Advanced Design of Pile Foundations Under Lateral Loading

This book presents models that capture the nonlinear response of piles subjected to lateral forces. Utilizing a consistent approach encompassing new mathematical models, it offers solutions presented as closed-form expressions and underpinned by the same set of 3–5 measurable soil-input parameters. These focus on nonlinear response of mono piles, anchored piles, pile groups and torsional piles, as well as passive piles subjected to soil movement induced in shearing, sliding slopes or excavation, and earthquake shaking. The models can also be used for pipelines and similar beam structures.

Solutions are provided in the form of design charts, with each parameter obtained using available test data and illustrated with real-world cases. The models reveal, for the first time, the mysterious mechanisms of amplification resulting from back-rotation, which have incurred the collapse of structures such as the Showa Bridge and Nicoll Highway, as well as the distortion of piles during earthquakes.

Advanced Design of Pile Foundations Under Lateral Loading is ideal for practicing foundation engineers and students at graduate level.

Wei Dong Guo is co-founder of Hans Innovation Group and former Associate Professor at the University of Wollongong, Australia. He is a Chartered Professional Engineer and is a Fellow of Engineers Australia by whom he was awarded the 2012 Warren Medal.

Advanced Design of Pile Foundations Under Lateral Loading

Wei Dong Guo

CRC Press
Taylor & Francis Group
Boca Raton London New York

CRC Press is an imprint of the
Taylor & Francis Group, an **informa** business

Cover image: Wei Dong Guo

First edition published 2024
by CRC Press
4 Park Square, Milton Park, Abingdon, Oxon, OX14 4RN

and by CRC Press
2385 NW Executive Center Drive, Suite 320, Boca Raton FL 33431

CRC Press is an imprint of Informa UK Limited

British Library Cataloguing-in-Publication Data
A catalogue record for this book is available from the British Library

ISBN: 978-1-032-32476-0 (hbk)
ISBN: 978-1-032-32477-7 (pbk)
ISBN: 978-1-003-31523-0 (ebk)

DOI: 10.1201/9781003315230

Typeset in Sabon
by SPi Technologies India Pvt Ltd (Straive)

Contents

Preface

We live in a world where people are no longer satisfied with merely knowing how to do things. We are constantly seeking ways to achieve our goals quickly and cost-effectively. This ambitious pursuit has led to significant advancements in communication and transportation, providing us with the conveniences we enjoy today. As we move forward, we are poised to witness even more efficient approaches to written communication through artificial intelligence (AI). We are also witnessing a gradual shift towards the practical use of cloud platforms. To leverage this technological revolution, the geotechnical profession must minimize computing data and acquire reliable input design parameters.

Utilizing mathematical models and expressions is the most effective way to handle big data. We advocate for a few methods (models) of design and input parameters that can address multiple problems. To accomplish our objectives, we have developed models and closed-form solutions for piles under vertical, lateral and torsional loading. The key findings are summarized in the book entitled *Theory and Practice of Pile Foundations*. Rigorous models and expressions are indispensable for our professional training and validation. As demonstrated in this book, they also reveal special mechanisms that often go unnoticed using AI and numerical approaches.

In the real world, piles are frequently subjected to lateral forces, as observed in pile foundations during earthquakes, stabilizing piles for landslide and piles adjacent to excavations or embankments, among other scenarios. Although various solutions exist for these issues, we have developed a consistent approach that encompasses new mathematical models and closed-from solutions, to address all of these problems by capitalizing on a set of five parameters. These models, for the first time, reveal the mysterious mechanisms of amplification resulting from back-rotation, which have incurred the collapse of structures such as the Showa Bridge and Nico Highway, as well as the distortion of piles during earthquakes.

This book presents models of the consistent approach that capture the nonlinear response of piles subjected to lateral forces. The models include load transfer models for lateral and torsional piles, a 3-layer model for rigid piles, as well as the BiP-$k_\theta(\eta)$ and BiP R-EP models for flexible piles. They are explained in detail for various scenarios involving lateral piles (Chapter 2), anchored piles (Chapter 3), slope stabilizing piles (Chapter 4), embankment piles (Chapter 5), piles subjected to lateral spreading (Chapters 6 and 7), piles near excavations (Chapter 8), cyclically loaded piles (Chapter 9), and torsional piles (Chapter 10). Other models are also presented to facilitate understanding, such as reducing the 3-layer model to a 2-layer model, interchangeability between the k_θ-model and η-model, and the reduction of the R-EP model to an R-E model. All of these models require the same set of a maximum of five soil-related input parameters.

The input parameters are derived from in situ tests, 1g model tests, centrifuge tests and numerical solutions. For example, when dealing with 'moving soil', the on-pile limiting force is theoretically reduced to 33% (uniform strength profile) or 67% (Gibson profile) of that for static soil. The nonlinearity of the *p-y* curve largely originates from the relative movement between the pile and the soil, which can be accounted for by using an ideal elastic *p-y* curve along with a soil movement factor. The rotational restraining stiffness becomes important for back-rotated piles. Examples illustrating these concepts are provided in pertinent chapter(s).

I would like to express my heartfelt gratitude for the invaluable opportunity to have worked in the Center of Excellence for Geotechnical Science and Engineering, supported by the Australian Research Council. The Center brought together numerous eminent experts (including four Rankine lecturers) from the University of Newcastle, the University of Western Australia, the University of Wollongong, and Australian industry, as well as overseas experts. I am particularly grateful for the complete freedom I was granted to pursue my research interests. It was during this time that I developed the 2- and 3-layer models.

I am indebted to my former students, particularly Drs Enghow Ghee, Hongyu Qin, Bitang Zhu, and many other Masters students, who provided invaluable support in conducting model tests. I would also like to extend my gratitude to the diligent copy-editing team from CRC, especially Tony Moore, Aimee Wragg and Aishwariya M. Shankar and others from Straive (SPi technologies).

Last but certainly not least, I must express my deepest appreciation to my wife, Xiaochun (Helen) Tang, for her unwavering support throughout this endeavour. My wife and our two daughters, Dina and Cindy, have endured unnecessary mental stress as a result of my continued work on developing the BiP models and completing this book during an extended freelance period. Additionally, I am overjoyed to welcome my newborn grandson, Greyson, into our lives five weeks ago. It is my fervent hope that Greyson will one day develop a passionate interest in our noble profession.

Wei Dong Guo
camsaweidguo@gmail.com

Notation

The following symbols are used in the book:

A_g, A_{gj} = constant for soil shear modulus profile, A_g for upper- and lower-layers;
A_L, A_r = coefficient for the force per unit length, $A_L = A_r d$ [FL^{-2-n}];
A_k = modulus factor;
A_p = cross-sectional area of a pile [L^2];
A_u, A_ω = average stress factor for displacement and rotation;
a, b, c = loading depths (Chapter 3);
c = loading depth, or cohesion (Chapter 7)
$C_{tj}(l_s)$ = value of the function $C_{tj}(z)$, at $z = l_s$ ($j = 1, 2$);
C_{to} = limiting value of function $C_t(z)$, as z approaches zero;
C_{t1o} = limiting value of function $C_{t1}(z)$, as z approaches zero;
d = diameter of an equivalent solid cylinder pile [L];
D_r = relative density;
e, e_p = eccentricity (= e_p), eccentricity of the load $H(H_g)$ above ground level [L];
e^- = depth of attachment for anchored piles [L];
e^+ = height (free-length) of load H above mudline; or $e = M_o/H$ for lateral piles [L];
$\bar{e}$ = e/l, normalized eccentricity;
E_p = Young's modulus of an equivalent solid cylinder pile [FL^{-2}];
E_s = Young's modulus of soil [FL^{-2}];
E_pI_p = flexural stiffness of a lateral pile [FL^2];
FPUL = net force per unit length [FL^{-1}];
FreH = free head, allowing translation and rotation at head-level;
FixH = fixed-head, allowing translation only at head level;
F_b = force per pile [F];
GL = ground-level;
$G(G_c)$ = shear modulus at depth z (at critical pile depth, l_c) [FL^{-2}];
G^*_{c1} = $A_{g1} r_0^{n_1}$, a specific value, not necessarily the modulus at depth r_o [FL^{-2}];
G_i, G_{si}, G_{avei} = initial shear modulus, shear modulus over sliding layer and stable layer, and an average shear modulus for each layer [FL^{-2}];
G_p = shear modulus of pile [FL^{-2}];
$G_s(\tilde{G}_s)$ = average soil shear modulus over the pile length, l [FL^{-2}];
G^* = average soil shear modulus over the critical length, l_c, and $G^* = (1 + 0.75\nu_s)G_s$ [FL^{-2}];
H = equivalent concentrated load at sliding depth (Chapter 3); shear force just about a liquefied layer or at GL, or lateral load at an eccentricity of 'e' above GL [F];
H_{av}, H_g = average load per pile in a group, and total load imposed on a group [F];

H_o	= lateral load H (capacity) at tip-yield or YRP state [F];
$\bar{H}_o$	= $H_o/(A_r d l^{1+n})$, normalized capacity;
$\bar{H}$	= $H\lambda^{1+n}/A_L$, normalized pile-head load;
I_p	= moment of inertia of an equivalent solid cylinder pile [L^4];
I_ϕ	= torsional influence factor;
J_p	= polar moment of inertia of the rod (pile);
k, k_o	= modulus of subgrade reaction for lateral piles, $k = k_o z^m$, for Constant ($m = 0$) and Gibson k ($m = 1$), respectively [FL^{-3}];
$K_i(\gamma)$	= modified Bessel function of second kind of ith order (Chapter 1);
K_a, K_{a1}	= coefficient of active earth pressure, and K_a of sliding layer;
k_G, k_T	= rotational stiffness of a pile-cap (or a non-liquefied layer), a stable layer (e.g. underlying a liquefied layer), respectively;
k_i	= coefficient for limiting resistance for sliding layer and stable layer;
K_p, K_{p2}	= coefficient of passive earth pressure and K_p of stable layer;
k_{nG}	= $k_G\lambda^3 / k = 4k_G / (E_p I_p \lambda) = \bar{k}_G / \bar{\omega}_r$, and $\bar{k}_G = k_G \omega_r \lambda^{2+n} / A_L$;
k_s	= kd, modulus of subgrade reaction, and that for piles in moving soil, $k_s < k$ [FL^{-2}];
$k_s^{profile}$	= k_s deduced from response profiles [FL^{-2}];
k_θ	= rotational stiffness along a passive pile including pile-cap, or a non-liquefied layer on liquefied layer;
k_θ^*, k_T^*	= stiffness k_θ and k_T that cause singularity and amplified response of piles;
$\bar{k}_\theta^*$	= a normalized singularity stiffness (NSS) $\bar{k}_\theta$ at which the pile response (e.g., displacement, rotation angle, bending moment and shear force) becomes infinitely large.
k_θ^{fA}, k_θ^{bA}	= value of k_θ for forward and backward rotation of pile A;
k_θ^S	= a sliding stiffness for a negligible value of maximum bending moment ;
l, l_m	= embedded pile length, and thickness of a upper moving soil layer, respectively [L];
l_c	= critical length beyond which extra embedment would not affect the pile response [L];
l^*, l_1^*, l_2^*	= critical lengths for rigid piles ($l \le l^*$), rigid in upper-layer ($l \le l_1^*$), or lower-layer ($l \le l_2^*$) [L];
l_e	= effective pile length, or length of elastic, lower segment [L];
l_{exc}	= depth of excavation [L];
l_s	= on-pile loading depth (for stepwise calculation), depth of upper layer (Chapter 10)[L];
LFP	= net limiting force per unit length profile [FL^{-1}];
LIFPULD	= linearly increasing (net) force per unit length with depth;
m	= K_{p2}/K_{a1}, ratio of subgrade modulus of stable layer k_{s2} over that of the upper layer k_s;
m_i	= $1/(2 + n_i)$, for upper- and lower- layers;
$M(z)$, $M_i(z)$	= bending moment at depth z [FL];
$M(\bar{x})$	= bending moment at the normalized depth $\bar{x}$ [FL];
$\bar{M}_i(\bar{z})$	= $M_i(z)/(A_r l^3)$, normalized bending moment at depth z, M_i = bending moment;
M_L	= $\bar{M}_L p_s l^2$, calculation parameter [FL];
M_m, M_{mi}	= maximum bending moment within a pile, and M_m in sliding and stable layers [FL];
M_o	= $k_\theta \omega_r$, bending moment at GL, or pile and cap connection, including P–Δ effect [FL];

M_o, M_T	= constraint moments at the top and bottom of a liquefied layer, respectively [FL];
M_o^{yA}, M_o^{yB}	= yielding bending moments of pile A and B at GL or pile and cap connection [FL];
$\bar{M}_m$	= $M_m\lambda^{2+n}/A_L$, or $M_m/(A_r d l^{2+n})$, normalized maximum bending moment;
$\bar{M}_o$	= $M_o\lambda^{2+n}/A_L$, normalized M_o;
M_y	= yielding bending moment of pile body [FL];
n	= power to the depth z, $n = 0$ (constant k or p_u), $n = 1$ (Gibson k or p_u) (Chapter 3), or to the depth of α_g(or α_o) + z (Chapters 2 and 10);
n_e	= equivalent power for shear modulus distribution, at $\alpha_g \neq 0$;
n_j	= power to depth of shear modulus distribution, for upper- and lower-layers;
N_g	= $A_L/(s_u d^{1-n})$ (clay), and $A_L / \left(\gamma_s' d^{2-n}\right)$ (cohesionless soil), $N_g = N_p$ ($n = 0$);
N_g^{FreH}, N_g^{FixH}	= N_g deduced from measured load-displacement of free- and fixed- head piles, respectively;
N_p	= $p_u/(s_u d)$, gradient correlated soil strength to the p_u; or fictitious tension of the membrane tied together the springs around pile shaft;
$N_{p\alpha}$	= N_p under an angle α of loading (anchored piles), or for a batter (angle α) piles);
P	= vertical load on passive piles during model tests [F];
p_1, p_2	= ultimate FPUL over sliding and stable layer, respectively [FL^{-1}] (Chapter 3);
P_h	= $H + l_s p_s$, calculation parameter [F];
p_m	= p-multipliers used to reduce modulus and FPUL for individual piles in a group;
p_s	= $p_b l_m / l$, on-pile FPUL in sliding layer due to soil movement ($\geq w_s^* = p_s/k_s$) [FL^{-1}];
p_u	= limiting (maximums) FPUL along a laterally loaded, free-head pile [FL^{-1}];
$p_{ub}(p_b)$	= limiting FPUL at the base of a free-head pile during passive loading, and $p_b = \alpha p_{ub}$ for rotationally restrained piles [FL^{-1}];
p_{ui}	= limiting FPUL p_u on a pile in sliding, and stable layer, respectively [FL^{-1}];
$p(\bar{x}), Q(\bar{x})$	= net FPUL, and shear force at the normalized depth $\bar{x}$;
$p(z)$, $p_i(z)$	= net on-pile FPUL at the depth z, and $p(z)$ for sliding (i =1), transition (i = 2), and stable layer (i = 3), respectively) [FL^{-1}];
q	= $p_s/(\alpha d)$, surcharge pressure of an embankment loading;
$Q_m(Q_{mi})$	= maximum shear force in a passive pile within sliding or stable layer) [F];
$Q(z)$, $Q_i(z)$	= shear force at depth z [F];
r_o	= pile radius [L];
s	= pile centre-to-centre spacing [L];
$S_u\left(\tilde{S}_u\right)$	= undrained shear strength of soil (average s_u over a maximum slip depth) [FL^{-2}];
t	= wall thickness of a pipe pile [L];
$T(l_s)$	= torque at the interface depth of l_s of the upper- and lower- layers;
T_t	= torque at ground level;
u, u_g, u_t	= lateral displacement of a rigid pile, u at GL and pile-head level, respectively [L];
u^*	= local threshold u^* above which pile-soil relative slip occurs [L];
w, $w(x)$, $w(z)$	= lateral deflection of a pile, w in the plastic, and w in elastic zone, respectively [L];
w_h	= pile-displacement at depth of z_m (i.e., joint of two segments) [L];
w_f, w_s, w_e	= frame movement in model pile tests, w_s (= $w_s^* = p_s/k_s$), effective soil movement [L];

w_f^*	= soil movement at which maximum pile response occurs [L];
w_g, w_g^s	= displacement and sliding displacement of pile at GL [L];
w_i	= initial frame movement, within which pile response is negligible [L];
w_L	= a fictitious pile-displacement at GL projected along the low segment (L);
w_s	= soil movement in model pile tests [L];
$\bar{w}_g$	= $w_g k_s / p_s$ (rigid pile) or $w_g k \lambda^n / A_L$ (flexible pile) normalized pile-displacement at GL;
$\bar{w}_g$	= $w_g k_s / (A_r l^n)$, normalized displacement (Chapter 3);
w_p	= local limiting deflection beyond which the FPUL stays at p_u[L];
w_t	= pile deflection at loading level [L];
$w(\bar{x}), w'(\bar{x})$, $w(z), w'(z)$	= deflection and rotation at depth z [L] or the normalized depth $\bar{x}$;
x	= $z - z_m$, depth measured from z_m (R-E model) [L];
$x, x_p\, \bar{x}, \bar{x}_p$	= depth below GL, slip depth of plastic state, $\bar{x} = \lambda x, \bar{x}_p = \lambda x_p$ (Chapter 2) [L];
YRP	= yield at rotation point;
$z, \bar{z}$	= depth [L] and normalized depth z/l, respectively; or depth below the slip depth x_p (i.e., $z = x - x_p$), and $\bar{z} = \lambda z$, respectively;
z_m, z_{mi}	= depth of maximum bending moment (i = 1, 2) [L];
z_{mt}	= depth for maximum shear force Q_m[L];
$z_o(z_1)$	= slip depth initiated from mudline (pile-base) [L];
$z_r, \bar{z}_r$	= depth of rotation point [L], and normalized $\bar{z}_r = z_z / l$;
z_o^y	= slip depth at tip-yield state [L];
α	= parameter for loading depth αb (Chapter 3); to cater for impact of soil movement or $\alpha = w_g / w_e$ (≈ the ratio of on-pile pressure over the embankment pressure);
α_c, β_c	= coupled elastic parameters;
α_e	= a factor for reducing the loading eccentricity;
α_g, α_{gj}	= an equivalent depth for shear modulus at GL, upper- and lower-layers [L];
α_N, β_N	= normalized stiffness factors;
$\alpha_o(\bar{\alpha}_o)$	= an equivalent depth for GL limiting force with $\bar{\alpha}_o = \alpha_o \lambda$;
α_r	= a partial factor to gain the A_L and G_s using those for free-head and fixed-head piles;
α_R	= a factor for accumulated response over cycles of loading;
α_s	= a factor for reducing modulus k_N (Chapter 9 only);
β	= c/l_m, effective depth of sliding layer c over the sliding depth l_m;
β_s	= $1.08\alpha_s$, a factor for quantifying depth of soil deformation (of $\beta_s l_{exc}$);
χ_1, χ_2	= equivalent base factor for upper-layer, base factor for lower-layer;
δu_N	= $u_{gN} - u_{gN}^e$, residual displacement for the Nth cycle [L];
Δu	= accumulated residual displacement after N cycles [L];
$\delta\omega_N$	= $\omega_N - \omega_N^e$, residual rotation for Nth cycle;
$\Delta\omega$	= total residual angle of rotation after N cycles;
$\delta\theta^s$	= an angle over sliding (with $M_m \approx 0$);
$\phi(\phi')$	= (effective) angle of internal friction of soil;
$\phi(\phi_t)$	= local (GL) angle of twist (Chapter 10);
ϕ_d'	= angle of dilatancy;
ϕ_i'	= effective angle of sliding interface;
$\phi(z)$	= angle of twist of pile at any depth, z (Chapter 10);
γ	= load transfer factor (Chapter 1);
$\gamma_s(\gamma_s')$	= (effective) unit weight of soil [FL^{-3}];
η	= ω_r / ω_{rL}, ratio of rotation angle of upper over lower segments;

η_g	= modulus ratio between upper- and lower- layers (= A_{g1}/A_{g2});
Δ_o, Δ	= initial lateral displacement and final lateral displacement (centrifuge tests) [L];
λ	= ratio of thickness of lower stable layer over sliding layer, or modification ratio in estimating k_s; or reciprocal of characteristic length;
λ_d	= 0.5–0.667, distribution factor for displacement due to cap-restraining stiffness;
λ_k	= 1–3, modification factor for modulus, high value for short piles;
ν_s	= Poisson's ratio of soil, taken as 0.25 for sand, otherwise 0.4;
π_t, π_{tj}	= non-dimensional torsional stiffness factor, π_t for upper- and lower- layers;
π_t^*, π_t'	= critical stiffness for rigid piles ($\pi_t < \pi_t^*$), and flexible piles ($\pi_t > \pi_t'$);
$\pi_{tj}^*(\pi_{tj}')$	= critical stiffness for rigid (flexible) piles, for upper- and lower-layers;
σ_v'	= effective overburden stress [FL^{-2}];
$\tau_o(\tau_f)$	= (limiting) shaft friction on pile-soil interface [FL^{-2}];
ω	= angle of change between pile-cap and the pile at GL;
ω_{cap}	= rotation angle of a rigid pile cap;
ω_g	= rotation angle of pile at GL;
ω_r, ω_{rL}	= rotation angle of upper segment and lower segment of pile;
ω_s	= rotation angle of soil movement profile;
$\bar{\omega}_g$	= $\omega_g k\lambda^{n-1}/A_L$, normalized mudline rotation;
$\bar{\omega}_r$	= $w'(z)k_s/(A_r l^{n-1})$, normalized rotation (Chapter 3);
$\bar{\omega}_r$	= $w'(z)k_s l/p_s$, normalized rotation angle; or $\bar{\omega}_r = \omega_r^m / \omega_r^{FH}$ (Chapter 8 only) measured angle of rotation over estimated angle of free-head pile at the same lateral displacement;
$\bar{\omega}_r^{max}$	= $\omega_r^{max}/\omega_r^{FH}$ (Chapter 6 only);
ω_r^{FH}	= angle of free-head pile at given lateral displacement w_g of 'fixed-head' pile (Chapter 7);
ω_r^m	= measured angle of pile at ground level (Chapter 7);
ζ_b	= H/H_m, level of cyclic load H over maximum load H_m (= H at tip-yield state)

Bar '-' for normalized parameters and variables. Depths c, z, l_m are all normalized by pile embedment length l.

Subscript '1' and '2' denote upper and lower layer; or '1' and 'N' for number of cyclic loading.

Subscript 'i = 1' and 'i = 2' denote for sliding and stable layers

Subscript 'j = 1' and 'j = 2' denote for upper- and lower-layers

Subscripts 'A' and 'B' refer to pile 'A' and 'B', respectively.

Subscripts 'f' and 'b' refer to 'forward' and 'backward' rotation;

Subscript 'g' for ground level; Subscript 's' for soil.

Subscript 'm' and 'FH' refer to 'measured' data and free-head respectively;

Subscripts 'max' refer to 'maximum';

Subscripts 's' and 'SY' refer to 'sliding', and 'sway' respectively;

Subscript 'y' refers to 'yield of pile at ground-level;

Superscripts '+' and '–' indicate a positive and negative value, respectively;

Chapter 1

Problems models and tests

1.1 INTRODUCTION

Pile foundations are employed to supporting buildings, bridges and other infrastructure. They are designed to withstand a range of forces, including axial, lateral and torsional loads, as is elaborated in various books. In practice, piles are susceptible to lateral soil movements exhibiting rotation (e.g., near tunnelling, excavation), translation (e.g., in a sliding slope, and lateral spreading during earthquake) or a combination of translation-rotation (e.g., next to embankment or pile driving operation) (see Figure 1.1). For instance, during the 2011 tsunami in Japan, piles were subjected to significant lateral spreading and torsion that uplifted them from the ground and caused buildings to overturn. Similarly, in the 2011 Christchurch earthquake, bridge piles were subjected to lateral spreading, resulting in significant back-rotation and permanent tilt. The piles need to sustain passive loading (from soil movement), etc.

The response of vertically loaded, passive piles is dominated by the sliding thickness ratio l_m/l (with l_m = thickness of sliding layer, and l = pile embedment). For instance, in the Christchurch earthquake, lateral spreading reached full embedment of piles ($l_m/l = 1$) underpinning Fitzgerald bridge and Avondale bridge; and major portion ($l_m/l = 0.87$) of the length for Gayhurst bridge. The piles supporting Anzac bridge and South Brighton bridge were subjected to spreading l_m/l of ~0.5 and 0.28–0.57 and suffered permanent back-rotation (tilt) of 5–6 and 7–8 degrees, respectively. The piles largely suffered two hinges at the pile-cap level and the depth of interface between liquefied and non-liquefied layer, respectively. However, highly rotated piles (of Anzac bridge) do not have any hinge at pile-cap level (Berrill et al. 2001; Haskell et al. 2013). These piles can exhibit sway, rotation and translational movement (Guo 2022b) owing to shifting profiles of spreading (Abdoun et al. 2003).

Design of these piles entails incorporating rotational constraints (e.g., from pile cap, base, or even moving soil), sliding depth and dragging. Over decades, sophisticated shaking table tests, centrifuge tests and numerical simulations have been conducted to examine the response of piles in various sliding soil, as well as in situ tests. Despite these efforts, some critical failure mechanisms have gone undetected, and inconsistent parameters prevailed across different cases, which lead to costly over-design or risky under-design. This disjointed effort has resulted in divergent outcomes, as is evident in the vast number of publications (over 31,587 journal papers to May 2017) on piles.

To improve productivity and develop resilient infrastructure, there is an urgent need for a united technical service platform that can capture nonlinear response of conventional and sustainable geo-beams including pipelines, capped piles, anchored piles, piled walls and soil

DOI: 10.1201/9781003315230-1

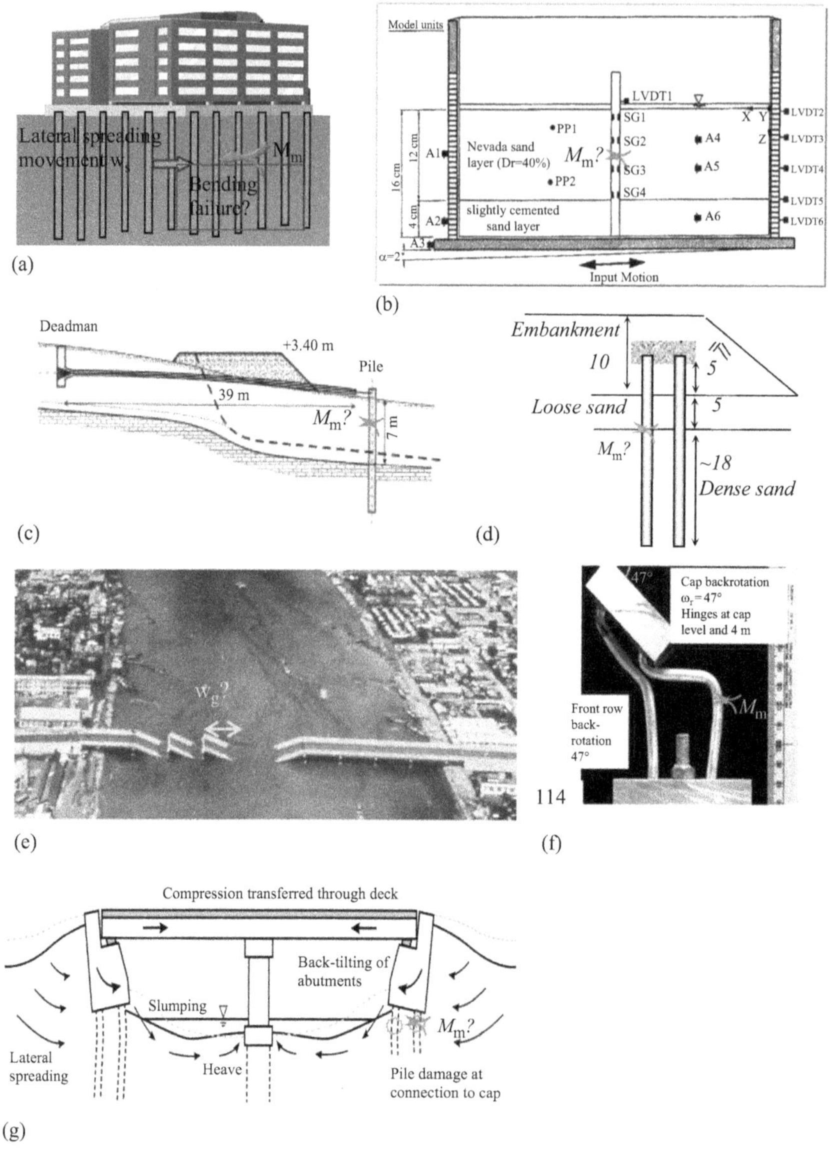

Figure 1.1 Problems resolved using 2-layer model: (a) piles under shaking (Seidel J.P. 2001); (b) centrifuge modelling (Dobry et al. 2003); (c) piles to stabilize embankment (Frank and Pouget 2008); (d) Armstrong et al. (2014); (e) failure from amplified w_g of P5 (show a bridge); (f) distortion from amplification (Knappett and Madabhushi 2009); (g) back-rotated abutment piles (Cubrinovski et al. 2014).

nails/anchors, under combined loading and in moving soil of diverse circumstances. This platform must be underpinned by a single set of parameters across all cases to empower easy access and efficient design without generating big data. Unfortunately, this need is currently not well recognized.

This book represents a crucial first step in developing united solutions for piles in the context of load transfer model. In particular, 2-/3-layer models and BiP models (BiP = **bi**-**p**ortion of a pile in sliding and stable layer, respectively) will be brought together to provide a consistent framework that can be applied across different scenarios. The united solutions are dominated by rotational restraints as elaborated individually.

1.2 LATERAL PILES WITH CAP RESTRAINT

1.2.1 Load transfer model

Pile-soil interaction for laterally loaded piles is simulated using a load transfer model. As shown in Figure 1.2a, the pile-soil interaction is characterized by a series of springs distributed along the shaft. The spring has a modulus of subgrade reaction k_s (= kd, d = pile outer diameter or width). Each spring induces force per unit length (FPUL) p of $k_s w$ (proportional to the local pile displacement, w), which has a limit value of p_u. A capped pile is restrained by a rotational stiffness k_G.

- As the lateral load H increases, pile slips from mudline down to a slip depth, x_p, at which the pile deflection $w(x_p)$ is equal to w_p (= p_u/k_s). Within the slip zone ($x = 0$–x_p), the FPUL p on each spring stays at p_u, which is uncoupled ($N_p = 0$) (see Figure 1.2d). Below the x_p ($x \geq x_p$), the p [= $k_s w(x)$] is proportional to the deflection $w(x)$ ($\leq w_p$). The k_s is determined using expressions in Table 1.1 (Guo and Lee 2001).
- The net FPUL p_u along the pile shaft is the sum of the passive soil resistance acting on the face of the pile in the direction of soil movement, and sliding resistance on the side of the pile, less any force due to active earth pressure on the rear face of the pile (Guo 2006). The p_u is described by parameters α_o, A_L, and n (see Table 1.2) and $p_u = A_L x^n$ ignoring the resistance at GL (Guo 2006), n = 0.5–0.7 (for a uniform strength profile), and 1.3–1.7 (for a linearly increasing strength profile, e.g., sand).
- Cap-rotation occurs when compression in piles located in front rows exceeds that in the back rows. The rotational stiffness of the cap k_G is taken as that of pile-head for a fully cast concrete cap with sufficient rigidity, and it is calculated using the axial stiffness and capacity of individual pile at diverse spacing (Mokwa and Duncan 2003).

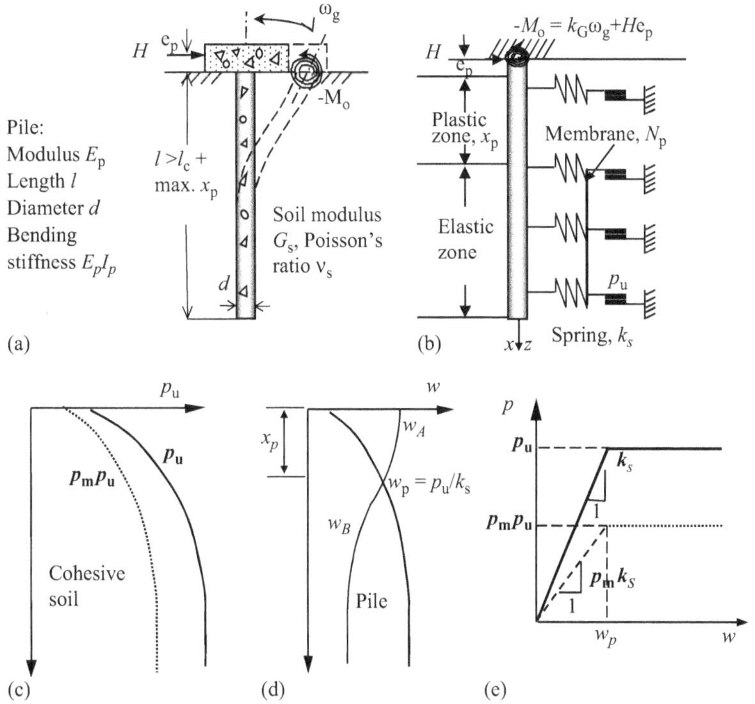

Figure 1.2 Schematic model for a laterally loaded capped pile: (a) a single pile; (b) schematic model; (c) limiting force profile (*LFP*); (d) pile deflection and w_p profiles; (e) *p-y*(*w*) curves for a single pile or a pile in a group. [Adapted from Guo, W. D., *J Eng Mech* **141**(9), 2015a.]

Table 1.1 Load transfer model (Guo and Lee 2001)

Parameter	$\alpha_c = \sqrt{\sqrt{\dfrac{k_s}{4E_pI_p}} + \dfrac{N_p}{4E_pI_p}}$	$\beta_c = \sqrt{\sqrt{\dfrac{k_s}{4E_pI_p}} - \dfrac{N_p}{4E_pI_p}}$
Simple k_s model	$\dfrac{k_s}{G_s} = 10.75\left(\dfrac{l}{d}\right)^{-0.45}$ (rigid piles)	$\dfrac{k_{s2}}{G_{s2}} = 8.78\left(\dfrac{E_p}{G_{s2}^*}\right)^{-0.087}$ (flexible piles)
Coupled model	$k_s = 1.5\pi G_s \begin{Bmatrix} 2\gamma_b K_1(\gamma_b)/K_o(\gamma_b) \\ -\gamma_b^2\left[\left(K_1(\gamma_b)/K_o(\gamma_b)\right)^2 - 1\right] \end{Bmatrix}$	$N_p = 0.25\pi d^2 G_s[(K_1(\gamma_b)/K_o(\gamma_b))^2 - 1]$
Load transfer parameters	$\gamma_b = k_1\left(\dfrac{E_p}{G_s^*}\right)k_2(2l/d)^{k3}$ $K_j(\gamma_b)$ = modified Bessel function of second kind of *j*th order. $G_s^* = (1 + 3\nu_s/4)G_s$; ν_s = soil Poisson's ratio. G_s = average shear modulus over the critical length l_c [$= 1.48d(E_p/G_s)^{0.25}$] $\approx 2.8/\alpha$.	Free-head: $k_1 = 2.14–3.8$, $k_2 = 0$, and $k_3 = -1$ (short piles); $k_1 = 1–2$, $k_2 = -0.25$, and $k_3 = 0$ (long piles). Fixed-head: $k_1 = 0.65$, $k_2 = -0.5$, and $k_3 = -0.04$ (long, fixed-head piles). Long fixed-head piles with $l > l_c + \max. x_p$), and x_p = length of plastic zone.
	k/G_s ratio is applicable for caps with rotational stiffness using an equivalent loading eccentricity e (= $e_p + k_r\omega_g/H$, ω_g/H is calculated for an average load level H and using k for e = 0).	

Table 1.2 Typical models for lateral and passive piles

Rigid pile (Guo 2008, 2013b)	*Lateral (flexible) pile (Guo 2006, 2015a)*	*Passive pile (Guo 2015b)*	*BiP models (Guo 2021a, 2022a)*
$p_u = A_r dz^n$ (n = 0 clay, n = 1.0 sand)	$p_u = A_L(x + \alpha_o)^n$ $A_L = \gamma_s' N_g d^{1-n}$ (n = 0.5–0.7, clay) $A_L = \gamma_s' N_g d^{2-n}$ (n = 1.3–1.7, sand)	$p_s = \alpha A_L z,\ p_{ub} = s_g\gamma_s' K_p^2 dl$ $A_L = p_{ub}/l$	$A_L = p_{ub}/z_m$
$A_r e^{-0.014\alpha}$ (reduced with loading angle α)	$p_m p_u$ (p_m = group interaction factor)	αp_u (α = soil movement and group interaction factor)	
$p = k_s u$, $k = k_o z^n$ $k_s = kd$ ($x = 0 - l$)	$p = k_s w(x)$ $k_s = kd$ ($x = 0 - l$)	$p_1 = k_s w(z) - p_s$, $k_s = kd$ ($z = 0 - l_s$), $p_2 = mk_s w(z)$, ($z = l_s - l$)	
$M_o = k_G\omega_r$	$M_o = k_G\omega_g$	$M_o = k\theta\omega_r$	$M_m = k\theta\omega_r$
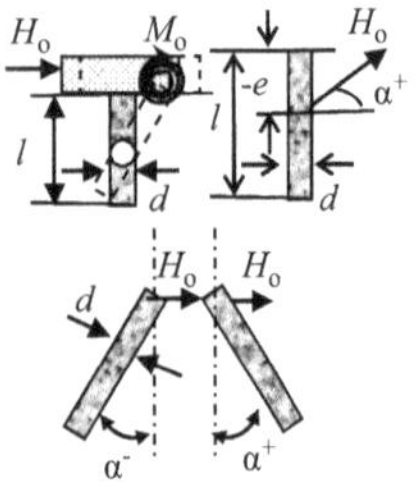	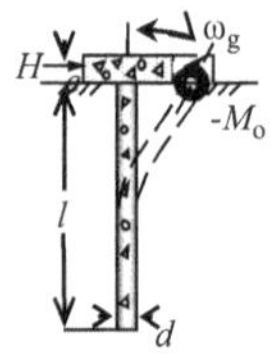	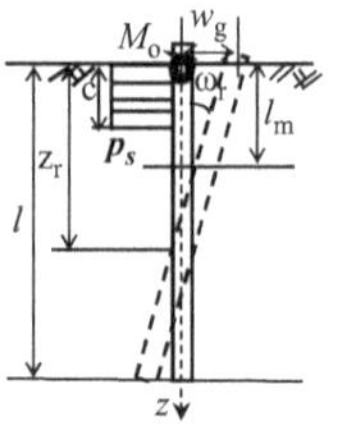	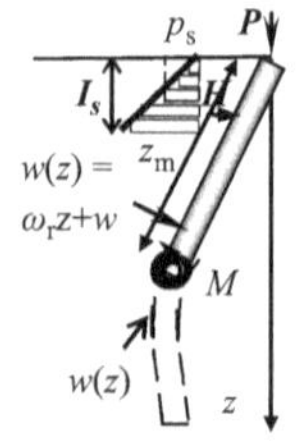
Chapter 3	Chapter 2	Chapter 4	Chapters 4 and 8

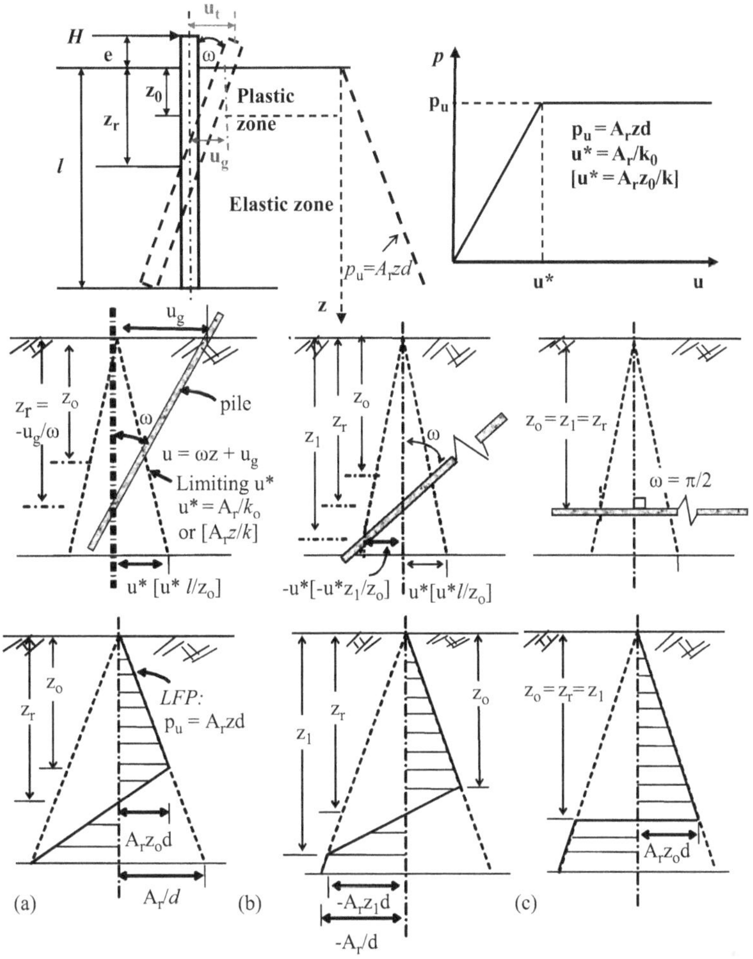

Figure 1.3 Schematic limiting force profile, on-pile force profile, and pile deformation: (a) tip yield state; (b) post-tip yield state; (c) impossible yield at rotation point (YRP). [Adapted from Guo, W.D., *Can Geotech J* **45**(5), 2008.]

The impact of semi-fixity (rotation restraint on piles) is empirically considered in relevant code [e.g., Chinese JGJ design code (JGJ 1994)]. It is more accurately calculated using elastic-plastic solutions (Guo 2015b) (see Chapter 2). A high stiffness k_G reduces the on-pile FPUL p_u value (Guo 2009) by a mobilization ratio N_g, which is also seen from rotation to translation of straight and raked piles (Chapter 3).

1.2.2 Solutions for lateral piles

A pile is deemed as rigid provided $E_p/G_s \geq (E_p/G_s)_c$, otherwise at $E_p/G_s < (E_p/G_s)_c$, it is defined as flexible, where $(E_p/G_s)_c = 0.832(l/d)^4$ and E_p is Young's modulus of an equivalent solid cylindrical pile of outer diameter d and embedded length l, and G_s is the soil shear modulus (Guo and Lee 2001). Guo (2008) develops closed-form solutions for rigid piles (see Table 1.2 and Figure 1.30); and for flexible piles (Guo 2015a, see Chapter 2) with any head restraints, which reduces to free-head (Guo 2006) through to capped head (Guo 2009). Importantly, these solutions reveal a coupling relationship between group interaction factor (p_m) and the cap-rotational stiffness (k_G). Moreover, the free-head solutions are successfully extended to capture response of structural nonlinear piles (Guo 2012) and slope-stabilizing piles (Guo 2013c).

1.2.3 Input parameters

The solutions (Guo 2015a) allow the parameters (k_s and p_u) to be deduced in consistent manner, which offer k_s/G_{s1} ratio against Table 1.1; as well as the p_{ub} (i.e., p_u at pile-tip level) with $p_{ub} = s_g\gamma_s'K_p^2dl$ in which γ_s' = (effective) unit weight, K_p = coefficient of passive earth pressure K_p over the pile embedment; s_g = 0.5–2.5 (an average of 1.29) varying with pile installation methods. The parameters α_o, A_L, and n (for p_u) were deduced using an extensive set of measured response of piles and hinged piles under static and cyclic loading, comprising 70 free-head piles and ~30 capped piles in layered soil (e.g., Guo 2012; Guo 2013a). The p_u may be estimated using Hansen's solution (Hansen 1961; Guo and Zhu 2011; Guo 2013a) based on angle of internal friction and cohesion of a subsoil.

Lateral load may exert at some eccentricity about GL. It may dis-align with the centroidal of cross-section, and incur rotation around the pile axis. The solutions for torsional piles are provided in Chapter 10, where the impact of the loading on input soil parameters is explored compared to vertical loading.

1.3 RESTRAINT AT MIDDLE DEPTH OF ANCHORED PILES

Chapter 3 provides closed-form solution for (i) rigid piles under a concerned lateral load (H-) or a uniformly distributed (p-based) over a desired area; (ii) anchored piles by simply using a negative eccentricity in the solutions of laterally loaded rigid piles (Guo 2008, 2012); and (iii) cap restrained rigid piles (Guo 2013b). Impact of rotational restraint (via anchoring depth and cap-rotation stiffness) is quantified on the pile response through the established expressions for the N_g (see Table 1.2). Response amplification of back-rotated piles is newly identified.

1.4 RIGID PILES SUBJECTED TO RESTRAINT OF SLIDING LAYER

1.4.1 2-layer and 3-layer model

Figure 1.4a indicates a rotationally restrained, rigid pile (with embedment of l) is subjected to a moving layer (of a thickness l_m) underlying by a stable, lower layer (of λl_m in thickness) namely 2-layer model. As with laterally loaded piles, the pile-soil interaction is modelled by a series of springs distributed along the pile shaft (Guo 2008). The spring now has a constant modulus of subgrade reaction k_s (= kd) and mk_s for the sliding and stable layer, respectively. It has a net FPUL p_s (= k_sw_s) at $w_s > w_s^*$ (= p_s/k_s) over a sliding depth c on the pile surface; otherwise below the transition depth (with $w_s < w_s^*$), the on-pile resistance $p_i(z)$ only relies on the local (elastic) pile displacement (see Table 1.2).

Guo (2016) reveals moving layer may rotationally restrain free-head piles. This effect is lumped into rotational restraining springs k_G and k_T at the pile-top and tip, respectively (see Figure 1.4c) and an ultimate FPUL value of αp_{ub}. A series of piece-wise, uniform p_s (= $\alpha p_{ub}l_m/l$, see Figure 1.4 for all symbols) over the on-pile 'sliding' depth c (= αl_m) offer a linear profile of soil movement w_s. The αp_u value well accommodate the impact of soil movement and its profile.

Example 1.1 Model for piles subjected to lateral spreading

A non-liquefied layer may cause dragging on a lateral spreading layer, which may be encapsulated as a head, rotational stiffness (thus moment M_G), a concentrated thrust H at the top of underlying layer, and a moment M_o due to loading eccentricity of H (Dobry et al. 2003; Brandenberg et al. 2005a). The lower layer on the pile is captured using a rotational constraint M_T (= $k_T\omega_r$). Under the impact of a non-uniform soil movement and pile-pile interaction (i.e., p_s), the modelling of the pile-soil interaction during lateral spreading thus becomes resolving the 2-layer model in Figure 1.4c.

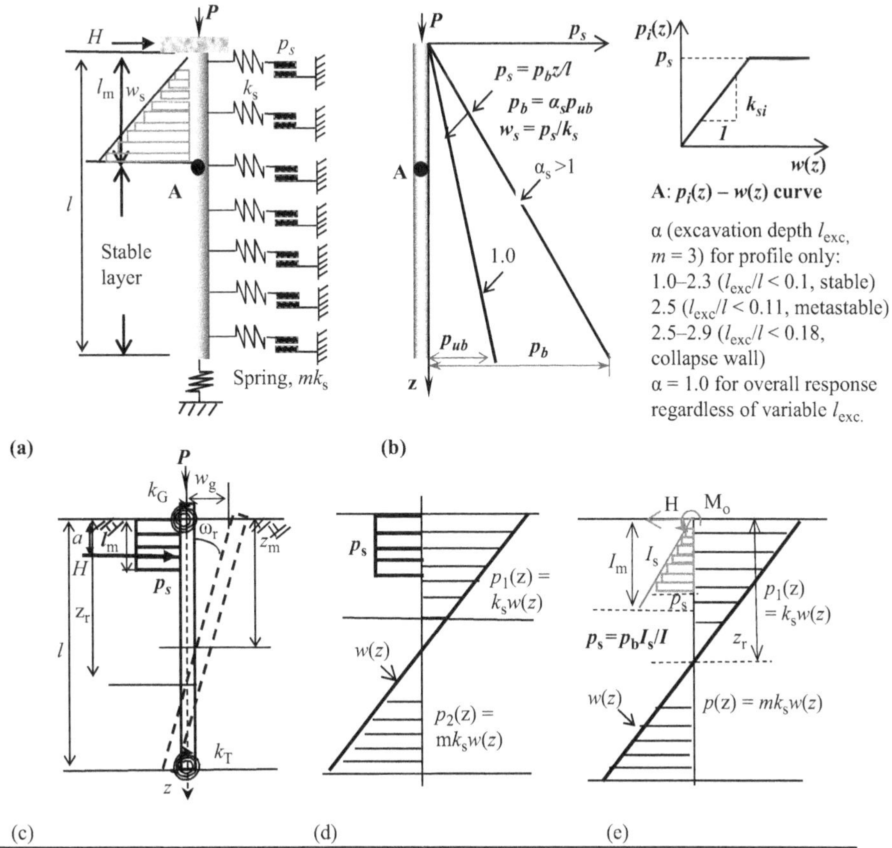

The pile-soil interaction is modelled by a series of springs distributed along the rigid pile shaft, with rotational springs k_G and k_T at GL and pile-tip, respectively.

l = pile embedment; l_m (l_s) = thickness of moving layer over a stable layer of λl_m in thickness; k_s = modulus of subgrade reaction (moving layer) and mk_s for stable layer; w_s (= p_s/k_s) = soil movement w_s converted into a uniform, force per unit length (FPUL) to an on-pile depth of c ($\leq l_s$);
p_s = ($p_b l_m/l$, $p_b = \alpha p_{ub}$) on-pile pressure over a depth of l_m; p_{ub} = limiting FPUL of free-head piles at pile-tip level; α = soil movement factor.
ω_r = rotation angle; and w_g = deflection at GL; $w(z) = \omega_r z + w_g$, pile deflection at depth z and $w(z_r) = 0$. $p(z) = k_s w(z)$, on-pile resistance per unit length in sliding layer, and $p(z) = mk_s w(z)$ in stable layer, respectively.
M_o, H = moment and shear force at pile-cap connection; P = vertical loading on each pile.

Figure 1.4 2-layer models for rigid passive pile: (a) pile-soil system; (b) p_s applied & $p(z)$ induced; (c) model with k_G & k_T; (d) pressure at a given l_m; (e) nonlinear model. [Adapted from Guo, W. D., *Int J Numer Anal Meth Geomech* **40**(14), 2016.]

Soil movement incurs rotation of the rigid pile. The pile, with a displacement $w(z) = \omega_r z + w_g$ at depth z below GL, rotates to an angle ω_r, a mudline displacement w_g, about a rotation depth z_r ($= -w_g/\omega_r$) (see Figure 1.4c). The net on-pile FPUL $p(z)$ is given by $p(z) = k_s w(z) - p_s$ in sliding layer, and $p(z) = mk_s w(z)$ in stable layer, respectively. The 2-layer model will be discussed in Chapter 4.

In practice, piles may be dragged by sliding soil. A dragging layer is introduced between the sliding and stable layers, resulting in a 3-layer model. Model tests (Guo 2012) indicate the on-pile FPUL $p(z)$ profile on a passive pile to the depth z_m (of maximum bending moment) resembles that along the entire length of a laterally loaded pile. Further down, the $p(z)$ profiles vary approximately linearly from the negative $p(z_m)$ at $z = z_m$ to the positive $p(l)$ at the pile-base. This offers a 3-layer model (see Figure 1.5a, Guo 2016) described by a subgrade modulus k_s in the upper layer (depth of $0 \sim l_m$), a linearly increasing modulus from k_s to mk_s in the transition layer (from depth l_m to z_m), and a modulus mk_s in the lower layer (in-depth $z_m \sim l$), respectively. The net on-pile FPUL $p(z)$ is now given by $k_s w(z) - p_s$ in sliding layer, $k_s w(z)[1 + (m - 1)(z - l_m)/(z_m - l_m)]$ in transition layer, and $mk_s w(z)$ in stable layer, respectively.

Closed-form expressions for rotationally restrained piles in the 2-layer and 3-layer models were developed (Guo 2014, 2015b, 2016), and are presented in Chapters 4 and 5, respectively. For instance, Chapter 5 shows how the impact of soil movement is incorporated through the new factor α (Guo 2014, 2015b, Guo et al. 2017). Chapter 7 shows the formation of triangular or uniform pressure profile due to diverse degrees of head-restraint piles.

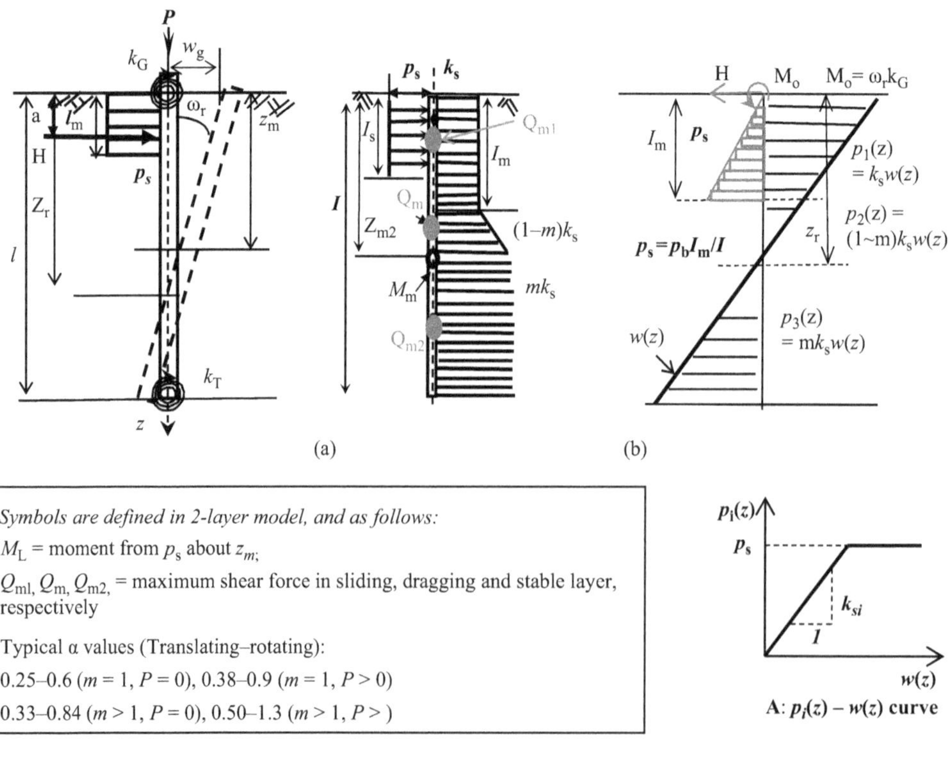
P
k_G
w_g
ω_r
a
l_m
z_m
H
p_s
Z_r
l
k_T
z
p_s k_s
l_s
Q_{m1}
l_m
Q_m
l
Z_{m2}
$(1-m)k_s$
M_m
mk_s
Q_{m2}
(a)
H
M_o
$M_o = \omega_r k_G$
l_m
p_s
$p_1(z) = k_s w(z)$
$p_2(z) = (1-m)k_s w(z)$
z_r
$p_s = p_b l_m / l$
$p_3(z) = mk_s w(z)$
$w(z)$
(b)
Symbols are defined in 2-layer model, and as follows:
M_L = moment from p_s about z_m;
Q_{ml}, Q_m, Q_{m2} = maximum shear force in sliding, dragging and stable layer, respectively
Typical α values (Translating–rotating):
0.25–0.6 ($m = 1, P = 0$), 0.38–0.9 ($m = 1, P > 0$)
0.33–0.84 ($m > 1, P = 0$), 0.50–1.3 ($m > 1, P >$)
$p_i(z)$
p_s
k_{si}
1
$w(z)$
A: $p_i(z) - w(z)$ curve

Figure 1.5 3-layer model for rigid passive pile: (a) pile and modulus model; (b) p_s and k_s profile for nonlinear response. [Adapted from Guo, W. D., *Int J Numer Anal Meth Geomech* **40**(14), 2016.]

1.4.2 Determination of model parameters

The model parameters p_{ub} and k_s for lateral piles are modified herein by the factor m and α to incorporate the impact of soil movement and rotating piles (of free-head, $k_\theta = 0$) and translating piles (of fixed-head, $k_\theta > 10E_pI_p$, E_pI_p = flexural stiffness of a pile, E_p = Young's modulus, Guo 2015a). Note that fixed-head piles may rotate as well at $l_m > 0.5l$. In principle, items G1–G6 should be consulted.

G1: Using associated α at $P = 0$ or $q = 0$ (otherwise see G3), the limiting FPUL p_{ub} (of a free-head pile) is estimated as follows

- $p_{ub}/(s_u dl) = (9\text{–}12)\ \alpha/c$ along with $\alpha = 1/3$ for a deep sliding, clay layer ($l_m > 0.5l$), to incur a limiting p_s of $(3\text{–}4)s_u dl/c$ to the loading depth c; $p_{ub}/(s_u dl) = (1\text{–}1.4)(9\text{–}12)\alpha/c$ (with $\alpha = 0.33\text{–}0.6$) for a shallow depth with ~40% increase in the resistance for sliding clay overlying stable sand layer.
- $p_s = p_b z/l$ and p_b ($= \alpha p_{ub}$ at pile-base), depending on rotational stiffness k_θ; and α = 1.39–1.5 (lateral spreading), 0.25–1.3 (sliding slope), and 0.3–1.4 (vertically loaded piles). p_{ub}/d is taken as the overburden stress at the pile-tip for piles subjected to lateral spreading (JRA 2002).

(JRA 2002)

G2: The limiting FPUL reduces by a multiplier factor p_m due to pile-pile interaction as suggested for laterally loaded piles (Guo 2012)

$$p_m = 1 - a(12 - s/d)^b \tag{1.1}$$

where s = pile centre-to-centre spacing; $a = 0.02 + 0.25\text{Ln}(n_r)$; $b = 0.97(n_r)^{-0.82}$, n_r = row number. The p_m is incorporated into the next factor α (see G3) due to soil movement.

G3: The α value reflects variation of p_s with rotational restraint and vertical load P (or embankment surcharge q), which needs to be checked against the following:

- Single, free-head piles ($l_m < 0.5l$): (i) $\alpha = 0.25\text{–}0.6$ (normal) for undefined sliding interface ($m = 1$) and $P = 0$; (ii) p_{ub} ($P \neq 0$, or $q \neq 0$) = ~$1.5p_{ub}$ ($P = 0$ or $q = 0$) noted in model tests [see $p(z)$ profiles]; and (iii) $\alpha \approx w_g/w_s$, the ratio of pile displacement over soil movement w_s.
- Capped piles: $\alpha = 0.25\text{–}0.6$ in general and $\alpha = 0.25$ for translating piles at $l_m < 0.5l$, otherwise $\alpha \approx 0.6\text{–}1.0$ for deep sliding and $P \neq 0$.
- Piles under embankment surcharge q ($p_s = \alpha dq$, Guo 2014): $\alpha = 0.50\text{–}0.72$ to include increasing impact of 'surcharge' and underlying stiff layer.

Example 1.2 α **values**

Tests and numerical analysis (Guo 2015b) indicate that (i) p_{ub} is mobilized at highly rotated piles, and drops to $0.25p_{ub}$ for translating piles; and (ii) The p_s increases by ~40% (Guo 2006; Yang and Jeremic 2002) from underlying stiff layer ($m > 1$), and by ~50% from vertical load P or embankment surcharge q (but with a reduced m). A stiff layer ($m > 1$) may lift α to 0.33–0.35 [= 0.25 × (1.3–1.4)] (2-layer clay), and 0.8–0.84 [= 0.6 × (1.32 – 1.4)] (2-layer sand); and further to 0.5–0.53 [= 1.5 × (0.33–0.35)], and 1.2–1.3 [= 1.5 × (0.8–0.84)] due to loading P and/or q, for translational and rotational movement, respectively. At a loading distance of $0.7l$ and a sliding depth of $0.286l$, the FPUL p_s is taken as $(\alpha p_b)(c/l)$ with $\alpha = 0.6$ in predicting profiles of on-pile response.

G4: The modulus ratio m (= k_{s2}/k_s) is taken as K_{p1}/K_{a2} at 'ultimate state', assuming limiting pressures on piles in either clay layer (Viggiani 1981) or sand layer (Muraro et al. 2014), respectively; as well as $m = (0.6\text{–}1.3)K_{p1}/K_{a2}$ and $m = 2.1K_{p1}/K_{a2}$ for a sliding sand overlying a stable, soft rock.

G5: Unless provided, the modulus of subgrade reaction k_s for a sliding sand is estimated as

$$k_s = 51d\lambda_k \frac{k_{p1}}{k_{a1}}\left(\frac{l_{2c}}{l_m}\right)^2 \frac{p_{u2}}{p_{u1}} \tag{1.2}$$

where d = pile diameter in m, k_s in kPa, λ_k = 1–3 and high for short piles; $l_{2c}/l_m = 1\text{–}1.3(l_m < l/3)$ and $l_{2c} = l_m$ ($l_m > 0.5l$); and $p_{u2}/p_{u1} \leq 3$ (p_{u1}, p_{u2} are limiting on-pile FPUL in sliding and stable layer, respectively). Equation (1.2) is purported for large deformation (typically > 0.5 m) and small k_s, regardless of the soil properties.

Example 1.3 Modulus of subgrade reaction k_s

The k_s for a model pile (d = 0.05 m) in sand (ϕ = 38°) was deduced as ~45 kPa during passive loading (Guo and Qin 2010). Using Equation (1.2), for instance, the modulus k_{s2} for an embankment pile with d = 430 mm (Stewart et al. 1994), is estimated as ~387 kPa (= 45 × 430/50, at ϕ = 38°). This in turn allows k_s of the overlying clay layer to be estimated as ~22 kPa (= ~387/m, m = 17.7). This k_s value gained using Equation (1.2) is consistent with these deduced from piles subjected to lateral spreading (Ishihara and Cubrinovski 1998, 2004), such as k_s = 25.2–72 kPa (Pile No. 1, Niigata FCH Building), k_s = 64 kPa (PC-pile building in Higashi-Nada, Kobe), and k_s = 23 kPa (Tank TA72 at Mikagehama Island).

G6: The normalized rotational stiffness $\overline{k_\theta}$ is less than 0.3 from the capped model tests

Guo (2015b, 2016) demonstrates that soil movement during landslide, earthquake shaking, simple shear and embankment loading leads to a progressively increasing p_s on piles that resembles a series of stepwise uniform increases, regardless of the final sliding depth. The nonlinear pile response is underpinned by the same set of parameters as those for lateral piles (Guo 2015b, 2016), as discussed further in Chapters 5 and 6.

1.4.3 α ratio for p_u between active and passive piles

Numerical analyses (e.g., Chow 1996; Poulos 1995) normally stipulate a p_s of $(2.8\text{–}4)s_u d$ for piles in sliding clay (s_u = undrained shear strength of clay) (Viggiani 1981; Guo 2016). This assumption implies $p_s = \alpha p_{u1}$ using α = 1/3 for p_{u1} of laterally loaded piles (Randolph and Houlsby 1984). This factor α is corroborated theoretically in Chapter 4. Without explicitly acknowledging the factor α, (i) direct use of p–y curves has incurred inconsistent predictions of passive piles (Chen et al. 2002; Frank and Pouget 2008; Smethurst and Powrie 2007), and an order of magnitude different p_s values for piles in liquefied sand (Franke and Rollins 2013); and (ii) A factor of safety of 1.5 in design actually becomes 2.5 for sliding resistance Q_m for a given ultimate bending capacity (= M_m), slope-stabilizing pile. The significance of the α values is explored extensively in Chapter 5.

1.4.4 Model solutions and features

On-pile FPUL is also described by $p_s = A_{L1}z$ (z = depth measured from GL) with A_{L1} (= p_b/z_m) and p_b at rotation depth z_m. As developed previously (Guo 2015b, 2016), the pile response is calculated for elastic interaction (via the modulus k_s and m) under a uniform FPUL being equal to the limit p_s (for the soil movement) over the sliding depth l_m. The piece-wise 'elastic response' together constitutes the nonlinear response of the piles. An extensive experimental and numerical analysis has been conducted over the past decades on response of piles subjected to lateral spreading, embankment movement, and embankment piles. The study reveals the difficulty in selecting input parameters to gain successful numerical predictions of the pile response, and the dominant impact of the p_u profile and pile-soil relative stiffness on the predictions (Guo 2012). Chapters 4, 5, 7 and 8 elaborate issues concerning slope-stabilizing piles, embankment loading, lateral spreading and excavation loading, as highlighted next.

1.5 APPLICATION OF 2-/3-LAYER MODELS

1.5.1 Slope-stabilizing (rigid, flexible) piles

Piles are designed to arrest landslide (Fukumoto 1972, 1976; Poulos 1995; Guo 2013c) using numerical methods (Springman 1989; Stewart et al. 1994), which are occasionally corroborated against instrumented data. A lateral soil movement w_s may exert a 'uniform' on-pile FPUL p_s over a sliding depth l_m (see Figure 1.6a) and a sliding thrust H_s at the depth l_m (see Figure 1.6b). The pile response may be estimated analytically for (i) the FPUL profile of p_s (sliding layer) and mp_s (stable layer) with m = 3–4 (see Figure 1.6a) (Ito and Matsui 1975; De Beer and Carpentier 1977; Viggiani 1981; Chmoulian 2004); and as shown in Figure 1.6b for (ii) the thrust H_s assuming a uniform profile (Fukuoka 1977) or an inverse triangular profile of moving soil (Cai and Ugai 2003) by elastic solutions.

Unfortunately, the early solutions are not related to magnitude of the soil movement (see Figure 1.6). Guo (2003, 2013c) proposed to gain a fictitious load on a passive pile for each magnitude of soil movement (w_s) using plastic (depth 0 ~ l_m) and elastic-plastic (l_m ~ l) solutions (i.e., *P-EP* mtd) (l = pile embedment). The pile is subsequently analysed using the ep solutions for a laterally loaded pile. The model and solutions well capture nonlinear response of two infinitely long piles, and six upper-rigid (in sliding layer) and low-flexible (in stable layer) piles (Guo 2012) against measured data. Nevertheless, 'flexible' piles turn to 'rigid' in sliding layer (upper rigid and lower flexible) during lateral spreading, for which use of a concentrated load is not justified. The 2-layer and 3-layer models should be used to avoid overestimating bending moment (see Chapter 4). The **bi-portion** (of upper-rigid and low-flexible) pile needs to be modelled using BiP models (Guo 2021a, Chapters 4 and 8).

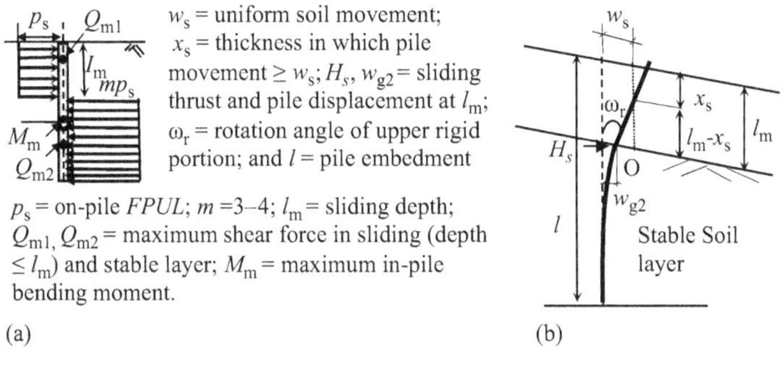

Figure 1.6 Models for a passive pile: (a) On-pile pressure model; (b) Equivalent load model (Guo 2013c). [Adapted from Guo, W. D., *Int J Numer Anal Meth Geomech* **47**(6), 2022a.]

All the available methods attempt to ensure the piles (in resisting landslide) within serviceability and limit states, but do not explicitly warrant a sufficient bending capacity of piles over a large displacement (i.e., ductile response). The limit states should be hinged on a level of coupled displacement and bending capacity (see Chapter 4).

1.5.2 Embankment loading (with restraining effect)

Piles underpinning bridge foundations are subjected to embankment loading (including lateral spreading). The pile behaviour has been modelled extensively numerically (Stewart et al. 1993; Jeong et al. 1995; Goh et al. 1997; Karim 2013). The accuracy depends on gaining reliable *LFPUL* p_u. The $p_u/(\sigma_v d)$ ratio ($\sigma_v = \gamma_s' z$, and z = depth), for instance, varies widely from unity (*JRA* 2002; González et al. 2009) to 0.26 (a single layer) or to 0.32 (2-layer soil) based on centrifuge tests on piles (Imamura et al. 2004). The ratio $p_u/(\sigma_v d)$ is close to S_r/σ_v ratio of 0.3–0.34 (S_r = residue strength for liquefied sand, Alba and Ballestero 2006; Miwaa et al. 2006; Mesri 2007). It is equal to the ratio α inflicted by rotational restraints (Guo 2016) between passive piles over laterally loaded piles. More importantly, the embankment induces rotational restraint on the piles.

1.5.3 Lateral spreading (limitation of *p–y* curves)

Significant research effort has been conducted over decades towards designing piles subjected to lateral spreading due to earthquake, including sophisticated shaking table tests, centrifuge tests and powerful numerical simulations (Jakrapiyanun 2002; Boulanger et al. 2003; Kagawa et al. 2004; Motamed and Towhata 2010; Juirnarongrit and Ashford 2006) among many others.

Several *p–y* curves (p = FPUL, y = local pile displacement) for liquefied soil are suggested, such as those using an average *p*-multiplier (Brandenberg et al. 2005b), an average residual strength (Seed and Harder 1990; Wang and Reese 1998; Olson and Stark 2002; Idriss and Boulanger 2007), and a dilation-based liquefaction model (Rollins et al. 2005). These *p–y* curves offer values of on-pile FPUL (thus p_u) that differ by up to an order of magnitude. This makes it difficult to gain reliable design (Chen et al. 2002; Frank and Pouget 2008). While useful through modifying residual strength and stiffness of the *p–y* curve for some circumstances (Franke and Rollins 2013), the prevalent *p–y* curve based methods are not suitable for ‘rigid' piles subjected to lateral spreading against 3D numerical analysis (McGann et al. 2011), as is well recognized for rigid piles underpinning wind-turbine foundations.

Given the difficulty in gaining reliable input parameters, simple methods have been developed for piles subjected to lateral spreading, especially for those of very low stiffness and strength liquefied sand (Dobry et al. 2003). These solutions provide interesting results, but are limited to a uniform distributed p profile. The 2- or 3-layer model and closed-form solutions offer promising outcome (see Chapters 3–8).

1.5.4 Response amplification

The 2-layer and 3-layer solutions (Guo 2015b, 2016) together with a set of five input parameters well capture the nonlinear response of piles in sliding soil induced by lateral spreading (Dobry et al. 2003), sliding sand (Guo and Ghee 2004), sliding slope (Frank and Pouget 2008), embankment loading (Guo 2016; Guo et al. 2017), excavation loading, and shaking embankment loading (Guo 2021a). In particular, the solutions well capture the 3–5 times higher bending moment induced by translational movement (Guo and Ghee 2004) than rotational movement, as well as the response of back-rotated, failed piles during earthquake (Guo 2021b), which are not seen in any other simulations. The models well interweaves all types of study together, and orchestrates a unitedly capturing of nonlinear response of the piles cost-effectively, as seen in Figure 1.1 and published by research groups worldwide. The solutions reveal a new mechanism of response amplification mathematically, which causes distortion / failure of piled bridges during lateral spreading (e.g., Showa bridge collapse, a mystery over six decades. Guo 2020a, b). This discovery is unlikely detected using numerical models or AI methods owing to its rare occurrence.

1.6 FLEXIBLE PILES UNDER PASSIVE LOADING (BIP MODELS)

Piles in sliding soil may behave either flexible in both sliding and stable layers (i.e., deep sliding), or rigid in sliding layer and flexible in stable layer (i.e., normal sliding) (Guo 2003, 2013c). In the case of deep sliding, the pile response can bc modelled using the elastic-plastic solutions. In the case of normal sliding, a sufficiently large displacement of piles (i.e., ductile failure) should be allowed prior to bending failure. This permits a plastic interaction between pile and soil in stable layer. To simulate flexible piles subjected to mixed (sliding-rotational) modes of soil movement, each pile is divided into upper and lower portions, which are joined together by a rotational spring (k_θ) at the depth of maximum bending moment z_m.

1.6.1 BiP models

The pile rotates rigidly about the rotation depth z_r ($\approx z_m$) to an angle ω_r and a GL deflection w_g (e.g., Figure 1.7b). It has a displacement $w(z)$ (= $\omega_r z + w_g$) at depth z ($<z_m$). Each stepwise

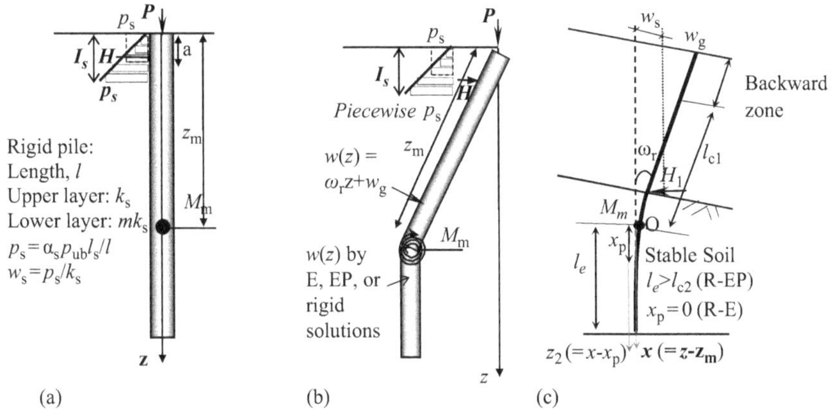

Symbols are defined in 2-layer model, and as follows:
Low section: E-model, EP model
I_c and I'_c = coefficients from elastic theory of lower segment; $k_\theta = -1/I'_c$;
$M_m = w_h/I_c$; w_h = displacement at depth z_m; M_L = moment from p_s about z_m.
Rigid low section: $w(z) = \omega_{rL} z + w_L$, w_L and ω_{rL} = projected displacement and rotation of the lower segment
BiP–k_θ model: $M_m = M_L + k_\theta(\omega_{rL} - \omega_r)$; M_L = moment from p_s about z_m.
BiP–η model: $\eta = \omega_r/\omega_{rL}$
H = anchor force at a depth a below GL; $(1-n)k$ = subgrade modulus of sliding, dragging layer; l_s = sliding depth.
α_s (excavation depth l_{exc}, m = 3) for profile only: 1.0–2.3 ($l_{exc}/l < 0.1$, stable);2.5 ($l_{exc}/l < 0.11$, metastable); 2.5–2.9 ($l_{exc}/l < 0.18$, collapse wall); α_s = 1.0 for overall response regardless of variable l_{exc}.

Figure 1.7 Nonlinear R-EP (R-E) models (Guo 2020b) for a passive pile: (a) loading p_s and M_m; (b) displacement of bi-portion; (c) effective length l_{c1}, l_e and slip depth x_p below M_m. [Adapted from Guo, W. D., *Int J Numer Anal Meth Geomech* **47**(6), 2022a.]

uniform p_s (over a depth 0 ~ l_s) together with the lateral force H (at a depth a below GL, Figure 1.7b) incur a bending moment M_L and a total passive moment of $M_L + k_\theta(-\omega_r)$ about the depth z_m. The maximum bending moment M_m (= $k_\theta\omega_r$) at z_m is correlated with a 'total' rotational stiffness k_θ along its embedment l. The bi-portions of a 'flexible' pile above and below depth z_m allow the development of analytical solutions (Guo 2021a) for a rigid low portion with a constant k_θ (BiP–k_θ) or a constant rotation ratio (η) between the bi-portions (BiP–η model), or an elastic low portion (R–E model, Figure 1.7c). Each mode is underpinned by five input parameters of modulus of soil k_s and k_{s2} (= nk_s below depth z_m), and pile-soil interaction of p_{ub}, α_s and k_θ. The BiP–k_θ, BiP–η, and R–E models and solutions (Chapter 8) well capture pile response including elastic rebound and brittle failure (at a small displacement), and reveal amplified response from stiffness singularity.

Among the five input parameters, the rotational stiffness k_θ is stipulated as a constant (Dobry et al. 2003; Guo 2015b) as is gained theoretically (Chapter 8). During the soil sliding, the ratio of the rotation angles of upper segment (ω_r) over the lower one (ω_{rL}) of the pile may stay as a constant η (termed as BiP–η model). Plastic hinge(s) should occur at a sufficiently high bending moment. The associated z_m should exceed $0.5l$, except for localized bending weakness. An explicit expression between the constant η and the stiffness k_θ is established in Chapter 8. The expression is useful to identify response amplification. The use of BiP models to analyse slope-stabilizing piles is provided in Chapter 4.

1.6.2 Excavation loading

Basement excavation work normally incurs soil movements behind retaining structure. Liquefaction may induce large flow of ground displacement, especially for loose sand. In either case, the soil movement/flow may inflict unacceptable large values of bending moment and deflection in existing pile foundations (Ishihara 1993; Miwaa et al. 2006; Haskell et al. 2013), which can incur catastrophic damage (Whittle and Davis 2006; Chai et al. 2014). To assess the safety of existing piles, centrifuge tests were conducted to study pile response when subjected to nearby excavation (Leung et al. 2000; Ong et al. 2003; Bourne-Webb et al. 2011), among many others. Nevertheless, none of the existing approaches provide integrated parameters for modelling piles in all cases, and the nonlinear response induced by mixed modes of soil movement. As mentioned early, the BiP models well capture ductile and brittle failure of piles next to excavation (Chapter 8).

1.7 1G TESTS FOR PASSIVE AND CYCLIC LOADING

To model the response of the rigid piles, Guo and Ghee (2004) developed a new shear apparatus (see Figure 1.8a). The square shear apparatus has an internal dimension of 1 × 1 m^2 in plan and 0.8 m in height. The upper part of the box is made of a series of 25-mm-thick square laminar steel frames underlying a 400-mm-high fixed timber box. Horizontal force was applied laterally (via the lateral jack) on a loading block to translate the aluminium frames of the upper portion of the shear box (and the adjacent sand). The loading block was made to a uniform (U), an inverse triangular (T, with a loading angle θ of 16.7°) (as shown in Figure 1.8a) and an arc (A) shape. The blocks generates a U, T or A profile of soil movement (thus referred to as U, T or A profiles) at the loading location (of a distance s_b from the centre of the shear box), respectively, but an unknown sand movement across the shear box and around the test pile. The impact of the s_b is quantified previously (Guo 2012).

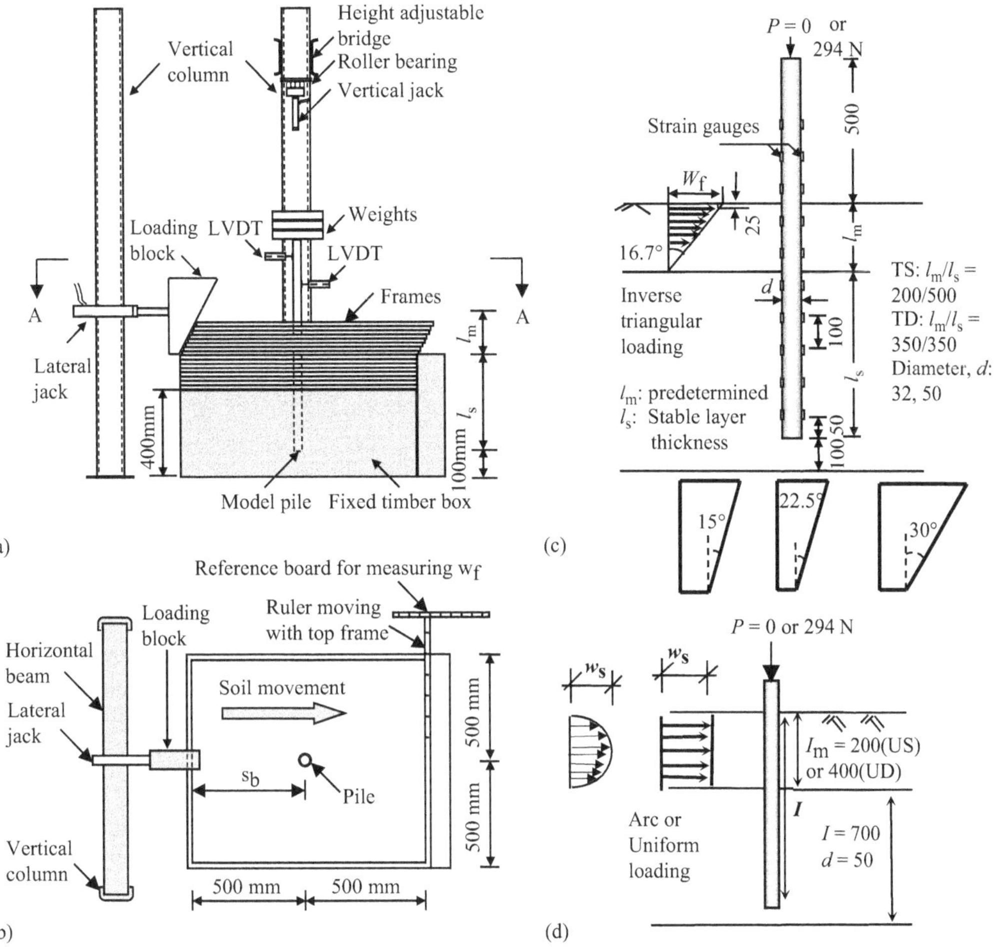

Figure 1.8 Schematic diagram of shear box: (a) elevation view; (b) A-A Plan view; (c) single piles with inverse triangular movement; (d) uniform or Arc movement. [Adapted from Guo, W. D., H.Y. Qin and E. H. Ghee, *Int J Geomechanics* **17**(4), 2017.]

1.7.1 Model pile and sand properties

The model piles tested, referred to as d_{32} or d_{50} piles, were all made of aluminium tube with 1,200 mm in length. The d_{32} piles have an outside diameter d of 32 mm, a wall thickness t of 1.5 mm, and E_pI_p (calculated bending stiffness) of 1.28 × 10^6 kN mm^2; whereas the d_{50} piles have d = 50 mm, t = 2.0 mm, and E_pI_p = 5.89 × 10^6 kN mm^2. The d_{50} and d_{32} piles were tested to model rigid and flexible piles, respectively. The pile was instrumented with 10 pairs of strain gauges at an interval of 100 mm. Prior to the testing, the strain gauges were calibrated by exerting a transverse load in the middle of the pile that was clamped at both ends. Various magnitudes of the load were applied, and measured voltages were compared with calculated strains to establish a strain/voltage relationship. A calibration factor was obtained for each gauge, allowing a measured strain to be converted to an actual strain. To protect from damage, the gauges were covered with 1 mm epoxy and wrapped by tapes. Two dial gauges (LVDTs) were mounted around pile cap at a height of 495 and 445 mm above GL, respectively (Figure 1.8a) to allow a measured accuracy of 0.00002 radians.

The model ground was medium-grained quartz (Queensland) sand with a uniformity coefficient C_u and coefficient of curvature of C_c of 2.92 and 1.15, respectively. The sand was discharged into the shear box through a rainer hanging over the box to achieve a reasonably uniform density of the sand within the shear box. The falling height of sand was chosen as 600 mm, resulting in a uniform relative density of about 89%, and a unit weight of 16.27 kN/m^3. The angle of internal friction was 38° as evaluated from direct shear tests. The d_{32} or d_{50} pile was instrumented with strain gauges, and jacked into a depth of 0.7 m in the model sand. It may be subjected to an axial load P of 294 N, which is ~10% the maximum jack-in resistance (during the pile installation) of 3.8 kN (d_{32}) and 5.4 kN (d_{50}), respectively. The load was applied on the pile-head by using a connector located 500 mm above the sand surface (Figure 1.8a).

1.7.2 Test results

Each test provides readings of ten pairs of strain gauges (along the pile length), two LVDTs (for measuring displacements at the pile-head level and pile rotation), and the force on the lateral jack under each frame movement. Numerical (using the trapezoidal rule) integration of $M(z)$ offer the rotation profile $\omega_r(z)$, which is integrated to gain the displacement $w(z)$ (incorporating the measured GL rotation ω_r and cap-level displacement w_{gc}). The differentiation of $M(z)$ offers the shear force $Q(z)$ and on-pile force per unit length $p(z)$. They in turn allow for maximum shear force Q_m, maximum bending moment M_m and its depth x_m, rotation angle ω_r, and pile-head deflection w_g. A cubic polynomial (by least squares) is used to fit five successive sets of equally spaced measured bending moment data, which is then differentiated at the central point to calculate the soil reaction $p(z)$ by imposing zero moment and shear force at the pile-tip. The calculation procedure has been compiled into a purposely designed spreadsheet program using Microsoft Excel VBA (Guo and Qin 2010).

1.7.3 Lateral cyclic loading

In the lateral cyclic loading tests, the apparatus of Figure 1.8 was modified and used. A vertical jack is used to install piles into the storage box. A triangular steel frame with an aluminium pulley was manufactured and clamped on the vertical columns to support the lateral loading system. The pile was loaded laterally by means of weights added to a loading pan carried by flexible wire acting over the pulley and attached to the pile at an eccentricity above the ground surface. A hydraulic jack was placed underneath the loading pan. Extending the stroke of the hydraulic jack lift the weights up thus unloading, while releasing the jack loaded the pile. Careful attention was paid when releasing the jack to ensure the pile was gradually loaded at a rate of about 2.0 mm/min without impact. Thus, any inertia effect and rate effect are negligible.

1.8 FUTURE RESEARCH

Guo (2012) presented compact, closed-form solutions (underpinning by ~5 soil related parameters) to capture nonlinear response of hundreds of piles under active (vertical, or lateral) loading right up to failure. This book provides the same type of united solutions (2-/3-layer, BiP, R-E and R-EP models by Guo 2014 through 2022a, b) for modelling piles in sliding soil and based on the parameters for laterally loaded piles.

The 2-/3-layer models should be developed further to account for impact of rotationally restrained (grouted, screw, branched) piles on both external loading and soil movement; as well as to model geo-beams in sliding soil (embankment, slope, etc.). The amalgamation of the analyses will significantly enhance cost-effective design of geo-beams and excogitate novel numerical approaches. The 'restrained' piles raise vertical and lateral capacity against conventional piles, and reduce carbon emissions (by 50–70% for piling).

The solutions largely start with a 'discovered' (rigorous) model, and concluded with concise, compact expressions that are handy for practical use. The development of convergent (models) solutions for challenging problems (*initiatives*) requires synergic (experimental, numerical and analytical) skills and puissance in modelling, evolved from integrating dedication, physical perspicacity, in-depth cross-field knowledge and skills. The solutions allow input parameters to be deduced consistently against test data, and/or numerical results. This effectively eradicates the root of inconsistency among diverse methods, and inaccuracy vexed by many parameters. The study, in foreseeable future, will allow a geo-technician to do complicated design by keying a few parameters (via cloud) at a fraction of the current cost.

Chapter 2

Laterally loaded pile groups

2.1 INTRODUCTION

Piles are commonly cast into a pile cap that restrains their rotation (Mokwa and Duncan 2003; Guo 2009), while still allowing for horizontal translation during lateral loading. However, at working load levels, cap-rotation occurs owing to relaxation of soil resistance beneath the pile-cap or around the piles, insufficient cap-restraint and possible cracking of the piles (Ooi et al. 2004; Guo 2009). This can lead to translation and rotation, resulting in up to fourfold difference in resistance for rigid piles. To ensure reliable design, especially for wind-turbine foundations and those structures subjected to earthquake loading, this difference must be properly considered.

To simulate lateral pile-soil interaction, a load transfer model is adopted with a series of independent springs-sliders distributed along the shaft and a membrane to incorporate couple interaction among the springs (e.g., Figure 1.2b, Chapter 1). The slider is characterized by the profile of limiting force per unit length, FPUL (p_u profile) to a slip depth x_p, and the spring has a modulus of subgrade reaction k_s. Elastic-plastic closed-form solutions (Guo 2012) were developed for free-head rotation (FreH piles) and fixed-head (no rotation, FixH piles) to predict pile behaviour. The FixH solutions are also used to predict response of pile groups by incorporating the pile-pile interaction factor p_m (= p-multipliers ≤ 1.0) (Brown et al. 1998) (see Figures 1.2c and e) to use a reduced limiting FPUL $p_m p_u$ and modulus $p_m k_s$.

Fixed-head solutions using numerical finite element and finite difference methods (Ooi et al. 2004) generally overestimate the maximum bending moment and underestimate the deflection against measured data for capped piles (Duncan et al. 2005). Free-head solutions, on the other hand, offer an incorrect depth of maximum bending moment and overestimate the head-deflection. In contrast, the use of p-multiplier, while less rigorous, provides a more reliable and efficient prediction of overall pile response than numerical methods (Guo 2009; Guo 2012). Nevertheless, to obtain reliable solutions, the ~4 times difference in resistance due to cap-rotational stiffness k_G, must be quantified to conform with the p_m. To meet more stringent design requirements, Guo (2015a) has developed closed-form solutions in the framework of the load transfer model for piles with a cap stiffness k_G. These solutions provide the impact of the stiffness k_G in non-dimensional charts and the key correlation between the stiffness k_G and the p_m.

2.2 OVERALL SOLUTIONS FOR A SINGLE PILE

2.2.1 Load transfer model

In Chapter 1, Figure 1.2b shows a load transfer model to simulate the pile-soil system under a pile-head load H using both the uncoupled (with $N_p = 0$, plastic zone) and coupled ($N_p \neq 0$,

 DOI: 10.1201/9781003315230-2

elastic zone) (Guo 2006). The pile-head has a rotational stiffness k_G, and a ground-level (GL) bending moment M_o of $k_G\omega_g$, where ω_g = pile-head rotation angle in radian, at zero loading eccentricity of $e_p = 0$. The pile-soil system retains the essential feature of free-head ($k_G = 0$) and fixed-head piles ($k_G > 10E_pI_p$, E_pI_p = flexural stiffness of a pile), as recaptured in Chapter 1. The net limiting FPUL, p_u along the pile shaft (see Figure 1.2c) is described by (Guo 2006)

$$p_u = A_L\left(x+\alpha_o\right)^n \tag{2.1}$$

$$A_L = \tilde{s}_u N_g d^{1-n} \text{ (cohesive soil), and } A_L = \gamma_s' N_g d^{2-n} \text{ (cohesionless soil)} \tag{2.2}$$

where d = pile diameter; x = depth below GL; α_o = an equivalent depth to include the p_u at GL; n = 0.5–0.7 for a uniform strength profile (e.g., clay), and 1.3–1.7 for a linearly increasing strength profile (e.g., sand); $\tilde{s}_u$ = average undrained shear strength s_u of the soil over the maximum x_p; N_g = gradient to convert clay strength or sand density to the limiting p_u; γ'_s = effective unit weight of the overburden soil (i.e., dry weight above water table and buoyant weight below). The parameters α_o, N_g, and n were deduced using measured response (Guo 2013a). The p_u from Equation (2.1), being less than $9.14–11.94s_ud$ for clay (Randolph and Houlsby 1984), is only effective up to the maximum x_p, and may be estimated using angle of internal friction and cohesion of a subsoil (Hansen 1961; Guo 2013a).

The uncoupled and coupled load transfer models (Hetenyi 1946; Guo and Lee 2001) allow the governing equations for the pile (see Figure 1.2) to be obtained as

$$E_pI_pw^{IV}(x) = -p_u \quad \left(\text{Elastic zone, } 0 \le x \le x_p\right) \tag{2.3}$$

$$E_pI_pw^{IV}(z) - N_pw''(z) + k_sw(z) = 0 \quad \left(\text{Plastic zone}, x_p \le x \le l, \text{ or } 0 \le z \le l - x_p\right) \tag{2.4}$$

where $w(x)$ = pile deflection at depth x; $w^{IV}(x)$ = 4th derivative of $w(x)$ with respect to x; I_p and E_p = moment of inertia and Young's modulus of an equivalent solid cylindrical pile, respectively. As with free-head and fixed-head piles (Guo 2006), response of the lateral pile is presented against the depth x (measured from GL) in the upper plastic zone, and a depth z (= $x-x_p$, measured from the slip depth x_p) in the lower elastic zone, respectively. The values of N_p and k_s are calculated using the average modulus G_s of the soil over the effective length l_c (see Table 1.1, Chapter 1).

2.2.2 Pile-head rotational stiffness k_G

The rotational stiffness of the pile-cap k_G is taken as that of pile-head for a fully cast concrete cap with sufficient rigidity. It is worth noting that the rotational stiffness k_G affects the magnitude of the p_u [and the multiplier N_g, see Equation (2.2)] (Guo 2009). The superscripts 'FreH'and 'FixH' are used to denote free-head and fixed-head piles. Elastic theory for a laterally loaded rigid pile provides the limit state of $p_mN_g^{FixH*} = N_g^{FreH*}$ for $4w_g^{FixH} = w_g^{FreH}$ and $p_mk^{FreH} = k^{FixH}$. Note that the w_g^{FreH}/w_g^{FixH} ratio exceeds far greater than 4 once plastic deformation is induced (Guo 2013b). The N_g with a stiffness k_G is determined using $N_g^{FixH*} = p_mN_g^{FreH}$. The bending moment M_o at GL may be measured for a lateral load H with a loading eccentricity e_p above GL. Owing to semi-fixity, the moment M_o is a fraction of the moment (M_o^{FixH}) induced on a pile-head with translation-only (FixH) pile, e.g., $M_o/M_o^{FixH} = 0.4$ in the Chinese JGJ design code (JGJ 1994). These ratios will be used later to determine the stiffness k_G.

2.2.3 Elastic-plastic solutions

Equations (2.3) and (2.4) were resolved using the same approach as for free-head piles (Guo 2006). The bending moment $-M_o = E_pI_pw''(0) = k_G\omega_g + He_p$, and the shear force $-Q(0) = H$ were enforced at the pile-head level ($x = 0$). Note that the ω_g is negative to offer a counter moment against He_p. The elastic-plastic solutions obtained are provided in Tables 2.1 and 2.2, for the response profiles and head response, respectively. These solutions involve the reciprocal

Table 2.1 Expressions for response profiles of a semi-fixed-head pile (Guo 2015a)

(a) Response profiles in plastic zone $\left(\bar{x} \le \bar{x}_p\right)$:

$$w\left(\bar{x}\right) = \frac{4A_L}{k_s\lambda^n}\left\{-F\left(4,\bar{x}\right)+F\left(4,0\right)+F\left(3,0\right)\bar{x}+\left[k_{nG}\bar{\omega}_g+\bar{H}\bar{e}_p+F\left(2,0\right)\right]\frac{\bar{x}^2}{2}+\left[F\left(1,0\right)+\bar{H}\right]\frac{\bar{x}^3}{6}\right\}+\frac{\omega_g}{\lambda}\bar{x}+w_g$$

$$w'\left(\bar{x}\right) = \frac{4A_L\lambda}{k_s\lambda^n}\left[-F\left(3,\bar{x}\right)+F\left(3,0\right)+\left[k_{nG}\bar{\omega}_g+\bar{H}\bar{e}_p+F\left(2,0\right)\right]\bar{x}+\left[F\left(1,0\right)+\bar{H}\right]\frac{\bar{x}^2}{2}\right]+\omega_g$$

$$-M\left(\bar{x}\right) = E_pI_pw''\left(x\right) = \frac{A_L}{\lambda^{2+n}}\left[-F\left(2,\bar{x}\right)+F\left(2,0\right)+\left(F\left(1,0\right)+\bar{H}\right)\bar{x}+k_{nG}\bar{\omega}_g+\bar{H}\bar{e}_p\right]$$

$$-Q\left(\bar{x}\right) = E_pI_pw'''\left(x\right) = \frac{A_L}{\lambda^{1+n}}\left[-F\left(1,\bar{x}\right)+F\left(1,0\right)+\bar{H}\right]$$

where $k_G = k_G\omega_g\lambda^{2+n}/A_L = k_{nG}\bar{\omega}_g$; $F\left(m,\bar{x}\right) = \left(\bar{x}+\bar{\alpha}_o\right)^{n+m}/(n+m)\ldots(n+2)(n+1)$ $(1 \le m \le 4)$; $\bar{\alpha}_o = \lambda\alpha_o$; $F\left(0,\bar{x}\right) = \left(\bar{x}+\bar{\alpha}_o\right)^n$; and $\bar{e}_p = \lambda e_p$. Note both ω_g and $\bar{\omega}_g$ are used in the $w\left(\bar{x}\right)$ expression.

(b) Response profiles in elastic zone $\left(\bar{x} > \bar{x}_p,\ \text{or}\ \bar{z} > 0,\ \bar{z} = \lambda z = \lambda\left(x - x_p\right)\right)$:

$$w\left(\bar{z}\right) = e^{-\alpha_N\bar{z}}\left[C_5\cos\left(\beta_N\bar{z}\right)+C_6\sin\left(\beta_N\bar{z}\right)\right]$$

$$w'\left(\bar{z}\right) = \lambda e^{-\alpha_N\bar{z}}\left[\left(-\alpha_NC_5+\beta_NC_6\right)\cos\left(\beta_N\bar{z}\right)+\left(-\beta_NC_5-\alpha_NC_6\right)\sin\left(\beta_N\bar{z}\right)\right]$$

$$-M\left(\bar{z}\right) = E_pI_pw''\left(z\right) = E_pI_p\lambda^2e^{-\alpha_N\bar{z}}\left\{\begin{matrix}\left[\left(\alpha_N^2-\beta_N^2\right)C_5-2\alpha_N\beta_NC_6\right]\cos\left(\beta_N\bar{z}\right)+\left[2\alpha_N\beta_NC_5+\left(\alpha_N^2-\beta_N^2\right)C_6\right] \\ \sin\left(\beta_N\bar{z}\right)\end{matrix}\right\}$$

$$-Q\left(\bar{z}\right) = E_pI_pw'''\left(z\right) = E_pI_p\lambda^3e^{-\alpha_N\bar{z}}\left\{\begin{matrix}\left[-\alpha_N\left(\alpha_N^2-3\beta_N^2\right)C_5+\beta_N\left(3\alpha_N^2-\beta_N^2\right)C_6\right]\cos\left(\beta_N\bar{z}\right) \\ +\left[-\left(3\alpha_N^2-\beta_N^2\right)\beta_NC_5-\alpha_N\left(\alpha_N^2-3\beta_N^2\right)C_6\right]\sin\left(\beta_N\bar{z}\right)\end{matrix}\right\}$$

where

$$C_5 = \frac{2A_L}{k_s\lambda^n}\left\{\left(1-2\alpha_N^2\right)\left[F\left(2,\bar{x}_p\right)-F\left(2,0\right)-k_{nG}\bar{\omega}_g-\bar{H}\bar{e}_p\right]-\alpha_NF\left(1,\bar{x}_p\right)+\left[\alpha_N-\left(1-2\alpha_N^2\right)\bar{x}_p\right]\left[F\left(1,0\right)+\bar{H}\right]\right\}$$

$$C_6 = \frac{2A_L}{k_s\beta_N\lambda^n}\left\{\begin{matrix}\alpha_N\left(2\alpha_N^2-3\right)\left\{-F\left(2,\bar{x}_p\right)+F\left(2,0\right)+\left[F\left(1,0\right)+\bar{H}\right]\bar{x}_p+k_{nG}\bar{\omega}_g+\bar{H}\bar{e}_p\right\} \\ +\left(\alpha_N^2-1\right)\left[-F\left(1,\bar{x}_p\right)+F\left(1,0\right)+\bar{H}\right]\end{matrix}\right\}$$

$$\alpha_N = \sqrt{1+N_p/\sqrt{4E_pI_pk_s}},\ \beta_N = \sqrt{1-N_p/\sqrt{4E_pI_pk_s}}$$

Table 2.2 $\bar{H}$, $\bar{w}_g$ and $\bar{M}_o$ of a semi-fixed-head pile (Guo 2015b)

(a) Normalized pile-head load, $\bar{H}$

$$\bar{H} = \frac{F(2,\bar{x}_p) - F(2,0) + \alpha_N\left[F(1,\bar{x}_p) - F(1,0)\right] + 0.5F(0,\bar{x}_p) - \bar{x}_p F(1,0)}{\alpha_N + \bar{x}_p + \bar{e}_p} - \frac{k_{nG}\bar{\omega}_g}{\alpha_N + \bar{x}_p + \bar{e}_p}$$

The $\bar{H}$ is deduced from the following relationship obtained for the depth, x_p (z = 0):

$$w_p^{IV} + \alpha_N w_p''' + 2\lambda^2 w_p'' = 0$$

where w''_p, w'''_p, and w_p^{IV} are values of 2nd, 3rd, and 4th derivatives of $w(x)$ with respect to depth z.
Given $\bar{x}_p = 0$, the minimum head load to initiate slip is obtained.

(b) Normalized ground-line deflection, $\bar{w}_g$

$$\bar{w}_g = 4\left[F(4,\bar{x}_p) - F(4,0) - \bar{x}_p F(3,\bar{x}_p)\right] - 2\left(2\alpha_N^2 - 1\right)\left[F(2,\bar{x}_p) - F(2,0) - k_{nr}\bar{\omega}_g\right] - \left(2k_{nr}\bar{x}_p + 1\right)\bar{x}_p\bar{\omega}_g$$
$$- 2\alpha_N F(1,\bar{x}_p) - 2F(2,0)\bar{x}_p^2 + 2\left[F(1,0) + \bar{H}\right]\left[\frac{-1}{3}\bar{x}_p^3 + \left(2\alpha_N^2 - 1\right)\bar{x}_p + \alpha_N\right] + 2\vec{e}_p\bar{H}\left[2\alpha_N^2 - 1 - \bar{x}_p^2\right]$$

The w_g is deduced from $w(x)$.

(c) Normalized rotation at ground level, $\bar{\omega}_g$

$$\bar{\omega}_g = \frac{4\left(\bar{x}_p + \alpha_N + \bar{e}_p\right)\left[F(3,\bar{x}_p) - F(3,0)\right] - 2\left(\bar{x}_p^2 - 2\alpha_N^2 + 1 + 2\bar{x}_p\bar{e}_p\right)F(2,\bar{x}_p)}{2k_{nG}\left(\bar{x}_p^2 + 2\alpha_N\bar{x}_p + 2\alpha_N^2 - 1\right) + \bar{x}_p + \alpha_N + \bar{e}_p}$$
$$- \frac{2\left(\bar{x}_p^2 + 2\alpha_N\bar{x}_p + 2\alpha_N^2 - 1\right)F(2,0) + 2\left[\bar{x}_p\left(\alpha_N\bar{x}_p + 2\alpha_N^2 - 1\right) + \bar{e}_p\left(2\alpha_N\bar{x}_p + 2\alpha_N^2 - 1\right)\right]F(1,\bar{x}_p)}{2k_{nG}\left(\bar{x}_p^2 + 3\alpha_N\bar{x}_p + 2\alpha_N^2 - 1\right) + \bar{x}_p + \alpha_N + \bar{e}_p}$$
$$- \frac{\left[\bar{x}_p^2 + 2\alpha_N\bar{x}_p + 1 + 2\left(\alpha_N + \bar{x}_p\right)\right]F(0,\bar{x}_p) - 2\bar{e}_p\left(\bar{x}_p^2 + 2\alpha_N\bar{x}_p + 2\alpha_N^2 - 1\right)F(1,0)}{2k_{nG}\left(\bar{x}_p^2 + 3\alpha_N\bar{x}_p + 2\alpha_N^2 - 1\right) + \bar{x}_p + \alpha_N + \bar{e}_p}$$

(d) Normalized ground-level bending moment

$$-\bar{M}_o = \bar{H}\bar{e}_p + k_{nG}\bar{\omega}_g$$

Note: The constants C_i are determined using the compatibility conditions of $Q(\bar{x})$, $M(\bar{x})$, $w'(\bar{x})$, and $w(\bar{x})$ at the normalized slip depth, $\bar{x}_p\left[\bar{x} = \bar{x}_p\right]$ or $\bar{z} = 0$. Elastic solutions validated for $N_g < 2(k_s E_p I_p)^{0.5}$ is ensured by $L > L_c$ + max. x_p

of a characteristic length $\lambda\left[= \sqrt[4]{k_s/(4E_p I_p)}\right]$, the on-pile FPUL $p = p_u$ at $x \le x_p$ (see Figure 1.2c, Chapter 1), the normalized ground-level resistance $\bar{\alpha}_o (= \lambda\alpha_o)$, and the two coupled parameters α_N and β_N in the elastic zone ($\alpha_N = \beta_N = 1$ for uncoupled springs with $N_p = 0$). The response profiles are dominated by the normalized depths $\bar{x} (= \lambda x)$ and $\bar{z} (= \bar{x} - \bar{x}_p, \bar{x}_p = \lambda x_p)$, respectively, for plastic and elastic zones. The solutions provide the normalized pile-head oad $\bar{H}\left(= H\lambda^{n+1}/A_L\right)$, the ground-level deflection w_g (via $\bar{w}_g = w_g k_s \lambda^n / A_L$) and the rotation ω_g (via $\bar{\omega}_g = \omega_g k_s \lambda^{n-1} / A_L$) for the normalized stiffness k_{nG} [$= k_G\lambda^3/k_s$].

2.2.4 Simplified expressions

Table 2.2 provides the dimensionless expressions of $\bar{H}$, $\bar{w}_g$, and $\bar{\omega}_g$, which largely reflect the consequence of the mobilization depth (via $\bar{x}_p$) of the limiting FPUL p_u along a laterally loaded pile. When on-pile resistance at ground-line and coupled interaction are negligible (thus $\alpha_o \approx 0$ and $N_p \approx 0$), these expressions can be simplified to

$$\bar{\omega}_g = \frac{\bar{x}_p^n}{2(\bar{x}_p+1)^2 k_{nG} + 1 + \bar{x}_p + \bar{e}_p}\left\{\begin{array}{l} \dfrac{4\bar{x}_p^3(\bar{x}_p+\bar{e}_p+1)}{(n+1)(n+2)(n+3)} + 2\bar{x}_p^2\dfrac{(1-\bar{x}_p^2-2\bar{x}_p\bar{e}_p)}{(n+1)(n+2)} \\ -2\bar{x}_p\dfrac{\bar{x}_p(1+\bar{x}_p)+(1+2\bar{x}_p)\bar{e}_p}{n+1} - (1+\bar{x}_p)(1+\bar{x}_p+2\bar{e}_p) \end{array}\right\} \quad (2.5)$$

$$\bar{H} = \frac{0.5\bar{x}_p^n\left[(n+1)(n+2)+2\bar{x}_p(2+n+\bar{x}_p)\right]}{(\bar{x}_p+1+\bar{e}_p)(n+1)(n+2)} - \frac{k_{nG}\bar{\omega}_g}{\bar{x}_p+1+\bar{e}_p} \quad (2.6)$$

$$\begin{aligned}\bar{w}_g = &-\bar{\omega}_g\bar{x}_p\left(\frac{2k_{nG}\bar{x}_p(2\bar{x}_p+3)}{3(\bar{x}_p+\bar{e}_p+1)}+1\right) \\ &+\frac{4\bar{x}_p^{4+n}}{(n+1)(n+2)(n+3)(n+4)} - \frac{\left[2\bar{x}_p^2+(2\bar{x}_p+n+1)(n+2)\right](\bar{x}_p+3\bar{e}_p)\bar{x}_p^{2+n}}{3(1+\bar{x}_p+\bar{e}_p)(n+1)(n+2)} + \bar{x}_p^n\end{aligned} \quad (2.7)$$

where $k_{nG} = k_G\lambda^3/k_s = 4k_G/(E_pI_p\lambda) = \bar{k}_G/\bar{\omega}_g$. The maximum bending moment is likely equal to the moment M_o at GL (with $\bar{M}_o = \bar{H}\bar{e}_p + k_{nG}\bar{\omega}_g$) for semi-FixH piles. It is important to note that the stiffness k_G has a significant effect on the solutions, but it is often difficult to estimate accurately. Some numerical program manuals suggest determining the k_G based on the upper structure behaviour, but this may not always be feasible. Overall, the solutions in Tables 2.1 and 2.2 share similar features with the FreH and FixH solutions (Guo and Lee 2001; Guo 2006), but with different N_g and n (thus A_L) values.

Example 2.1 Estimation of stiffness k_G

Assuming a design ratio $M_o/M_o^{FixH} = 0.4$ (JGJ 1994), the k_G is deduced using Equations (2.5) and (2.6) for the elastic case, which in turn is simplified as Equation (2.8) for $e_p = 0$.

$$\frac{\bar{\omega}_g k_{nG}}{\bar{\omega}_g k_{nG}^{FixH}} = \frac{\omega_g k_{nG}}{\omega_g k_{nG}^{FixH}} p_m^{0.25(n-1)} = \frac{M_o}{M_o^{FixH}} p_m^{0.25(n-1)} \quad (2.8)$$

The FixH condition is enforced using a $k_{nG}^{FixH} > 50$ (i.e., k_{nG} for a fully fixed-head). The k_{nG} values are determined for $n = 0.7$ and 1.7, which are suitable for most piles in clay and sand, respectively. Using Equation (2.8) (taking $\bar{x}_p = 0.0001$) and $M_o/M_o^{FixH} = 0.4$, the k_{nG} at $n = 0.7$ is deduced as 0.275–0.333 with k_{nG}/p_m = 0.275/0.2, 0.288/0.3, 0.298/0.4, 0.306/0.5, 0.313/0.6, 0.319/0.7, 0.324/0.8, 0.329/0.9, and 0.333/1.0, respectively. Similarly, the k_{nG} at $n = 1.7$ is obtained as 0.333–0.564, with k_{nG}/p_m = 0.564/0.2, 0.488/0.3, 0.443/0.4, 0.412/0.5, 0.389/0.6, 0.371/0.7, 0.356/0.8, 0.344/0.9, and 0.333/1.0, respectively. These values may be used for initial design based on the JGJ code. Note that $k_{nG} = 0.5$ for n = 0.7–1.7 is deduced using Equation (2.9) for elastic case and $w_g/w_g^{FixH} = 1.25$ (JGJ 1994), although this is yet to be confirmed.

$$\frac{\bar{w}_g}{\bar{w}_g^{FixH}} = \frac{w_g}{w_g^{FixH}} p_m^{0.25n} \quad (2.9)$$

2.2.5 Amplification

The solution presented has a singularity at the normalized stiffness $k_{nG}{}^*$:

$$k_{nG}^* = -\frac{1+\overline{x}_p+\overline{e}_p}{2\left(\overline{x}_p+1\right)^2} \tag{2.10}$$

The value of the stiffness is affected by the normalized loading eccentricity and slip depth.

2.3 PARAMETRIC ANALYSIS

The closed-form (CF) solutions for capped piles can be reduced to those of free-head piles at k_{nG} = 0 (Guo 2006) and to those of the fixed-head piles at k_{nG} > 50 (Guo 2009). These solutions have shown good agreement with both finite element approach and experimental results. The nonlinear response of capped piles was examined for a rotational stiffness k_G = (0–10)$E_pI_p\lambda$ at the typical n = 0.7 and 1.7, respectively. Figure 2.1a shows the normalized load $\left(\overline{H}\right)$ –displacement $\left(\overline{w}_g\right)$ curves at GL for n = 0.7 and 1.7, while Figure 2.1b illustrates normalized displacement $\left(\overline{w}_g\right)$ –bending moment $\left(\overline{M}_o\right)$ curves. The normalized load $\left(\overline{H}\right)$ –bending moment $\left(\overline{M}_o\right)$ curves at GL are depicted in Figure 2.2a, and normalized displacement $\left(\overline{w}_g\right)$ –head-rotation angle $\left(\overline{w}_g\right)$ curves are shown in Figure 2.2b. The profiles of non-dimensional displacement, slope, bending moment and shear force, for example, at $\overline{x}_p = 1$ are illustrated in Figures 2.3a and 2.4a through d for n = 0.7 and 1.7, respectively.

For piles in clay (n = 0.7), Figure 2.5a shows the normalized load $\left(\overline{H}\right)$ –displacement $\left(\overline{w}_g\right)$ curves and normalized $\overline{H}$ –bending moment $\left(\overline{M}_o\right)$ curves; while Figure 2.5b presents the normalized load $\overline{H}$ -bending moment $\overline{M}_o$ and moment $\left(\overline{M}_o\right)$ –normalized rotation angle $\left(\overline{\omega}_g\right)$. The results indicate that pile response is amplified at a negative stiffness (back-rotation of pile head) of –(0.2–0.4). The impact of rotational stiffness on pile response is presented in the following example.

Example 2.2 A Pile in 10-pile group in clay (Matlock et al. 1980)

Matlock et al. (1980) performed lateral loading tests on piles at a site in Harvey, Louisiana. The tests comprised a single pile, as well as two circular groups consisting of 5-pile and 10-pile each, respectively in a 2.4-m-deep pit. The site consisted of a highly plastic grey-clay, with occasional thin layers of peat, sand or silt within a depth of 2.4–18 m. The plasticity index was 77–100 (at a depth 0–1.2 m below the test pit) and 100–185 (1.2–2 m below), respectively, which allow the angle of the soil friction ϕ (drained case) to be estimated as 12 degree (BSI 1985), and a cohesion c as 1 kPa (Guo 2013a). This analysis is focused on the 10-pile group.

The tubular steel piles were installed in a circle (see insert of Figure 2.6) at a centre-to-centre spacing of 1.8 pile diameters. Each pile was 13.4 m in length, 168 mm in outside diameter, 7.1 mm in wall thickness, and had a bending stiffness of 2.326 MN-m^2. The piles were driven 11.6 m into a uniform soft clay with an undrained shear strength s_u of 20 kPa. To simulate FixH restraints, the group was subjected to deflection loading at two support levels, 0.305 and 1.83 m above ground-line. The measured curves of average load per pile (H_{av}) versus pile deflection (w_t) and w_t versus M_o for a pile in the group are plotted in Figures 2.6b and c, respectively. In addition, the bending moment profiles were measured and are plotted in Figure 2.6d for four typical values of ground-level deflection w_g.

The response of a capped pile in the group was predicted using free-head and fixed-head solutions with n = 0.85, p_m = 0.2 (for either solution), and G_s = 13s_u (fixed-head

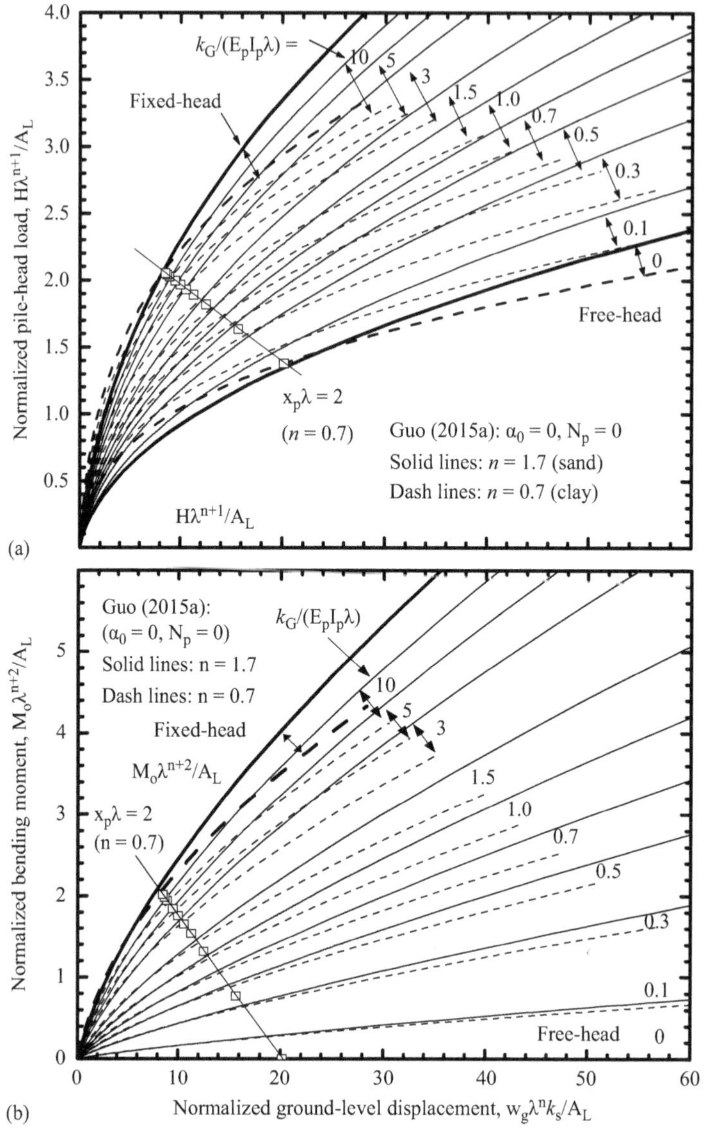

Figure 2.1 Nonlinear response of capped piles (n = 0.7 and 1.7): (a) displacement-load; (b) displacement-bending moment. [Adapted from Guo, W. D., *J of Engrg Mechanics* **141**(9), 2015a.]

solutions) or $33s_u$ (free-head solutions). The respective fictitious profile of limiting force per unit length p_u is shown in Figure 2.6a. Note that the semi-fixed-head constraints and less interaction from the special (circle) layout of the group results in the correlation of $G_s^{FixH} \approx 0.39G_s^{FreH}$, otherwise $G_s^{FixH} = p_m G_s^{FreH}$. The predicted H_{av}–w_g curves are plotted in Figure 2.6b, which agree remarkably well with the measured H_{av}–w_t curve (thus $w_t \approx w_g$). However, the predicted M_o values using fixed-head solutions are much higher than the measured values (see Figures 2.6c and d) (Guo 2009), even after deducting the moment $H_{av}e_p$ caused by loading eccentricity above GL e_p.

The solutions of caped piles are then used to examine impact of rotational stiffness k_G on the modulus G_s, the limiting FPUL p_u, and the response of a pile in the group. This is based on revised modulus G_s and p_u using a partial factor α_r (rather than the p_m) determined iteratively. The α_r is initially taken as the ratio of M_o/M_o^{FixH} (JGJ 1994).

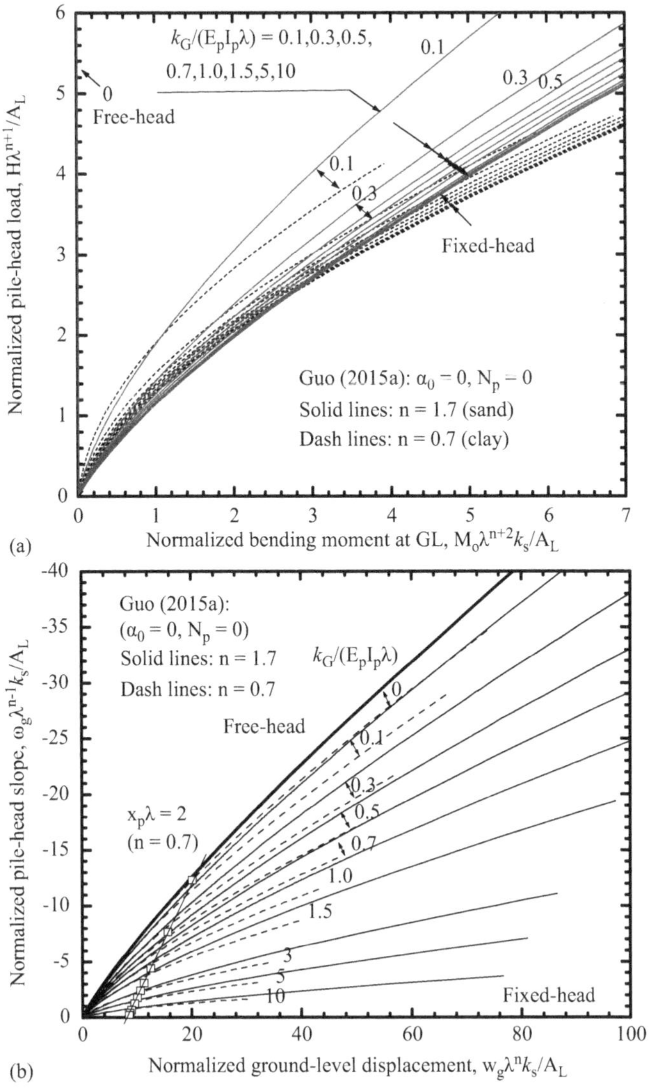

Figure 2.2 Nonlinear response of capped piles (n = 0.7 and 1.7): (a) load-bending moment; (b) displacement-rotation.

First, the modulus G_s was estimated using $\alpha_r = 0.42$ and $G_s^{FixH*} + \alpha_r(G_s^{FreH*} - G_s^{FixH*})$. Given $G_s^{FixH*} = 13s_u$ and $G_s^{FreH*} = 33s_u$, the G_s was estimated as $25s_u$ [$= 13s_u + 0.42 \times (33\text{–}13)s_u$, or $p_m = 0.76 = G_s/G_s^{FreH*}$], which was translated to $k_s = 1.59 \times 10^3$ kPa ($= 3.18G_s$, $s_u = 20$ kPa), and $\lambda = 0.643$/m {$= [k_s/(4\times2.326)]^{0.25}$}.

Second, the limiting p_u profile was estimated using Hansen's expression for free-head piles with $c = 1$ kPa and $\phi = 12°$ and is plotted in Figure 2.6a. As with G_s, the A_L is estimated as $2.26A_L^{FixH*}$ using $A_L^{FixH*} + \alpha_r(A_L^{FreH*} - A_L^{FixH*})$, and $A_L^{FreH*} = 4A_L^{FixH*}$, or $p_m = 0.565 = A_L/A_L^{FreH*}$. Note that this p_m value differs from that for shear modulus owing to the special group layout (Guo 2009). The A_L is calculated as 5.22 kN/m$^{1.85}$ ($= 1 \times 20 \times 0.1681^{-0.15} \times 0.2$) (Guo 2009), which was reduced to 5.142 kN/m$^{1.85}$ to account for the use of $e_p = 0$ against $e_p > 0$. With $n = 0.85$, and $A_L = 5.142$ kN/m$^{1.85}$, the p_u profile was obtained

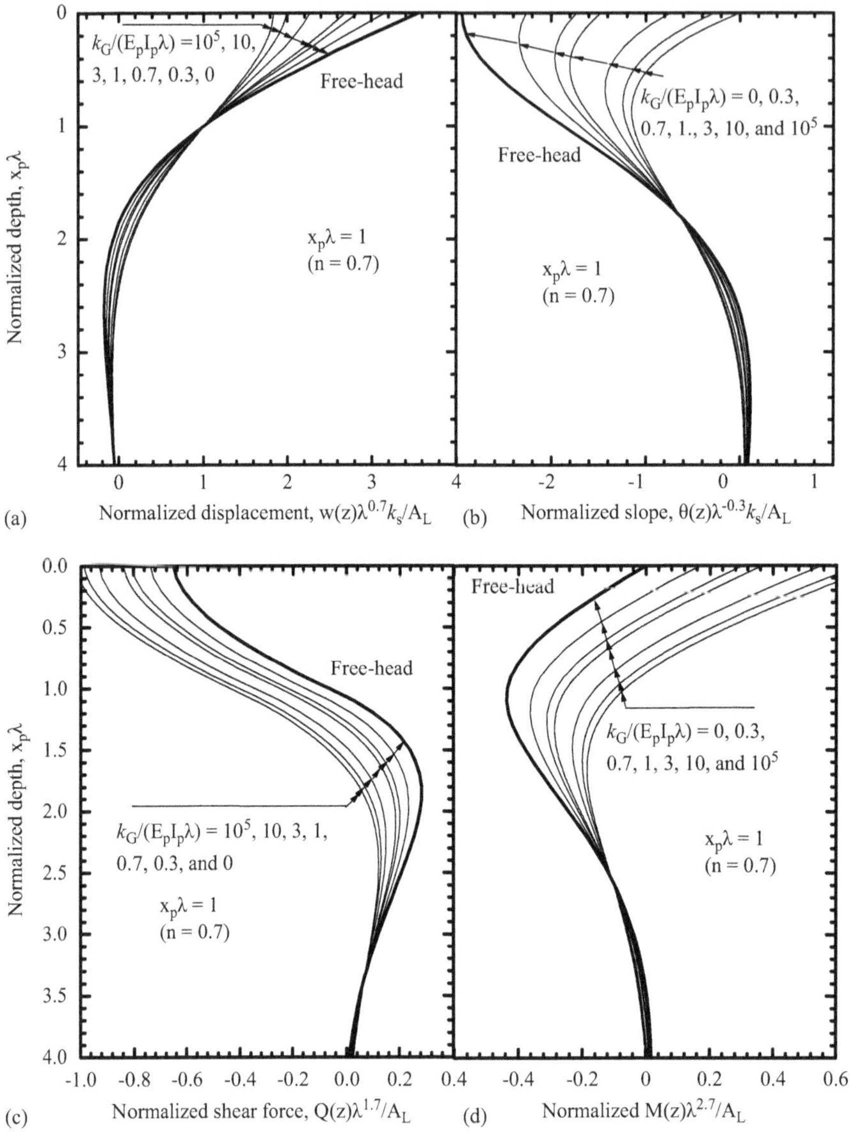

Figure 2.3 Non-dimensional profiles ($x_p\lambda = 1, n = 0.7$): (a) displacement; (b) slope; (c) bending moment; and (d) shear force. [Adapted from Guo, W. D., *J of Engrg Mechanics* **141**(9), 2015a.]

and is plotted in Figure 2.6a as Guo LFP, which matches well with the Hansen LFP. The difference between Bogard and Matlock's p_u profile and the Guo LFP and Hansen LFP has limited impact on the pile response (with a maximum x_p of 22.14d).

Third, the measured load H, ground-level bending moment M_o and displacement w_g were normalized using A_L, n, k, and λ to obtain normalized load $\bar{H}$, moment (thus $\bar{\omega}_g$ with a known $\bar{e}_p$), and displacement $\bar{w}_g$. The three measured values were then used to iteratively resolve the three unknown k_{nG}, A_L (thus α_r) and $\bar{x}_p$ (via k_s) using Equations (2.5)–(2.7).

If the calculated α_r value is outside acceptable difference from the assumed value, a new α_r is stipulated (resulting in new G_s, A_L, and k), and the three steps are repeated. The calculation can be readily done using a professional mathematical program (e.g., Mathcad™).

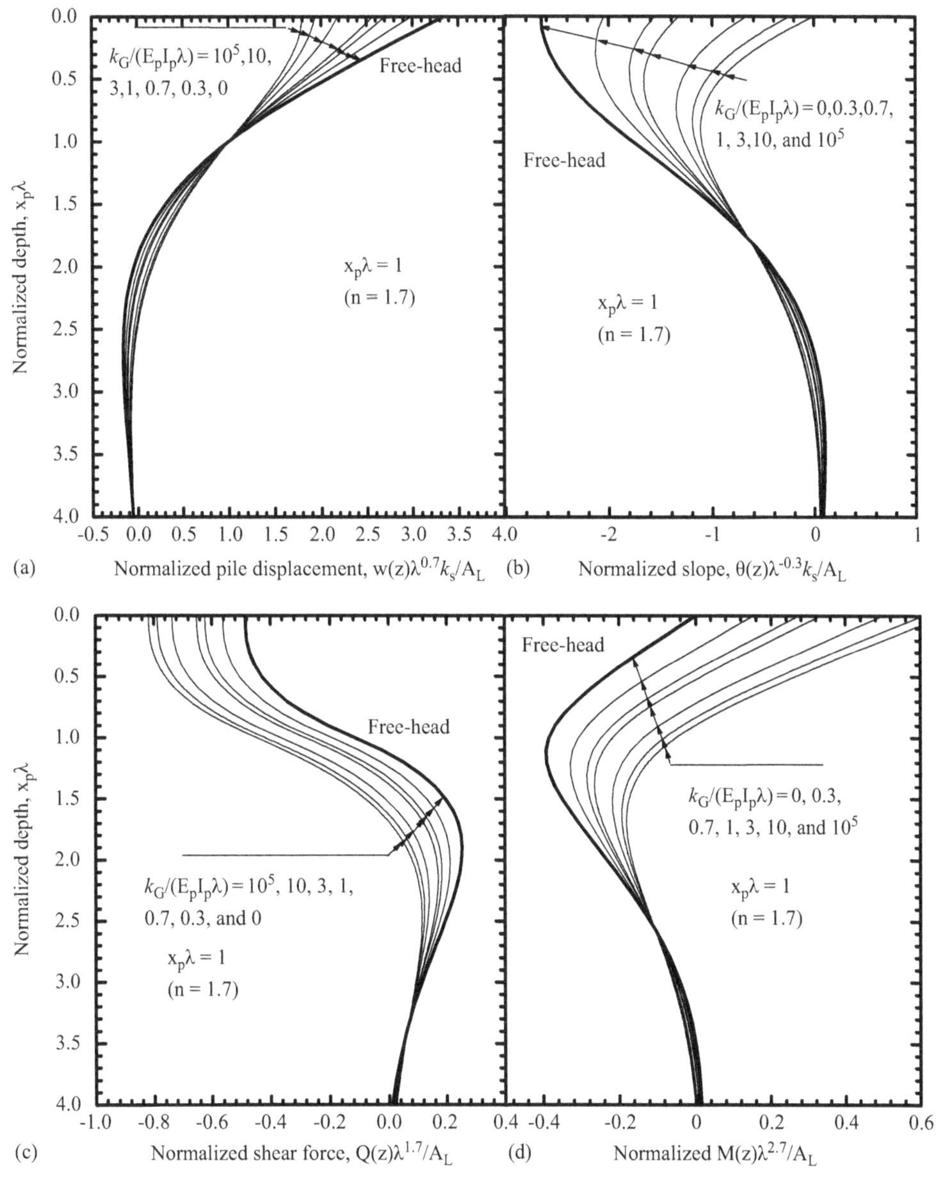

Figure 2.4 Non-dimensional profiles (n = 1.7): (a) displacement; (b) slope; (c) bending moment; (d) shear force. [Adapted from Guo, W. D., *J of Engrg Mechanics* **141**(9), 2015a.]

This iterative calculation is illustrated using the design chart for n = 0.7 (note n = 0.85 for the current pile) with three steps: (i) normalizing a pair of measured load H and displacement w_g using A_L, n, and k_s gained for free-head piles (e.g., $\bar{H} = 1.56$ at $\bar{w}_g = 10 = 10$, see italic values in Table 2.3); (ii) Using the pair $\bar{H}$ and $\bar{w}_g$ to ascertain $k_G/(E_pI_p\lambda)$ = 1.1 (or k_{nG} = 0.275) in Figure 2.1a, for k_G = 1.643 MNm (= $1.1E_pI_p\lambda$); and (iii) estimating α_r using $\bar{H}(\text{measured}) = \bar{H}^{\text{FixH}^*} + \alpha_r\left(\bar{H}^{\text{FreH}^*} - \bar{H}^{\text{FixH}^*}\right)$. For instance, at $\bar{w}_g = 10$, $\bar{H}^{\text{FixH}^*} = 1.0$ and $\bar{H}^{\text{FreH}^*} = 2.2$ (see Figure 2.1a), and $\bar{H}(\text{measured}) = 1.5$, the α_r is obtained as 0.42 from 1.5 = 1.0 + α_r(2.2–1). The deduced values are k_{nG} = 0.275 and α_r = 0.42 for the measured $\bar{H}$ and $\bar{w}_g$. The process may be repeated for other measured pairs of $\bar{H}$ and $\bar{w}_g$, and similar k_{nG} and α_r should be obtained (Guo 2013a).

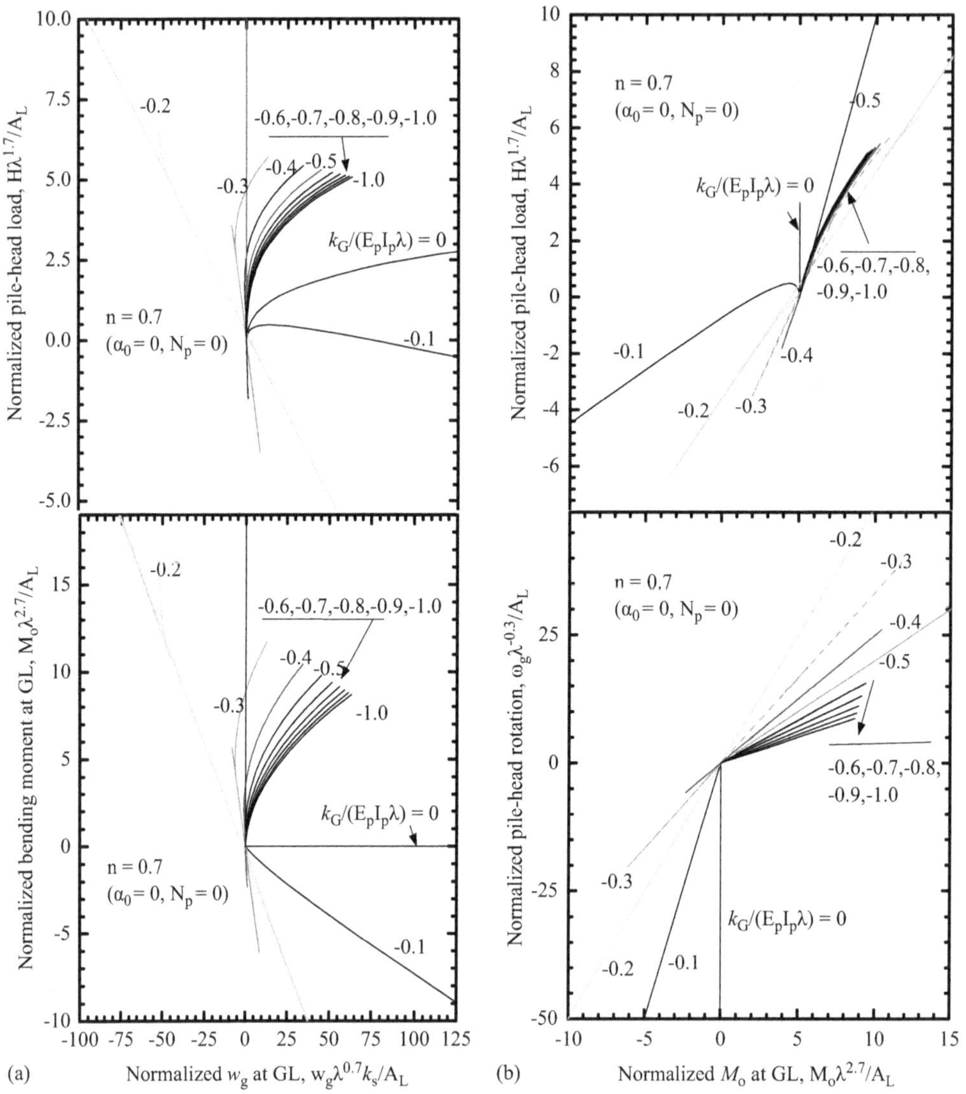

Figure 2.5 Nonlinear response of capped back-rotated piles (n = 0.7): (a) load/moment-displacement; (b) displacement/rotation angle-bending moment.

Having obtained p_u (n = 0.85), G_s and k_{nG} (= 0.275), the pile response is readily predicted. Especially, bending moment profiles for a w_g at GL of 9.4, 23.5, 47.3 and 83.1 mm are provided for e_p = 0 (zero loading eccentricity). For w_g = 9.4 mm, the normalized slip depth $\bar{x}_p$ was estimated as 0.863 (x_p = 1.342 m). (i) The $\omega_g k_s \lambda^{n-1}/A_L$ was calculated using Equation (2.5).

$$\frac{\omega_g k\lambda^{-0.15}}{A_L} = \frac{-0.863^{0.85}}{1.863\times 2\times 0.275+1}\left(\frac{2\times 0.863^2\left[0.863+3+0.85\right]}{2.85\times 3.85}+1+0.863\right) \qquad (2.11)$$
$$= -1.091$$

Likewise, the values of $H\lambda^{1.85}/A_L$ = 0.686 and $w_g k_s \lambda^{0.85}/A_L$ = 1.997 were obtained using Equations (2.6) and (2.7), respectively, along with $-M_o\lambda^{2.85}/A_L$ = 5.427. (ii) With $k_s = 1.59\times10^3$

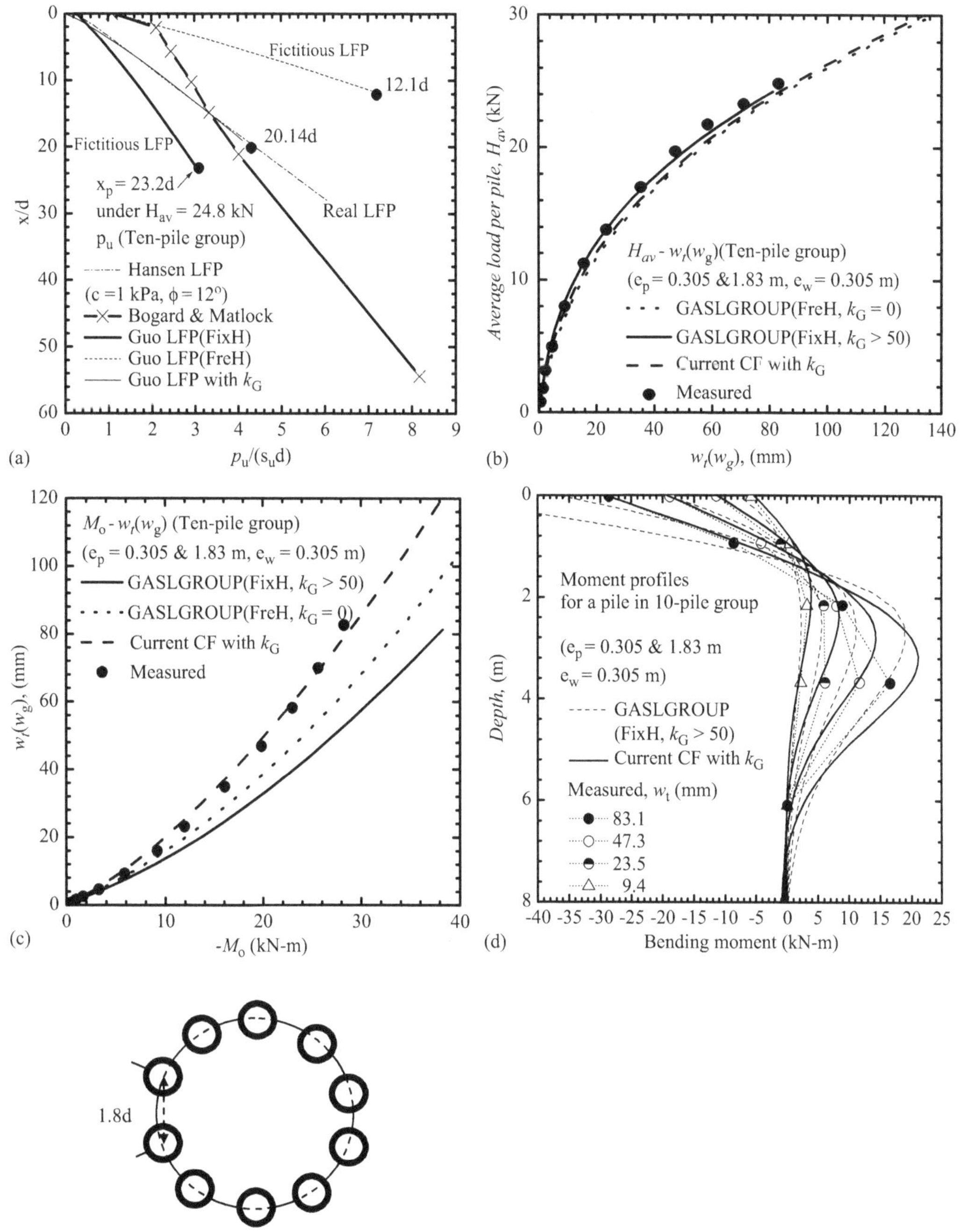

Figure 2.6 Predicted versus measured (Matlock et al. 1980) response of a pile in 10-pile group: (a) p_u profile; (b) w_g-H_{av} curves; (c) -M_o-w_g curves; (d) moment profiles. [Adapted from Guo, W. D., *J of Engrg Mechanics* **141**(9), 2015a.]

kPa and $A_L = 5.142$ kN/m$^{1.85}$, the GL rotation angle ω_g and H were obtained as -3.3035×10^{-3}, and 7.981 kN at $w_g = 9.4$ mm; and (iii) With expressions in Table 2.1, $F(1,0) = F(2,0) = 0$, constants $C_5 = 4.154 \times 10^{-3}$ and $C_6 = -1.574 \times 10^{-3}$ were obtained, respectively. The pile deflection at depth x (plastic zone) or z (elastic zone) is calculated respectively as

Table 2.3 Key response and parameters for a pile in 10-pile group (Guo 2015b)

w_g (mm)	$\bar{x}_p$	$\bar{H}/\bar{w}_g$	H (kN)	$-\omega_g$ (×10^{-3})	$C_5/-C_6$ (×10^{-3})	$-M_o/-M_o^{FixHa}$ (kNm)	M_o/M_o^{FixH}
9.4	0.863	0.686/1.997	7.981	3.3035	4.154/1.574	5.43/12.29	0.442
23.5	1.34836	1.132/4.992	13.179	6.938	6.069/4.259	11.40/28.81	0.396
47.3	1.78227	1.559/10.05	18.616	12.0	7.964/7.462	19.43/51.5	0.377
83.1	2.17641	2.082/17.65	24.23	17.9	9.118/10.975	29.40/79.10	0.371

Note:
[a] Values of M_o^{FixH} is cited from Guo (2009).

$$w(x) = \frac{5.142}{E_pI_p}\left[\frac{-x^{4.85}}{4.85\times3.85\times2.85\times1.85} + \frac{1.33017x^3 - 2.7135x^2}{5.142} - (3.3035x - 9.4)\times10^{-3}](m)\right] \tag{2.12a}$$

$$w(z) - e^{-0.643(x-1.342)}\{4.154\cos[0.643(x-1.342)] - 1.574\sin[0.643(x-1.342)]\}\times10^{-3}(m) \tag{2.12b}$$

The profile of bending moment $M(\bar{x})$ or $M(\bar{z})$ is obtained as

$$-M(x) = -\frac{x^{2+n}A_L}{(n+2)(n+1)} + Hx + k_r\omega_g = -\frac{5.142x^{2.85}}{1.85\times2.85} + 7.98x + 1.645\times10^3\times(-3.3035\times10^{-3}) \tag{2.13a}$$

$$\begin{aligned}-M(\bar{z}) &= 2E_pI_p\lambda^2e^{-\bar{z}}[-C_6\cos(\bar{z}) + C_5\sin(\bar{z})] \\ &= 1.923\times10^3 e^{-0.643(x-x_p)}\left\{\begin{matrix}-C_6\cos[0.643(x-x_p)] \\ +C_5\sin[0.643(x-x_p)]\end{matrix}\right\}\end{aligned} \tag{2.13b}$$

The calculated bending moment profile is plotted in Figure 2.6d as a solid line.

As with $w_g = 9.4$ mm, the normalized slip depth $\bar{x}_p$ was estimated as 1.3484, 1.7823, and 2.1764 for $w_g = 23.5, 47.3$ and 83.1 mm respectively. The H, ω_g, C_5 and C_6 values (see Table 2.3) were obtained; and the bending moment profiles obtained are plotted in Figure 2.6d. The predicted moment profiles agree well with the respective measured data rather than the fixed–head predictions (via the program GASLGROUP). Note that the profile of shear force $Q(\bar{x})$ or $Q(\bar{z})$ (not shown herein) can be predicted using the expressions in Table 2.1.

Importantly, this capped pile has a moment ratio M_o/M_o^{FixH} of 0.371–0.442 at different GL deflection w_g (9.4, 23.5, 47.3, and 83.1 mm, respectively, see Table 2.3). The average of these ratios (0.406) is quite close to 0.4 specified by the JGJ code, despite the special group layout, cap configuration and nonlinear response. This consistency is further evidenced between the deduced values of k_{nG}/p_m = 0.275/0.565–0.76 (for M_o/M_o^{FixH} = 0.371–0.442) and the theoretical values of 0.31/0.565 (for M_o/M_o^{FixH} = 0.4), vindicating the reliability of the capped-pile solutions and the values of k_{nG} and p_m values.

At a normalized stiffness of k_{nG} = 0.275, if the pile-cap is subjected to sway (forward- and back-rotation), the pile response will be amplified as highlighted in Figure 2.7. The finding is critical.

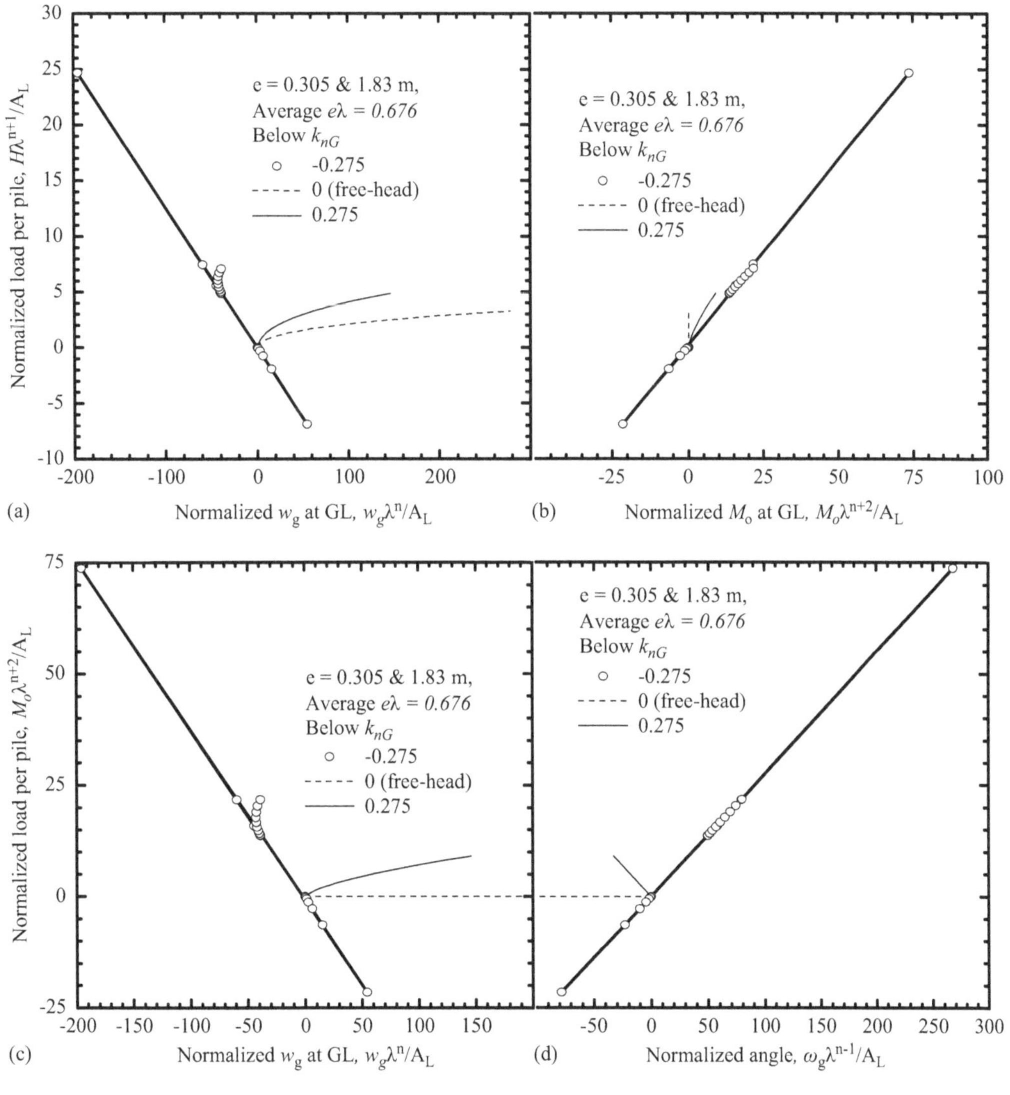

Figure 2.7 Response of 10 pile group to head sway: (a) load-displacement; (b) load-bending moment; (c) displacement-bending moment; (d) rotation-bending moment.

2.4 SUMMARY

In this chapter, closed-form solutions for semi-fixed-head pile-cap are presented, which provide an effective means of predicting the response of lateral pile groups. The solutions are essentially underpinned by limiting force per unit length $p_m p_u$, modulus of subgrade reaction $p_m k_s$, and p-multiplier p_m (to cater for pile-pile interaction, $p_m = 1$ for single piles). Non-dimensional response for free-head ($k_G = 0$) through to fixed-head ($k_G > 10E_pI_p$) piles is presented for any cap-rotational stiffness k_G. Use of the solutions is elaborated for a typical offshore pile group against measured response. The example study reveals that:

- The existing p_m values, which bear no link to the stiffness k_G, are inconsistent with 'p_m = 0.25' for capped piles at the limiting state of elastic solutions. This raises doubts about the accuracy of available solutions and highlights the need for compatible values of k_G and p_m.
- The compatible normalized stiffness k_{nG} is equal to 0.275–0.333 (n = 0.7) and 0.333–0.564 (n = 1.7) for the p_m at design level of GL bending moment specified in the JGJ code.

It is important to consider the coupled k_G and p_m when designing capped piles, and these values should be employed in pertinent design methods

Chapter 3

Anchored piles

3.1 INTRODUCTION

Offshore exploration has propelled analytical, numerical and experimental investigation into bearing capacity of anchored piles (Reese 1973; Senpere and Auvergne 1982; Vivatrat et al. 1982; Doyle et al. 2004; Erbrich 2004; Young et al. 2009). Several methods have been developed to determine the capacity, including the strength mobilization (SM) method (Bang and Cho 2001), finite element method (FEM) (Anderson et al. 2005) and plastic limit analysis (PLA) (Aubeny et al. 2003) among others. The study to date provides the evolution law of the capacity with depth of loading attachment e^- for a constant FPUL p_u with depth and a linearly increasing FPUL p_u (Gibson p_u). The p_u profile was obtained along rigid piles (Murff and Hamilton 1993), caissons (Aubeny et al. 2003) and footings (Yun and Bransby 2007) in cohesive soil based on FEM and/or PLA analyses. The 'upper bound' p_u of 52 free-head piles (e^+) was deduced using closed-form solutions and measured data (Guo 2013a). Note that the symbol e is taken as negative (e^-) for depth of attachment to distinguish it from the positive (e^+) loading above GL. The p_u reduces by ~4 times reduction from free-head to fixed-head conditions (Guo 2005, 2009, 2012). The study also provides ratio of rotation depth z_r/l and a normalized capacity $H_o/(s_u dl)$ (i.e., N_p) for anisotropic to isotropic strength profiles, gapping between rigid inclusion and soil, etc. To utilize these valuable data consistently, new solutions are developed (Guo 2013b).

In the anchor-pile system, an anchor may impose a patch or concentred load on rigid piles, which is resisted by the subgrade reaction of surrounding static and/or moving soil. To estimate the capacity, and profiles of shear force, bending moment and deflection, among others, Guo (2014) developed models and closed-form solutions based on a power-law increase with depth of limiting force per unit length (FPUL) p_u with depth for the anchor-pile in a single layer. The solutions allow for the determination of maximum bending moment M_m, shear force Q_m and their respective depths z_m and z_{mt}.

3.2 ELASTIC SOLUTIONS FOR ANCHORED PILES

Figure 3.1 shows a rigid pile is subjected to a patch FPUL p_u to a depth of c (αl_m) below GL. The on-pile p_u of $A_r d z^n$ may be uniform or linearly reduced with depth (referred to as the p-based model). In the context of load transfer model (Guo 2012), the p-based pile-soil interaction models are underpinned by the following hypotheses:

- The pile-soil interaction is modelled by a series of elastic springs along the pile shaft with no shear resistance at the pile tip. Each spring has a constant coefficient of subgrade reaction k_s over the entire length l of a single layer;

DOI: 10.1201/9781003315230-3

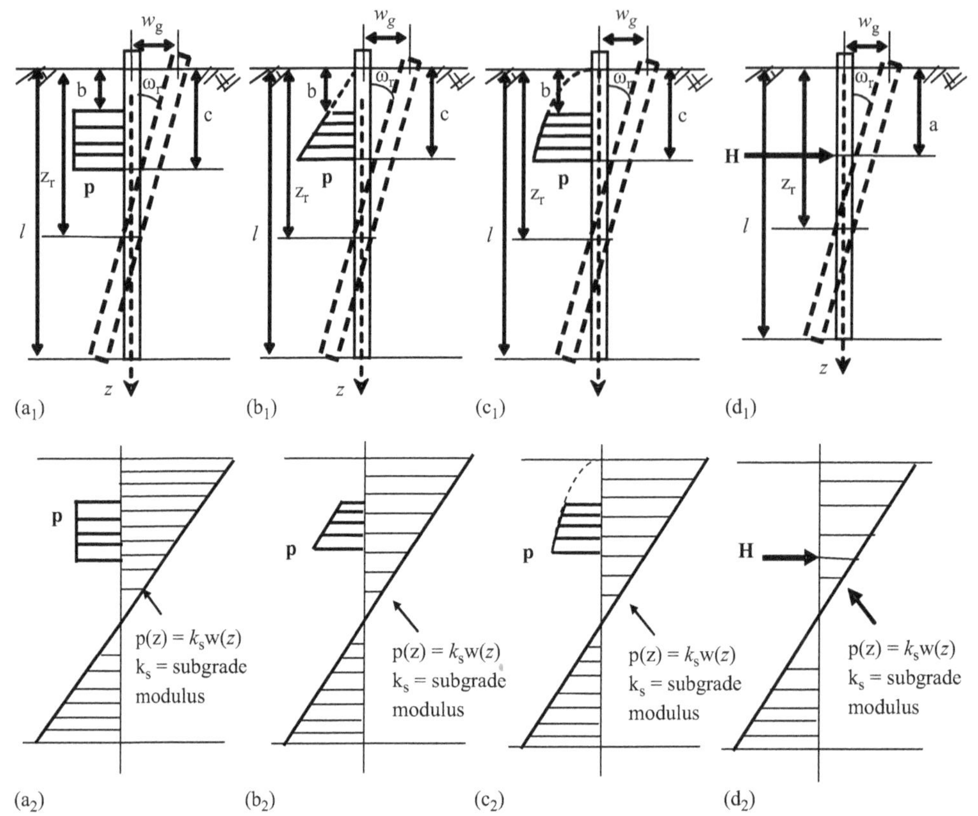

Figure 3.1 p-based models of passive piles in a single layer: (a_1) constant $p = A_r d$; (b_1) linear $p = A_r dz$; (c_1) parabolic $p = A_r dz^n$; (d_1) concentrated load H; the associated p applied and $p(z)$ induced in (a_2), (b_2), (c_2) and (d_2) respectively. [Adapted from Guo, W. D., *Int J Numer and Anal Meth in Geomech* **38**(18), 2014.]

- The pile rotates rigidly with a deflection $w(z)$ at depth z, which induces a FPUL $p(z)$ [$= k_s w(z)$] (with a specific upper limit).

$$w(z) = (\bar{\omega}_r \bar{z} + \bar{w}_g)\frac{A_r d l^n}{k_s} \tag{3.1}$$

where $\bar{\omega}_r = w'(z)k_s/(A_r d l^{n-1})$, normalized rotation; $\bar{w}_g = w_g k_s/(A_r d l^n)$, normalized displacement; $w'(z)$ = pile rotation (a constant along the rigid pile length), and w_g = pile-displacement at GL. The normalized $\bar{\omega}_r$ and $\bar{w}_g$ alter with the p profile (via A_r and n).

- The distributed loading p over the depth between b and c on a single pile (see Figures 3.1a–c) is described by $p = A_r dz^n$ (A_r = gradient, z = depth), which can be uniform (n = 0), linearly increasing (n = 1.0), or observe a power-law increase (n = 0–1.7).

3.2.1 Solutions for $p = A_r z^n$ (passive movement in a single layer)

Solutions for the pile under the distributed loading p were deduced using Equation (3.1) and force and moment equilibrium (Guo 2014). Tables 3.1–3.3 provide the results concerning the loading zone (from depths b to c with subscript i = 2) and non-loading zone

Table 3.1 M_m and z_m for piles with a constant *p* over depths *b* through *c* (Guo 2014)

Depth	*Expressions*
b–c	$\bar{M}_{m2} = \frac{\bar{\omega}_r \bar{z}_{m2}^3}{6} + \frac{(\bar{w}_g - 1)}{2}\bar{z}_{m2}^2 + \frac{\bar{b}}{2}\bar{z}_{m2} - \frac{\bar{b}^2}{2}$ (Loading zone)
	$\bar{z}_{m2} = -\frac{\bar{w}_g - 1 + \sqrt{(\bar{w}_g - 1)^2 - 2\bar{b}\bar{\omega}_r}}{\bar{\omega}_r}$ (Loading zone)
c–l	$\bar{M}_{m3} = \frac{\bar{\omega}_r \bar{z}_{m3}^3}{6} + \frac{\bar{w}_g}{2}\bar{z}_{m3}^2 + \frac{\bar{b} - \bar{c}}{2}\bar{z}_{m3} + \frac{\bar{c}^2 - \bar{b}^2}{2}$ (Non-loading zone)
	$\bar{z}_{m3} = -\frac{\bar{w}_g - \sqrt{\bar{w}_g^2 + 2(\bar{c} - \bar{b})\bar{\omega}_r}}{\bar{\omega}_r}$ (Non-loading zone)

Note: Single layer, constant k_s and constant *p* of $A_r d$ (over $z = b–c, n = 0$). $\bar{M}_{mi} = M_{mi} / (A_r d l^2)$, the values of $\bar{\omega}_r \left[= w¢(z) k_s / (A_r d l^{n-1})\right]$, and $\bar{w}_g \left[= w_g k_s / (A_r d l^n)\right]$ are determined using Table 3.3 with $n = 0$.

Table 3.2 M_m and z_m for piles with a linear *p* over depths *b* through *c* (Guo 2014)

Depth	*Expressions*
$b \sim c$	$\bar{M}_{m2} = \frac{(-1 + \bar{\omega}_r)\bar{z}_{m2}^3}{6} + \frac{\bar{w}_g}{2}\bar{z}_{m2}^2 + \left(\frac{\bar{b}^2}{2}\bar{z}_{m2} - \frac{\bar{b}^3}{3}\right)$ (Loading zone)
	$\bar{z}_{m2} = \frac{\bar{w}_g + \sqrt{\bar{w}_g^2 + \bar{b}^2(1 - \bar{\omega}_r)}}{1 - \bar{\omega}_r}$ (Loading zone)
$c \sim l$	$\bar{M}_{m3} = \frac{\bar{\omega}_r \bar{z}_{m3}^3}{6} + \frac{\bar{w}_g}{2}\bar{z}_{m3}^2 + \frac{\bar{b}^2 - \bar{c}^2}{2}\bar{z}_{m3} + \frac{\bar{c}^3 - \bar{b}^3}{3}$ (Non-loading zone)
	$\bar{z}_{m2} = \frac{-\bar{w}_g + \sqrt{\bar{w}_g^2 - (\bar{b}^2 - \bar{c}^2)(1 - \bar{\omega}_r)}}{\bar{\omega}_r}$ (Non-loading zone)

Note: Single layer, constant k_s and linear *p* of $A_r dz$ (over $z = b \sim c$). $\bar{M}_{mi} = M_{mi} / (A_r d l^3)$, the values of $\bar{\omega}_r \left[= w'(z) k_s / (A_r d l^{n-1})\right]$, and $\bar{w}_g \left[= w_g k_s / (A_r d l^n)\right]$ are determined using Table 3.3 with $n = 1$.

(depth 0 to *b* with $i = 1$, and *c* to *l* with $i = 3$). Table 3.1 provides the normalized maximum bending moment $\bar{M}_{mi}$ and its depth $\bar{z}_{mi}$ for a constant FPUL *p* over depths *b–c* (see Figure 3.1a), while Table 3.2 presents those for a linear increasing FPUL *p* ($n = 1$, see Figure 3.2b). Table 3.3 presents the profiles of displacement $w(z)$, shear force $Q(z)$, and bending moment $M(z)$ for the power-law increasing *p*, respectively via non-dimensional $\bar{Q}_i(\bar{z}) = Q_i(z) / (A_r d l^2)$, $\bar{M}_i(\bar{z}) = M_i(z) / (A_r d l^3)$, $\bar{z} = z / l$, $\bar{b} = b / l$ and $\bar{c} = c / l$, as well as the normalized angle of rotation $\bar{\omega}_r$ and displacement at GL $\bar{w}_g$. The subscript *i* of 2 and 3 denotes the loading and non-loading zones respectively, except for the concentrated load *H*-based model, whereas subscript *i* of 1 is reserved for the non-loading zone from GL to depth *b* (see Figure 3.1).

Table 3.3 Solutions for on-pile patch loading $p = A_r d z^n$ (Guo 2014)

Depth $\bar{z}$	*Expressions*
$0 \sim l$	$w(\bar{z}) = (\bar{\omega}_r \bar{z} + \bar{w}_g)\dfrac{A_r d l^n}{k_s}$ $\bar{\omega}_r = \dfrac{-6\left[2\bar{b}(n+1)-2-n\right]\bar{b}^{n+1} + 6\left[2\bar{c}(1+n)-2-n\right]\bar{c}^{n+1}}{(n+1)(2+n)}$ $\bar{w}_g = \dfrac{\left[6\bar{b}(n+1)-4(n+2)\right]\bar{b}^{n+1} - \left[6\bar{c}(n+1)-4(n+2)\right]\bar{c}^{n+1}}{(n+1)(2+n)}$
$0 \sim \bar{b}$	$\bar{Q}_1(\bar{z}) = \dfrac{1}{2}\bar{\omega}_r \bar{z}^2 + \bar{w}_g \bar{z},\ \bar{M}_1(\bar{z}) = \dfrac{\bar{\omega}_r \bar{z}^{2+n}}{6} + \dfrac{\bar{w}_g}{2}\bar{z}^2$
$\bar{b} \sim \bar{c}$	$\bar{Q}_2(\bar{z}) = \dfrac{1}{2}\bar{\omega}_r \bar{z}^2 + \bar{w}_g \bar{z} + \dfrac{1}{1+n}\left(\bar{b}^{n+1} - \bar{z}^{n+1}\right)$ $\bar{M}_2(\bar{z}) = \dfrac{\bar{\omega}_r \bar{z}^3}{6} + \dfrac{\bar{w}_g}{2}\bar{z}^2 + \left[\dfrac{\bar{b}^{n+1}}{n+1}\bar{z} - \dfrac{(1+n)\bar{b}^{n+2} + \bar{z}^{2+n}}{(n+2)(n+1)}\right]$
$\bar{c} \sim l$	$\bar{Q}_3(\bar{z}) = \dfrac{1}{2}\bar{\omega}_r \bar{z}^2 + \bar{w}_g \bar{z} + \dfrac{1}{1+n}\left(\bar{b}^{n+1} - \bar{c}^{n+1}\right)$ $\bar{M}_3(\bar{z}) = \dfrac{\bar{\omega}_r}{6}\bar{z}^3 + \dfrac{\bar{w}_g}{2}\bar{z}^2 + \left[\dfrac{\bar{b}^{n+1} - \bar{c}^{n+1}}{n+1}\bar{z} + \dfrac{\bar{c}^{2+n} - \bar{b}^{2+n}}{n+2}\right]$

Note: Single layer, constant k_s and $p = A_r d z^n$ ($z = 0 \sim c$). $\bar{Q}_i(\bar{z}) = Q_i(z)/(A_r d l^{1+n})$, $\bar{M}_i(\bar{z}) = M_i(z)/(A_r d l^{2+n})$, $\bar{\omega}_r = w'(z)k_s/(A_r d l^{n-1})$, $\bar{w}_g = w_g k_s/(A_r d l^n)$

3.2.2 Parametric analysis for a single layer

The solutions for the anchored piles in a single layer reveal salient features.

3.2.2.1 *Variations in* M_m *and* z_m

The moment M_m and its depth were obtained for a loading length of $(\alpha - 1)b$ from depths b to c ($= \alpha b$, see Figure 3.1) using the solutions. Figures 3.2a and b present the normalized moment $-M_m/(A_r d l^{2+n})$ and depth z_m/l, respectively assuming either uniform or linear FPUL p profiles, for full length of loading to depth l_m and to $\alpha = 1.5$, 2.0 and 3. The plots show that as the loading length of $(\alpha - 1)b$ increases, the normalized depth z_m/l shifts towards the pile tip at a higher rate for higher α values with limited difference between uniform and linear p profiles. Figures 3.2c and d illustrate the difference in the normalized moment M_m and depth z_m between non-loading and loading zones. It is noteworthy that the alteration from uniform to linear 'p' profiles reduces the normalized M_m by ~16 times in the non-loading zone, indicating that the loading zone dominates the design.

3.2.2.2 *Normalized force and bending moment*

Table 3.3 presents the solutions of normalized shear force $Q(z)/(A_r d l^{n+1})$ and bending moment $M(z)/(A_r d l^{n+2})$ for either a uniform FPUL p ($= A_r d$), or a linearly increasing FPUL p ($= A_r d z$)

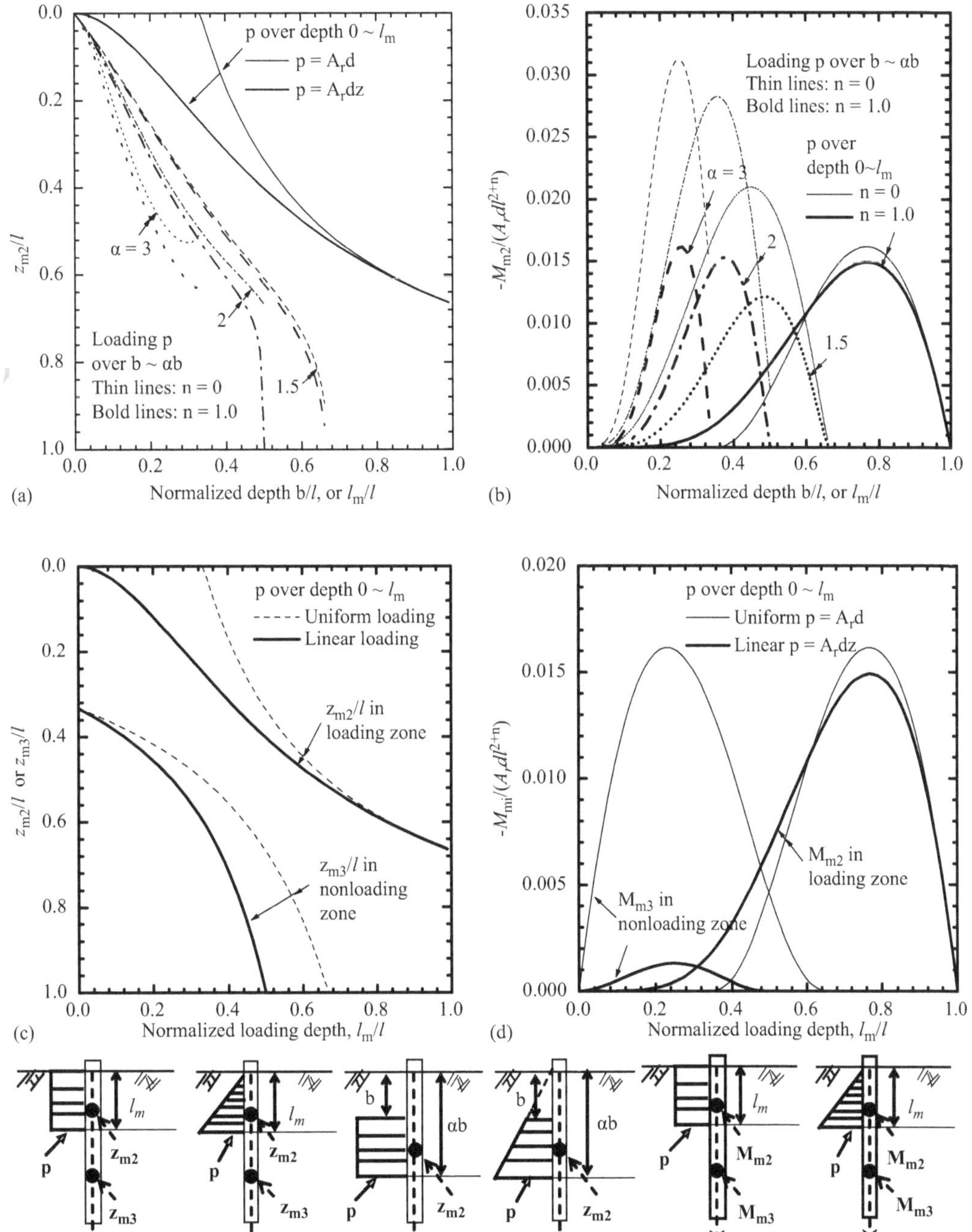

Figure 3.2 Response to normalized loading length (single layer): (a and b) z_{m2}/l; and (c, d) $M_{m2}/(A_r d l^{2+n})$. [Adapted from Guo, W. D., *Int J Numer and Anal Meth in Geomech* **38**(18), 2014.]

from GL to depth c. Meanwhile, Table 3.4 provides the $M(z)/Hl$ profile for a concentrated load H with uniform k_s. Figures 3.3a and b provide the normalized shear force $Q(z)/(pl^{n+1})$, and bending moment $M(z)/(pl^{2+n})$ for $n = 0$ and 1, respectively. The normalized force $Q_m/(pl^{n+1})$ and moment $M_m/(pl^{n+1})$ shift with the normalized loading length c/l ($c = l_m$), as previously depicted in Figure 3.2d for the normalized M_m.

Table 3.4 Solutions for piles under a concentrated load H (Guo 2014)

Depth $\bar{z}$	*Expressions*
$0 \sim \bar{a}$	$\bar{Q}_1(\bar{z}) = \left[3(2\bar{a}-1)\bar{z} + 2(-3\bar{a}+2)\right]\bar{z}$ $\bar{M}_1(\bar{z}) = \left[(2\bar{a}-1)\bar{z} + (-3\bar{a}+2)\right]\bar{z}^2$
$\bar{a} - 1$	$\bar{Q}_2(\bar{z}) = \left[3(2\bar{a}-1)\bar{z} + 2(-3\bar{a}+2)\right]\bar{z} - 1$ $\bar{M}_2(\bar{z}) = \left[(2\bar{a}-1)\bar{z} + \bar{a}\right](\bar{z}-1)^2$
$0 \sim 1$	$w(\bar{z}) = \left[3(2\bar{a}-1)\bar{z} + (-3\bar{a}+2)\right]\dfrac{2H}{k_s l}$
$\bar{a} \neq 0.5$	$\bar{Q}_{m1} = \dfrac{-(3\bar{a}-2)^2}{3(2\bar{a}-1)}$, $\bar{Q}_{m2} = \dfrac{-(3\bar{a}-1)^2}{3(2\bar{a}-1)}$, $\bar{Q}_{m1} = \dfrac{2(3\bar{a}-2)}{3(2\bar{a}-1)}$ $\bar{M}_{m1} = \dfrac{-4(3\bar{a}-2)^3}{27(2\bar{a}-1)^2}$, $\bar{M}_{m2} = \dfrac{9\bar{a}-4}{27(2\bar{a}-1)^2} + \bar{a}$, $\bar{z}_{m2} = \dfrac{-1}{3(2\bar{a}-1)}$
	$\bar{Q}_m = 0.5$, $\bar{M}_{m1} = \bar{M}_{m2} = 1/8$

Note: Single layer, constant k_s, with H at depth a; $\bar{Q}_i(\bar{z}) = Q_i(z)/H, \bar{M}_i(\bar{z}) = M_i(z)/(Hl)$.

It is worth noting that the H-based solution is only recommended for flexible piles, and is included here for comparison purposes. Among the distributed loading profiles ($p = A_r dz^n$), a uniform p seems to be more critical than other cases, particularly for partial loading at depth b to ab (as in anchors), and full loading to depth l_m. The magnitude of p should increase proportionally with the sliding depth (l_m or ab), as shown in an example study later on.

3.3 ELASTIC-PLASTIC SOLUTIONS FOR ANCHORED PILES

The load transfer model describing the pile-soil interaction is shown in Figure 3.4. The plastic zone (depth $z < z_o$) is described by $p_u = A_r dz^n$, while the elastic zone by $p = kdu$ where $k = k_o z^n$. Elastic-plastic solutions for the model were developed to capture response of a rigid pile (Guo 2008) for three typical pairs of p_u and k profiles: (i) constant p_u and constant k; (ii) Gibson p_u and constant k; and (iii) Gibson p_u and Gibson k. The response is presented in non-dimensional form such as normalized capacity $\bar{H}_o = H_o / \left(A_r dl^{1+n}\right)$, normalized maximum bending moments $\bar{M}_m = M_m / \left(A_r dl^{2+n}\right)$, and applied bending moment (at GL) $\bar{M}_o = M_o / \left(A_r dl^{2+n}\right)$, etc.

In the elastic zone, between depths z_o and z_1 (Figures 3.1 d_1 or d_2), the on-pile FPUL is given by $p = kdu$, where the kd is the gradient of the p-u curve with $k = k_o z^n$ (see Table 3.5). A Gibson k ($n = 1$) profile is characterized by the gradient k_o. The slip (plastic) zone(s) develop once the p attains the $p_u = A_r dz^n$, including the zone $0 \sim z_o$ prior to the tip-yield state, or both zones of $0 \sim z_o$ and $z_1 \sim l$ after the tip-yield state. The A_r is the gradient of the limiting FPUL p_u profile, with $A_r = N_p s_u$ and $n = 0$ for constant p_u; s_u is an average undrained shear strength over the pile embedment. The A_r for Gibson p_u ($n = 1$) can be estimated, for instance, by employing Hansen's plasticity solutions using frictional angle and effective unit weight of the subsoil.

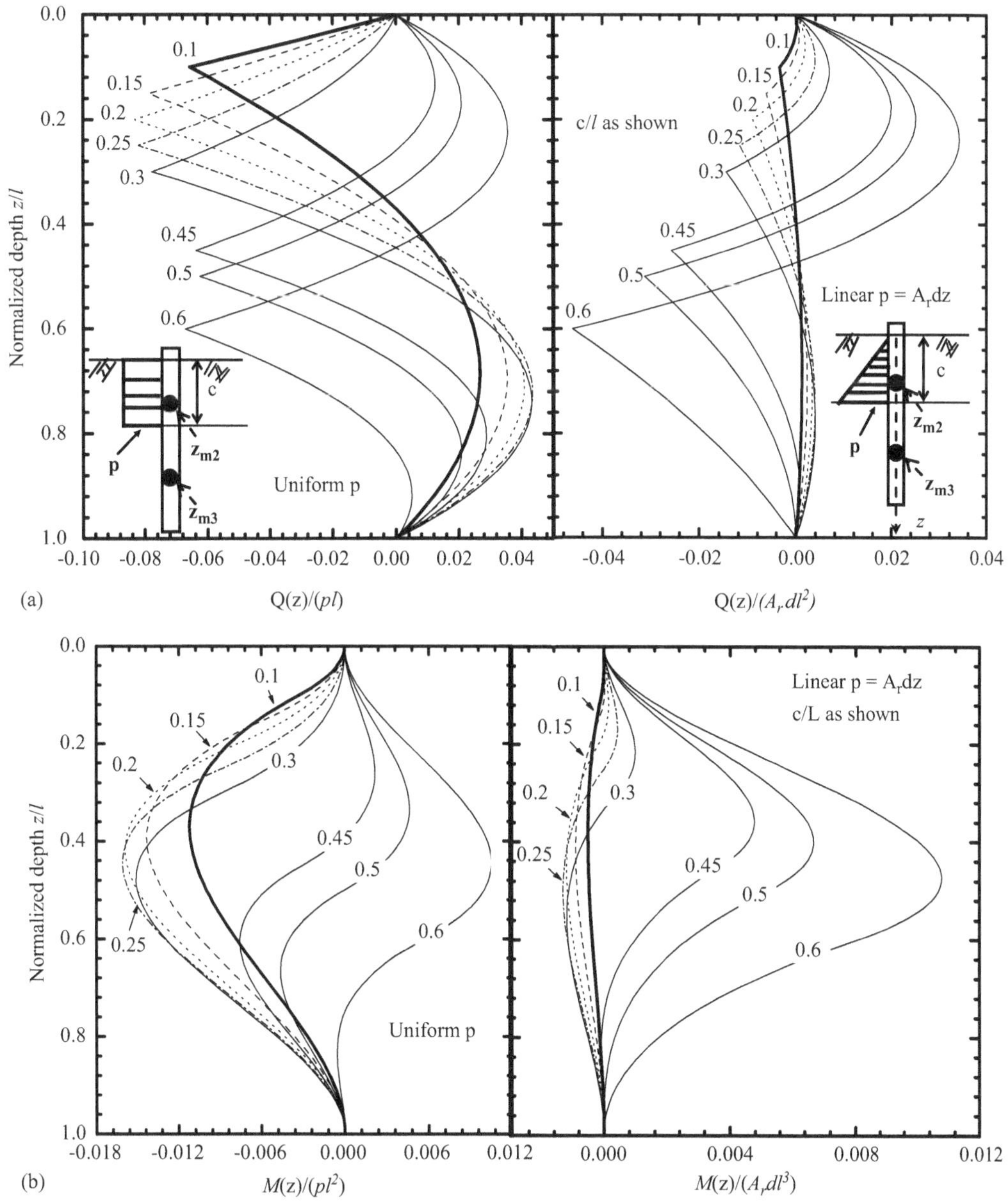

Figure 3.3 (a) Normalized shear force; (b) normalized bending moment with depth (*p* over depths 0 through *c*). [Adapted from Guo, W. D., *Int J Numer and Anal Meth in Geomech* **38**(18), 2014.]

At the transition (slip) depth z_o from the plastic to the elastic state, the limiting FPUL p is equal to the p_u, at which the displacement threshold u^* (see Figure 3.4) is deduced as

$$u^* = N_p s_u / k \left(\text{Constant}\, p_u \text{ and } k\right) \tag{3.2}$$

$$u^* = A_r / k_o \left(\text{Gibson}\, p_u \text{ and } k\right), \text{or } u^* = A_r z / k \left(\text{Gibson}\, p_u \text{ and constant } k\right) \tag{3.3}$$

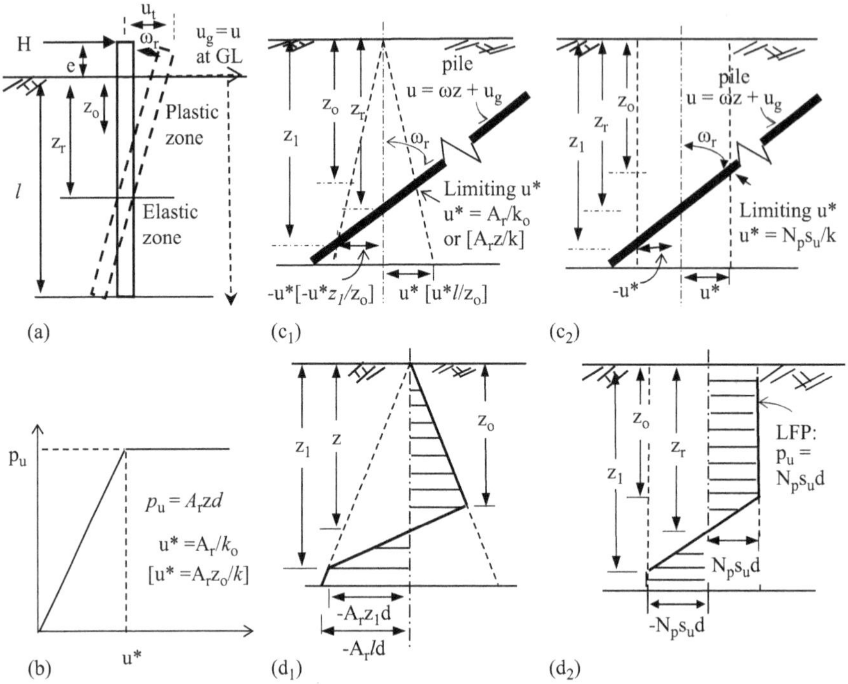

l = embedded length; d = outer diameter of the pile; e = loading eccentricity above ground line (GL); H = lateral load; u_g = pile displacement at ground line; ω_r = angle of rotation (in radian); z = depth from ground line; z_0 = depth of slip; z_r = depth of rotation point; p = soil resistance per unit length; p_u = ultimate limiting force per unit length; A_r = gradient of limiting force profile; u = pile displacement; u^* = local threshold u above which pile soil relative slip is initiated; k, k_0 = modulus of subgrade reaction, $k = k_0 z^m$, m = 0, and 1 for constant and Gibson k.

Figure 3.4 Schematic analysis for a rigid pile (Guo 2013b): (a) pile-soil system; (b) load transfer model; (c_i) pile displacement features; (d_i) p_u (LFP) profiles (i = 1 and 2 for Gibson p_u and constant p_u respectively). [Adapted from Guo, W. D., *Computers and Geotechnics* **54**(2), 2013b.]

The pile-tip threshold displacement $-u^*$ is depicted in Figures 3.4 c_1 and c_2. The on-pile FPUL is a consequence of the p_u mobilized by a linear deflection u (= $\omega_r z + u_g$) of the pile rotating about a depth z_r (= $-u_g/\omega_r$) to an angle ω_r (note that ω_r is in radian, and u_g is deflection at GL). For instance, the Gibson profile is characterized by $p_u = A_r z_o d$ and u^*[= $\omega z_o + u_g$, or Equation (3.3)] at the slip depth z_o, and by $p_u = -A_r z_1 d$ and $-u^*$ (= $\omega z_1 + u_g$,) at the slip depth z_1, respectively. On the other hand, the constant profile is described by $p_u = N_p s_u d$ and $u^* = N_p s_u / k$, at the depth z_o, and by $p_u = -N_p s_u d$ and $-u^* = N_p s_u / k$ at the depth z_1. The dashed lines in Figures 3.4 c_1 and c_2 represent the profiles of both p_u and u^*.

The free-head solutions has been reformulated to determine the capacity H_o at both the tip-yield and yield at rotation point (YRP) states, as well as to calculate load, displacement, rotation and maximum bending moment response of anchored piles. However, it is important to note that when using rigid-pile solutions to predict the response of flexible piles (Meyerhof et al. 1983), bending failure should be assessed based on the maximum bending moment in the piles including the applied bending moment (Guo 2012).

Table 3.5 Solutions for a rigid pile at pre-tip and tip-yield states (Guo 2013b)

Gibson p_u (n = 1) and constant k (n = 0)	*Gibson p_u and Gibson k (n = 1)*
$u = \omega_r z + u_g$, $z_r/l = -u_g/(\omega_r l)$, $kd = \frac{3\pi\tilde{G}}{2}\left\{2\gamma_b \frac{K_1(\gamma_b)}{K_o(\gamma_b)} - \gamma_b^2\left[\left(\frac{K_1(\gamma_b)}{K_o(\gamma_b)}\right)^2 - 1\right]\right\}$	
where $K_i(\gamma_b)$ (i = 0, 1) is modified Bessel functions of the second kind, and of order i. $\gamma_b = k_1(r_o/l)$; $k_1 = 2.14 + \bar{e}/(0.2 + 0.6\bar{e})$, increases from 2.14 at e = 0 to 3.8 at e = ∞.	
$p = kdu$, $p_u = A_r dz$, $u^* = A_r z_o/k$	$p = k_o dzu$, $p_u = A_r dz$, $u^* = A_r/k$
$\bar{H} = \frac{0.5\bar{z}_o\left[(\bar{z}_o - 1)^2 - 6\bar{k}_G(\bar{z}_o - 2)\right]}{(\bar{z}_o - 1)^2(2 + \bar{z}_o + 3\bar{e}) + 6\bar{k}_G}$	$\bar{H} = \frac{1}{6}\frac{(1 + 2\bar{z}_o + 3\bar{z}_o^2)(\bar{z}_o - 1)^2 + 36\bar{k}_G}{\left[(2 + \bar{z}_o)(2\bar{e} + \bar{z}_o) + 3\right](\bar{z}_o - 1)^2 + 12\bar{k}_G}$
$\bar{u}_g = \frac{(2 + 3\bar{e} + 6\bar{k}_G)\bar{z}_o}{(2 + \bar{z}_o + 3\bar{e})(\bar{z}_o - 1)^2 + 6\bar{k}_G}$	$\bar{u}_g = \frac{3 + 2(2 + \bar{z}_o^3)\bar{e} + \bar{z}_o^4 + 12\bar{k}_G}{\left[(2 + \bar{z}_o)(2\bar{e} + \bar{z}_o) + 3\right](\bar{z}_o - 1)^2 + 12\bar{k}_G}$
$\varpi_r = \frac{\bar{z}_o\left[\bar{z}_o^2 - 3 + 3\bar{e}(\bar{z}_o - 2)\right]}{(2 + \bar{z}_o + 3\bar{e})(\bar{z}_o - 1)^2 + 6\bar{k}_r}$	$\bar{\omega}_r = \frac{-2(2 + 3\bar{e})}{\left[(2 + \bar{z}_o)(2\bar{e} + \bar{z}_o) + 3\right](\bar{z}_o - 1)^2 + 12\bar{k}_r}$
$\bar{z}_m = \sqrt{2\bar{H}}$ $\bar{M}_m = \bar{H}(\bar{e} + 2\bar{z}_m/3) + \bar{k}_G\bar{\omega}_r$ $(z_m \le z_o)$	$\bar{z}_m = \sqrt{2\bar{H}}$ $(z_m \le z_o)$ $\bar{M}_m = \bar{H}(\bar{e} + 2\bar{z}_m/3) + \bar{k}_G\bar{\omega}_r$ $(z_m \le z_o)$
At tip-yield state $\bar{z}_*^3 + 3\bar{e}\bar{z}_*^2 + \left[3\bar{k}_G - (2 + 4.5\bar{e})\right]\bar{z}_*$ $+ (1 + 1.5\bar{e} + 3\bar{k}_G) = 0$	 $(\bar{z}_* - 1)\left[\bar{z}_*^3 + (2\bar{e} + 1)(\bar{z}_*^2 + \bar{z}_*) - (1 + \bar{e})\right]$ $+ 12\bar{k}_G = 0$
$\bar{H} = H/(A_r dl^2)$, $\bar{u}_g = u_g k/A_r l$, $\bar{\omega}_r = \omega_r k/A_r d$, $\bar{M}_m = M_m/(A_r dl^3)$, and $\bar{k}_G = k_G/(kl^3)$	$\bar{H} = H/(A_r dl^2)$, $\bar{u}_g = u_g k_o/A_r$, $\bar{\omega} = \omega k_o l/A_r$, $\bar{M}_m = M_m/(A_r dl^3)$, $\bar{M}_o = M_o/(A_r dl^3)$, and $\bar{k}_G = k_G/(kl^4)$

Note: $\bar{z}_o = z_o/l$, $\bar{z}_r = z_r/l = -\bar{u}_g/\bar{\omega}$, $\bar{z}_m = z_m/l$, and $\bar{e} = e/l$. The solutions for free-head piles (Guo 2012) are used to predict response profiles by replacing $M_o = He$ with the counterparts of $M_o = He + k_G\omega$. For example, the $M(z)$ for Gibson p_u and constant k is as follows:

$$\bar{M}_1(\bar{z}) = \bar{H}(\bar{e} + \bar{z}) + \bar{k}_G\bar{\omega}_r - \frac{1}{6}\bar{z}^3$$

$$\bar{M}_2(\bar{z}) = \bar{H}(\bar{e} + \bar{z}) + \bar{k}_G\bar{\omega}_r - \left(0.5\bar{z}\bar{z}_o^2 - \frac{\bar{z}_o^3}{3}\right) - \frac{\bar{\omega}_r}{6}\left(\bar{z}^3 - 3\bar{z}\bar{z}_o^2 + 2\bar{z}_o^3\right) - 0.5\bar{u}_g(\bar{z} - \bar{z}_o)^2 \quad (z_o < z \le l, z_1 \approx l)$$

3.3.1 Ultimate capacity

Assuming a constant p_u (n = 0) and a constant k with depth, the normalized capacity $\bar{H}_o$ at the tip-yield state or the YRP state (Guo 2012) is given respectively by:

$$\bar{H}_o = \frac{H_o}{N_p s_u dl} = \frac{-3\bar{e} + (9\bar{e}^2 + 6\bar{e} + 3)^{0.5}}{3(1 + \bar{e}) + (9\bar{e}^2 + 6\bar{e} + 3)^{0.5}} \text{ (Tip-yield state)} \tag{3.4}$$

$$\bar{H}_o = \frac{H_o}{N_p s_u dl} = 2\left[-\bar{e} + \left(\bar{e}^2 + \bar{e} + 0.5\right)^{0.5}\right] - 1 \text{ (YRP state)} \tag{3.5}$$

where $\bar{e} = e / l$ and $e = M_o/H_o$ or loading eccentricity. The normalized moment $\bar{M}_m$ at the tip-yield through the YRP state is given by

$$\bar{M}_m = \bar{M}_o + 0.5\bar{H}_o^2 \tag{3.6a}$$

where $\bar{M}_o = \bar{H}_o\bar{e}$ with tip-yield state $\bar{H}_o$ determined from Equation (3.4); and $\bar{M}_o$ at YRP state is related to $\bar{H}_o$ (depending on loading direction), e.g., for the fourth quadrant, by

$$\bar{M}_o = -\left[\frac{1}{4}\left(\bar{H}_o + 1\right)^2 - \frac{1}{2}\right] \tag{3.6b}$$

Stipulating a Gibson p_u ($n = 1$) and a constant k, the capacity $\bar{H}_o$ at the tip-yield state or the YRP state (Guo 2008) is given respectively by

$$\bar{H}_o = \frac{H_o}{A_r dl^2} = \frac{0.5\left[-\left(3\bar{e} + 1\right) + \left(9\bar{e}^2 + 12\bar{e} + 5\right)^{0.5}\right]}{3\left(1 + \bar{e}\right) + \left(9\bar{e}^2 + 12\bar{e} + 5\right)^{0.5}} \text{ (Tip-yield state)} \tag{3.7}$$

$$\bar{H}_o = \frac{H_o}{A_r dl^2} = 0.25\left(\sqrt[3]{A_0} + \sqrt[3]{A_1} - \bar{e}\right)^2 - 0.5 \text{ (YRP state)} \tag{3.8}$$

where

$$A_j = \left(-\bar{e}^3 + 2 + 3\bar{e}\right) + (-1)^j\left[\left(2 + 3\bar{e}\right)\left(-2\bar{e}^3 + 2 + 3\bar{e}\right)\right]^{0.5} \quad (j = 0, 1) \tag{3.9}$$

The normalized moment $\bar{M}_m$ at the tip-yield though the YRP state is given by

$$\bar{M}_m = \bar{M}_o + \frac{1}{3}\left(2\bar{H}_o\right)^{1.5} \tag{3.10a}$$

where $\bar{M}_o = \bar{H}_o\bar{e}$ with tip-yield state $\bar{H}_o$ gained from Equation (3.7), and the $\bar{M}_o$ is correlated with the $\bar{H}_o$ at YRP state, e.g., for the fourth quadrant, by

$$\bar{M}_o = -\left[\frac{2}{3}\left(\bar{H}_o + 0.5\right)^{1.5} - \frac{1}{3}\right] \tag{3.10b}$$

Equation (3.4) or (3.7) and Equation (3.5) or (3.8) allow calculation of the normalized capacity of lateral piles (with $\bar{e} \geq 0$) for the tip-yield and the YRP states, respectively. The capacities for piles with an l/d of 4–20 were obtained using the Broms' solutions (Broms 1964) underpinned by an 'around-pile' gap to a depth of 1.5d, and were normalized by $N_p = 9$. All the normalized capacities well bracketed by the current solutions of constant p_u ($n = 0$) and Gibson p_u($n = 1$) (Guo 2012, 2013b). Note that both solutions neglect the longitudinal resistance along the shaft and transverse shear resistance on the pile-tip.

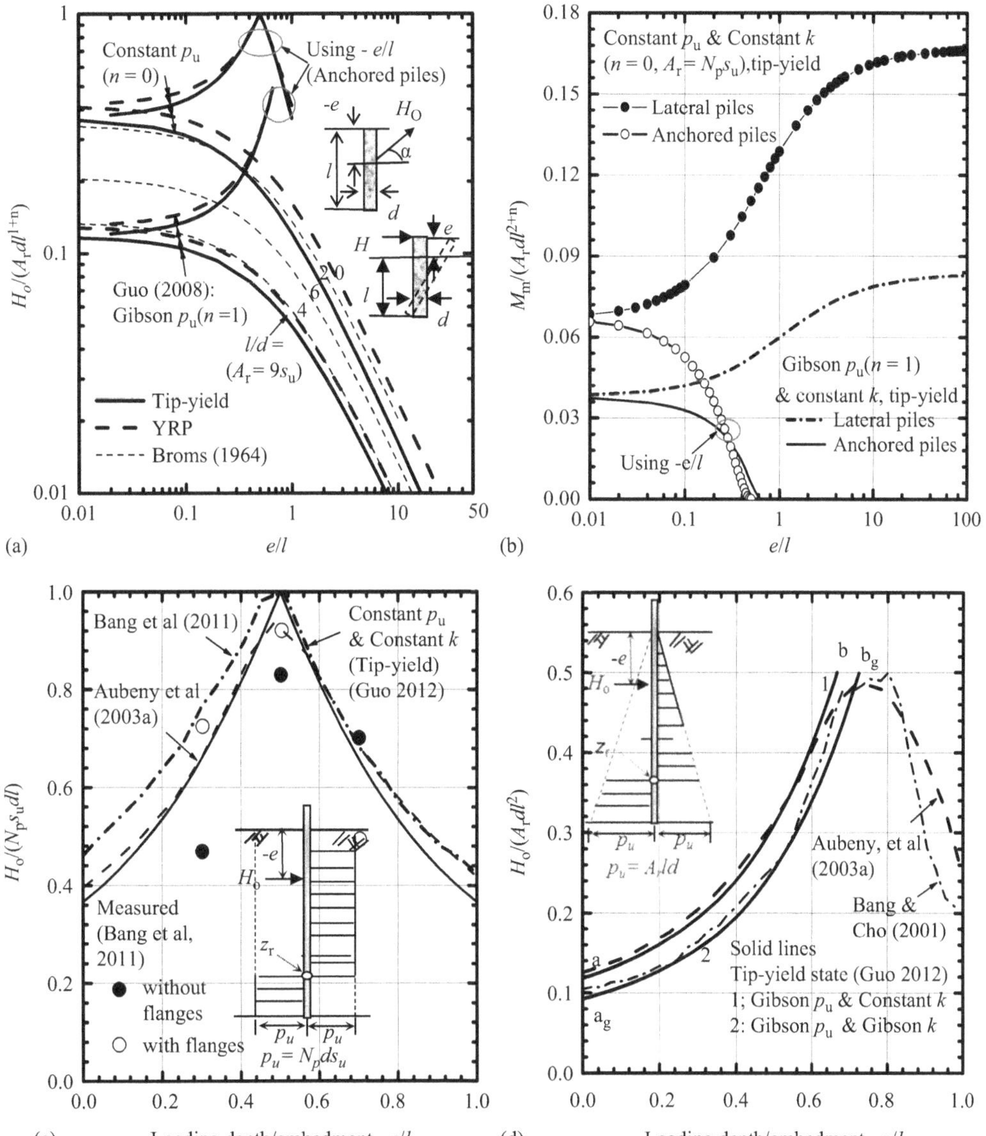

Figure 3.5 Lateral and anchored piles: (a) normalized capacity; (b) normalized maximum bending moment and measured data (α = 0); (c) in clay; (d) in sand. [Adapted from Guo, W. D., *Computers and Geotechnics* **54(2)**, 2013b.]

3.3.2 Capacity of anchored piles and loading depth

Guo (2012, 2013b) demonstrates that lateral piles (e^+) at YRP state can achieve a normalized capacity of $\bar{H}_o$ = ~0.414 by rotating about a depth of (0.5–0.707)l for constant $p_{u;}$ or rotate about a depth of (0.707–0.794)l to reach $\bar{H}_o$ = ~0.113 for Gibson p_u. The capacity increases as loading attachment (rotation) shifts downwards below GL (with e < 0 in anchored piles), as is evident Figure 3.5 concerning three pairs of the p_u and k profiles, at the tip-yield and the YRP states:

- The normalized capacity $\bar{H}_o$ at tip-yield state reduces by a maximum of 25% from constant k to Gibson k. It is slightly smaller than the $\bar{H}_o$ at the YRP state. The $\bar{H}_o$ approaches unity as $\bar{e}$ approaches −0.5(constant p_u) or −0.667 (Gibson p_u), and the pile shifts from rotation to translation.
- The capacity increases by ~2.73 (= 1/0.414) times (constant p_u) and by 3.84 (= 0.5/0.13) times (Gibson p_u) as the attachment depth is shifted to 0.5l and (0.667–0.725)l below GL, respectively. This increase is physically associated with the shift of the on-pile resistance along two-sides of the pile (divided by rotation depth) to one-side (translation) for a specified p_u of $N_p s_u d$ or $A_r zd$.

Example 3.1 Capacity against measured data

The capacity $\bar{H}_o$ at the tip-yield state is compared with published results in Figures 3.5c and d for constant p_u and Gibson p_u, respectively.

- In Figure 3.5c, the $\bar{H}_o$ follows a similar increase with the normalized depth to $-\bar{e} = 0.5$ and decrease afterwards to the PLA analysis (with/without gapping between caisson and surrounding soil). Nevertheless, it is ~15% smaller than the SM prediction on caissons (Bang and Cho 2001). The $\bar{H}_o$ agrees fairly well with the measured capacities (centrifuge tests) for three e/l ratios of anchored piles with flanges (Bang et al. 2011).
- The $\bar{H}_o$ agrees well (Figure 3.5d) between the current solution (Gibson p_u and Gibson k) and the prediction by SM (Bang and Cho 2001); as well as between the current solution (Gibson p_u and constant k) and the prediction by PLA (Aubeny et al. 2003).

Example 3.2 Capacity for various rotational depth

The ultimate lateral-moment loading capacity of anchored piles and lateral piles is governed by the same p_u profile. Equations (3.5) and (3.6b) yield a concave portion of the $\bar{M}_o - \bar{H}_o$ curve in the fourth quadrant of Figure 3.6a for a constant p_u, which overlaps the loading capacity locus for lateral piles (Guo 2012). Similarly, Equations (3.8) and (3.10b) offer the same portion of locus in Figure 3.6b for a Gibson p_u. Compared to lateral piles, an anchored pile attains a higher capacity (i.e., $\bar{H}_o = 0.414 - 1.0$ for constant p_u) for rotating about a depth z_r either within its body at $-\bar{e} = 0 – 0.33$ (e.g., $\bar{z}_r = z_r / l = 0.791 – 0.833$ for $\bar{e} = -0.25$, and 0.865–1.055 for $\bar{e} = -0.35$), or outside its body as $-\bar{e}$ approaches 0.5 [e.g., $\bar{z}_r = 5.3 - 83.7\left(\bar{e} = -0.499, \text{and} \propto (-0.5)\right)$, respectively]. These diverse rotation depths resemble the mechanisms revealed in the SM method characterized by four sets of complex expressions (Bang and Cho 2001). The normalized capacity or moment (see Figure 3.5) is independent of the slenderness ratio l/d, gapping development, rotation-translation mode, and loading angle, but the normalizer N_p (or A_r) is dependent of these factors, as discussed later.

The normalized moment $\bar{M}_m$ at the tip-yield state [gained using Equations (3.6a) and (3.10a), respectively] is shown in Figure 3.5b. Indeed, the p_u profiles have limited effect on the $\bar{M}_m$ in anchored piles using $-\bar{e} = 0.25 – 0.49$ [with $\bar{M}_m = 0$ at $-\bar{e} = 0.5$ (constant p_u) or 0.667 (Gibson p_u)], compared to their remarkable effect on the $\bar{M}_m$ in lateral piles with e^+ (= 0.25–0.667).

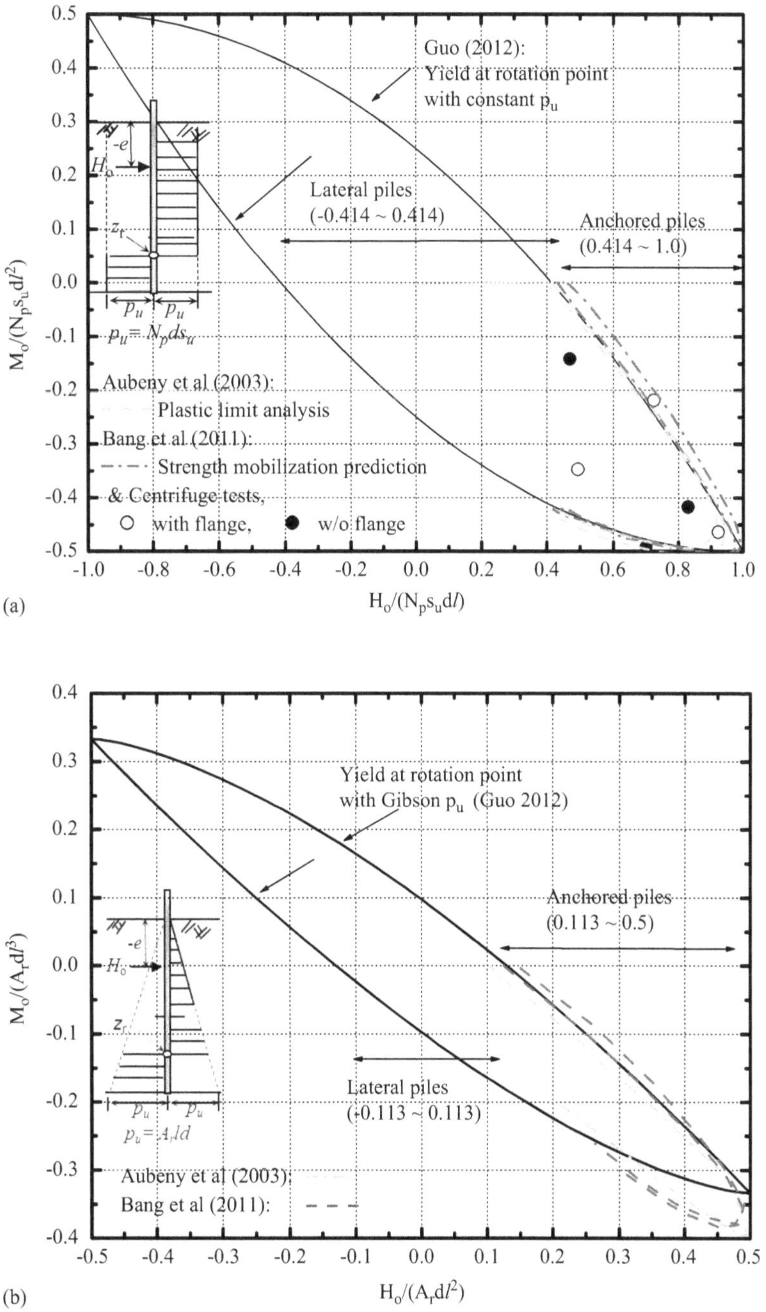

Figure 3.6 Comparison of normalized capacity $\bar{H}_o$ and normalized bending moment loci between lateral piles and anchored piles (a) constant p_u; (b) Gibson p_u.

3.4 SOLUTIONS FOR PILES WITH ROTATING CAP

Ideal elastic-plastic p–$y(u)$ curves were used to develop closed-form solutions for nonlinear response of capped lateral piles (Guo 2013b) in the context of load transfer model. The bearing capacity of rigid piles with a rotating cap is obtained for both constant and Gibson p_u profiles, which is compared with experimental data, numerical (FEM, PLA) results, and SM (rotation-dependent) solutions. These comparisons reveal the pile response (i.e., capacity, displacement and rotation) to p_u profiles (with constant k), loading eccentricity, and yielding states (at tip and rotation point), and rotational stiffness of pile-cap. Additionally, expressions are established for the resistance factor N_p (thus p_u) for cohesive soil.

Piles are largely cast into a pile-cap, which restrains the pile-head rotation with a restraining bending moment M_o. This moment normally raises with rotation angle ω_r at a constant pile-cap rotational stiffness k_G with $M_o = k_G\omega_r$ (Mokwa and Duncan 2003). As with free-head piles (Guo 2008), the model stipulates a constant k [= $p/(du)$], and a constant p_u (= $N_p s_u d$), and $u = \omega_r z + u_g$ with unknown rotation ω and displacement u_g at GL The pile response is related to the rotational stiffness, k_G, as well as the eccentricity, e, of the head load, and the slip depth, z_o (see Figure 3.4a), as outlined below:

$$\bar{H} = \frac{0.5\left[(1+2\bar{z}_o)(\bar{z}_o-1)^2+12\bar{k}_G\right]}{(\bar{z}_o-1)^2(2+\bar{z}_o+3\bar{e})+6\bar{k}_G} \tag{3.11}$$

$$\bar{u}_g = \frac{2+3(1+\bar{z}_o{}^2)\bar{e}+\bar{z}_o{}^3+6\bar{k}_G}{(2+\bar{z}_o+3\bar{e})(1-\bar{z}_o)^2+6\bar{k}_G} \tag{3.12}$$

$$\bar{\omega}_r = \frac{-3(1+2\bar{e})}{(2+\bar{z}_o+3\bar{e})(1-\bar{z}_o)^2+6\bar{k}_G} \tag{3.13}$$

$$\bar{z}_r = \frac{-u_g}{\omega_r l} \tag{3.14}$$

$$\bar{z}_m = \bar{H},\ \bar{M}_m = \bar{H}(\bar{e}+0.5\bar{z}_m)+\bar{k}_G\bar{\omega}_r, \tag{3.15}$$

$$\bar{z}_*^3+3\bar{e}\bar{z}_*^2-1.5(1+2\bar{e})\bar{z}_*+0.5(1+12\bar{k}_G)=0 \tag{3.16}$$

where $\bar{H} = H/(N_p s_u dl)$, the normalized pile-head load; $\bar{u}_g = u_g k/(N_p s_u)$, the normalized groundline displacement; $\bar{\omega} = \omega kl/(N_p s_u)$, the normalized rotation angle; $\bar{M}_m\left[= M_m/(N_p s_u dl^2)\right]$ and $\bar{M}_o\left[= M_o/(N_p s_u dl^2)\right]$, the normalized maximum bending moment, and moment at GL; $\bar{k}_G = k_G/(kl^3)$, the normalized rotational stiffness of the pile-cap; $\bar{z}_o = z_o/l$, the normalized slip depth; $\bar{z}_r$ and $\bar{e}$ were defined earlier. Table 3.5 provides the solutions for a Gibson p_u (= $A_r dz$) with a constant k (i.e., $p = kdu$, thus $u^* = A_r z_o/k_o$), or with a Gibson k (i.e., $p = k_o dzu$, thus $u^* = A_r/k_o$).

The impact of the stiffness $\bar{k}_G$ (= 0, 0.05, 0.1, 0.2, 0.4, 0.8, 1.5, 3.0 and 10) on typical responses of $\bar{H}-\bar{u}_g$ – and $\bar{H}-\bar{\omega}_r$ were obtained for $n = 0$ and $n = 1$, and are plotted in Figures 3.7a and b, respectively. Their salient features are outlined in the following examples.

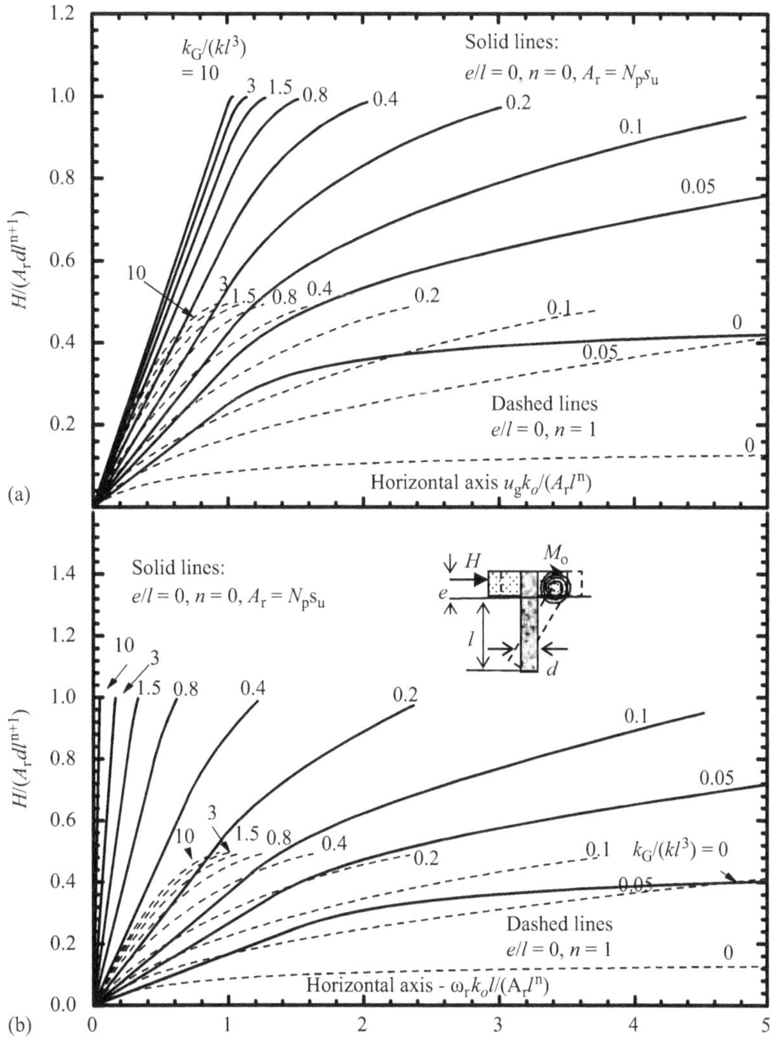

Figure 3.7 Response of (a) pile-head load $\bar{H}$ and groundline displacement $\bar{u}_g$; (b) $\bar{H}$ and rotation $\bar{\omega}$ (n = 1 for Gibson p_u and constant k, and n = 0 for constant p_u and constant k). [Adapted from Guo, W. D., *Computers and Geotechnics* **54(**2), 2013b.]

Example 3.3 Response of capped piles

- Rigid piles with fixed-head ($\bar{k}_G > 10$) experience a maximum lateral load at a unit normalized displacement ($\bar{u}_g = 1$, see Figure 3.7a).
- Increasing the stiffness k_G [see Equation (3.13)] can reduce the rotation at a constant p_u and k. As k_G approaches infinity, the pile becomes fully fixed-head ($\omega = 0$), and the normalized load $\bar{H}$ reaches a maximum of 1.0 (the normalized capacity) as the slip depth extends from GL to the pile tip $(\bar{z}_o \to 1)$.
- For Gibson p_u and constant k, increasing stiffness k_G reduces the rotation. However, a fully fixed-head theoretically occurs at a normalized slip depth of 1.732–2.0 [using $\bar{z}_o = -1.5\bar{e} + 0.5\left(9\bar{e}^2 + 24\bar{e} + 12\right)^{0.5}$]. This means that fixed-head is practically unattainable given $\bar{z}_o \leq 1$, as noted for Gibson p_u and Gibson k as well (see Table 3.5). The normalized load (or capacity) $\bar{H}$ reaches a maximum of 0.5 (the fixed-head capacity) as the slip depth extends to the entire pile length $(\bar{z}_o \to 1)$.

Equation (3.16) and the expressions in Table 3.5 are used to estimate z_*/l for the tip-yield state. For instance, given $k_G/(kl^3) = 0.02$ for constant p_u and constant k, the tip-yield occurs at z_*/l of $0.49(\bar{e} = 0)$, and $0.462(\bar{e} = 0.03)$), respectively. At large k_G values, the yield between pile and soil occur almost simultaneously over the entire pile length. However, fully fixed-head conditions are theoretically impossible for laterally loaded rigid piles with a constant p_u through a Gibson p_u, as rotation and/or cracking may be induced beforehand.

Example 3.4 Capped piles covering free-head and fixed-head cases

The new solutions cover the free-head (k_G = 0) to fixed-head cases. Setting k_G = 0, for instance, Equations (3.11) and (3.16) reduce to

$$\bar{H} = \frac{0.5(1+2\bar{z}_o)}{2+\bar{z}_o+3\bar{e}} \tag{3.17}$$

$$\bar{z}_* = -(1.5\bar{e}+0.5)+0.5(3+6\bar{e}+9\bar{e}^2)^{0.5} \tag{3.18}$$

Substituting Equation (3.18) into Equation (3.17) yields Equation (3.4). Likewise, with a Gibson p_u and a constant k, the expressions in Table 3.7 offer

$$\bar{H} = \frac{0.5\bar{z}_o}{2+\bar{z}_o+3\bar{e}} \tag{3.19}$$

$$\bar{z}_* = -(1.5\bar{e}+0.5)+0.5(3+12\bar{e}+9\bar{e}^2)^{0.5} \tag{3.20}$$

Substituting Equation (3.20) into Equation (3.19) yields Equation (3.7). Similarly, the solutions for a Gibson p_u and a constant k can be deduced using the expressions in Table 3.5.

Example 3.5 Response for $k_G = 0$

To estimate free-head response, the abovementioned solutions for constant p_u and these in Table 3.5 for Gibson p_u were adopted by setting k_G = 0. The normalized responses of $\bar{H}$, $\bar{u}_g$, $\bar{\omega}_r$ and $\bar{M}_m$ obtained for a few typical e/l ratios in the range of 3 to −0.5 (n = 0) or 3 to −0.667 (n = 1) are shown in Figure 3.8. For anchored piles with a constant p_u, Figure 3.8 shows that loading at the middle depth of embedment $(\bar{e} = -0.5)$ results in pure translation of the pile (ω_r = 0), zero maximum bending moment ($\bar{M}_m$ = 0, see Figure 3.5b), and a limited elastic displacement $\bar{u}_g = 1$ at $\bar{H} = 1$ (not shown fully in the figure). Slip from the pile base may commence shortly after or even simultaneously with the incipient of the top-slip z_o (see Figure 3.4). This slip causes a limited impact on the capacity, as evidenced by the small gap between the current and the previous solutions (see Figure 3.5d) and the discrepancy between the tip-yield and the YRP states (Guo 2013b). Similar features are noted for a Gibson p_u, albeit the translation occurring at $\bar{e} - 0.667$ (constant k) to −0.725 (Gibson k). Furthermore, the $\bar{M}_m$ is doubled by using $p_u = 2A_r dz$ (e.g., at tip-yield state and $\bar{e} = 0.01$, $\bar{M}_m = 0.07728$ in Figure 3.8, compared to $\bar{M}_m = 0.03864$ in Figure 3.5b), and it occurs prior to reaching the tip-yield state. These findings for anchored piles are as expected, but their validity should be confirmed by checking displacement performance when available.

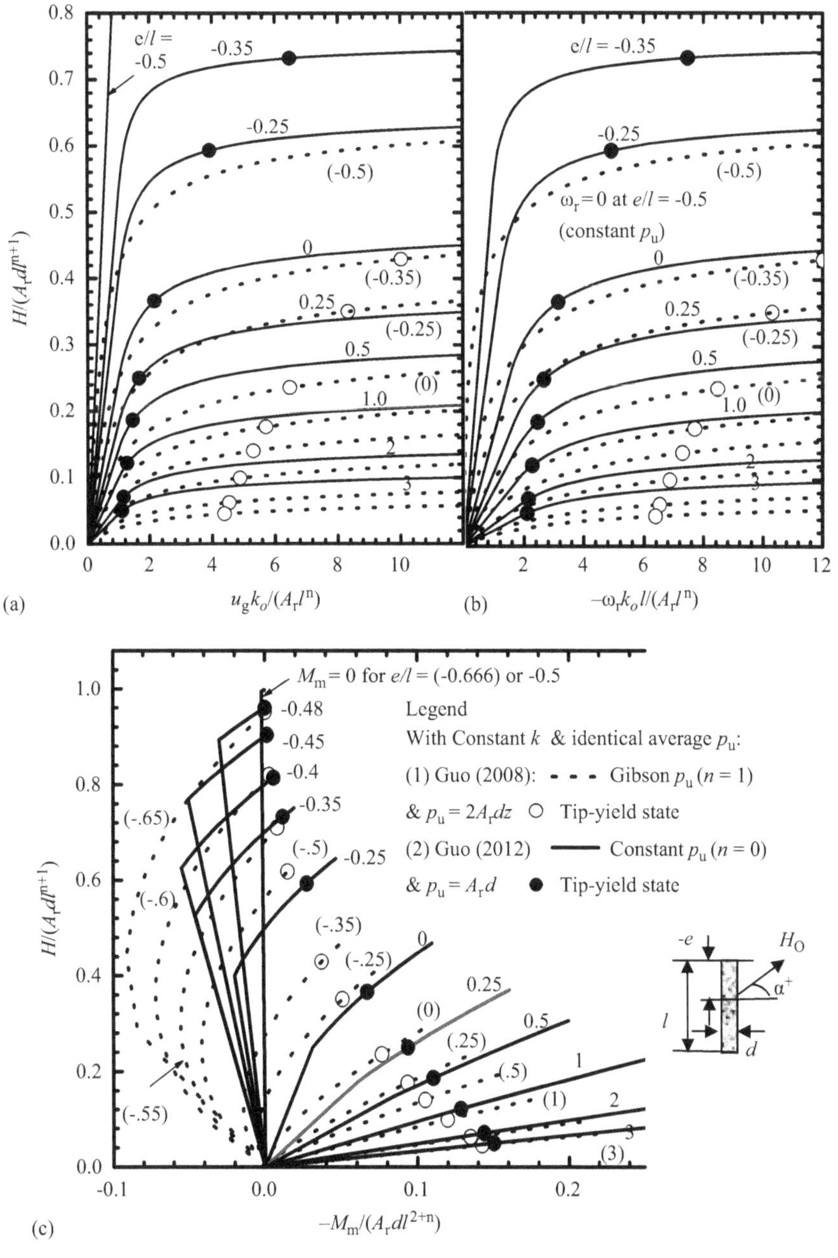

Figure 3.8 Normalized response for typical ratios of *e*/*l*: (a) pile-head load $\bar{H}$ and mudline displacement $\bar{u}_g$; (b) $\bar{H}$ and rotation $\bar{\omega}$; (c) $\bar{H}$ and maximum bending moment $\bar{M}_{max}$. [Adapted from Guo, W. D., *Computers and Geotechnics* **54**(2), 2013b.]

3.4.1 Impact of eccentricity and rotation stiffness

Figure 3.9 shows the impact of normalized rotational stiffness k_{nG} [= $k_G/(kl^{3+n})$ = 0.1–1.0] on normalized pile-head load and displacement for e/l = 0. For a constant k and normalized eccentricity e/l (= 0.1–1.0), the normalized displacement and rotation are presented in Figures 3.9a and b for a constant p_u at k_{nG} = 0.3, and in Figure 3.9d for a Gibson p_u at k_{nG} = 0.1. Figures 3.10a and b present the same results for Gibson k and Gibson p_u.

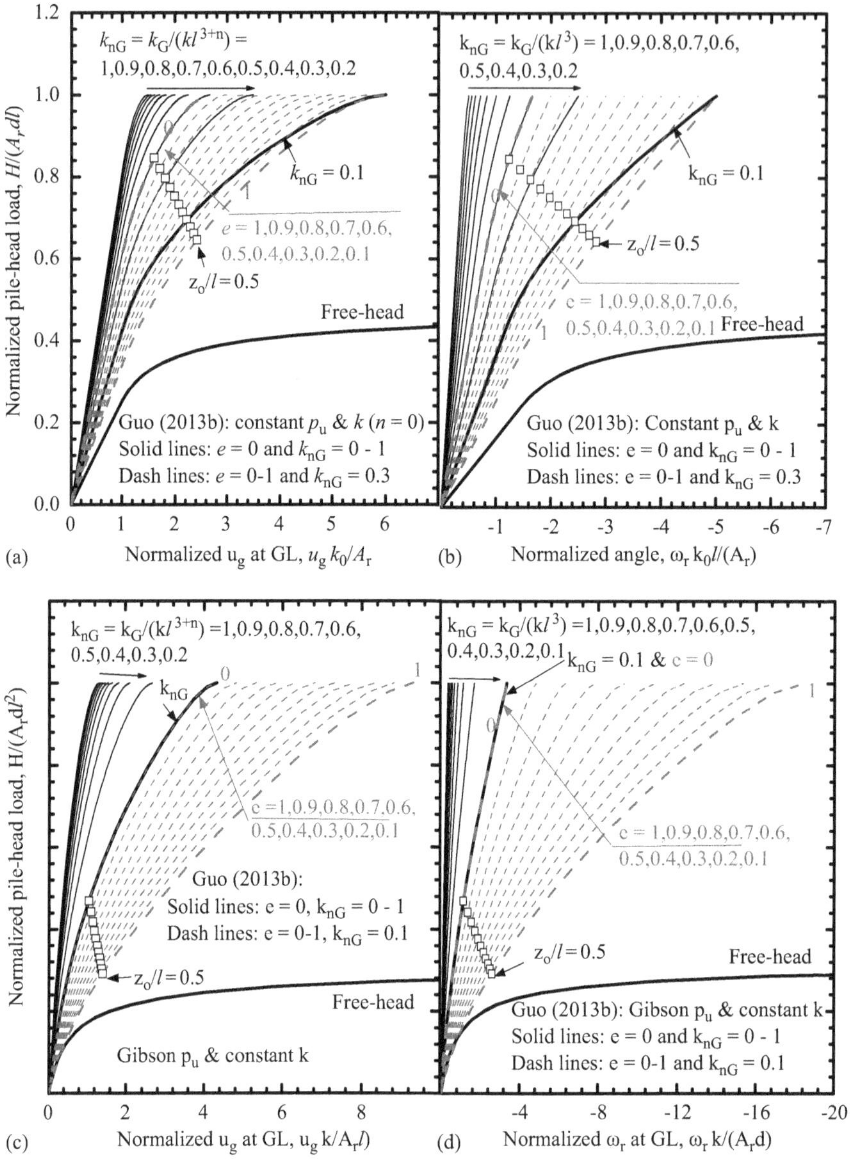

Figure 3.9 Response of capped piles (constant k): (a)(b) constant p_u; (c)(d) Gibson p_u.

In reality, the pile-cap may rotate in the opposite direction to pile body (i.e., back-rotation), resulting in a negative rotational stiffness k_G. At a special value of normalized stiffness $k_G^*/(kl^{3+n})$, the denominator becomes singular [see Equation (3.11)]:

$$\bar{k}_G^* = -\left(2 + \bar{z}_o + 3\bar{e}\right)\left(1 - \bar{z}_o\right)^2 / 6 \tag{3.21}$$

For $\bar{k}_G$ values of −0.1 through −0.5, the normalized response (e.g., $\bar{H}, \bar{u}_g, \bar{\omega}_r$) is amplified (see Figures 3.10c and d), which is otherwise limited for $\bar{k}_G$ of −0.6 through −1.0.

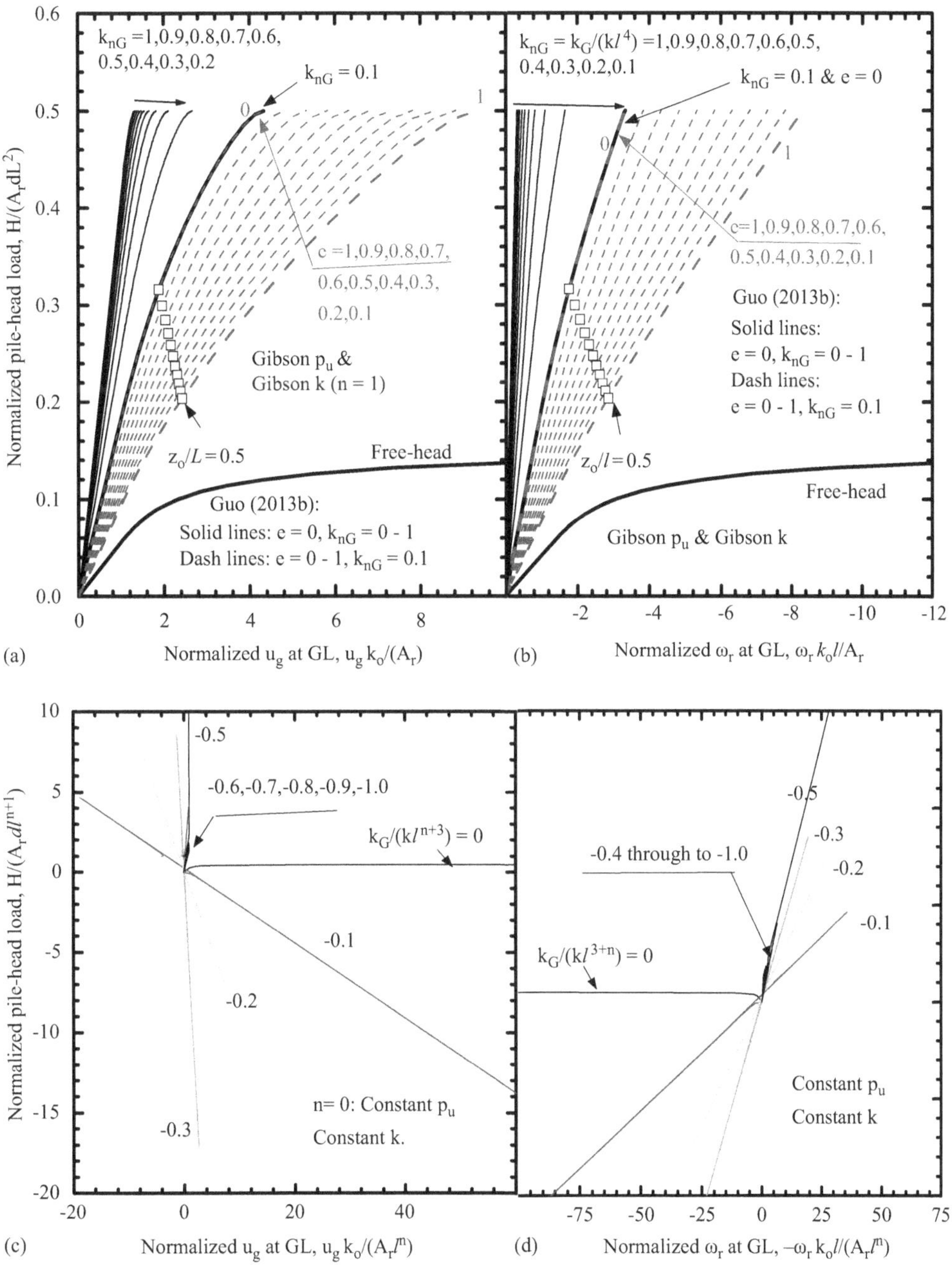

Figure 3.10 Response of capped piles: (a and b) Gibson p_u & Gibson k; (c and d) Constant p_u & Constant k.

Furthermore, Figures 3.11a, b and c present the impact of varying *e/l* of 0.1–1.0 at k_{nG} = 0.1 on normalized displacement and rotation angle for constant k with a constant p_u and Gibson k or constant k with Gibson p_u, respectively. Figure 3.11d shows the impact on normalized load and bending moment at GL for the same conditions.

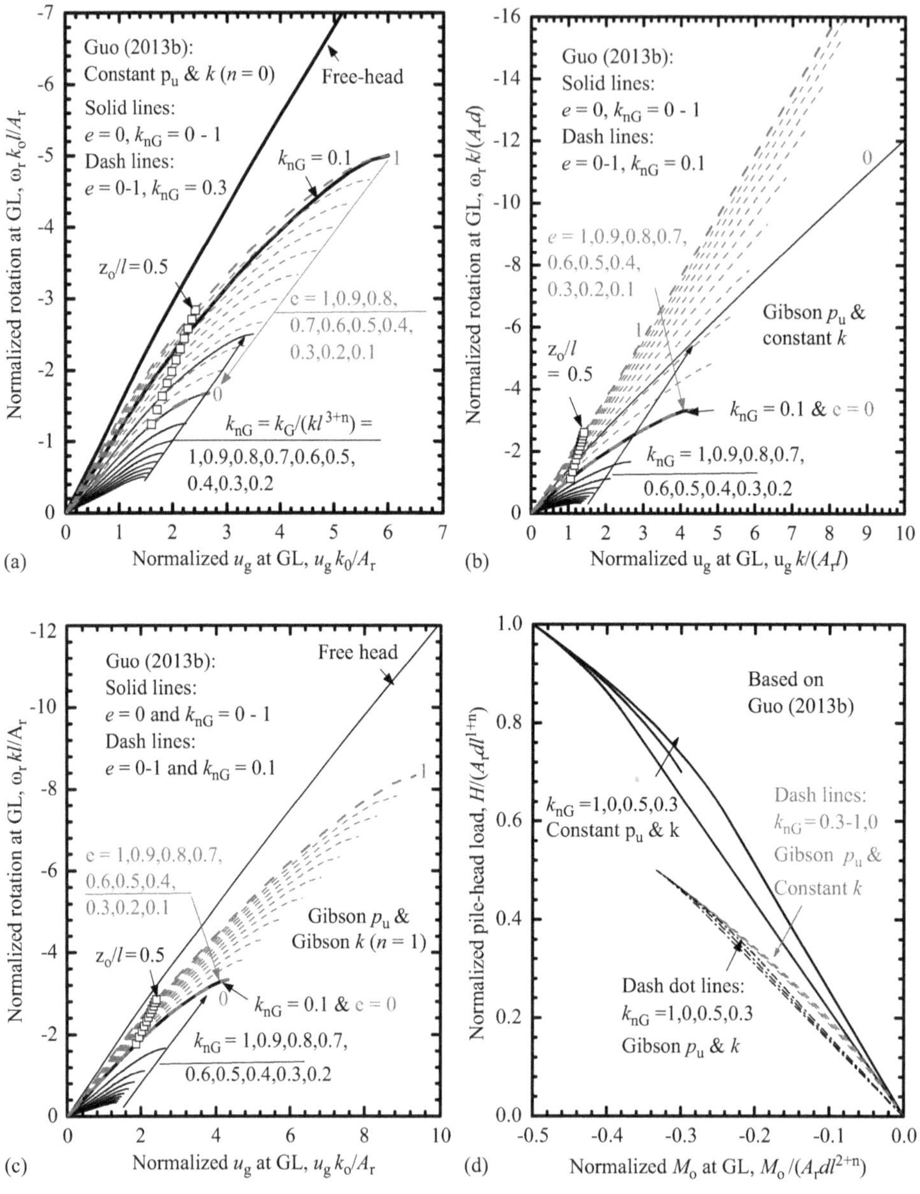

Figure 3.11 Response of capped piles: (a–c) rotation; (d) bending moment at GL.

3.4.2 Response profile from both eccentricity and rotation stiffness

To predict the response profiles of semi-fixed head piles (with $\bar{k}_G > 0$), the existing solutions for free-head piles with $M_o = He$ (Guo 2012) are utilized by replacing the M_o with $M_o = He + k_G\omega_r$. Specifically, the $M(z)$ and $H(z)$ for a constant p_u (constant k) at the pre-tip yield state are given by

$$\bar{M}_1(\bar{z}) = \bar{H}(\bar{e} + \bar{z}) + \bar{k}_G\bar{\omega}_r - 0.5\bar{z}^2 \tag{3.22a}$$

$$\bar{H}_1(\bar{z}) = \bar{H} - \bar{z} \quad (0 < z \leq z_o) \tag{3.22b}$$

$$\bar{M}_2(\bar{z}) = \bar{H}(\bar{e} + \bar{z}) + \bar{k}_G\bar{\omega}_r - (\bar{z}_o\bar{z} - 0.5\bar{z}_o^2) - \frac{\bar{\omega}_r}{6}(\bar{z}^3 - 3\bar{z}\bar{z}_o^2 + 2\bar{z}_o^3) - 0.5\bar{u}_g(\bar{z} - \bar{z}_o)^2 \tag{3.23a}$$

$$\bar{H}_2(\bar{z}) = \bar{H} - \bar{z}_o - 0.5\bar{\omega}(\bar{z}^2 - \bar{z}_o^2) - \bar{u}_g(\bar{z} - \bar{z}_o) \quad (z_o < z \leq l) \tag{3.23b}$$

where yielding may occur simultaneously over the pile length and the tip-yield state is excluded herein. Using Equations (3.21), (3.22a and b) and (3.23a and b) for $\bar{z}_o = 0.4$ and nine typical values of normalized rotation stiffness, shear force and bending moment profiles were obtained and are plotted in Figures 3.12a$_1$ and b$_1$. The same profiles were obtained for Gibson p_u (constant k) using the expressions in Table 3.5 and are plotted in Figures 3.12a$_2$ and b$_2$.

3.5 FACTORS N_p AND $N_{p\alpha}$ FOR COHESIVE SOIL

The normalizer N_p [e.g., in Equation (3.4)] or A_r (see Table 3.5) is useful across footing, lateral piles (caissons), anchored piles and pipelines (Guo 2012). Under combined lateral-moment loading (loading angle), piles may rotate (largely observed on lateral piles), or shift to translation (observed in anchored piles), depending on the restraints at the pile-head and base, depths of stiff layers and the loading positions. For a typical $\bar{k}_G$, the N_p value may be interpolated from the extreme values of free-head (denoted superscript 'FreH', $\bar{k}_G = 0$) (Guo 2006, 2008) and fixed-head (superscript 'FixH', say, $\bar{k}_G > 10$) (Guo 2009), with $N_p^{FreH} = 4N_p^{FixH}$.

Using the normalized capacities for free-head cases (see Table 3.6), the N_p values for a uniform shear strength profile were deduced using the solutions for a constant p_u (Guo 2013b). These solutions take into account the impact of pile-movement mode, gapping and slenderness ratios. The N_p may be estimated using Equations (T3.7a) and (T3.7b and c) (see Table 3.7) for pure rotation and rotation-translation (R-T), respectively.

Example 3.6 N_p from PLA and upper bound solution

For l/d = 2–10 and e = 0, the N_p was deduced as 10.14–11.6 (no gapping) and 5.92–8.33 (with gapping), respectively. The former range was deduced using Equation (3.5) at YRP state for rotating mode with an ultimate $H_o/(s_u dl)$ of $0.414N_p$ = 4.2–4.8 (PLA on caissons). As rotation mode evolves into gapping, the N_p decreases to $0.414N_p$ of 2.45–3.5 (PLA on caissons). The no-gapping N_p of 10.14–11.6 (also see Table 3.6) falls within the upper bound solution of 9.14–11.94 for translating piles (Randolph and Houlsby 1984). It is independent of the mobilization of limiting force on both sides of a rotating pile (see Figure 3.4 d$_2$, the current solutions) or on one side of a translating pile (the upper bound approach). The gapping N_p of 5.92–8.33 agrees well with 5.6–8.6 (Aubeny et al. 2003) and represents 40–80% the no-gapping N_p.

For short piles with l/d = 0.08–1 or footings (Yun and Bransby 2007), the N_p for $\bar{e}$ = –0.5 (translational movement) was deduced as 1.0–6.1 using $N_p = H_o/(s_u dl)$ at the YRP state; while the N_p for $e = \infty$ (rotational movement) was gained as 2.8–9.8 based on $0.25N_p$ = 0.7–2.45 [= $M_o/(s_u dl^2)$ at YRP state, Guo 2012].

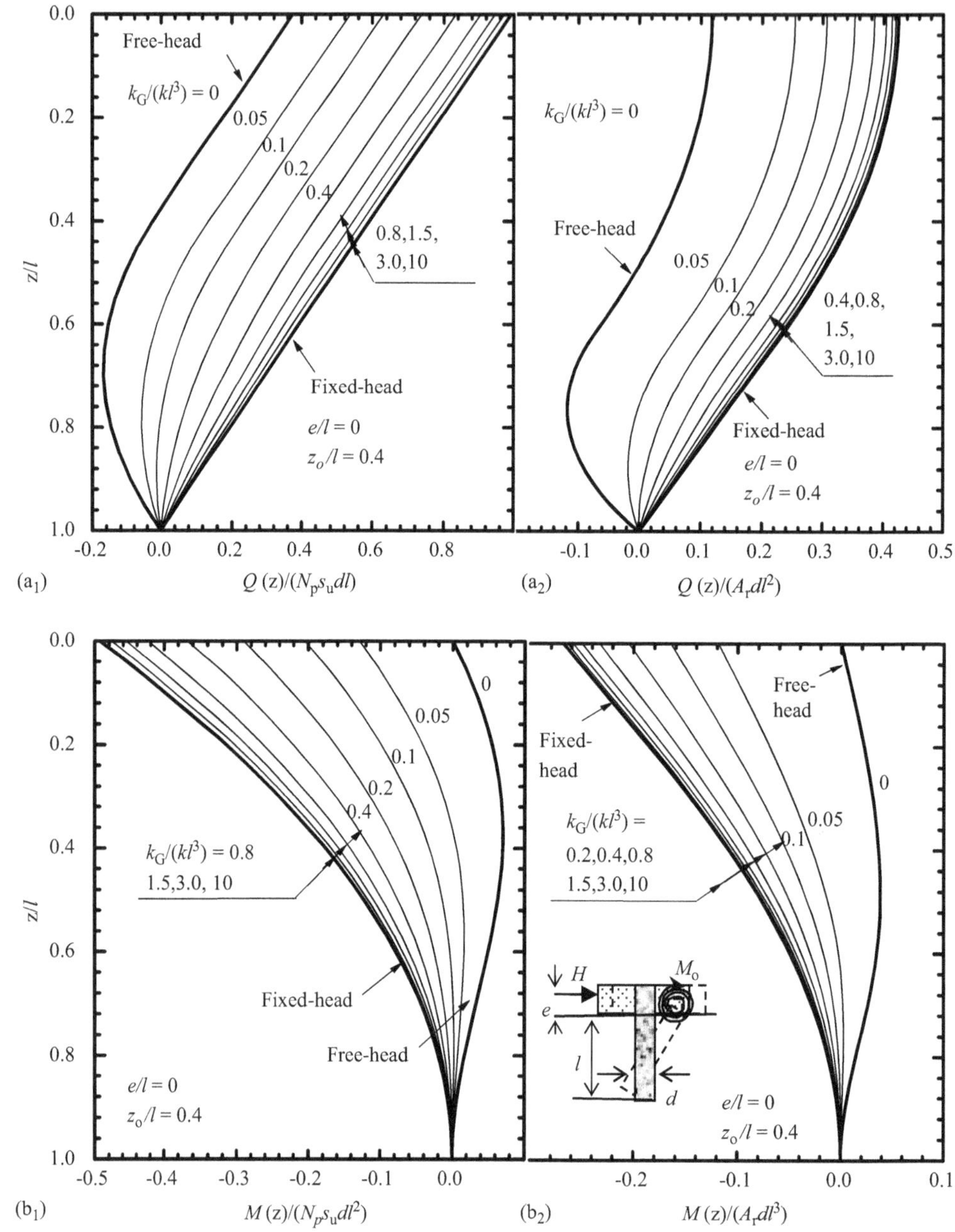

Figure 3.12 Normalized profiles of $\bar{H}(z)$ and $\bar{M}(z)$ for typical normalized k_G at $z_o/l = 0.4$ (constant k): (a_1)–(b_1) constant p_u; (a_2)–(b_2) Gibson p_u. [Adapted from Guo, W. D., *Computers and Geotechnics* **54(**2), 2013b.]

Table 3.6 Normalized capacities and N_p values deduced using Constant p_u and constant k (Guo 2013b)

l/d	*0.08*	*0.2*	*0.5*	*1.0*	*2.0*	*4.0*	*6.0*	*8.0*	*10*
$H_o/(N_p s_u dl)$	1.0	2.0	3.6	6.1	4.2	4.5	4.75	4.8	4.8
N_p	1.0	2.0	3.6	6.1	10.14	10.87	11.47	11.59	11.59
References	Yun and Bransby (2007), translation				Aubeny et al. (2003), rotation no gapping				
l/d	0.08	0.2	0.5	1.0	2.0	4.0	6.0	8.0	10
$M_o/(N_p s_u dl^2)$[a] $H_o/(N_p s_u dl)$[b]	0.7[a]	0.925[a]	1.3[a]	2.45[a]	2.45[b]	2.8[b]	3.1[b]	3.3[b]	3.45[b]
N_p	2.8	3.7	5.2	9.8	5.92	6.76	7.49	7.97	8.33
References	Yun and Bransby (2007), rotation				Aubeny et al. (2003), rotation with gapping				

[a] Indicates $M_o/(N_p s_u dl^2)$.
[b] Indicates $H_o/(N_p s_u dl)$.

Table 3.7 Solutions for a rigid pile at pre-tip and tip-yield states (Guo 2013b)

Expressions and conditions	*Equation*
$N_p = 2 + 8.9 \arctan(0.9l/d)$ (Pure rotation)	(T3.7a)
$N_p = 1 + 7.5 \arctan(0.8l/d)$ ($R - T$ mode without gapping)	(T3.7b)
$N_p = 1 + 5 \arctan(\varepsilon l/d)$, $\varepsilon = 0.7 - 1.0$ ($R - T$ mode with gapping)	(T3.7b)
$N_p\alpha = N_p \exp(-0.014\alpha)$	(T3.7d)
$N_{p\alpha} = N_{p\alpha=90°} + \left(N_{p\alpha=0°} - N_{p\alpha=90°}\right)\left(\dfrac{90° - \alpha}{90°}\right)^2$	(T3.7e)
$N_p = [2 + 8.9 \arctan(0.9l/d)] \exp(-0.014\alpha)$	(T3.7f)
$N_p = [1 + 7.5 \arctan(0.8l/d)] \exp(-0.014\alpha)$ ($R - T$ mode without gapping)	(T3.7g)
$N_p = [1 + 5 \arctan(\varepsilon l/d)] \exp(-0.014\alpha)$ $\varepsilon = 0.7 - 1.$ ($R - T$ mode with gapping)	(T3.7h)

Example 3.7 N_p from rotation

Table 3.7 provides expressions for estimating N_p for various cases. Figure 3.13a shows the plot of Equations (T3.7a, b and c) of Table 3.7 with $\varepsilon = 0.7$. Equation (T3.7a) fits well with those deduced from rotation of footing under moment loading as well as with $N_p = 15$ from cone penetration (CPT) tests. CPT tests force rotational soil movement around the cone tip, and may induce a higher N_p than 15 (Yu et al. 2000), as is confirmed by analytical solutions for a cone angle of 30°–180° (Houlsby and Martin 2003). Equation (T3.7b) closely matches with N_p values deduced from footing and caissons under lateral loading without gapping, while Equation (T3.7c) is based on values for footing, caissons under lateral loading with gapping and anchors.

Upper bound solutions (Murff and Hamilton 1993) were obtained for rough-surface piles in uniform and non-uniform shear strength profiles, as well as for smooth surface piles in the same strength profiles. These solutions are plotted in Figure 3.13a and confirm the current variation law of the N_p with l/d ratio, despite some scatter for $l/d < 1$. The latter may be due to using the pile-length l in place of the embedment depth of anchors or pipelines when determining the N_p values using empirical expressions deduced numerically for vertical anchors without gapping (Merifield 2011) and with gapping (Yu et al. 2011) as well as deduced from centrifuge tests for pipelines with gapping (Oliveira, et al. 2010).

Example 3.8 N_p variation to loading angle

Lateral loading is pragmatically applied at an angle of α + (>0) from the horizon on anchored piles, or horizontally on a group of rake piles with a batter angle of α (see the insert of Figure 3.13b). Measured capacities were normalized using centrifuge tests for anchored piles with α = 0 (Kim et al. 2006), and laboratory 1g model tests (Rao and Veeresh 1994) for rake piles. The normalized capacities were taken as the ratios of $N_{p\alpha}/N_p$ [$N_{p\alpha} = p_u/(s_u d)$, involving an inclined loading or batter angles], and are plotted in Figure 3.13b. The figure demonstrates a starkly similar variation in the $N_{p\alpha}/N_p$ (thus p_u) with the angle α between the anchored and battered piles, regardless of loading eccentricity e^+ or e^-.

The impact of batter angle α on lateral piles is well mimicked by factoring the p_u for 'vertical piles', as seen in centrifuge tests on pile groups (Zhang et al. 1999) and finite element modelling on single piles (Rajashree and Sitharam 2001). The factor $N_{p\alpha}$ of Equation (T3.7d) (see Figure 3.13b) is equally sufficiently accurate for anchored piles, which are governed by the same limiting p_u. Assuming an identical variation law to inclined anchor in clay (Yu et al. 2011), the $N_{p\alpha}$ may follow Equation (T3.7e). For convenience, Equations (T3.7a, b, c) are rewritten as Equations (T3.7f, g and h) by dropping the subscript 'α'. The N_p of Equations (T3.7f, g and h) may be used in Equations (3.4)–(3.6) to calculate response of anchored piles under inclined loading, as the impact of the loading position on the $N_{p\alpha}$ is limited (Anderson et al. 2005). The N_p value may be reduced further to account for the shadowing effect to model capped piles.

Example 3.9

Aubeny et al. (2003) conducted plasticity limit analysis (PLA) and finite element analyses to gain lateral capacity of caissons (with length-to-diameter ratios of 2, 6 and 10) under various loading angle. The normalized $N_{p\alpha}/N_p$ ratios for a uniform shear strength profile are plotted in Figure 3.13b for $-\bar{e}$ = 0, 0.25, 0.5, 0.75 and 1.0. The ratio for each $\bar{e}$ hardly reduces at α = 0–15°, but drops rapidly at α > 75°, which is not seen in the cited test data. This phenomenon requires further experiment investigation. To count for the faster reduction in $N_{p\alpha}$ with α the factor -0.014α in Equation (T3.7d) may be replaced with -0.018α.

It is common practice to estimate the p_u utilizing N_p and undrained shear strength. For Gibson profile, the p_u may be obtained using plasticity solutions (Brinch Hansen 1961, Guo 2013a) and cohesion and frictional angle of soil (Guo 2006) especially for piles in layered soil. The gradient A_r may then be obtained by fitting the p_u profile, allowing for use of the corresponding solutions, as is noted for flexible piles in stiff clay (Guo 2008). Further investigation into the factor A_r is warranted for anchors, anchored piles in Gibson p_u profile.

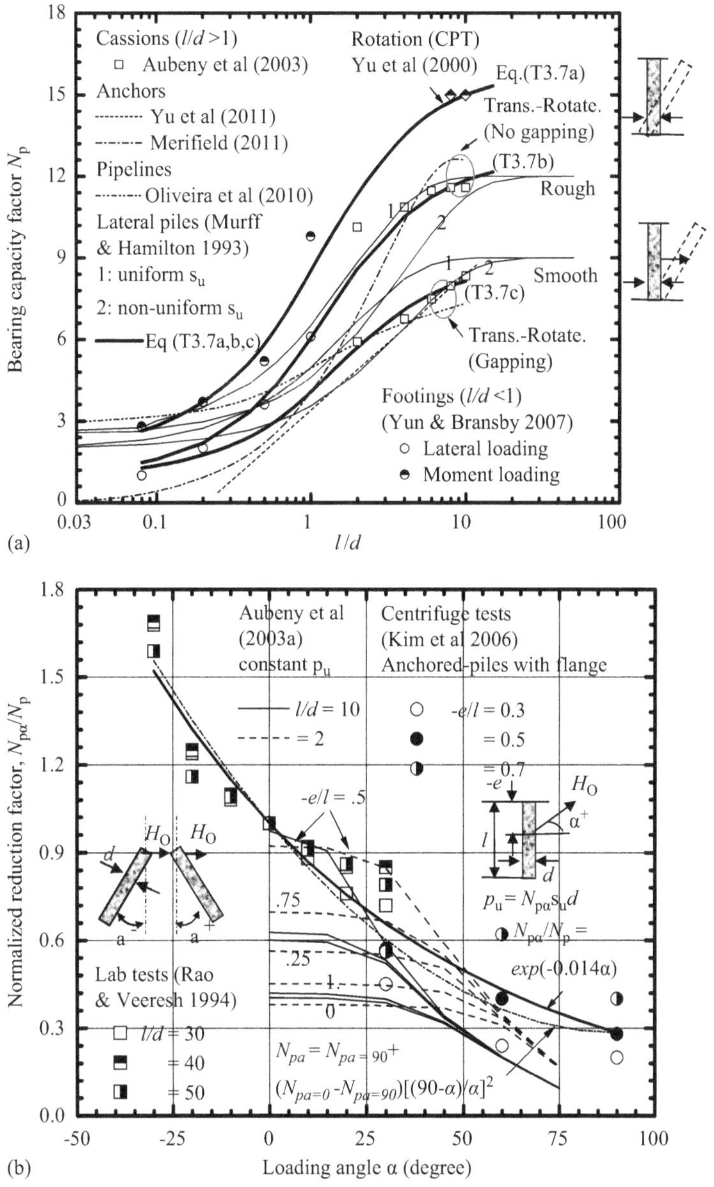

Figure 3.13 (a) Variation of N_p with gapping and movement (rotation, translation) modes; (b) Variation of $N_p\alpha/N_p$ with loading angle. [Adapted from Guo, W. D., *Computers and Geotechnics* **54**(2), 2013b.]

3.6 SUMMARY

This chapter presents comprehensive set of solutions for capturing the response of rigid piles under different loading conditions, which include

- Elastic solutions for capturing the response of rigid anchored piles under various loading types, such as a power-law distributed loading p (via the gradient A_r) over a loading depth ($b \sim \alpha b$, c, or l_m); or a uniform p to a sliding depth l_m.

- Explicit expressions for estimating loading capacity of free-head and capped piles, which encompass the solutions for lateral loading capacity H_o at tip-yield state and the YRP state, the maximum bending moment M_m, and the lateral capacity H_o–M_o (applied moment) locus for a constant FPUL p_u or a linear increasing FPUL p_u (Gibson p_u) with depth.
- Expressions for estimating the resistance factor N_p (= p_u/s_u), which include the impact of rotation or translation, slenderness ratio, gapping and loading angles on lateral piles, anchors, caissons and pipelines.

The rigid-pile solutions are adopted together with a negative loading eccentricity to capture response of anchored piles. This approach may be extended to model piles in multi-layered soil, as was accomplished previously for flexible piles. Under elastic-plastic conditions, anchored piles are subjected to large bending moment. The predicted capacity is on conservative side for $l/d < 3$ owing to ignoring base resistance. The displacement and A_r value gained (for Gibson p_u) are yet to be corroborated through test data.

Chapter 4

Slope stabilizing piles

2-layer model

4.1 INTRODUCTION

Piles are subjected to soil movement in stabilizing sliding slope, supporting bridge abutments and providing lateral pressure barrier during pile driving or excavation operation. In earthquake-prone zone, vertically loaded piles are susceptible to lateral spreading. To simulate the behaviour of slope stabilizing piles, elastic solutions are proposed using measured thrust and gradient of soil movement with depth (Fukuoka 1977; Cai and Ugai 2003). These solutions are compelled to be coupled with the magnitude of soil movement, as are other methods (Chmoulian 2004; De Beer and Carpentier 1977; Ito and Matsui 1975; Viggiani 1981). Theoretically, thrust can be estimated for each movement (Guo 2012). Nevertheless, during lateral spreading, 'flexible' piles become 'rigid', for which the use of the concentrated load is not justified.

In Chapter 1, Figure 1.5 depicts a 3-layer model for piles in sliding soil. This model accounts for the impact of pile-slide relative position and the evolution of the on-pile force per unit length (FPUL) p_s which reflects the pile response as revealed in experiment (Guo 2003, 2008; Dobry et al. 2003; He et al. 2009) to the distance, stiffness and profile of soil movement. The chapter also introduces BiP (R-E, R-EP) models that capture the nonlinear response of rotationally restrained piles subjected to soil movement (sliding soil) over stable layer.

4.2 2-LAYER MODEL AND SOLUTIONS

Guo (2014, 2015) developed the 2-layer model and its solutions to estimate the response of rigid piles (see Figure 1.4, Chapter 1). The pile is subjected to a moving layer (of thickness l_m and a subgrade reaction k_s) embedded in a stable layer (of λl_m and mk_s). The model considers the moment from rotational restraint (= $k_\theta \omega_r$) over pile embedment, including the head-constraint moment M_G (= $k_G \omega_r$) and the tip-constraint moment M_T (= $k_T \omega_r$), as well as the lateral shear force H at the head level and bending moment M_o. Additionally, an on-pile FPUL p_s (= $\alpha p_{ub} l_m / l$), which incorporate the factor α of soil movement profile and loading distance from the pile(s), is considered. The resistance FPUL $p(z)$ is proportional to the subgrade modulus k_s of the sliding layer and mk_s of the stable layer, respectively, and is confined to the upper limit of the p_s at l_m. The model is resolved using force and bending moment equilibrium about the pile (see Table 4.1).

The solutions to the model are presented in Table 4.2, which encompass the normalized depth $\bar{z}_{m2}$ of the maximum bending moment M_{m2}, the maximum shear force Q_{m2}, the shear

DOI: 10.1201/9781003315230-4

Table 4.1 Governing equations for 2-layer model

The horizontal force and bending moment equilibrium of the rigid pile (see Figure 1.4c and d, Chapter 1) are given by Equation (T4.1a) and (T4.1b), respectively
$\int_{o}^{l_m}(\bar{\omega}_r s+\bar{w}_g)p_s ds+\int_{l_m}^{l}(\bar{\omega}_r s+\bar{w}_g)mp_s ds-\int_{0}^{c}p_s ds=H$ (T4.1a)
$\int_{o}^{l_m}(\bar{\omega}_r s+\bar{w}_g)p_s(l-s)ds+\int_{l_m}^{l}(\bar{\omega}_r s+\bar{w}_g)mp_s(l-s)ds$ $-\int_{0}^{c}p_s(l-s)ds=pl^2\bar{k}_\theta\bar{\omega}_r+Hl+M_o$ (T4.1b)
Equations (T4.1a) and (T4.1b) allow the $\bar{\omega}_r$ and $\bar{w}_g$ to be determined as Equations (4.2) and (4.3). The expressions for the $Q(z)$ and $M(z)$ are provided in Table 4.2. By $Q_i'(z) = 0$, the depth z_{mti} of maximum shear force Q_{mi} is determined; whereas with $M_1'(z) = 0$, the depth z_{mi} of the maximum bending moment M_{mi} is gained.

Table 4.2 2-layer theoretical model for response profiles (Guo 2015b)

Depth	$z \le l_m$	$z > l_m$
$w(z)$	$w(z)=(\bar{\omega}_r\bar{z}+\bar{w}_g)p/k_s$	
$p(z)$	$p_1(z)=(\bar{\omega}_r\bar{z}+\bar{w}_g-1)p_s$	$p_2(z)=(\bar{\omega}_r\bar{z}+\bar{w}_g)mp_s$
Shear force $Q(z)$	$\frac{Q_1(z)}{p_s l}=0.5\bar{\omega}_r\bar{z}^2+\bar{w}_g\bar{z}-\bar{z}-\bar{H}$	$\frac{Q_2(z)}{p_s l}=05\left[(1-m)\bar{l}_m^2+m\bar{z}^2\right]\bar{\omega}_r$ $+\left[(1-m)\bar{l}_m+m\bar{z}\right]\bar{w}_g-\bar{c}-\bar{H}$
Bending moment $M(z)$	$\frac{M_1(z)}{p_s l^2}=-\bar{M}_o+\frac{1}{6}\bar{\omega}_r\bar{z}^3$ $+\frac{1}{2}(\bar{w}_g-1)\bar{z}^2-\bar{H}\bar{z}+\bar{k}_\theta\bar{\omega}_r$	$\frac{M_2(z)}{p_s l_2}=\left[\left(\frac{1}{2}\bar{l}_m^2\bar{z}-\frac{1}{3}\bar{l}_m^3\right)(1-m)+\frac{m}{6}\bar{z}^3+\bar{k}_G\right]\bar{\omega}_r$ $-\bar{H}\bar{z}+\left[\left(\bar{l}_m\bar{z}-0.5\bar{l}_m^2\right)(1-m)+0.5m\bar{z}^2\right]\bar{w}_g$ $-0.5\bar{c}(2\bar{z}-\bar{c})-\bar{M}_o$
z_m	$\bar{z}_m=\frac{-1}{\bar{\omega}_r}\left\{(\bar{w}_g-1+\left[(\bar{w}_g-1)^2+2\bar{\omega}_r\bar{H}\right]^{0.5}\right\}$ $(0\le z\le l_m)$ $\bar{z}_m=\bar{z}_{m2}=\frac{-1}{\bar{\omega}_r}\left\{-\bar{w}_g+\left\{\frac{m-1}{m}\bar{\omega}_r^2\bar{l}_m^2+2\left[\frac{m-1}{m}\bar{l}_m\bar{w}_g+\frac{\bar{H}+\bar{l}_m}{m}\right]\bar{\omega}_r+\bar{w}_g^2\right\}^{05}\right\}$ $(l_m<z\le l)$	
Q_{m1} Q_{m2}	$\bar{z}_{mt1}=(1-\bar{w}_g)/\bar{\omega}_r,\ Q_{m1}=Q_1(z_{mt1})$ $(0\le z\le l_m)$ $\bar{z}_{mt2}=-\bar{w}_g/\bar{\omega}_r,\ Q_{m2}=Q_2(z_{mt2})$ $(l_m\le z\le l)$	

force $Q_i(z)$ and the bending moment $M_i(z)$ at depth z of $0 \sim c$, (measured from GL, with subscript 1) and at z of $c \sim l$ (with subscript 2). The main expressions are as follows:

(1) The pile-deflection at depth z, $w(z)$ is given by

$$w(z)=(\bar{\omega}_r\bar{z}+\bar{w}_g)p_s/k_s \tag{4.1}$$

where $\bar{\omega}_r\left[=w'(z)k_s l/p_s\right]$ and $\bar{w}_g\left(=w_g k_s/p_s\right)$ are given by

$$\bar{\omega}_r=\frac{-6\left[\left(2m\lambda+m\lambda^2+1\right)\left(\bar{c}+\bar{H}\right)\bar{l}_m-\left(m\lambda+1\right)\left(\bar{c}^2-2\bar{M}_o\right)\right]}{\left(1+4m\lambda+6m\lambda^2+m^2\lambda^4+4m\lambda^3\right)\bar{l}_m^3+12\left(1+m\lambda\right)\bar{k}_\theta} \tag{4.2}$$

$$\bar{w}_g=\frac{4\left(3m\lambda+3m\lambda^2+m\lambda^3+1\right)\left(\bar{c}+\bar{H}\right)\bar{l}_m^3+\left(6\bar{M}_o-3\bar{c}^2\right)\left(2m\lambda+m\lambda^2+1\right)\bar{l}_m^2+12\left(\bar{c}+\bar{H}\right)\bar{k}_\theta}{\left[\left(1+4m\lambda+6m\lambda^2+m^2\lambda^4+4m\lambda^3\right)\bar{l}_m^3+12\left(1+m\lambda\right)\bar{k}_\theta\right]\bar{l}_m} \tag{4.3}$$

where $\bar{k}_\theta=k_\theta/\left(k_s l^3\right)$, $\bar{c}=c/l$, $\bar{l}_m=l_m/l$, $\bar{H}=H/\left(p_s l\right)$, $\bar{M}_o=M_o/\left(p_s l^2\right)$, $\lambda=(l-l_m)/l_m$, m = ratio of subgrade modulus of stable layer over sliding layer, and k_θ = the total rotational stiffness along the pile, which consists of the top non-liquefied layer k_G (= M_o/ω_r), and the low non-liquefied layer k_T (= M_T/ω_r) (i.e., $k_\theta = k_G + k_T$). Note that while k_G and k_T values may differ, the associated angle of rotation ω_r is identical along the rigid pile.

(2) The maximum bending moment M_{m2} is given by

$$\begin{aligned}M_{m2}/\left(p_s l^2\right)=&\left[\frac{m}{6}\bar{z}_{m2}^3+\left(1-m\right)\bar{l}_m^2\left(\frac{\bar{z}_{m2}}{2}-\frac{\bar{l}_m}{3}\right)+\bar{k}_G\right]\bar{\omega}_r-\bar{z}_{m2}\bar{H}\\&-\bar{M}_o+\left[0.5m\bar{z}_{m2}^2+\left(1-m\right)\bar{l}_m\left(\bar{z}_{m2}-0.5\bar{l}_m\right)\right]\bar{w}_g-0.5\bar{c}\left(2\bar{z}_{m2}-\bar{c}\right)\end{aligned} \tag{4.4}$$

where $\bar{M}_{m2}=M_{m2}/\left(p_s l^2\right)$, $\bar{k}_G=k_G/\left(k_s l^3\right)$, and $\bar{z}_{m2}=z_{m2}/l$. The solutions are defined by four input parameters, namely m, k_s, p_s (via p_b), and k_θ, and characterized by the net resistance per unit length $p_1(z) = p - k_s w_1(z)$ within $z = 0 \sim l_m$, the loading depth $c = l_m$ ($<l$) for piles in a 2-layer soil, and $c < l_m$ for full-length ($l_m = l$) lateral spreading case.

The model can be applied to a wide range of soil movement profiles by altering p_b (= αp_{ub}) or α. By setting $c = l_m$ and $\bar{k}_\theta=0$, the normalized rotation $\bar{\omega}_r$, displacement $\bar{w}_g$, depth $\bar{z}_m\left(=z_m/l\right)$, bending moment $M_m/(p_s l_m l)$, and depth z_{mt2}/l (in stable layer) of normalized maximum shear force $Q_m/(p_s l_m)$ were obtained. In particular, the M_{m2} was obtained with $z > c$ (for non-loading zone). Figure 4.1 shows the normalized depth z_{mt}/l (of maximum shear force Q_m), depth z_m/l (of maximum bending moment M_m), force $Q_m/p_s l_m$ and moment $M_m/(p_s l_m l)$ at various normalized sliding depths l_m for m = 1, 1.5, 3 and 5.

Example 4.1 Impact of concentrated load (*H*) versus p_s

Table 3.4 in Chapter 3 presents compact solutions for rigid piles embedded in a soil with a uniform k_s, under a lateral load H at a depth 'a'. Figure 4.1 plots the calculated ratios of z_m/l, Q_m/H, and $M_m/(Hl)$ for a range of normalized location a/l ($a = l_m$). In particular, Figure 4.1c presents the positive M_m (i.e., M_m^+) as negative to facilitate comparison. The ratios of $M_m/(Hl)$ and $M_m/(Hl_m l)$ decrease from a maximum of 0.148 (at l_m/l = 0) to zero (at l_m/l = ~0.8), while the normalized depth z_m/l increases from 0.33 to 1.0 (see Figure 4.1a), which induces the normalized moment M_m (i.e., M_{m1}) in the loading zone. The ratios of $M_m/(Q_m l)$ are important to design slope-stabilizing piles. Figure 4.1d illustrates the profiles of normalized bending moment $M(z)/(Hl)$ for typical normalized depths of loading by pile embedment (a/l). The maximum moment-force ratio $M(z)/(Hl)$ in Figure 4.1d is similar to the maximum $M_m/(p_s l^2)$ in Figure 3.3b (Chapter 3), but the latter (based on distributed p_s = p) has a shape close to measured data.

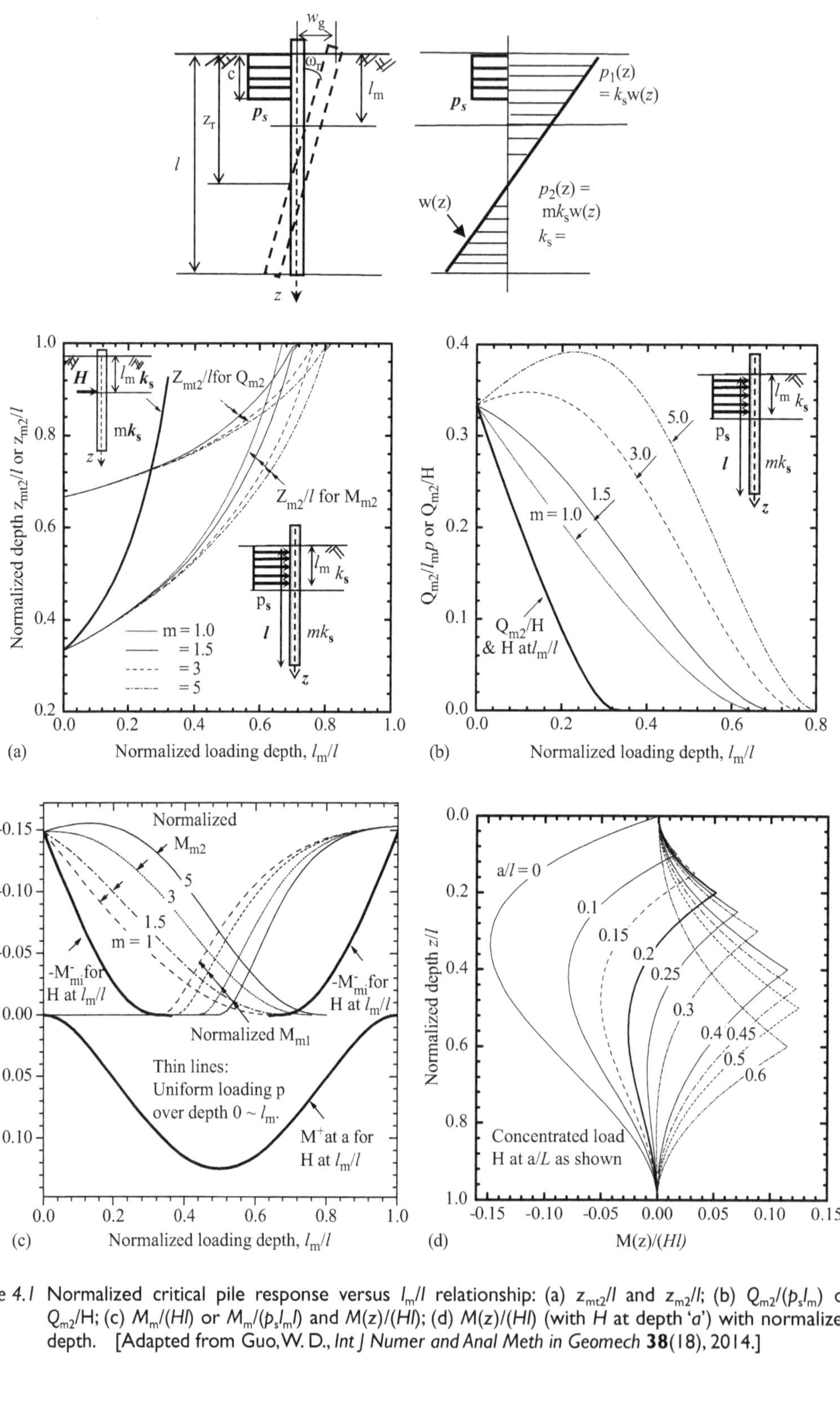

Figure 4.1 Normalized critical pile response versus l_m/l relationship: (a) z_{mt2}/l and z_{m2}/l; (b) $Q_{m2}/(p_s l_m)$ or Q_{m2}/H; (c) $M_m/(Hl)$ or $M_m/(p_s l_m l)$ and $M(z)/(Hl)$; (d) $M(z)/(Hl)$ (with H at depth 'a') with normalized depth. [Adapted from Guo, W. D., *Int J Numer and Anal Meth in Geomech* **38**(18), 2014.]

4.2.1 Features of 2-layer model

The 2-layer model by Guo (2014, 2015b) has several salient features such as

- The calculated on-pile pressure matches measured values for passive piles in clay (Viggiani 1981), and reveals elastic pile-soil interaction. The estimated maximum shear force, however, exceeds measured values for model piles (Guo 2014) in sliding sand (and on the safe side).
- The normalized maximum bending moment $M_m k_s/(p_s l_m)$ at various normalized displacements of $\bar{w}_g$ compares well with the boundary element solution (BEM) (Chen and Poulos 1997) at given a pile deflection w_g (= p_s/k_s) under uniform soil movement (Guo 2015b) (see Figures 4.2a and b). The trend of increasing p_s/k_s ratio for decreasing slenderness ratio l/d as noted in the correlation of $k_s/G_s \approx 10(l/d)^{-0.5}$ for laterally loaded rigid piles (Guo 2012) is also observed.
- The soil movement w_s encompasses the displacement w_g derived from rotation and translation. The differential displacement between the pile- top and base is equal to $w'(z)l$ with $w'(z) = \omega_r$. The net local pile displacement y (between pile and moving soil) at depth z is equal to $\omega_r z$ after deducting any translation component. The associated resistant FPUL $p(z)$ is equal to $\omega_r z k_s$ after deducting the translational resistance $w_g k_s$. The displacement $w(z)$ and the FPUL $p(z)$ constitute the p–y (w) curve at depth z.

Chapter 1 provides estimation of input parameters p_b, k_s, and m. Equations (4.1)–(4.4) and the expressions in Table 4.2 take into account the impact of pile cross-section shape and any vertical load P (see Figure 1.4c) on the pile through a modified value α on the FPUL p_b (see Chapter 1).

Example 4.2 Piles subjected to an inverse triangular profile of soil movement

Initially, solutions are obtained for a uniform soil movement ($w_g \approx w_s$). To accommodate an inverse triangular moving soil (i.e., IT w_s), for instance, the movement w_s and its depth l_m are modified to w_s/α and l_m/α, respectively, with $\alpha = 0.72$. The resulting normalized displacement versus loading depth curve is demonstrated in Figure 4.2b and closely matches that of the BEM solution for IT w_s movement. The reduced ratio of w_g/w_s is due to an increase in w_s (to w_s/α), and $w_g \approx w_s$ for rigid piles in a uniform soil movement. Guo (2015b) further demonstrates that the model tests under T-block (Chapter 1) have a measured effective soil movement (= pile displacement w_g) of $0.72w_s$ where $\alpha = 0.72$, $w_s \approx w_f$ −42 mm, with 42 mm of ineffective movement.

4.2.2 Non-dimensional analysis (H = 0, M_o = 0)

The parameters k_s and p_s (out of the four input parameters m, k_s, p_s via p_b, and k_θ) are used to normalize the pile response, which offer

- Normalized soil movement (= $\alpha w_g k_s/p_s$) versus sliding depths [$l_m/(\alpha l)$] (see Figure 4.3);
- Normalized pile-soil relative displacement (= $\omega_r k_s l/p_s$) versus soil displacement (= $\alpha w_g k_s/p_s$) (Figure 4.4);
- Normalized bending moment [= $M_m/(p_s l_m l)$] versus soil movement (Figure 4.5);
- Normalized thrust [= $Q_m/(p_s l_m)$] at sliding depth (Figure 4.6a) and at true depth (Figure 4.6b), respectively.

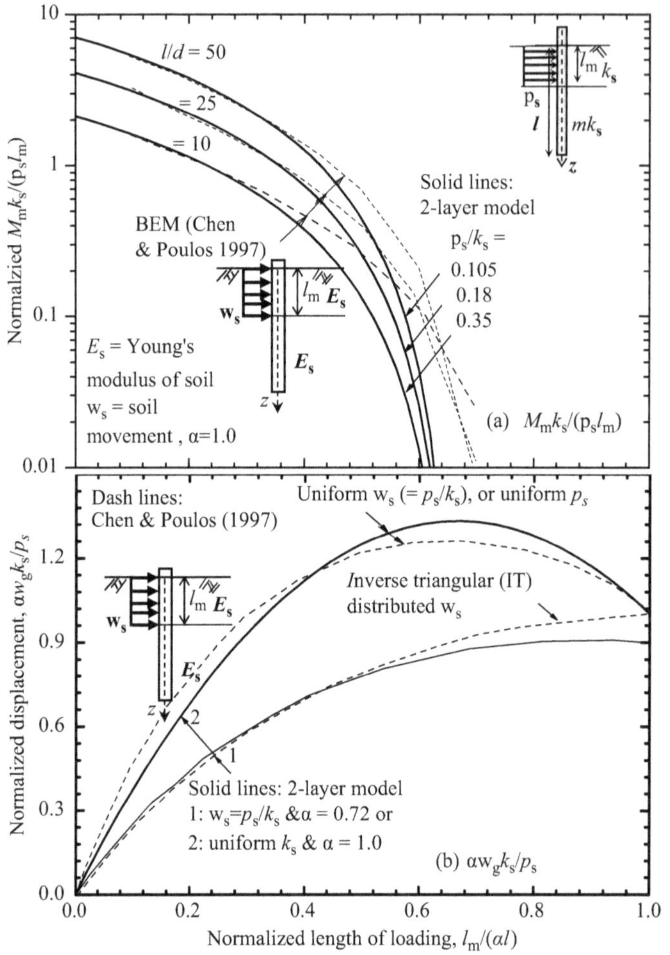

Figure 4.2 Comparison between current solution and BEM results (Chen and Poulos 1997) for uniform loading: (a) Normalized bending moment; (b) Normalized deflection. [Adapted from Guo, W. D., *Int J Numer and Anal Meth in Geomech* **38**(18), 2014.]

Guo (2015b) demonstrated that these normalized values are effective in capturing the effect of soil movement in various scenarios, including landslides, earthquake shaking, simple shear tests and embankment loading. Particularly, (i) a series of stepwise increased uniform p_s ($= p_{ub}z/l$) with depth z effectively capture the effect of soil movement, and (ii) the p_{ub} (thus p_s) reduces from rotating (free-head) to translating (fixed-head) piles, and exhibits a triangular to a uniform profile of on-pile pressure, respectively. To obtain the nonlinear pile response, such as the moment M_m and pile-displacement w_g (Figure 1.4c), for a specific l_m (i.e., uniform p_s in Figure 1.4d, Chapter 1), the 'elastic' model is repeatedly calculated for a series of l_m ($< l$) and the associated on-pile pressure p_s ($= p_b l_m/l$). Finally, the net pressure gradually increases to a maximum and reduces subsequently with the lateral movement, and a translational resistance may hold a residual bending moment if $k_T \neq 0$.

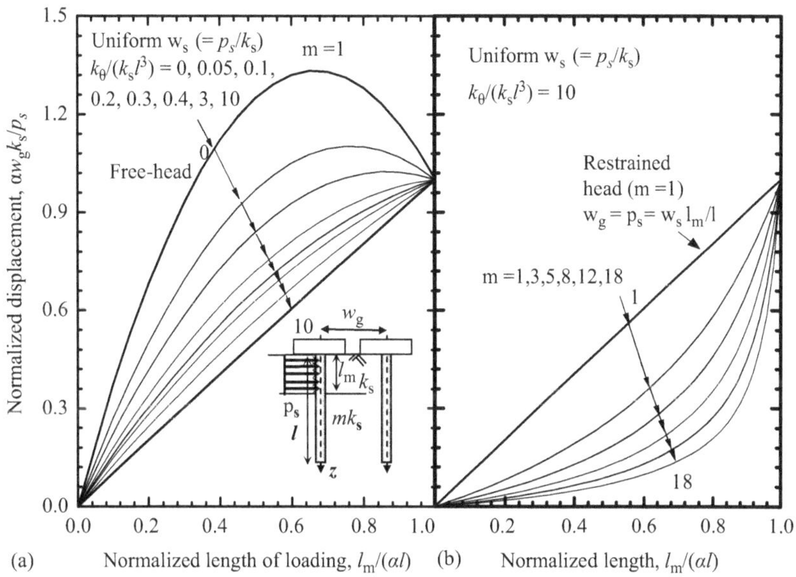

Figure 4.3 Normalized uniform soil movement w_g (constant k_s) with loading depth owing to (a) normalized cap stiffness (m = 1); (b) modulus ratio m and $\bar{k}_\theta = 10$. [Adapted from Guo, W. D., *Can Geotech J* **52**(7), 2015b.]

Example 4.3 Impact of modulus ratio m on $\alpha w_g/w_s$

Figures 4.3a and b demonstrate a linear correlation between $\alpha w_g/w_s$ and $l_m/(\alpha l)$ for a perfectly head-rotationally restrained pile. The induced average pressure over the pile embedment at a movement w_s is $w_s k_s l_m/(\alpha^2 l)$. The transitional movement w_g is equal to $w_s l_m/(\alpha^2 l)$ (i.e., the pressure over the k_s), or $\alpha w_g/w_s = l_m/(\alpha l)$. As the modulus ratio m increases, the base resistance becomes apparent, resulting in a significant reduction in the ratio $\alpha w_g/w_s$ (see Figure 4.3b).

Example 4.4 Impact of modulus ratio m on $-\omega_r l/w_g$

Figure 4.4 shows an upper limit ratio $-\omega_r l/w_g$ (pile-soil relative displacement over w_g) of 1.5 for $l_m/l < 0.5$. This ratio and its displacement mode are independent of loading properties, and are, therefore, identical to a laterally loaded rigid pile (Guo 2012). At high l_m/l values (>0.5), the normalized displacement $\bar{w}_g^*$ (* denotes the lower bound) displays an inverted mirror image of that for $l_m/l < 0.5$, as illustrated in the inset of Figure 4.4a. For $l_m/l < 0.5$, the $\bar{w}_g^*$ is thus equal to the normalized base displacement w_b/w_s. Therefore, $\bar{w}_g^* = w(l)/w_s = \bar{\omega}_r + \bar{w}_g$ is obtained as per Equation (4.1). For $l_m/l > 0.5$, '$\bar{w}_g^* = -\bar{\omega}_r/3$' (the lower bound) due to $-\omega_r l/w_g = 1.5$. The two bold extreme lines in Figure 4.4a intersect at the point $(\bar{w}_g = 2, -\bar{\omega}_r = 3)$, which implies $\bar{w}_g \le 2$ and $|\bar{\omega}_r| \le 3$ for any rigid piles. For a highly rotational restrained pile, the moment at the pile tip, $M_m \left(= k_T\omega_r = \omega_r \bar{k}_\theta k_s l^3\right)$, is equal to $p_s l_m l/2$ at a negligible displacement w_g/w_s (≈ 0). The normalized angle $-\bar{\omega}_r$ should be equal to $1/\left(2\bar{k}_\theta\right)$. For instance, if $\bar{k}_\theta = 10$, then $-\bar{\omega}_r = 0.05$ is obtained at $w_g/w_s = 0$, as illustrated in Figure 4.4c. As the m increases, the normalized maximum pile-displacement increases and converges towards $1/\left(2\bar{k}_\theta\right)$.

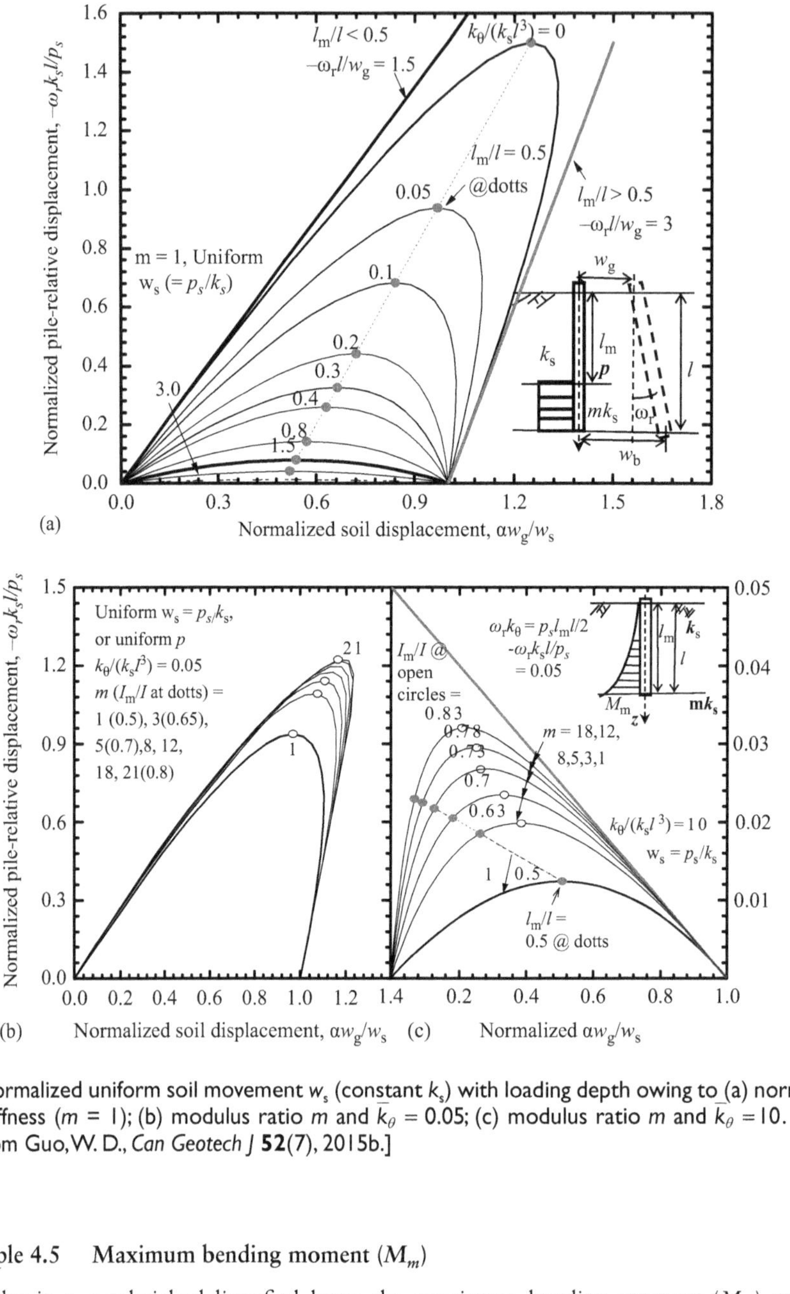

Figure 4.4 Normalized uniform soil movement w_s (constant k_s) with loading depth owing to (a) normalized cap stiffness (m = 1); (b) modulus ratio m and $\bar{k}_\theta = 0.05$; (c) modulus ratio m and $\bar{k}_\theta = 10$. [Adapted from Guo, W. D., *Can Geotech J* **52**(7), 2015b.]

Example 4.5 Maximum bending moment (M_m)

For piles in a sandwiched liquefied layer, the maximum bending moment (M_m) generally occurs at the depth l_m, which differs from that for a free-head laterally loaded pile. Irrespective of the head restrained conditions, the bending moment M_m was calculated using $z = l_m$ in $M_1(z)$ (see Table 4.2) for typical $\bar{k}_\theta$ and m. Figure 4.5 shows the resulting normalized M_m obtained. In particular, for a fixed-head pile $\left(\bar{k}_\theta = 10\right)$, the l_m/l (at m = 1) is 0.5, which yields the p_s distribution profiles shown in the insert of Figure 4.5a. The M_m at l_m then equals $p_s l^2/16$, or $M_m/(p_s l l_m) = 0.125$. For fully base-restrained piles, the normalized M_m increases by a factor of 2.6 from 0.124 (m = 1, $\bar{l}_m = 0.5$) to 0.32 (m = 18, $\bar{l}_m = 0.8$), and converges towards 0.5 with increasing m value (see Figure 4.5b). This is comparable

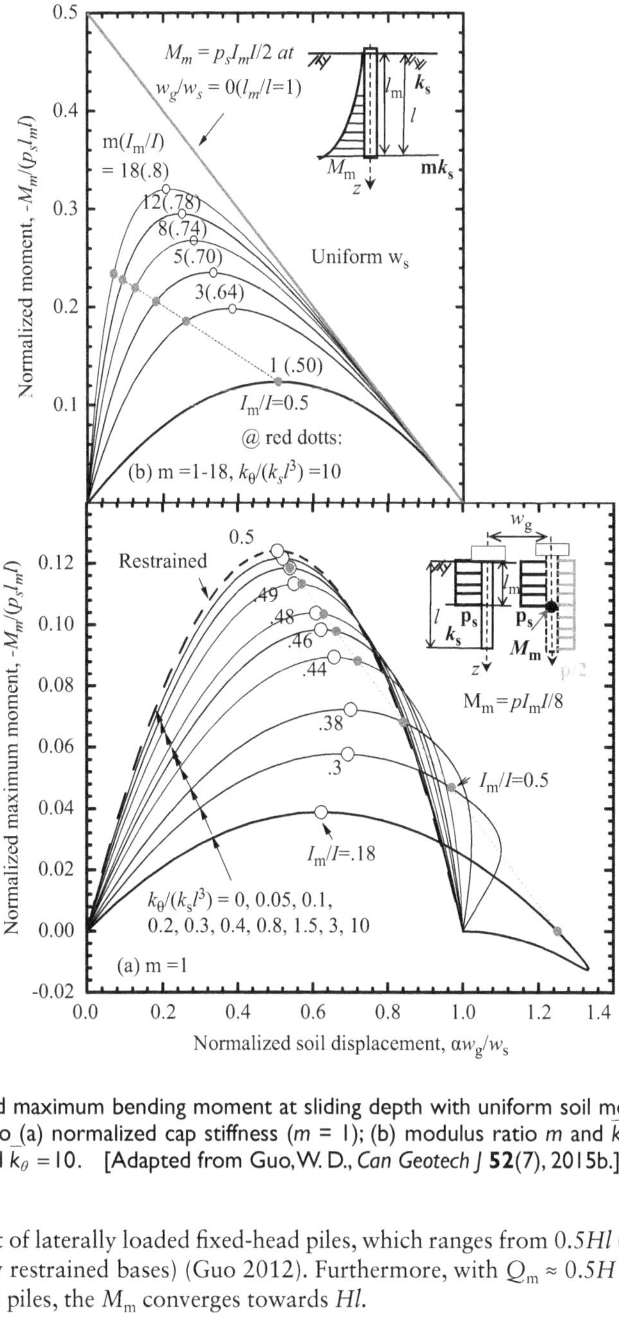

Figure 4.5 Normalized maximum bending moment at sliding depth with uniform soil movement w_s (constant k_s) owing to (a) normalized cap stiffness (m = 1); (b) modulus ratio m and $\bar{k}_\theta$ = 0.05; (c) modulus ratio m and $\bar{k}_\theta$ = 10. [Adapted from Guo, W. D., *Can Geotech J* **52**(7), 2015b.]

to the moment of laterally loaded fixed-head piles, which ranges from $0.5Hl$ (floating base) to $0.6Hl$ (fully restrained bases) (Guo 2012). Furthermore, with $Q_m \approx 0.5H$ for restrained head and base piles, the M_m converges towards Hl.

Example 4.6 Maximum shear force (Q_m)

The normalized (maximum) thrust $Q_m/(p_s l_m)$ at sliding level should not exceed 0.333 (Viggiani 1981; Guo 2014); see Figure 4.6. For lightly head-restrained (with $\bar{k}_\theta$ = 0.05) to fully fixed-head piles, the corresponding $\bar{l}_m$ should not exceed 0.4 to 0.7 (Guo 2015b). Nevertheless, the Q_m may attain a high value at a different depth from the l_m through forming a translation (or dragging) layer (indicated by a high m value, Figure 4.6b-right). In

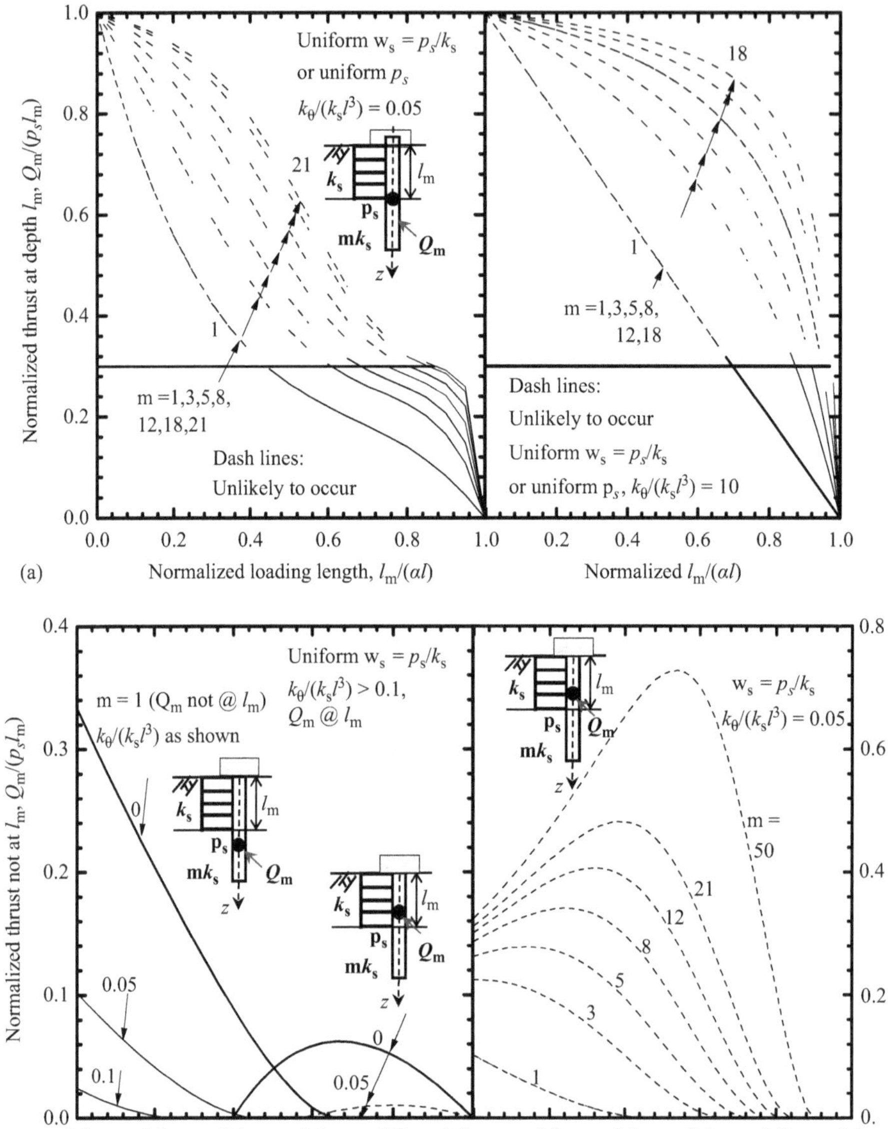

Figure 4.6 Normalized thrust $\bar{Q}_m$ at sliding depth l_m owing to (a-left) modulus ratio *m* and $\bar{k}_\theta = 0.05$; (a-right) modulus ratio *m* and $\bar{k}_\theta = 10$; and $\bar{Q}_m$ (not @ l_m) owing to (b-left) normalized cap stiffness ($m = 1$); and (b-right) the modulus ratio m ($\bar{k}_\theta = 0.05$). [Adapted from Guo, W. D., *Can Geotech J* **52**(7), 2015b.]

this case, the normalized $\bar{Q}_m$ reduces; see Figure 4.6b-left (for $m = 1$), with the increase in the normalized stiffness $\bar{k}_\theta$. On the other hand, at $\bar{l}_m = 0.4 - 0.7$, a much lower, normalized thrust $\bar{Q}_m$ may be induced, governed by Mode A in Figure 4.6b-left (Guo 2014).

Finally, Figures 4.3 through to 4.6 are for elastic response for a particular on-pile FPUL p_s. To obtain nonlinear response, as an example, the 2-layer model is adopted to simulate an in situ pile in sliding soil with $H \neq 0$, and $M_o \neq 0$ ($k_G = k_T = 0$).

4.3 MOMENT THRUST RATIO FOR RIGID PILES

To start with, assuming free loading at the pile-head level $\left(\bar{k}_\theta = 0,\ \bar{H} = 0, \text{ and } \bar{M}_o = 0\right)$, the 2-layer model and solutions presented in Table 4.2 offer the maximum bending moment M_{mi}, and shear force Q_{mi} for the sliding and stable layer (subscript i = 1 and 2, respectively).

$$Q_{m1} = p_s z_{mt1}\left[\frac{1}{2} z_{mt1}\frac{\bar{\omega}_r}{l} + \bar{w}_g - 1\right] \tag{4.5}$$

$$M_{m1} = p_s\left[\frac{1}{6}\frac{\bar{\omega}_r}{l} z_{m1}^3 + \frac{1}{2}\left(\bar{w}_g - 1\right) z_{m1}^2\right] \tag{4.6}$$

$$Q_{m2} = p_s\left\{\left[\frac{m}{2} z_{m2}^2 + \frac{1-m}{2} l_m^2\right]\frac{\bar{\omega}_r}{l} + \left[m z_{m2} + (1-m) l_m\right]\bar{w}_g - c\right\} \tag{4.7}$$

$$\begin{aligned} M_{m2} = p_s\Bigg\{&\left[\frac{m}{6} z_{m2}^3 + \frac{1-m}{2} l_m^2 z_{m2} + \frac{m-1}{3} l_m^3\right]\frac{\bar{\omega}_r}{l} \\ &+ \left[\frac{m}{2} z_{m2}^2 + (1-m) l_m z_{m2} + \frac{m-1}{2} l_m^2\right]\bar{w}_g - \frac{c}{2}\left(2 z_{m2} - c\right)\Bigg\} \end{aligned} \tag{4.8}$$

where z_{mt1}, z_{mt2}, z_{m1} and z_{m2} are provided in the table.

$$\bar{\omega}_r = \frac{-6cl}{l_m^3}\frac{\left(2m\lambda + m\lambda^2 + 1\right) l_m - \left(m\lambda + 1\right) c}{1 + 4m\lambda + 6m\lambda^2 + m^2\lambda^4 + 4m\lambda^3} \tag{4.9}$$

$$\bar{w}_g = \frac{c}{l_m^2}\frac{4\left(3m\lambda + 3m\lambda^2 + m\lambda^3 + 1\right) l_m - 3c\left(2m\lambda + m\lambda^2 + 1\right)}{1 + 4m\lambda + 6m\lambda^2 + m^2\lambda^4 + 4m\lambda^3} \tag{4.10}$$

Given $\lambda = 0$, and $l_m = l$, the following expressions are deduced

$$\bar{\omega}_r = 6cl\left(c - l_m\right)/ l_m^3;\ \bar{w}_g = -c\left(3c - 4l_m\right)/ l_m^2 \tag{4.11}$$

$$z_{m2} = l_m^2 / 3\left(-c + l_m\right);\ M_{m2} = \frac{p_s c m}{54}\frac{\left(3c - 2l_m\right)^3}{\left(c - l_m\right)^2} \tag{4.12}$$

Example 4.7 An in situ test pile in sliding soil ($H \neq 0$, $M_o \neq 0$)-Mirrored M_m

Frank and Pouget (2008) reported the response of a pipe pile installed downslope of a 'sliding' embankment over 16 years. The pile (l = 11.0 m, d = 0.915 m, and t = 19 mm) was instrumented and subjected to a trapezoidal profile of movement to a depth of 6.8 m. The soil properties of the site were characterized by s_u = 88 kPa, γ_s' = 17.0 kN/m^3 and ϕ' = 24.5°. During the 16-year test period, the pile was pulled back four times at a lateral load H and moment M_o (at ~0.5 m above GL). The response of the pile was measured and is plotted in Figures 4.7b–d, which include:

- The time-evolution of the maximum bending moment M_{m1} at a depth of 3.75m, as well as those estimated using $M_{m1} = 0.25 Q_{m1} l$ (Guo and Qin 2010) from the measured Q_{m1};

- Five profiles of FPUL p_s after each 'pulling back' and at year 1999;
- Four pile-deflection profiles before and after each pulling-back.

Figure 4.8 provides the measured bending moment profiles for each of the four pulling-back stages. Table 4.3 provides the measured M_{m1} and GL displacement w_g, given the applied bending moment M_o and force H. The displacement profiles in Figure 4.7 indicate that the pile was 'fixed-head' and rigid (linear displacement to sliding depth) prior to and after each pre-pulling-back event.

For the pile in the 'pre-pulling back' state, the 2-layer prediction was made using l = 11.0 m, d = 0.915 m, and $c = l_m$ = 6.8 m (λ = 0.618). The p_b (= p_u at l = 11.0 m) was estimated as 749.7 kN/m [= $0.75\gamma_s' K_p^2 dz$, with γ_s' = 17.0 kN/m^3, ϕ = 24.5°, and d = 0.915 m (see Figure 4.7c)]. The ultimate p_b increases with the pulling-backs (see Table 4.3). For instance, taking $p_b = 0.9p_u$, the 1986 pulling-back involves p_s (= $p_b l_m/l$) of 417.1 kN/m at the sliding level.

The set of M_o = −94 kNm and H = 0 before pulling back (see Table 4.3) were normalized to $\bar{M}_0 = -1.863\times10^{-3}\left[= -94/(417.1\times11)\right]$, and $\bar{H} = 0$, while $\bar{k}_\theta = 0$ and k_G = 0. By taking m = 4.5, and k_s (at large movement) = 2.86 MPa (<k for lateral loading), Equations (4.9) and (4.10) yielded $\bar{\omega}_r = -1.458$, and $\bar{w}_g = 1.251$. Using the expressions in Table 4.2, the profiles of displacement and bending moment were predicted, and are plotted in Figures 4.7d and 4.8a together with the measured profiles. The depths z_{m1} and z_{m2} of the maximum bending moment M_{m1} (in the sliding layer) and M_{m2} (in the stable layer) were estimated as 3.729 m $(\bar{z}_{m1} = 0.345)$, and 8.217 m $(\bar{z}_{m2} = 0.717)$. The corresponding moment M_{m1} [= $M_1(z_{m1})$, see Table 4.2] and M_{m2} [= $M_2(z_{m2})$, Equation (4.4)] were obtained as −345.23 kNm, and 659.23 kNm, respectively. For the 1986-pulling stage, using M_o = −209 kNm, H = 310 kN, m = 5.5 (high for large dragging), and k_s = 2.86 MPa, the predictions are also included in the figures.

Likewise, predictions were made for other three stages in 1988, 1992 and 1995, using the values of M_o, H, m, and k_s shown in Table 4.3. Figures 4.7d and 4.8 demonstrate that the measured bending moment and deflection profiles were well predicted for the pre-pulling back stages. However, the deflection profiles during the pulling-back stages (solid symbols) are not simulated herein, as they require a theory for a laterally loaded pile. The bending moment profiles depend solely on the ultimate on-pile pressure and can be estimated using the 2-layer model, regardless of the pile-soil relative stiffness and/or loading properties.

The bending moments M_{m1} (in the sliding layer) and M_{m2} (in the stable layer) fluctuate with the pile-head displacement w_g during the loading cycles (see Figures 4.8e and f). To predict the evolution of the maximum bending moments with the soil movement (with α = 0.588), and with the pile-head displacement over the 16 years, a simplified loading (M_o = 0, and H = 0 kN), along with m = 4.5, k_s = 2.4 MPa, and p_b = 900 kN/m (= 1.2 × 749.68 kN/m) were utilized. The resulting curves are plotted in Figures 4.9a and b, respectively. The predictions compare well with the measured data after swapping $-M_{m1}$ at a depth of 3.75 m with the M_{m2} at depth 8–9 m. Moreover, the predicted base displacement w_b versus moment M_m curve compares well with the measured w_g–M_m curve at 8–9 m (another swap). These swaps demonstrate the impact of deep sliding (l_m/l > 0.5) depicted in the insert of Figure 4.4a. Additionally, the impact of non-homogeneity m and any dragging ($\bar{k}_\theta > 0$) can be assessed against Figures 4.5a and b.

The prediction adopts a stepwise increasing LFPUL p_s (= $\alpha p_{ub} l_m/l$) with l_m to obtain nonlinear response of the pile. The parameter α of 0.59–0.72 was chosen to derive p_b (= αp_{ub}), the sliding depth of l_m/α and the movement w_s/α that incorporate impact of soil movement profile (source) w_s on the piles. Sliding resistance is ignored; otherwise, a transitional layer would be introduced into 2-layer model (Guo 2016), using slightly different values of k_s, m and p_{ub} (see Chapter 5).

Table 4.3 Calculated and measured (Frank and Pouget 2008) response of an in situ tested pile (Guo 2015b)

Year	*Applied*		*Measured*		*Input*		*Calculated*		
	$-M_o$ (kNm)	H (kN)	w_g (mm)	$-M_{ml}$ (kNm)	m	p_b/p_u	$-Q_{ml}/Q_{m2}$ (kN)	$-Q_m$@ l_m (kN)	$-M_{ml}$ (kNm)
1984.01.01	0	0	0	0				0	0
1986.11.4	94	0	17.8		4.5	0.9	99.4/302.8	565.3	345.2
1986.11.5	209	−310	−2.4	901.9	5.5		310.0/230.0	705.8	973.2
1988.11.10	154	0	16.5	536.9	3.9	1.1	133.0/302.	655.5	505.3
1988.11.11	262	−273	−0.9	1102.3	4.6		285.4/250.8	787.3	1050.0
1992.9.30	138	0	35.2	544.3	4.3	1.73	200.8/552.0	1063	656.8
1992.10.01	313	−321	−2.0	1473.5	4.3		375/405.0	1171	1403
1995.07.05	182	0	29.8	756.1	4.0	1.48	177.0/422.3	888.4	646.5
1995.07.06	295	−347	−1.8	1434.2	4.3		372.0/314.6	1023	1358
1999.07.20	184	—	21.7	932.7					

Note: l = 11.0 m, l_m = 6.8 m, d = 0.915 m, k_s = 2.86 MPa (1986, 1988 after pulling backs), and 2.5 MPa (1992, 1995, after pulling backs), and $p_s = p_b l_m / l$.

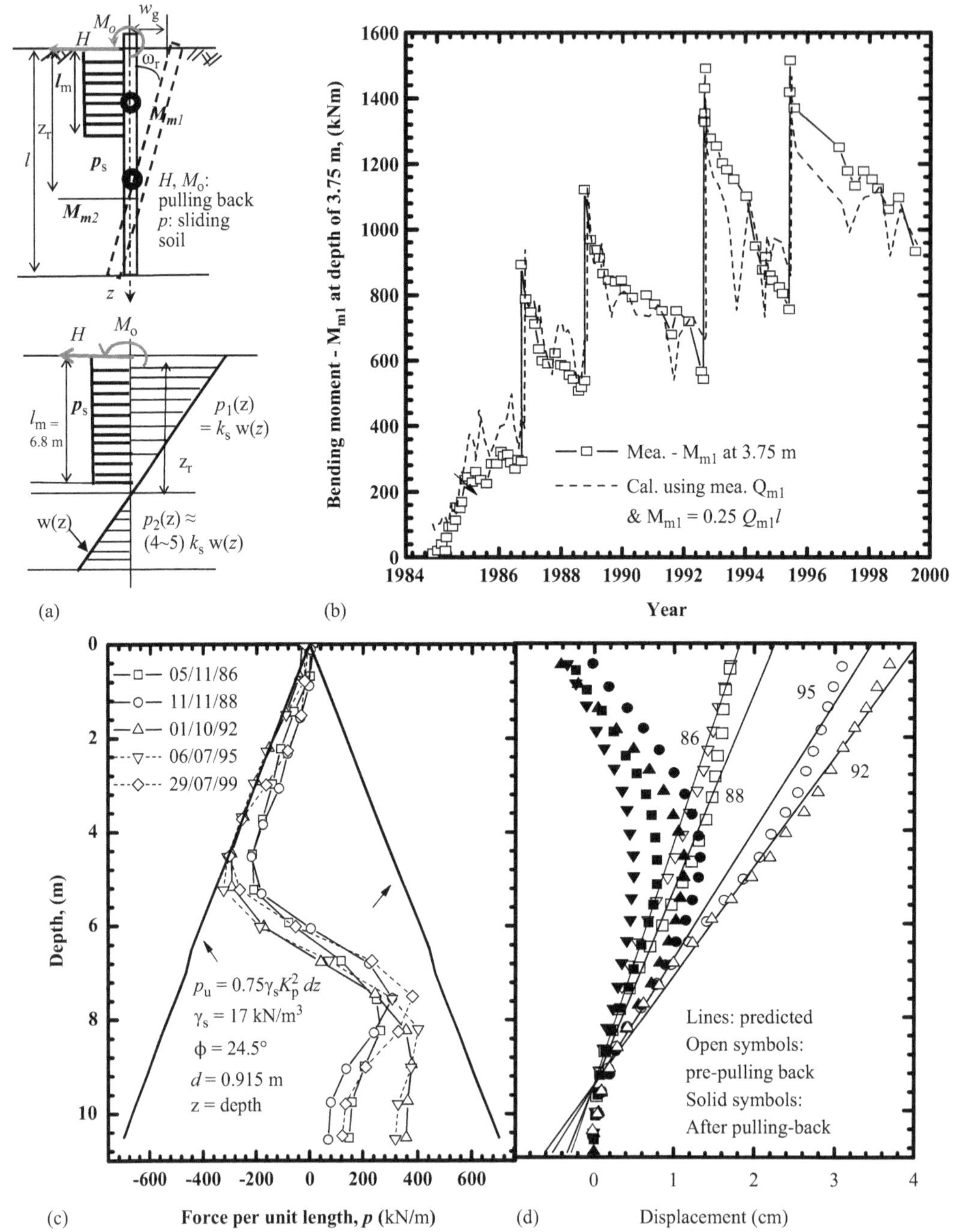

Figure 4.7 An in situ pile (Frank and Pouget 2008): (a) 2-model for the in situ pile; (b)-(d) evolution of the response of: (b) maximum bending moment at a depth of 3.75 m and pile-head shear load; (c) force per unit length p; (d) profiles of pile deflections. [Adapted from Guo, W. D., *Can Geotech J* **52**(7), 2015b.]

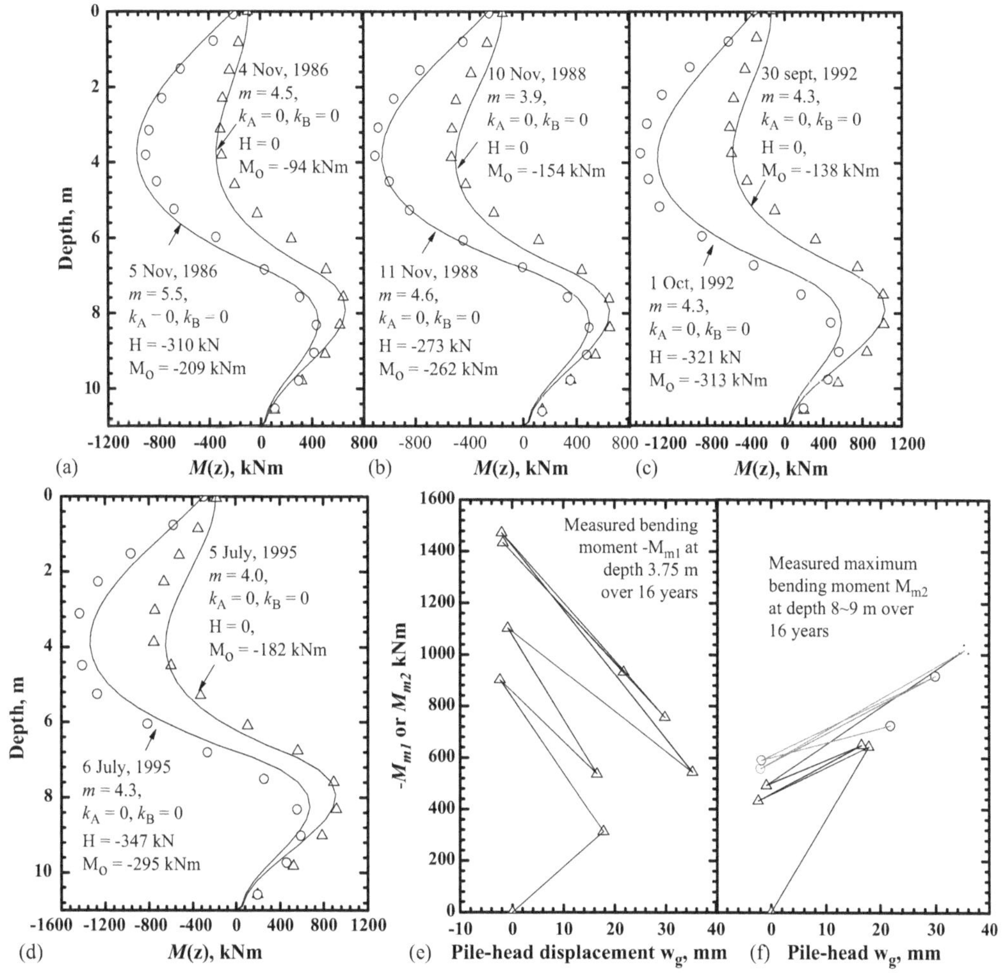

Figure 4.8 Measured (Frank and Pouget 2008) bending moment profiles at (a) 1986; (b) 1988; (c) 1992; (d) 1995 (with predictions using $p = p_b l_m/l$, l_m = 6.8 m and l = 11 m), and maximum bending moment; (e) $-M_{m1}$ at depth 3.75 m; and (f) M_{m2} at 8–9 m. [Adapted from Guo, W. D., *Can Geotech J* **52**(7), 2015b.]

4.3.1 p_{u1} for elastic interaction of passive piles

Given $H = 0$, and with $Q_1(z)$ in Table 4.2, the shear force Q_m was calculated as $Q_m(l_m)$ at sliding depth l_m, and as Q_m [= $Q_1(z_{mt1})$] at z_{mt1}. The calculation is repeated for each l_m (= 0–l) and m (= 1.5, 3 and 5). The normalized maximum shear force $Q_m/(p_s l_m)$ at l_m is plotted in Figure 4.10a (compared to Figure 4.6a); while Q_m [= $Q_1(z_{mt1})$] at z_{mt1} in Figure 4.10b (compared to Figure 4.6b).

Viggiani (1981) developed 'ultimate-state' solutions for passive piles with a limiting FPUL p_{ui} (= $k_i s_{ui} d$) in sliding layer (see Table 4.4). These solutions are categorized as modes A, B and C based on the sliding depth ratio λ, and a ratio m of limiting FPUL of stable layer to that of sliding layer. Figure 4.10 shows the $Q_m/(p_s l_m)$ plotted against l_m/l ratios using an average normalizer p_s of $3.33k_1 s_{u1} d$ (k_1 = 2.5–4.0, Viggiani 1981, Poulos 1995, Guo 2014). The ultimate-state $Q_m/(p_s l_m)$ ratios obtained agree with those obtained at a fixed l_m of the (elastic) 2-layer

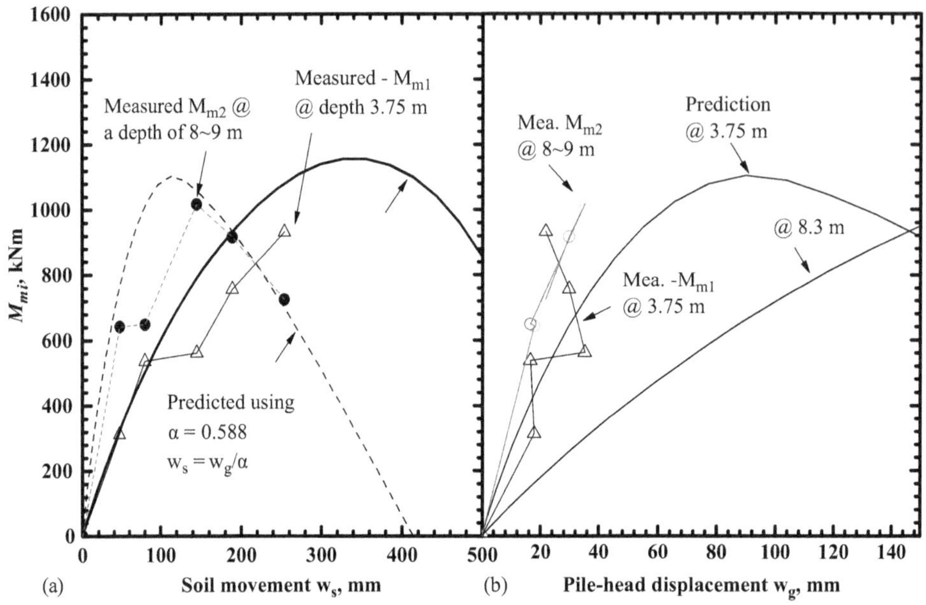

Figure 4.9 Predicted versus measured (Frank and Pouget 2008) (a) $w_s - M_{mi}$; and (b) $w_g - M_{mi}$. [Adapted from Guo, W. D., *Can Geotech J* **52**(7), 2015b.]

model. The ultimate-state is therefore dominated by 'elastic' interaction between 'flowing' soil and piles (see Guo and Qin 2010) with little effect of excess soil movement from 'plastic flow'.

Sliding at depth l_m may incur a different depth z_{mt} of maximum shear force. The 2-layer model (see Figure 4.10b) predicts a gradual shift of the depth z_{mt} (as observed in model tests by Guo and Qin 2010) to either the sliding layer (z_{mt1}, Mode A) or the stable layer (z_{mt2} Mode B). Using the expressions in Table 4.2, the calculated maximum shear force Q_{m1} at depth z_{mt1} ($\neq l_m$) [or $Q_m/(p_s l_m)$] and bending moment M_{m1} [or $M_{mi}/(pl_m l)$] are plotted in Figure 4.11 as Mode A. Similarly, at a depth z_{mt2} ($>l_m$) in the stable layer, the calculated maximum force Q_{m2}, and moment M_{m2} using Equations (4.7)–(4.8) are plotted in Figures 4.10b and 4.11 as Mode B.

4.3.2 $M_m/Q_m l$ ratio for slope stabilizing piles

Figure 4.12 shows the $M_m/(Q_m l)$ ratio for a concentrated loading H (H-based model, taking $a = l_m$) and for a 2-layered soil at m = 1–12. Importantly, the $M_m/(Q_m l)$ ratio reaches 0.3–0.45 when loading H at $l_m \approx 0.2l$ or the sliding l_m up to $0.5l$, which agrees with 0.33–0.39 for d_{50} (50 mm diameter) piles, and 0.35 ~ 0.38 for d_{32} piles in model tests on 'rigid' piles under passive loading (Guo and Qin 2010). For sliding depth $l_m/l > 0.5$, the M_m may occur at a depth symmetrical about $l_m/l = 0.5$ and with the same magnitude (Guo 2014).

The $M_m/(Q_m l)$ ratios for eight in situ test piles (Guo and Qin 2010) were calculated using measured data by taking l as a sum of l_m and the minimum pile length l_{c2} in stable layer and thickness of the stable layer. These ratios are plotted in Figure 4.12 against the normalized sliding depth l_m/l. Despite the pile flexibility, the reduction of the ratio with the sliding ratio l_m/l agrees with the 2-layer prediction. In contrast, the H-based solutions consistently

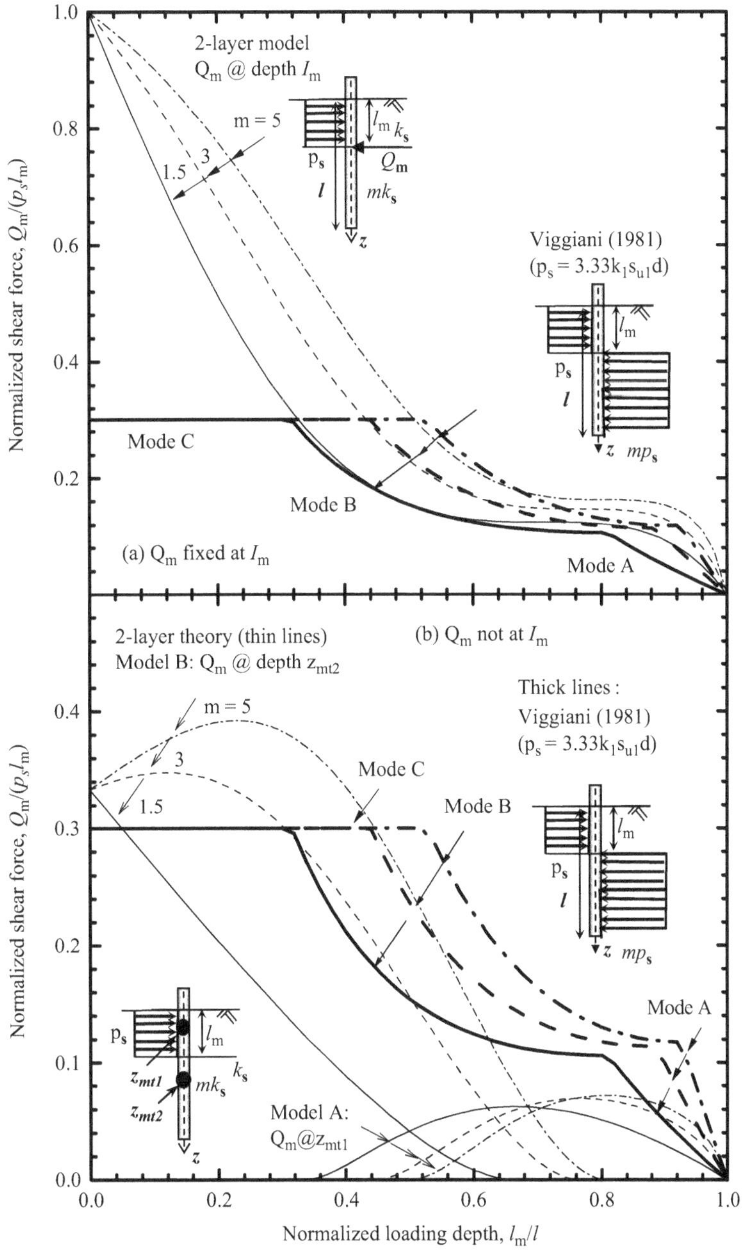

Figure 4.10 $Q_m/(p_s l_m)$ versus normalized length (p force per unit length): (a) Q_m fixed at depth l_m; (b) Q_m not fixed at depth l_m. [Adapted from Guo, W. D., *Int J Numer and Anal Meth in Geomech* **38**(18), 2014.]

underestimate the ratio of $M_m/(Q_m l)$. This suggests that the 2-layer model or the p_u-based elastic-plastic solutions (Guo 2013c) are more accurate for passive piles than solutions based on a thrust H. It is worth noting that the moment M_m and the sliding force Q_m are 'overestimated' assuming fully fixed-head for working piles (Guo 2016). Chapters 5 and 7 provide further analysis on the effect of semi-fixed head k_θ of $\pm(0 \sim 0.3)$.

Table 4.4 'Ultimate state' solutions (Viggiani 1981) for piles in 2-layered soil

Mode	*Expressions*
A	With $\lambda < \lambda_A = \left[\sqrt{\frac{2(1+m)}{m}} - 1\right] / (2+m)$ $\bar{Q}_A = Q_A / (k_1 c_1 d l_m) = m\lambda, \bar{M}_{2A} = M_{2A} / \left(k_1 c_1 d l_m^2\right) = 0.5 m \lambda^2$
B	With $\lambda > \lambda_A = \left[\sqrt{\frac{2(1+m)}{m}} - 1\right] / (2+m)$ $\bar{Q}_B = Q_B / (k_1 c_1 d l_m) = \sqrt{\left[\frac{(1+\lambda)m}{1+m}\right]^2 + \left[\frac{(m\lambda^2+1)m}{1+m}\right]} - \frac{(1+\lambda)m}{1+m}$, $\bar{M}_{1B} = M_{1B} / \left(k_1 c_1 d l_m^2\right) = \left(1 - \bar{Q}_B\right)^2 / 4, \bar{M}_{2B} = M_{2B} / \left(k_1 c_1 d l_m^2\right) = \frac{m}{4}\left(\lambda - \frac{\bar{Q}_B}{m}\right)^2$
C	With $\lambda > \lambda_c = \left[1 + \sqrt{2(1+m)}\right] / m$ $\bar{Q}_C = Q_C / (k_1 c_1 d l_m) = 1, \bar{M}_{1c} = M_{1c} / \left(k_1 c_1 d l_m^2\right) = 0.5$

Note: 2-layered, $\lambda = (l - l_m)/l_m$, $m = k_2 c_2/(k_1 c_1)$.

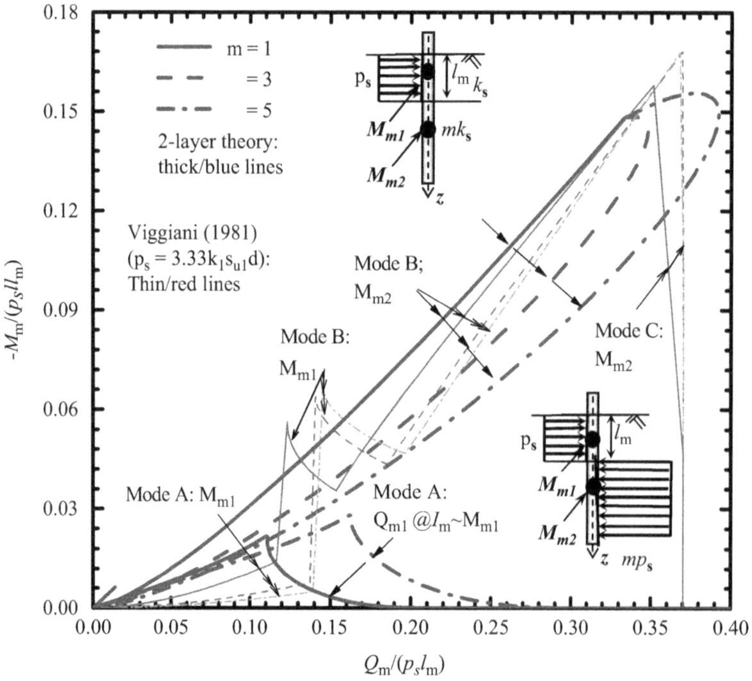

Figure 4.11 $M_m/(p_s l l_m)$ versus normalized shear force $Q_m/(p_s l_m)$. [Adapted from Guo, W. D., *Int J Numer and Anal Meth in Geomech* **38**(18), 2014.]

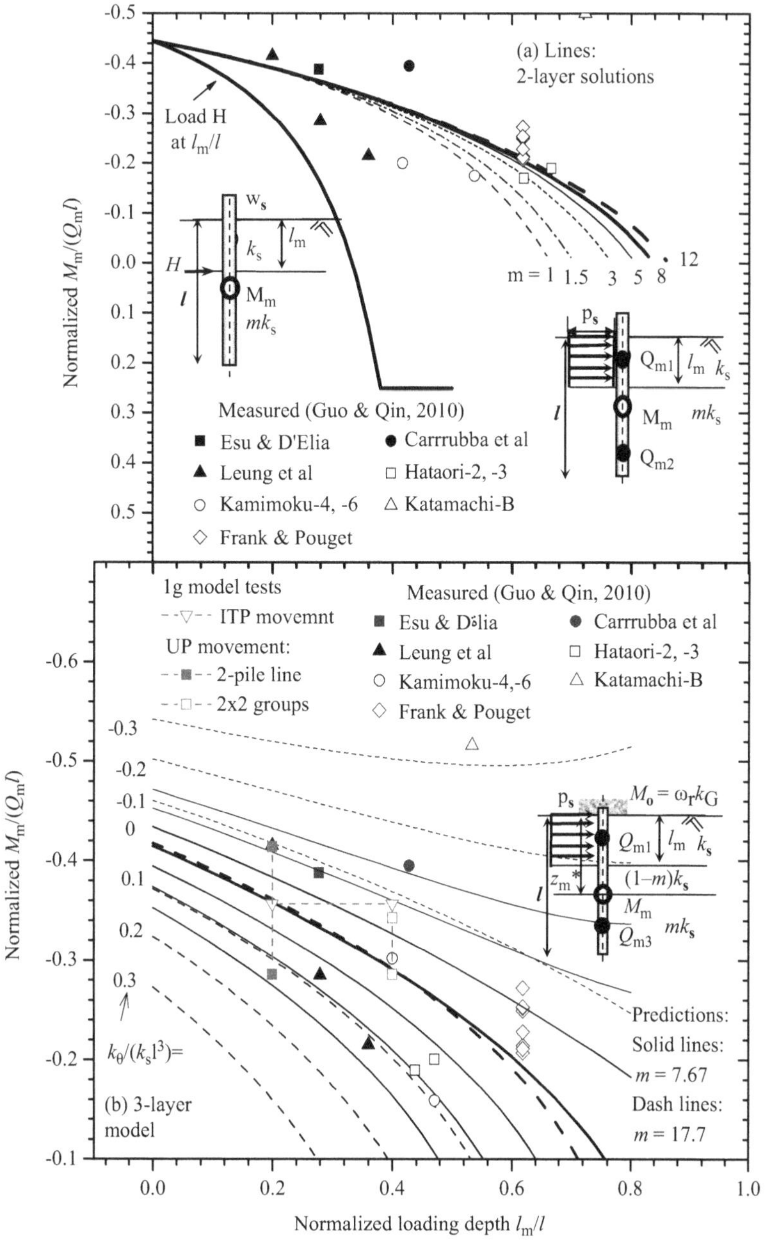

Figure 4.12 Predicted versus measured ratios of $M_m/(Q_m l)$: (a) *H*-based and 2-layer model (b) 3-layer model. [Adapted from Guo, W. D., *Int J Numer and Anal Meth in Geomech* **38**(18), 2014.]

4.4 RIGID – ELASTIC PILES IN STABILIZING SLOPE

Slope stabilizing piles may behave flexible (rather than rigid) in both sliding and stable layers (i.e., deep sliding), which can then be modelled using the elastic-plastic solutions (Guo 2012, 2013c). A large proportion of piles can be modelled as **bi-**portion (BiP), i.e., rigid in sliding layer (above the depth z_m of maximum bending moment M_m) and elastic in stable layer (i.e.,

Table 4.5 **R-E** and **R-EP** models for passive piles (Guo 2021a)

Depth z ($\le l_s$)	$w(z) = (\bar{\omega}_r\bar{z} + \bar{w}_g)p_s / k_s$	$p_l(z) = (\bar{\omega}_r\bar{z} + \bar{w}_g)p_s - p_s$
Shear force $Q(z)$	$Q_{lc}(z)/(p_s l) = 0.5\bar{\omega}_r\bar{z}^2 + \bar{w}_g\bar{z} - \bar{z} - \bar{H}$	
Bending moment $M(z)$	$\dfrac{M_{lc}(z)}{p_s l^2} = \dfrac{1}{6}\bar{\omega}_r\bar{z}^3 + \dfrac{1}{2}(\bar{w}_g - 1)\bar{z}^2 - \bar{H}(\bar{z} - \bar{a}) - (\bar{k}_\theta - \bar{k}_G)\bar{\omega}_r$	
$l_s \le z \le z_m$	$w(z) = (\bar{\omega}_r\bar{z} + \bar{w}_g)p_s / k_s$	$p_l(z) = (\bar{\omega}_r\bar{z} + \bar{w}_g)p_s$
Shear force $Q(z)$	$Q_l(z)/(p_s l) = 0.5\bar{\omega}_r\bar{z}^2 + \bar{w}_g\bar{z} - \bar{l}_s - \bar{H}$	
Bending moment $M(z)$	$\dfrac{M_l(z)}{p_s l^2} = \dfrac{1}{6}\bar{\omega}_r\bar{z}^3 + \dfrac{1}{2}\bar{w}_g\bar{z}^2 - \dfrac{1}{2}\bar{l}_s(2\bar{z} - \bar{l}_s) - \bar{H}(\bar{z} - \bar{a}) - (\bar{k}_\theta - \bar{k}_G)\bar{\omega}_r$	
Peak M_{ml} (Q_{ml}) in sliding layer, M_m at z_m, $Q(l_s)$ at l_s	$\dfrac{M_{ml}}{p_s l^2} = \dfrac{2}{3}\dfrac{(1 - \bar{w}_g)^3}{\bar{\omega}_r^2}$	$\dfrac{M_m}{p_s l^2} = \dfrac{1}{6}\bar{\omega}_r\bar{z}_m^3 + \dfrac{1}{2}\bar{w}_g\bar{z}_m^2 - \bar{M}_L$
	$\dfrac{Q_{ml}}{p_s l} = \dfrac{-1}{2}\dfrac{\bar{w}_g^2}{\bar{\omega}_r} - \bar{P}_h$	$\dfrac{Q(l_s)}{p_s l} = 0.5\bar{\omega}_r\bar{l}_s^2 + \bar{w}_g\bar{l}_s - (\bar{H} + \bar{l}_s)$
$z_m \le z \le l$[a]	Elastic solution (Table 4.7)	Elastic-plastic solution (Table 4.8)
Notes	$\bar{w}_g = w_g k_s / p_s, \bar{\omega}_r = \omega_r k_s l / p_s, \bar{M}_L = M_L/(p_s l^2), \bar{z}_m = z_m / l$	
Moment M_L and force	$\bar{M}_L = M_L/(p_s l^2)$ $= \bar{H}(\bar{z}_m - \bar{a}) + 0.5\bar{l}_s(2\bar{z}_m - \bar{l}_s)$	$\bar{H} = H/(p_b l_s) = H/(p_s l), \bar{z} = z/l$ $\bar{a} = a/l, \bar{l}_s = l_s/l, \bar{P}_h = \bar{H} + \bar{l}_s$
$\bar{k}_\theta, \bar{k}_G$	$\bar{k}_\theta = k_\theta/(k_s l^3)$	$\bar{k}_G = k_G/(k_s l^3)$

[a] In this zone, *R-E* and *R-EP* models follow different solutions.

normal sliding). The **R-E** model (see Figure 1.7b) and solutions are applicable to those piles as outlined next.

As with 2-later model, the pile rigidly rotates about the rotation depth z_r (= $-w_g/\omega_r$, and $z_r \approx z_m$) to an angle ω_r and a GL deflection w_g (e.g., Figure 1.4c, Chapter 1). The pile has a displacement $w(z)$ (= $\omega_r z + w_g$) at depth z (< z_m) (see Table 4.5). A 'stepwise' uniform p_s (over a depth 0 ~ l_s) together with the lateral force H (at a depth a below GL, Figure 1.4c) exert a bending moment $M_L = H(z_m - a) + 0.5p_s l_s(2z_m - l_s)$ and a total passive moment of $M_L + k_\theta(-\omega_r)$ about the depth z_m. The bending moment M_m (= $k_\theta\omega_r$) is correlated with a rotational stiffness k_θ along the pile embedment l (see Figure 1.7b). The stiffness k_T is negligible for flexible piles.

4.4.1 R-E model

The R-E model (Guo 2021a) provides the following pile response (see Figure 1.7, Chapter 1, and Tables 4.6 and 4.7): (i) The displacement (w_{g2}) and rotation (ω_r) at z_{me} (subscript '*e*' refers to R-E model); and (ii) The compatible M_m (= $k_\theta\omega_r = w_h/I_c$) at z_{me} based on the local displacement w_h and the parameter I_c, as well as importantly $k_\theta = -1/I'_c$ (Guo and Lee 2001). The expressions (see Table 4.5) are used to estimate the pile response at a sliding depth l_m, including

- Net on-pile FPUL $p(z)$ at depth z, in particular, $p(z) = p_s - k_s w(z)$;
- Bending moment $M(z)$ and shear force $Q(z)$;

- Maximum bending moment M_m and its depth z_{me};
- Maximum moment M_{m1} and shear force Q_{m1} in the sliding layer, and shear force $Q(l_m)$ at the sliding depth;
- Singularity stiffness k_θ*.

The fictitious tension N_p (linked together the springs distributed along pile shaft) has minimal impact on overall pile response. Taking $N_p = 0$ and $H = 0$, first, estimation of z_{me} and $k_{\theta e}$ (see Table 4.6) is simplified (Guo 2022a) as follows:

$$\bar{z}_{me}^3 - 3\left(0.5\bar{l}_s + \chi_c\right)\bar{z}_{me}^2 + 3\bar{l}_s\chi_c\bar{z}_{me} - 6\bar{k}_{\theta e} = 0 \tag{4.13}$$

where

$$\chi_c = \frac{1}{\lambda_2 l}\frac{\sinh^2\left[\lambda_2\left(l - z_{me}\right)\right] + \sin^2\left[\lambda_2\left(l - z_{me}\right)\right]}{\sinh\left[2\lambda_2\left(l - z_{me}\right)\right] + \sin\left[2\lambda_2\left(l - z_{me}\right)\right]} \tag{4.14}$$

$$-\bar{k}_{\theta e} = \frac{-k_{\theta e}}{k_s l^3} = \frac{0.5\chi_c}{\left(\lambda_2 l\right)^2}\frac{k_{s2}}{k_s} \text{ and } \frac{-k_{\theta e}}{E_p I_p \lambda_2} = 2\lambda_2 l\chi_c \tag{4.15}$$

Table 4.6 **R-E** (BiP) model for passive loading (Guo 2021a)

Normalized $\bar{z}_{me}$ ($= z_{me}/l$)	$\bar{z}_{me}^3 - 1.5\left(\frac{\bar{M}_L}{\bar{P}_h} + \frac{I_c}{I'_c l}\right)\bar{z}_{me}^2 + 3\frac{\bar{M}_L}{\bar{P}_h}\frac{I_c}{I'_c l}\bar{z}_{me} - 3\bar{k}_{\theta e} = 0$	
Normalized w_g and ω_r	$\bar{w}_g = \frac{2\left[\left(\bar{z}_{me}^3 + 6\bar{k}_{\theta e}\right)\bar{P}_h - 3\bar{z}_{me}^2\bar{M}_L\right]}{\left(-\bar{z}_{me}^3 + 12\bar{k}_{\theta e}\right)\bar{z}_{me}}$	$\bar{\omega}_r = \frac{-6\left(\bar{z}_{me}\bar{P}_h - 2\bar{M}_L\right)}{-\bar{z}_{me}^3 + 12\bar{k}_{\theta e}}$
Normalized w_h and $k\theta_e$	$\bar{w}_h = \frac{\left(-4\bar{z}_{me}^3 + 12\bar{k}_{\theta e}\right)\bar{P}_h + 6\bar{z}_{me}^2\bar{M}_L}{\bar{z}_{me}\left(-\bar{z}_{me}^3 + 12\bar{k}_{\theta e}\right)}$	$\bar{k}_{\theta e} = \frac{k_{\theta e}}{k_s l^3}, k_{\theta e} = -\frac{1}{I'_c}$
I_c and I_c'	$I_c = \frac{1}{E_p I_p \delta}\left(I(0) + \frac{\sqrt{k_{s2}N_p}}{E_p I_p}C(0)\right)$	$I'_c = \frac{1}{E_p I_p \delta}\left(I'(0) + \frac{\sqrt{k_{s2}N_p}}{E_p I_p}C'(0)\right)$

$I'(0)$, $C'(0)$ and δ are as follows:

$$I'(0) = \alpha_c\beta_c\left(\alpha_c^2 + \beta_c^2\right)^2\left[\beta_c\left(3\alpha_c^2 - \beta_c^2\right)\text{sh}\left(2\alpha_c l_e\right) - \alpha_c\left(\alpha_c^2 - 3\beta_c^2\right)\sin\left(2\beta_c l_e\right)\right],$$

$$C'(0) = 4\alpha_c^2\beta_c^2\left(\alpha_c^2 + \beta_c^2\right)\left[\text{sh}^2\left(\alpha_c l_e\right) - \cos^2\left(\beta_c l_e\right)\right],$$

$$\delta = \left(\alpha_c^2 + \beta_c^2\right)^2\left[\beta_c^2\left(3\alpha_c^2 - \beta_c^2\right)^2\text{sh}^2\left(\alpha_c l_e\right) - \alpha_c^2\left(\alpha_c^2 - 3\beta_c^2\right)^2\sin^2\left(\beta_c l_e\right)\right]$$

$$+\frac{\sqrt{k_{s2}N_p}}{E_p I_p}\alpha_c\beta_c\left(\alpha_c^2 + \beta_c^2\right)\left[\alpha_c\left(\alpha_c^2 - 3\beta_c^2\right)\sin\left(2\beta_c l_e\right) + \beta_c\left(3\alpha_c^2 - \beta_c^2\right)\text{sh}\left(2\alpha_c l_e\right)\right]$$

$$\alpha_c = \sqrt{\sqrt{\frac{k_s}{4E_p I_p}} + \frac{N_p}{4E_p I_p}}, \beta_c = \sqrt{\sqrt{\frac{k_s}{4E_p I_p}} - \frac{N_p}{4E_p I_p}}$$

α_c and β_c are based lower layer k_s ($= k_{s2}$), and $G_{s2} = nk_{s2}/3$. k_{s2} is estimated using simple model and coupled model provided in Table 1.1.

where $\lambda_2 = [k_{s2}/(4E_pI_p)]^{0.25}$, E_pI_p = bending stiffness, E_p, I_p = Young's modulus and moment of inertia of an equivalent solid cylinder pile). Second, the local displacement w_h (= M_mI_c) at z_{me} is given by

$$w_h = \frac{-M_m}{\sqrt{E_pI_pk_{s2}}}\lambda_2 l\chi_c \tag{4.16}$$

Third, the k_{s2} is given by the expression in Table 1.1 (Chapter 1) for non-zero loading eccentricity. And finally, the R-E model is applicable for $\lambda_2(l - z_{me}) < 2.5$; otherwise, it is capped at $\lambda_2 l\chi_c \approx 0.5$, $-k_{\theta e}/(E_pI_p\lambda_2) \approx 1.0$, and $w_h = -0.5M_m/(E_pI_pk_{s2})^{0.5}$.

The modelling is underpinned by five input parameters, including the soil modulus k_s and k_{s2} (= mk_s below depth z_m), as well as three pile-soil interaction parameters of p_{ub}, α and k_θ (see Figure 1.7b). In stabilizing landslide, a large displacement of ductile piles should be allowed, which can inflict **elastic**-plastic interaction of the piles in stable layer (i.e., BiP R-EP model).

4.5 RIGID – ELASTIC-PLASTIC MODEL

In the R-E model, the bending moment (M_m) may be larger enough to mobilize a (slip) depth x_p in stable layer (see Figure 1.7c). The M_m occurs at depth z_{mep} using **elastic**-plastic solutions (denoted by 'ep' or 'EP') of Chapter 2, for pile length l_e (= $l - z_{mep}$) ≤ length l_{c2} (see Figure 1.7, Chapter 1 and Table 4.7). Above the depth z_{mep}, the fictitious 'tension' vanishes (i.e., $N_p = 0$), and FPUL p_{s2} attains

$$p_{s2} = A_{L2}x^{n_2} \tag{4.17}$$

where x = depth measured from the depth z_{mep}; n_2 = power to the depth x; and A_{L2} = gradient of the p_{s2} profile (Guo 2013c, and Chapter 1). Below the depth x_p, the FPUL is proportional to the pile deflection $w(x)$ or $w(z_2)$. The shear force $Q(z_{mep})$ and rotation angle ω_r (i.e., $M_m = k_{\theta ep}\omega_r$) are given by Equations (2.6) and (2.5) (Chapter 2), which warrant

$$Q(z_{mep})\lambda_2^{n_2+1} / A_{L2} = \bar{H} \tag{4.18}$$

Table 4.7 Elastic solutions for lower portion ($\mathbf{z} \geq \mathbf{z}_{me}$) (Guo 2012)

$w(x), \omega(x), M(x), Q(x)\ p(x)$	$w(x) = \dfrac{M_m}{E_pI_p\delta}\left(I(x) + \dfrac{\sqrt{k_{s2}N_p}}{E_pI_p}C(x)\right)$	$\omega(x) = w'(x), M(x) = -E_pI_pw''(x)$, $Q(x) = E_pI_pw'''(x), p(x) = E_pI_pw^{IV}(x)$
x, x', l_e	$x = z - z_{me}$ (from z_{me} downwards)	$x' = l_e - x, l_e = l - z_{me}$

$I(x)$, and $C(x)$ for $w(x)$

$$I(x) = \alpha_c(\alpha_c^2+\beta_c^2)^2(\alpha_c^2-3\beta_c^2)\sin(\beta_c l_e)[-\beta_c \text{sh}(\alpha_c x)\cos(\beta_c x') + \alpha_c \text{ch}(\alpha_c x)\sin(\beta_c x')]$$
$$+\beta_c(\alpha_c^2+\beta_c^2)^2(3\alpha_c^2-\beta_c^2)\text{sh}(\alpha_c l_e)[-\beta_c \text{sh}(\alpha_c x')\cos(\beta_c x) + \alpha_c \text{ch}(\alpha_c x')\sin(\beta_c x)]$$

$$C(x) = 2\alpha_c\beta_c(\alpha_c^2+\beta_c^2)^2\text{sh}(\alpha_c l_e)[-\beta_c \text{ch}(\alpha_c x')\cos(\beta_c x) + \alpha_c \text{sh}(\alpha_c x')\sin(\beta_c x)]$$
$$+2\alpha_c\beta_c(\alpha_c^2+\beta_c^2)^2\cos(\beta_c l_e)[\beta_c \text{sh}(\alpha_c x)\cos(\beta_c x') - \alpha_c \text{ch}(\alpha_c x)\sin(\beta_c x')]$$

The determination of δ, α_c and β_c is explained in Table 4.6.

$$\bar{\omega}_r = \frac{\omega_r k_{s2}\lambda_2^{n_2-1}}{A_{L2}} = \frac{-\bar{x}_p^{n_2}}{2(\bar{x}_p+1)k_{nr}+1}\left[\frac{2\bar{x}_p^3+2(n_2+3)\bar{x}_p^2}{(n_2+2)(n_2+3)}+(1+\bar{x}_p)\right] \tag{4.19}$$

where $k_{nr} = k_{\theta ep}/(4E_pI_pk_{s2}) = k_{\theta ep}\lambda_2^3/k_{s2}$. As $Q(z_{mep}) = 0$ at M_m, Equations (4.18) and (4.19) allow the k_{nr} to be deduced as

$$k_{nr} = \frac{k_{\theta ep}\lambda_2^3}{k_{s2}} = \frac{-(n_2+3)}{4(n_2+2)}\frac{2\bar{x}_p^2+(n_2+2)(2\bar{x}_p+1+n_2)}{2\bar{x}_p^3+(n_2+3)\left[\bar{x}_p^2+(\bar{x}_p+1+n_2)(\bar{x}_p+1)\right]} \tag{4.20}$$

and $k_{nr}/\bar{k}_{\theta e} = (\lambda_3 l)^3(k_{\theta ep}k_s)/(k_{\theta e}k_{s2})$. The compatibility conditions at z_{mep} warrant $w_{g2} = w_h$ ($= \omega_r z_{mep} + w_g$) and ω_r using Equations (2.7) (Chapter 2), (4.19) and Table 4.6, which offer

$$\frac{\left(-4\bar{z}_{mep}^3+12\bar{k}_{\theta ep}\right)\bar{P}_h+6\bar{z}_{mep}^2\bar{M}_L}{\bar{z}_{mep}\left(-\bar{z}_{mep}^3+12\bar{k}_{\theta ep}\right)}\frac{p_s}{k_s}\frac{k_{s2}\lambda_2^{n_2}}{A_{L2}} = -\bar{\omega}_r\bar{x}_p\left(\frac{2k_{nr}\bar{x}_p(2\bar{x}_p+3)}{3(\bar{x}_p+1)}+1\right)$$
$$+\frac{4\bar{x}_p^{4+n_2}}{(n_2+1)(n_2+2)(n_2+3)(n_2+4)}-\frac{\left[2\bar{x}_p^2+(2\bar{x}_p+n_2+1)(n_2+2)\right]\bar{x}_p^{3+n_2}}{3(1+\bar{x}_p)(n_2+1)(n_2+2)}+\bar{x}_p^{n_2} \tag{4.21}$$

$$\frac{-6\left(\bar{z}_{mep}\bar{P}_h-2\bar{M}_L\right)}{-\bar{z}_{mep}^3+12\bar{k}_{\theta ep}}\frac{p_s}{k_s l}\frac{k_{s2}\lambda_2^{n_2-1}}{A_{L2}} = \frac{-\bar{x}_p^{n_2}}{2(\bar{x}_p+1)k_{nr}+1}\left[\frac{2\bar{x}_p^3+2(n_2+3)\bar{x}_p^2}{(n_2+2)(n_2+3)}+(1+\bar{x}_p)\right] \tag{4.22}$$

By solving Equations (4.20), (4.21) and (4.22) together, the z_{mep}, x_p and $k_{\theta ep}$ values are obtained. The calculation is quite straightforward, despite more complex than direct resolution of Equation (4.13) to gain z_m. Subsequently, response profiles are estimated using the expressions in Table 4.8 (Guo 2021b), and in Table 4.5 (identical to R-E model) for the lower and upper portions, as exemplified later.

Table 4.8 Elastic-plastic solutions for lower portion ($z \geq z_{mep}$) (Guo 2015a)

	Under the moment M_m [i.e., $Q(z_{mep}) = 0$]: Plastic zone ($x \leq x_p$)
$w(x)$	$w(x) = \frac{A_{L2}}{E_pI_p}\left[\frac{-x^{n_2+4}}{(n_2+4)(n_2+3)(n_2+2)(n_2+1)}+\frac{k_\theta\omega_r}{A_{L2}}\frac{x^2}{2}\right]+\omega_r x+w_{g2}$
	Elastic zone [$x > x_p$, and $\bar{z}_2 = \lambda_2 z_2 = \lambda_2(x - x_{p2})$, with $z_2 = x - x_{p2}$]
$w(\bar{z}_2)$	$w(\bar{z}_2) = e^{-\bar{z}_2}\left[C_5\cos(\bar{z}_2)+C_6\sin(\bar{z}_2)\right]$
	$C_5 = \frac{2A_{L2}\lambda_2}{k_{s2}}\left[\frac{-x_p^{n_2+2}\lambda_2}{(n_2+2)(n_2+1)}-\frac{x_p^{n_2+1}}{n_2+1}+\frac{\lambda_2 k_\theta\omega_r}{A_{L2}}\right]$
	$C_6 = \frac{2\lambda_2^2}{k_{s2}}\left[\frac{A_{L2}x_p^{n_2+2}}{(n_2+2)(n_2+1)}-k_\theta\omega_r\right]$

Note:

- $x \leq x_p$: $\omega(x) = w'(x)$, $-M(x) = E_pI_pw''(x)$, $-Q(x) = E_pI_pw'''(x)$
- $x > x_p$: $\omega(\bar{z}_2) = w'(\bar{z}_2)$, $-M(\bar{z}_2) = E_pI_pw''(\bar{z}_2)$, $-Q(\bar{z}_2) = E_pI_pw'''(\bar{z}_2)$
- Expressions are identical to those for free-head piles, but for different C_5 and C_6 values
- Elastic solutions for $N_p < 2(k_{s2}E_pI_p)^{0.5}$ is validated by $l - z_{mep} > l_{c2}$ + maximum x_p, and $l_{ci} = 1.48d(E_p/G_{si})^{0.25}$.

4.6 R-EP AND R-E MODELS

The ratio $k_{\theta e}/(E_pI_p\lambda_2)$ was estimated using the k_{s2} and N_p expressions provided in Table 1.1 (Chapter 1, Guo 2012) for relative stiffness E_p/G_{si}^* (G_{si}^* = modified shear modulus of the soil) with respect to (i) the coupled model; (ii) the uncoupled model (N_p = 0) but the coupled k_{s2}; and (iii) the simple (Winkle) k_{s2} (N_p = 0). Figure 4.13 indicates that Equation (4.15) underestimates the $k_{\theta e}/(E_pI_p\lambda_2)$ ratio by (5–15)% for 'flexible piles' compared to the coupled model. Nonetheless, it remains sufficiently accurate to assess potential amplification of w_g and ω_r (see Table 4.6) around the normalized singularity stiffness $\bar{k}_\theta^*$ of $\bar{z}_m^3/12$. Back-rotated piles involves a negative normalized stiffness $\bar{k}_{\theta e}\left\{= -k_\theta / \left[k_{s2}\left(l - z_{me}\right)^3\right]\right\}$. Figures 4.14a, b and c shows the variations of the positive and negative $\bar{k}_{\theta e}$ or $k_{\theta e}/(k_{si}l^3)$ with the normalized depths z_{me}/l and z_m/l for k_{s2}/k_s = 1 and 3 and a normalized characteristic 'length' $\lambda_2 l$ (of 0.3–3.6). The normalized back-rotation stiffness $\left(-\bar{k}_{\theta e}\right)$ of the piles ($\lambda_2 l$ = 0.3–4.5) largely increases with the z_{me}/l ratio (from a sliding depth l_s of 0.02l to l_m, see Figure 4.14a). The $-\bar{k}_{\theta e}$ versus z_m/l curve of a pile may cross the singularity $\bar{k}_\theta^* - z_m/l$ curve, leading to response amplification, which is also evident in Figures 4.14b and c for k_{s2}/k_s = 3.

The R-EP model predicts a reducing 'elastic' stiffness with evolving plastic depth (x_p). The reduction is accurately captured using Equation (4.20) (Guo 2022a). Figure 4.15 shows the modelling of three instrumented piles concerning the normalized maximum bending moment $M_m/(p_b l_m z_m)$, pile-head displacement $w_g k_s/p_s$ and rotational angle $\omega_r k_s l/p_s$ at GL. The R-EP model promotes prolonged displacements (ductile) behaviour, preventing the moment capacity from dropping quickly to zero (brittle failure, R-E model) around a unit normalized displacement and rotation. The predictions of the two models are exemplified next.

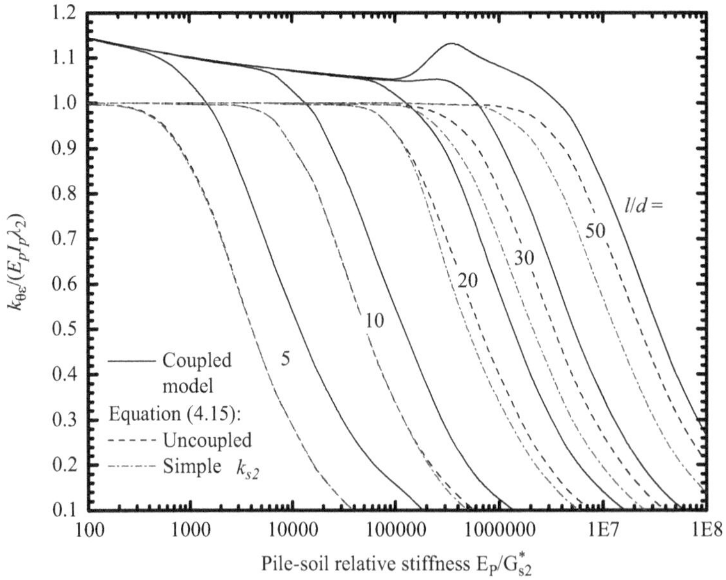

Figure 4.13 Normalized rotational stiffness $k_{\theta e}/(E_pI_p\lambda_2)$ versus pile-soil relative stiffness E_p/G_{s2}^*. [Adapted from Guo, W. D., *Int J Numer and Anal Meth in Geomech* **38**(18), 2014.]

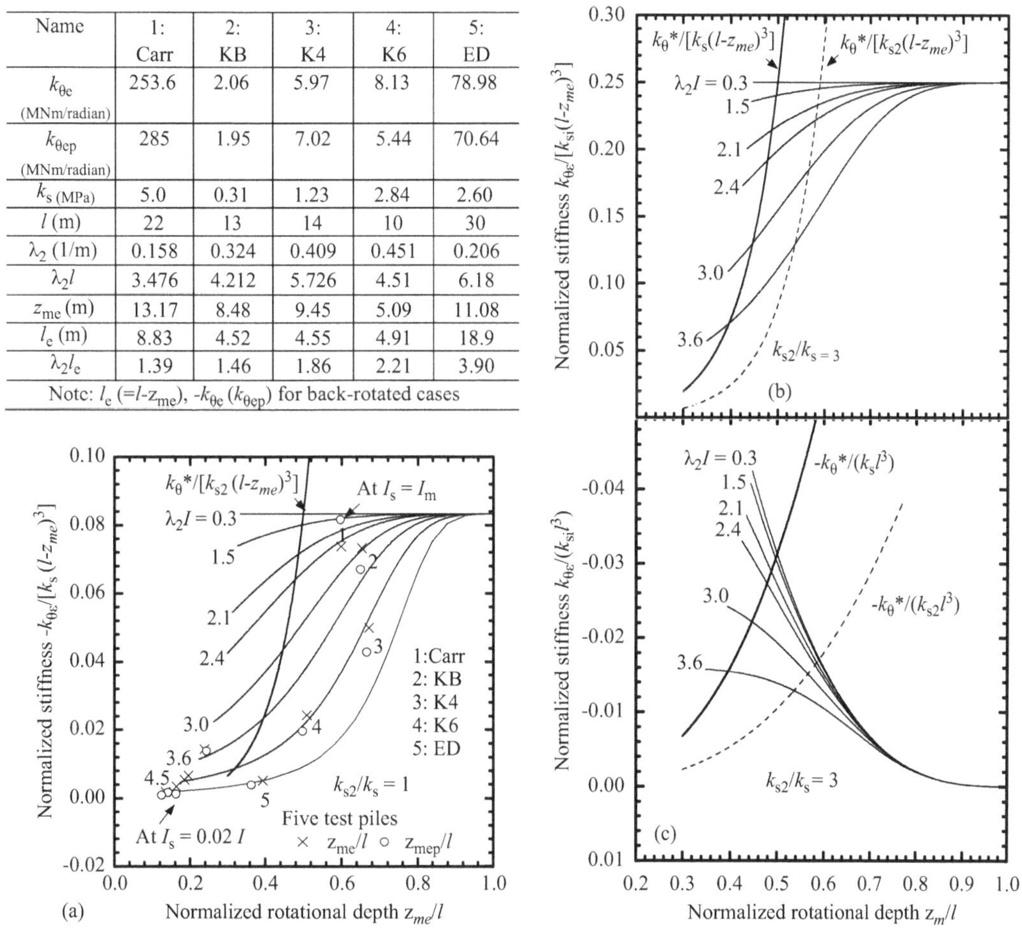

Name	1: Carr	2: KB	3: K4	4: K6	5: ED
$k_{\theta e}$ (MNm/radian)	253.6	2.06	5.97	8.13	78.98
$k_{\theta ep}$ (MNm/radian)	285	1.95	7.02	5.44	70.64
k_s (MPa)	5.0	0.31	1.23	2.84	2.60
l (m)	22	13	14	10	30
λ_2 (1/m)	0.158	0.324	0.409	0.451	0.206
$\lambda_2 l$	3.476	4.212	5.726	4.51	6.18
z_{me} (m)	13.17	8.48	9.45	5.09	11.08
l_e (m)	8.83	4.52	4.55	4.91	18.9
$\lambda_2 l_e$	1.39	1.46	1.86	2.21	3.90

Note: l_e ($=l-z_{me}$), $-k_{\theta e}$ ($k_{\theta ep}$) for back-rotated cases

Figure 4.14 Normalized rotational depth, back-rotation stiffness and singularity stiffness (a) (b) k_{s2}/k_s = 1 and 3; (c) $k_{\theta e}/(k_{si}l^3)$ and k_{s2}/k_s = 3. [Adapted from Guo, W. D., *Int J Numer and Anal Meth in Geomech* **38**(18), 2014.]

4.6.1 Response prediction using R-EP (R-E) models

To predict the response of piles using the R-EP model, the following steps are taken:

- Determine the parameters k_{si}, p_{ub} ($= p_b$ with $\alpha = 1$); loading properties H and a; l_{c2} and k_{s2}, and coupled parameters α_c and β_c for a 'guessed' z_{mep}.
- Determine z_{mep}, x_p and k_θ for a selected sliding depth l_m by resolving Equations (4.20), (4.21) and (4.22).
- Estimate the values of w_g, M_{m1}, M_m ($= k_\theta \omega_r$), Q_{m1}, $Q(l_s)$, and ω_r, and obtain the response profiles at l_m using expressions in Tables 4.5 and 4.8.
- Repeat the previous three steps for a range of l_m (<0.8l) and FPUL p_s ($= p_b l_m/l$, or $p_s = p_b z/z_m$, p_b at rotation depth z_m) to obtain nonlinear response.

The calculation of the R-E model follows the same steps but with a new value of $k_{\theta e} = -1/I'_c$, and the z_{me} estimated using the expressions in Tables 4.6 and 4.7.

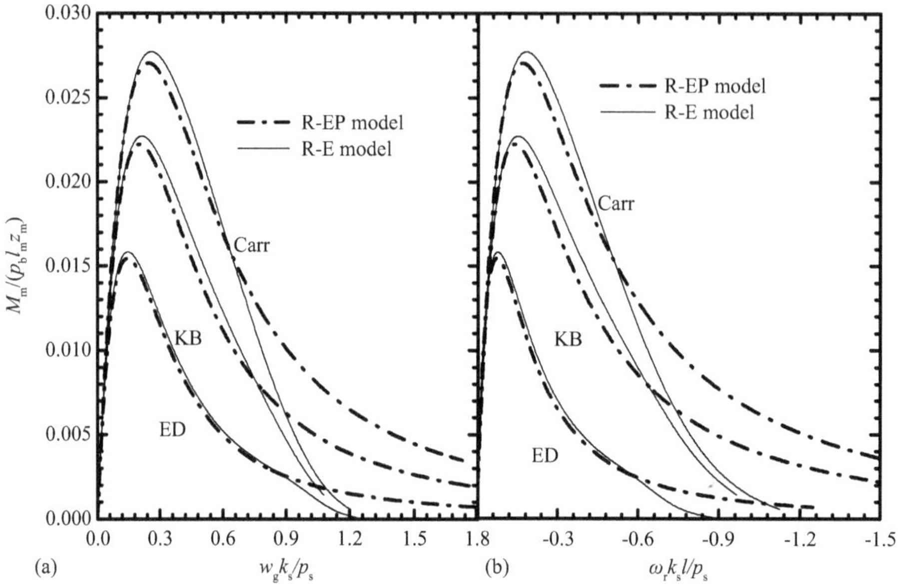

Figure 4.15 Ductile and brittle behaviour of normalized bending moment versus (a) pile deflection; and (b) angle of rotation. [Adapted from Guo, W. D., *Int J Numer and Anal Meth in Geomech* **38**(18), 2014.]

Examples 4.8 Modelling an Instrumented Pile

Reinforced concrete piles (d = 0.79 m, l = 30 m, and E_pI_p = 360 MNm^2) were installed to arrest a sliding slope (Esu and D'Elia 1974). The profiles of shear force $Q(z)$, bending moment $M(z)$, and deflection $w(z)$ along the pile were measured at a uniform w_s = 110 mm to a sliding depth of 7.5 m (256 days after installation). The maximum moment was scaled up by 68% to yielding (i.e., 0.87 MNm, see Figure 4.16c), as was the force Q_{m2} (see Figures 4.17a and b) (Guo 2022a). The 'ultimate' profiles of $Q(z)$, $M(z)$, and $w(z)$ are plotted in Figure 4.16, together with boundary element analysis (Chen and Poulos 1997) using a Young's modulus E_s of 0.533z MPa (n_2 = 1.0), $p_{s1} = 3ds_{u1}$, and $p_{s2} = 8ds_{u2}$ ($s_{u1} = s_{u2}$ = 40 kPa).

As an example, the R-EP model prediction was made using the following parameters: $H = 0$, $a = 0$ (no anchors), p_{ub} = 1,244 kN/m (= 3.6 × 40 × 0.79z_m, $z_m \approx$ 11 m confirmed late), A_{Li} = 113.1 kN/m^{ni+1} (= p_{ub}/z_m, n_1 = 1, and n_2 = 0.81); k_s = 2.6 MPa ($\approx 0.9E_s$, E_s = 2.93 MPa), α = 1.04 with p_s/k_s = 0.114 m and w_s = 0.11 m, l_{c2} = 10.1 m (taking $G_{s2} = k_{s2}/3$, and $E_p = 1.88 \times 10^7$ kPa), as well as coupled parameters λ_2 = 0.206/m [= (2.6/4 × 360)$^{0.25}$], α_c = 0.211/m, and β_c = 0.202/m. Note that the k_{s2} was taken as k_s by using a reduced modulus of nG_{s2} (n = 0.813) and N_p = 2.72 MN.

At l_m = 7.8 m (including a 0.3 m dragging layer) and p_s = 323.5 kN/m, Equations (4.20)–(4.22) were resolved, resulting in x_p (slip depth) = 0.521 m (= 0.107/0.206, $\bar{x}_p$ = 0.107), z_{mep} (z_m) = 10.898 m, and $-k_{\theta ep}$ = 70.64 MNm/radians (= 0.238 × $4E_pI_p\lambda_2$, k_{nr} = -0.238), respectively. Using Table 4.6, the following were obtained: (i) M_L = 17.66 MNm $\left(\bar{M}_L = 0.061\right)$ and P_h = 2.523 MN $\left(\bar{P}_h = 0.26\right)$; (ii) $\bar{w}_g = 1.203$ and ω_r = -0.011 radians $\left[= -2.684 \times 323.5/(2{,}600 \times 30),\ \bar{\omega}_r = -2.684\right]$, and (iii) w_h (displacement at z_m) = 0.028 m, and M_m = 786.37 kNm (= $k_\theta \omega_r$, Figure 4.16c). These values were subsequently substituted into the solutions (Tables 4.5 and 4.8) to gain response profiles (Guo 2020a), which are plotted in Figures 4.16b–d, showing good agreement well with measured data.

Using the R-EP model, the λ_2, α_c and β_c values were identical to those for the R-E model, while $-k_\theta$ = 78.98 MNm/radians (= $1/I'_c$) and z_{me} = 11.079 m [from Equation (4.13)]. As x_p raises from 0 to 0.521 m, the stiffness k_θ decreases by 10.6% from 78.98 to 70.64 MNm/

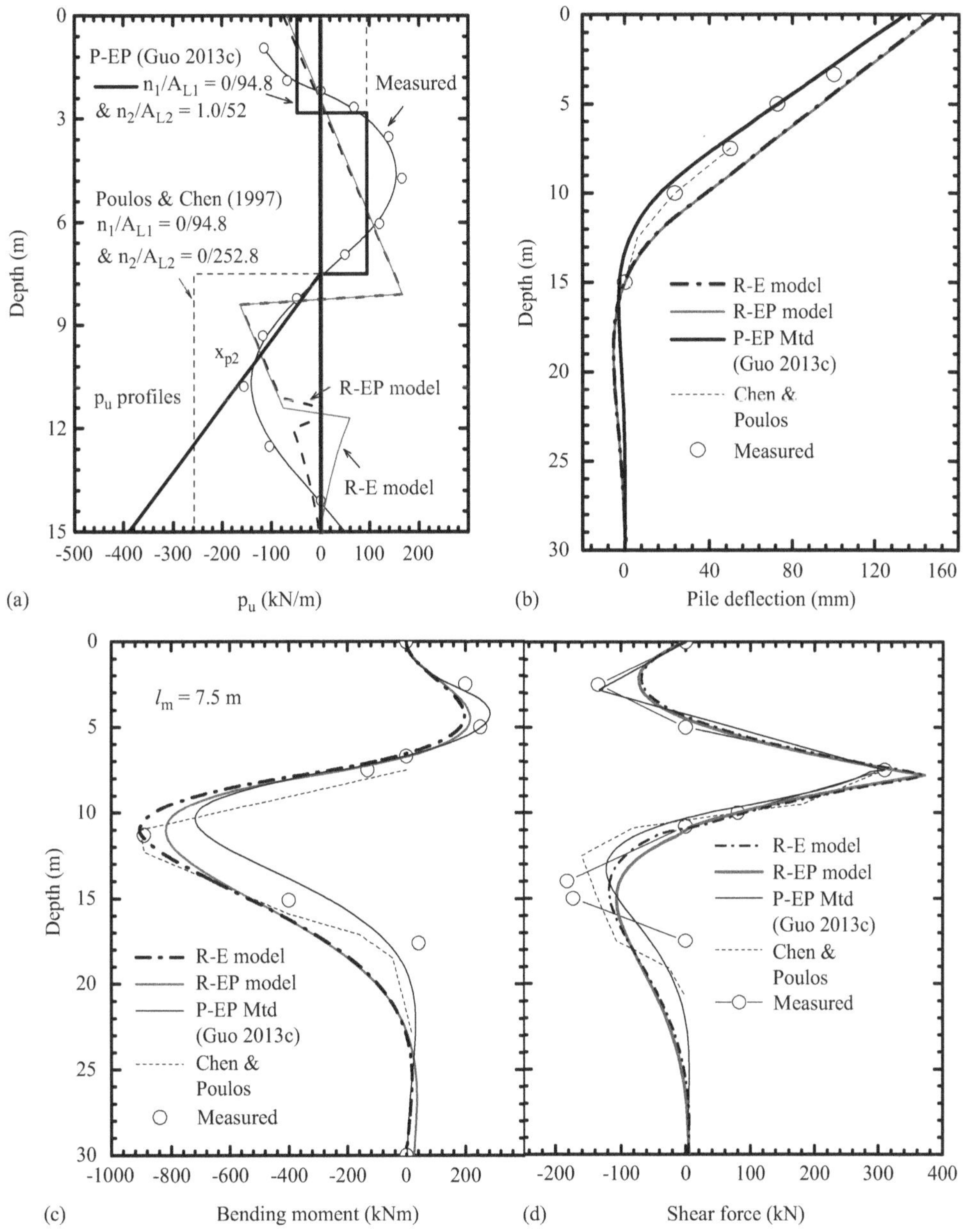

Figure 4.16 Predicted versus measured (Esu and D'Elia 1974) responses (Case ED): (a) limiting force per unit length; (b) pile deflection; (c) bending moment; and (d) shear force. [Adapted from Guo, W. D., *Int J Numer and Anal Meth in Geomech* **38**(18), 2014.]

radians; while the decrease was underestimated as 4.7% from 74.13 to 70.64 MNm/radians using $N_p = 0$ in Equations (4.15) and (4.20). Nevertheless, using $N_p = 0$ and a simple k_{s2} of 2.619 MPa, the R-EP model predictions are barely different from those gained using the coupled N_p with $M_L = 17.65$ MNm, $\bar{P}_h = 0.26, \bar{w}_g = 1.2038, x_p = 0.521$ m, $w_h = 0.0282$ m, $\omega_r = -0.011$ radians, $-k_\theta = 70.08$ MNm/radians, and $M_m = 782.83$ kNm.

Estimations were made for the displacement w_g, the moments M_{m1} (sliding layer), M_{m2} (stable layer), the force Q_{m1}, $Q(l_m)$, and the rotation ω_r for a series of l_m values (thus p_s). The resulting force Q_{m1}, $Q(l_m)$ and moment values M_{m1}, M_{m2} were plotted against the displacement w_g and the rotation ω_r in Figures 4.17a and b, and 4.17c and d, which agree well with the measured data. Using $k_s = k_{s2}$, all BiP models (see Figure 4.17c) predict similar linear correlations between the moment M_m (or M_{m2}) and displacement w_g up to 50% of the failure moment. The elastic R-E model and the rigid lower portions of the k_θ-model (Guo 2020a) both predicted a 'brittle' collapse, whereas the R-EP model predicted a ductile M_m with large displacement without failure.

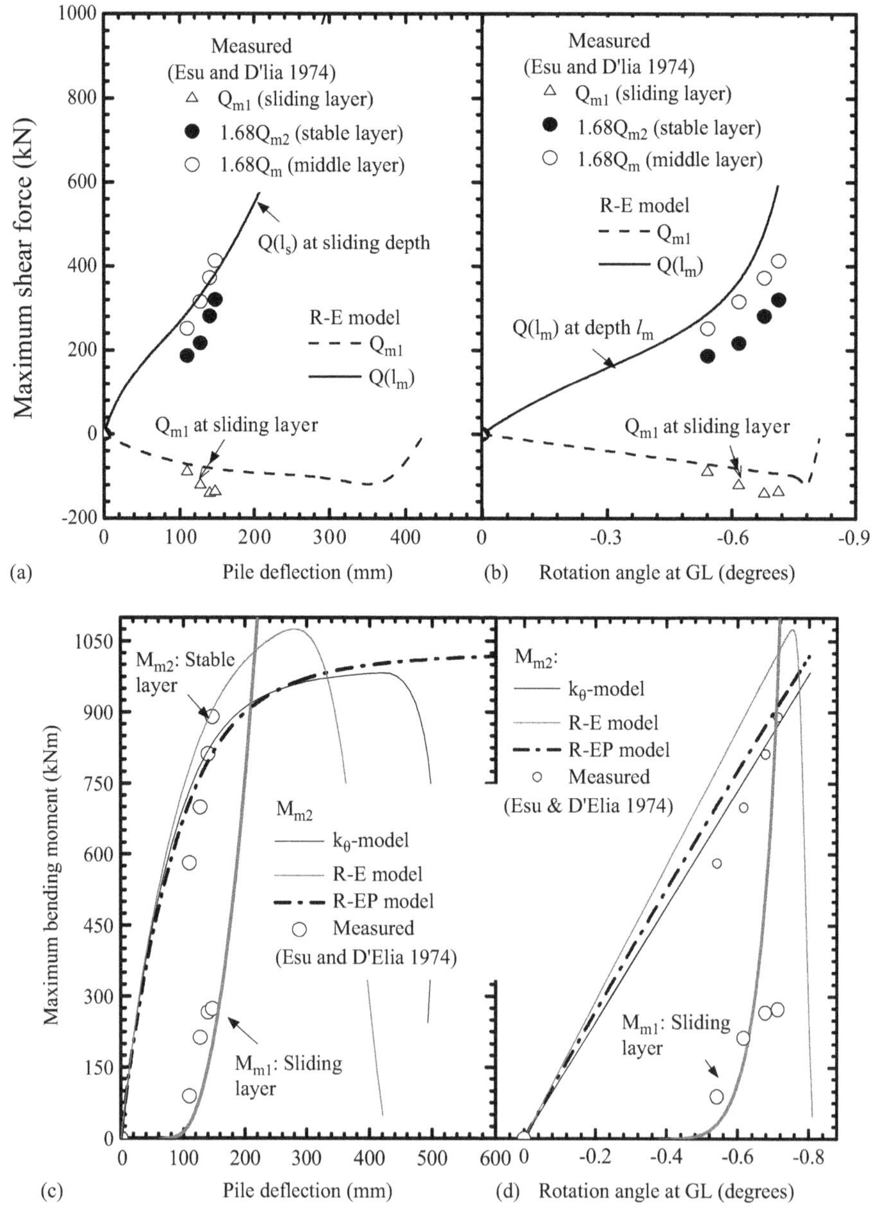

Figure 4.17 Predicted versus measured (Esu and D'Elia 1974) pile responses: (a) $w_g - Q_{m1}[Q(l_m)]$; (b) $\omega_r - Q_{m1}[Q(l_m)]$; (c) $w_g - M_{m2}(M_{m1})$; and (d) $\omega_r - M_{m2}(M_{m1})$; for Case ED. [Adapted from Guo, W. D., *Int J Numer and Anal Meth in Geomech* **38**(18), 2014.]

4.6.2 Structural nonlinearity induced by cracking

When subjected to a large displacement, pile body may exhibit nonlinear structural behaviour. In such case, the deformation may be estimated by reducing bending rigidity E_pI_p beyond cracking load/soil movement, as proposed for rock socketed piles (Guo and Zhu 2011).

Examples 4.9 Rigid displacement incurred by landslide

The deformation from structural nonlinearity depends on the distributed bending moment $M_{c1}(z)$ (Table 4.5) over a sliding depth l_m, and the rotational restraint k_0 as well as cap-restraint k_G. Both restraints are ignored to gain maximum displacement difference (Δw) along the rigid portion of the pile in sliding layer (Guo 2022a)

$$\Delta w = \left\{ -\frac{\bar{\omega}_r}{30}\left(\frac{l_m}{l}\right)^5 + \frac{1}{8}\left(\bar{w}_g - 1\right)\left(\frac{l_m}{l}\right)^4 + \bar{H}\left[\frac{1}{3}\left(\frac{l_m}{l}\right)^3 - \bar{a}\frac{1}{2}\left(\frac{l_m}{l}\right)^2\right]\right\} \frac{p_s l^4}{E_p I_p} \tag{4.23}$$

The rigid, upper portion of the pile should have an identical rotation angle to the angle ω_r of the low portion at the depth z_m, regardless of crack evolution. At a sliding depth l_m of 7.5–7.8 m, for instance, the pile ED has $\bar{w}_g = 1.203$, $\bar{\omega}_r = -2.684$, and E_pI_p of 3.6×10^5 kNm^2, the Δw is estimated using Equation (4.23) as 0.136–0.162 m (i.e., an average of 149 mm). This displacement agrees well with the calculated displacement of 150 mm (see Figure 4.16b) and incorporates the pile rigidity (E_pI_p).

4.6.3 Cracking moment M_{cr} and effective flexural rigidity E_cI_e

The crack moment and effective flexural rigidity of a concrete pile are estimated in the same manner as that for rock-socketed piles. The crack incepts at maximum bending moment M_{max} (known as cracking moment M_{cr}) of a concrete pile (ACI 1993):

$$M_{cr} = k_r\sqrt{f'_c}\, I_g / y_r \tag{4.24}$$

where f_c' = characteristic compressive strength of concrete (kPa); k_r = 16.7–62.7 for rock-socketed, concrete piles (Guo 2012); I_g = moment of inertia of the pile cross-section about centroidal axis neglecting reinforcement; and y_r = the distance between extreme (tensile) fibres and the centroidal axis. The effective moment of inertia I_e reduces with increase in M_{max} (ACI 1993)

$$I_e = \left(\frac{M_{cr}}{M_{max}}\right)^3 I_g + \left[1 - \left(\frac{M_{cr}}{M_{max}}\right)^3\right] I_{cr} \tag{4.25}$$

where I_{cr} = moment of inertia of the cracked transformed section, at which the bending moment attains the ultimate value, M_{ult}. Assuming a constant Young's modulus of concrete E_c, the effective rigidity E_pI_p (= E_cI_e) degenerates from the elastic value E_cI_g to the ultimate cracked value $(EI)_{cr}$ (= E_cI_{cr}). To estimate M_{ult} and I_{cr}, bending theory (moment-curvature method), stress block method (Guo and Zhu 2011, ACI. 1993, Whitney 1937), or nonlinear stress-strain relationships for concrete and steel (Reese 1997) can be used, considering E_cI_{cr} = M_{ult}/φ_{ult} and φ_{ult} = ultimate curvature. The study on eight instrumented piles with cracks offers the k_r values in Equation (4.24), a crack curvature φ_{cr} (= M_{cr}/E_cI_{cr}) of 0.1427–0.503

and an ultimate φ_{ult} of 1.386–26.925. While these parameters may spread widely, the degeneration of rigidity follows the rule stipulated by Equation (4.25).

4.6.4 Procedure for analysing structurally nonlinear piles

Structural nonlinearity of piles can be modelled using piece-wise (increment) elastic approach for linear elastic piles, as with rock-socketed piles (Guo and Zhu 2011). For each loading increment, the variable rigidity E_cI_e replaces the constant rigidity E_pI_p of linear elastic piles. First, the maximum moment M_{max} and deflection w_g at each soil movement w_s (l_m) level are estimated for elastic piles. The M_{max} is taken as that induced in nonlinear pile. Next, the incremental response of linear piles is obtained by calculating:

- The cracking moment M_{cr} using Equation (4.24);
- The ultimate moment M_{ult} and the cracked rigidity E_cI_{cr} in terms of pile size, rebar strength and layout, and concrete strength;
- The soil movement w_s at the onset of crack by taking $M_{max} = M_{cr}$;
- The effective E_cI_e using Equation (4.25) and values of E_cI_g, E_cI_{cr}, M_{cr} and M_{max}; and finally
- The moment M_{max} and deflection w_g at w_s by using the calculated E_cI_e to replace E_pI_p in the R-E and R-EP models.

This process is repeated for a set of w_s (thus M_{max}) to gain response of cracking piles. An detailed example of this procedure is presented by Guo and Zhu (2011), which is not replicated herein.

Over the length of plastic hinge l_p, an additional displacement (Δw_c) may occur, in addition to the deflection Δw caused by a curvature φ_{cr}. In free-moving soil, the pile behaves as a cantilever with $\Delta w_c = (\varphi_{ult} - \varphi_{cr})l_p(l_s - 0.5l_p)$ (Park and Paulay 1975). Limited studies indicate the length l_p of plastic zone is (1.2–1.5)d (Chai and Hutchinson 2002) and (0.1–1.5)d (Huang et al. 2021).

The ED pile can be compared to a rock-socketed pile SN2 (d = 0.8 m, l = 34 m) with an E_pI_p of 790 MNm^2 (Guo 2012). The SN2 pile exhibits structural nonlinearity with an ultimate moment M_{ult} of 1.89 MN-m at a load H of 863 kN. The bending curvature rises from 0.5886×10^{-3} (φ_{cr}) to 16.32×10^{-3} (φ_{ult}). Without reinforcement detail, the ED pile is assumed to have the same φ_{cr} and φ_{ult} values, and a plastic hinge length l_p of 0.79 m ($\approx d$). This results in a displacement Δw_c of 131.8 mm [= 0.79 × (11.0 – 0.5 × 0.79)(16.32 – 0.586)]. Therefore, the ED pile may be able to sustain a head displacement (= $\Delta w + \Delta w_c$) of 280.8 mm (= 149.0 + 131.8) before collapsing along the trajectory portrayed by the R-E model (see Figure 4.17). Clearly, the structural nonlinearity of pile body may limit the maximum displacement of 'ductile' behaviour.

4.6.5 Comments

The normalized stiffness $\bar{k}_{\theta e}$ of pile ED is determined based on the measured sliding depth l_m. Likewise, piles K4 and K6 (Fukumoto 1976), KB (Fukumoto 1972), and Carr (Carrubba et al. 1989) were analysed, and their results are presented in Figures 4.14, and 4.15. Figure 4.14 indicates that the associated normalized depths (z_{me}/l, z_{mep}/l) for ED pile (number 5, see the upper-left table) fall in between the $\bar{k}_{\theta e}$ versus z_m/l curves for $\lambda_2 l$ of 6.18, and $\lambda_2 l_e$ of 3.9 (not shown) with length l_e (= l–z_{me}) of pile portion in stable layer. At a small sliding length l_s of $0.02l$, the z_{me}/l ratio is over 50% lower, resulting in a slightly reduced normalized $\bar{k}_{\theta e}$. For stiff piles like Carr (point 1), this reduction (variation) in $\bar{k}_{\theta e}$ can be extremely large. Should the sliding depth gradually increase from $0.02l$ to $\bar{l}_m$, the large range of $\bar{k}_{\theta e}$ values would cultivate a high probability of crossing the singularity $\bar{k}_\theta^* - z_m / l$ curve, and incurring response amplification (plastic hinge). Fortunately, the sliding depth varies narrowly between top and bottom of a dragging layer, and back-rotation of piles may only occur naturally for a highly stiff topsoil layer. To prevent the 'çross', the $\bar{k}_\theta$ can be adjusted away from the $\bar{k}_\theta^*$ curve together with curbing the back-rotation.

The R-E and R-EP models are strictly intended for rigid movement in sliding layer and infinite length in stable layer. Otherwise, the models may predict an incorrect non-zero bending moment at pile-tip level of short piles. In the case of a large soil movement, the rigid upper portion of the pile may be carried by moving soil, resulting in a larger rotation than that of the pile (ω_r) and ultimately leading to a lack of compatibility of displacement and rotation at the depth of maximum bending moment. Given pile-soil relative stiffness $E_p/G_{s2}{}^*$, the normalized rotational stiffness $k_{\theta e}/(E_p I_p \lambda_2)$ may be roughly estimated using the curves in Figure 4.13. The rotational depth z_{me} can be obtained using Equation (4.13) or Figure 4.14. Subsequently, the pile response can be estimated using available elastic or elastic-plastic solutions (see Tables 4.5, 4.7 and 4.8) for piles subjected to the rotational moment M_m at depth z_{me}, as with the presented example (pile ED).

4.7 SUMMARY

This chapter provides models to predict the nonlinear response of piles in sliding soils for different scenarios. The pile may exhibit rigid behaviour in both the sliding and stable layers (2-layer model), rigid in sliding layer and elastic in stable layer (R-E model), and rigid in sliding layer and elastic-plastic in stable layer (R-EP model). The models capture impact of rotational restraints (at pile-cap, sliding depth or pile-tip), mirrored behaviour of deep sliding, plasticity and ductility of the pile on pile response including bending moment-to-thrust ratio. The R-E and R-EP models also provide expressions for estimating rotational stiffness. These models use a common set of parameters (LFPUL, modulus k_s and stiffness k_θ) and can accurately predict the response of both rigid piles and bi-portion piles in moving slopes.

Chapter 5

Embankment piles and shear tests

3-layer model

5.1 INTRODUCTION

Piles may be subjected to moving soil as induced by lateral spreading during an earthquake, creep slope, embankment loading or bulging from excavation. Their design entails incorporating the impact of sliding depth, pile-soil relative stiffness, rotational constraints and possible dragging from soil movement (Guo 2003, 2013c). Guo and Ghee (2005a,b) conducted extensive tests using the new shear apparatus (Chapter 1) to gain the impact of these parameters on piles. Guo (2014) developed a 2-layer model (see Chapter 4) to capture the nonlinear response of the piles induced by soil movement w_s (= p_s/k_s, k_s = modulus of subgrade reaction of sliding layer) over a sliding depth l_m. In particular, the model adopts a stepwise uniform pressure p_s to simulate a linearly increasing net force per unit length (*FPUL*) p_s (= $p_b l_m/l$), in which $p_b = \alpha p_{ub}$, p_{ub} = *FPUL* at base level of a lateral pile, and l = pile embedment (Chapter 1). The soil movement w_s (> $w_s^* = p_s/k_s$) exerts rotational restraint, acting as a 'thick' pile-cap, on free-head piles (Guo 2012) to enforce a *FPUL* of αp_{ub} (α = 0.25–1.5).

The factor α is sufficiently accurate to incorporate the impact of soil movement profiles, underlying stiff layer and vertical and/or embankment loading on the p_b. To incorporate effect of 'dragging layer', this chapter presents a 3-layer model and solutions underpinned by the same set of input parameters of the 2-layer model, which are m (modulus ratio of stable over sliding layer), k_s, p_b, and k_θ (cap rotational stiffness) (Chapter 4). The impact of dragging effect is explored together with further justification of the parameters at design and ultimate states. Example solutions are provided concerning free-head piles, pile groups, and piles subject to static and lateral spreading embankment loading.

5.2 3-LAYER MODEL AND SOLUTIONS

5.2.1 Model solutions and features

Chapters 1 and 4 provide the 3-layer model to simulate pile under passive loading p_s. The pile-displacement at depth z, $w(z)$ (= $\omega_r z + w_g$) is recast into

$$w(z) = (\bar{\omega}_r \bar{z} + \bar{w}_g) p_s / k_s \tag{5.1}$$

where $\bar{\omega}_r \left[= w'(z) k_s l / p_s\right]$, $\bar{w}_g \left(= w_g k_s / p_s\right)$, and $\bar{z} = z/l$. The 3-layer model is deduced using Equation (5.1), the force equilibrium and bending moment equilibrium of the pile-soil system, as well as the pile-soil displacement compatibility (see Figure 1.5, Chapter 1). The derivation

DOI: 10.1201/9781003315230-5

process resembles that of 2-layer model presented in Chapter 4. For instance, the net *FPUL* $p_2(z)$ on the piles in the transition layer is described by

$$p_2(z) = \left(\bar{\omega}_r \bar{z} + \bar{w}_g\right)\left[(m-1)\frac{\bar{z} - \bar{l}_m}{\bar{z}_m - \bar{l}_m} + 1\right] p_s \tag{5.2}$$

where

$$\bar{w}_g = \frac{-6\left(1 + \bar{z}_m\right)\left(\bar{H} + \bar{l}_m\right)}{(m-1)\left(-2\bar{l}_m^2 + \bar{z}_m \bar{l}_m + 3\bar{l}_m + \bar{z}_m^2\right) - 3(1+m)\bar{z}_m} \tag{5.3}$$

$$\bar{\omega}_r = \frac{12\left(\bar{H} + \bar{l}_m\right)}{(m-1)\left(-2\bar{l}_m^2 + \bar{z}_m \bar{l}_m + 3\bar{l}_m + \bar{z}_m^2\right) - 3(1+m)\bar{z}_m} \tag{5.4}$$

where $\bar{l}_m = l_m / l$, normalized sliding depth, $\bar{H} = H/(p_s l)$, normalized lateral shear force (at head-level) and $\bar{z}_m = z_m / l$, normalized depth of maximum bending moment, which is given by

$$\bar{z}_m^2 + \bar{B}\bar{z}_m + \bar{C} = 0 \tag{5.5}$$

where

$$\bar{B} = \frac{3 + \left[3(1+m) + \bar{l}_m(1-m)\right]\bar{C}_1}{(1-m)\bar{C}_1} \tag{5.6}$$

$$\bar{C} = \frac{m - (m-1)\bar{l}_m\left[3\left(\bar{C}_1 + 1\right) + \bar{l}_m\left(\bar{l}_m - 2\bar{C}_1 - 3\right)\right]}{(1-m)\bar{C}_1} \tag{5.7}$$

$$\bar{C}_1 = \frac{\bar{k}_\theta \bar{\omega}_r + \bar{H} + \bar{l}_m\left(1 - 0.5\bar{l}_m\right)}{-\left(\bar{H} + \bar{l}_m\right)} \tag{5.8}$$

where $\bar{k}_\theta = k_\theta / \left(k_s l^3\right)$, normalized rotational stiffness; and k_θ is the total rotational stiffness along the pile, $k_\theta = k_G + k_T$ in Figure 1.5, Chapter 1. The $\bar{z}_m$ value is calculated iteratively using Equation (5.5) and the $\bar{k}_\theta$-dependent coefficients $\bar{B}$ and $\bar{C}$, or can be directly estimated (without iteration) using Equation (5.5) at $\bar{k}_\theta = 0$. Modern mathematical software such as Mathcad™ can be used for this purpose, as illustrated later for free-head ($k_G = k_T = 0$) and capped piles ($k_G \neq 0$). Table 5.1 provides other expressions for the normalized shear force $Q_i(z)/(p_s l)$, normalized bending moment $M_i(z)/(p_s l^2)$ and on-pile *FPUL* $p_i(z)$ for the upper, transitional, and lower (stable) layer, respectively. These expressions allow for estimating the normalized maximum bending moment $M_{m2}/(p_s l l_m)$ and shear force $Q_{m3}/(p_s l)$ [M_{m2} and Q_{m3} = maximum of $M_2(z)$ and $Q_3(z)$]. It should be stressed that

Table 5.1 3-layer solutions for passive piles (Guo 2016)

Depth z	Sliding layer with k_s, $z \le l_m$, $p_1(z) = (\bar{\omega}_r \bar{z} + \bar{w}_g) p_s$, $p(z) = p_1(z) - p_s$
Q(z)	$Q_1(z)/(p_s l) = 0.5\bar{\omega}_r \bar{z}^2 + \bar{w}_g \bar{z} - \bar{z} - \bar{H}$
Moment M(z)	$\dfrac{M_1(z)}{p_s l^2} = \dfrac{1}{6}\bar{\omega}_r \bar{z}^3 + \dfrac{1}{2}(\bar{w}_g - 1)\bar{z}^2 - \bar{H}\bar{z} + \bar{k}_\theta \lambda(\bar{z})\bar{\omega}_r$
Depth	k_s transition layer, $l_m \le z \le z_m$
p(z)	$p(z) = p_2(z) = (\bar{\omega}_r \bar{z} + \bar{w}_g)\left[(m-1)\dfrac{\bar{z} - \bar{l}_m}{\bar{z}_m - \bar{l}_m} + 1\right] p_s$
Shear force Q(z)	$\dfrac{Q_2(z)}{p_s l} = \dfrac{1}{6(\bar{l}_m - \bar{z}_m)}\left\{\begin{array}{l}(1-m)\left[3\bar{w}_g(\bar{z}^2 + \bar{l}_m^2) + \bar{\omega}_r(2\bar{z}^3 + \bar{l}_m^3)\right] \\ -3(\bar{z}_m - m\bar{l}_m)(2\bar{w}_g + \bar{z}\bar{\omega}_r)\bar{z}\end{array}\right\} - \bar{H} - \bar{l}_m$
Bending moment M(z)	$\dfrac{M_2(z)}{p_s l^2} = \dfrac{1}{12(\bar{l}_m - \bar{z}_m)}\left\{(1-m)\left[2\bar{w}_g(\bar{z}^3 + 3\bar{l}_m^2 \bar{z} - \bar{l}_m^3) + \bar{\omega}_r(\bar{z}^4 + 2\bar{l}_m^3 \bar{z} - \bar{l}_m^4)\right]\right.$
	$\left. -2(\bar{z}_m - m\bar{l}_m)(3\bar{w}_g + \bar{\omega}_r \bar{z})\bar{z}^2\right\} - \bar{H}\bar{z} - (\bar{z} - 0.5\bar{l}_m)\bar{l}_m + \bar{k}_\theta \lambda(\bar{z})\bar{\omega}_r$
Depth z	Layer with mk_s, $z_m \le z \le l$, $p(z) = p_3(z) = (\bar{\omega}_r \bar{z} + \bar{w}_g) m p_s$
Shear force Q(z)	$\dfrac{Q_3(z)}{p_s l} = \left[\dfrac{1-m}{2}(\bar{l}_m + \bar{z}_m) + m\bar{z}\right]\bar{w}_g + \left[\dfrac{1-m}{6}(\bar{z}_m^2 + \bar{z}_m \bar{l}_m + \bar{l}_m^2) + \dfrac{m}{2}\bar{z}^2\right]\bar{\omega}_r - \bar{H} - \bar{l}_m$
Bending moment M (z)	$\dfrac{M_3(z)}{p_s l^2} = \dfrac{1-m}{12}\left\{2(\bar{\omega}_r \bar{z} - \bar{w}_g)(\bar{z}_m^2 + \bar{l}_m \bar{z}_m + \bar{l}_m^2) + \left[6\bar{w}_g \bar{z} - \bar{\omega}_r(\bar{l}_m^2 + \bar{z}_m^2)\right](\bar{l}_m + \bar{z}_m)\right\}$
	$+\dfrac{m}{6}\bar{z}^2(3\bar{w}_g + \bar{\omega}_r \bar{z}) - \bar{H}\bar{z} - (\bar{z} - 0.5\bar{l}_m)\bar{l}_m + \bar{k}_\theta \lambda(\bar{z})\bar{\omega}_r$
	$\bar{z}_m = z_m/l = 0.5\left(-\bar{B} - \sqrt{\bar{B}^2 - 4\bar{C}}\right)$, where $\beta = \dfrac{0.5\bar{l}_m^2}{\bar{H} + \bar{l}_m}$,
	$\bar{B} = \bar{l}_m - \dfrac{3[1 - (1-\beta)(m+1)]}{(1-\beta)(m-1)}$ $\bar{C} = (-2\bar{l}_m + 3)\bar{l}_m + \dfrac{(1-\bar{l}_m)^3}{1-\beta} + \dfrac{1 + 12\bar{k}_\theta}{(m-1)(1-\beta)}$
	$w(z) = (\bar{\omega}_r \bar{z} + w_g) p_s/k_s$, $\dfrac{w_g k_s}{p_s} = \bar{w}_g = \dfrac{-6(1+\bar{z}_m)(\bar{H} + \bar{l}_m)}{(m-1)(-2\bar{l}_m^2 + \bar{z}_m \bar{l}_m + 3\bar{l}_m + \bar{z}_m^2) - (3(1+m)\bar{z}_m}$,
	$\dfrac{\omega_r k_s l}{p_s} = \bar{\omega}_r = \dfrac{12(\bar{H} + \bar{l}_m)}{(m-1)(-2\bar{l}_m^2 + \bar{z}_m \bar{l}_m + 3\bar{l}_m + \bar{z}_m^2) - (3(1+m)\bar{z}_m}$, $\bar{k}_\theta = k_\theta/(k_s l^3)$, $\bar{l}_m = l_m/l$,
	$\bar{H} = H/(p_s l)$, $\lambda(\bar{z}) = (2\bar{z}^3 - 3\bar{z}^2 + 1)\dfrac{k_G}{k_\theta} + (2(1-\bar{z})^3 - 3(1-\bar{z})^2 + 1)\dfrac{k_T}{k_\theta}$, $k_\vartheta = k_G + k_T$

- The net *FPUL* $p(z)$ in sliding layer is equal to $p_1(z) - p_s$;
- As the rotational stiffness k_θ (via $k_G = k_\theta$) increases, the depth z_m advances towards the pile base ($z_m \rightarrow l$) to reduce thickness of the stable layer;
- A large sliding depth l_m renders the two depths z_m and l_m converge together (i.e., the 3-layer profile reduces to a 2-layer profile);
- The sliding depth at the pile location reduces to c ($= l_m/\beta$, $\beta \ge 1$);

- With rotational restraint lumped at the pile-head ($k_\theta = k_G$) and/or the tip ($k_\theta = k_T$), the net pile displacement with depth should be revised as $w(z) + \lambda_d z M_m/k_\theta$ or $w(z) + \lambda_d(l - z)M_m/k_\theta$, respectively, in which $\lambda_d = 0.667$ for a linearly increasing *FPUL* p_s with depth ($\lambda_d = 0.5$ for uniform p_s);
- The non-dimensional response (e.g., $\bar{\omega}_r$ and $\bar{w}_g$) depends on m, $\bar{l}_m$, $\bar{H}$, and $\bar{k}_\theta$ as does the $\bar{z}_m$. The soil movement of p_s/k_s only comes into play when obtaining the dimensional response.

5.2.2 Equivalent p_s (limiting *FPUL*)

The 3-layer model solutions can be readily applied to piles in groups by multiplying the pile-pile interaction factor p_m to the moduli and the factor α to *FPUL*. This results in $p_m k_s$, $p_m m k_s$ and αp_s. The p_s (= $\alpha p_{ub} l_m/l$) is always stipulated as linearly increase with sliding depth. These solutions are contemplated for rigid piles with free displacement at the pile base; otherwise, the previous solutions (e.g., Poulos and Davis 1980; Guo 2022a, 2013c) should be consulted. The input parameters m, p_{ub}, k_s, α, and k_θ are determined as described in Chapter 1. The reduction of p_{ub} from active to passive loading via factor α is substantiated through examples.

5.2.3 Comments on 3-layer model ($k_G = k_T = 0$, and $H = 0$)

Both the 3-layer and the 2-layer models warrant force and bending moment equilibrium, and pile-soil displacement compatibility under the 'external' *FPUL* p_s. While the 2-layer model is readily obtained (without iteration to gain the depth of maximum bending moment z_m), the 3-layer model is recommended for modelling piles with dragging.

5.2.3.1 *α = 1/3 for a uniform p_u and dragging resistance*

Chapter 4 introduces a normalized k_i [= $p_{ui}/(ds_{ui})$] of (9.1–11.9)α for sliding layer and $\alpha \approx$ 1/3 for uniform p_u. The 2-layer (Chapter 4) and 3-layer models are adopted to gain the shear force (or thrust) Q_m at the sliding depth l_m and maximum bending moment M_m. Figure 5.1 shows the normalized thrust $Q_m/(p_s l_m)$ ($p_s = 3.33s_{u1}d$) for l_m/l = 0–1. For m = 3, 5, 10 and 15, the 3-layer model (including dragging effect) predicts a higher thrust than the 2-layer model (Figure 5.1a). The predicted $Q_m l/M_m$ ratios (gradient of the curves) fall within 2.5–6 (thin, solid lines) as gained from field and laboratory tests on piles in sliding soil or adjacent to excavation (Guo and Qin 2010). Interestingly, the two models (see Figure 5.1b) predict barely different $M_m/(p_s l l_m)$ values for $Q_m/(p_s l_m) < 0.4$ (m = 1–10). Given a definite bending capacity, the 2-layer and 3-layer models predict a similar resistance Q_m, but the former offers a deeper sliding depth by about 20% (see Figure 5.1a). This suggests that the use of the larger gradient of *LFPUL* proposed for laterally loaded piles has led to over-design or under-design.

5.2.3.2 *α = 0.6–0.74 for linear FPUL*

A linear increasing *FPUL* with depth (*LIFPULD*) p_s (= $A_r z$) along piles in sand or stiff clay was used to obtain numerically the normalized sliding thrust of $Q_m/(p_s l_m)$ ($p_s = 0.5A_r l_m$ and $A_r = 3\gamma_s' K_{p1} d$) for piles at 'ultimate state' (Muraro et al. 2014) at K_{p2}/K_{p1} = 1, 2 and 3, respectively and typical sliding depth ratios of l_m/l. The shear force (thrust) Q_m was estimated using the 2-layer model for $m = K_{p1}/K_{a2}$, $2K_{p1}/K_{a2}$ and $3K_{p1}/K_{a2}$, respectively (see Guo 2016), and then normalized by $p_s = 0.37A_r l_m$ (= $0.5\alpha A_r l_m$, α = 0.74) to 'match' with the 'ultimate-state'

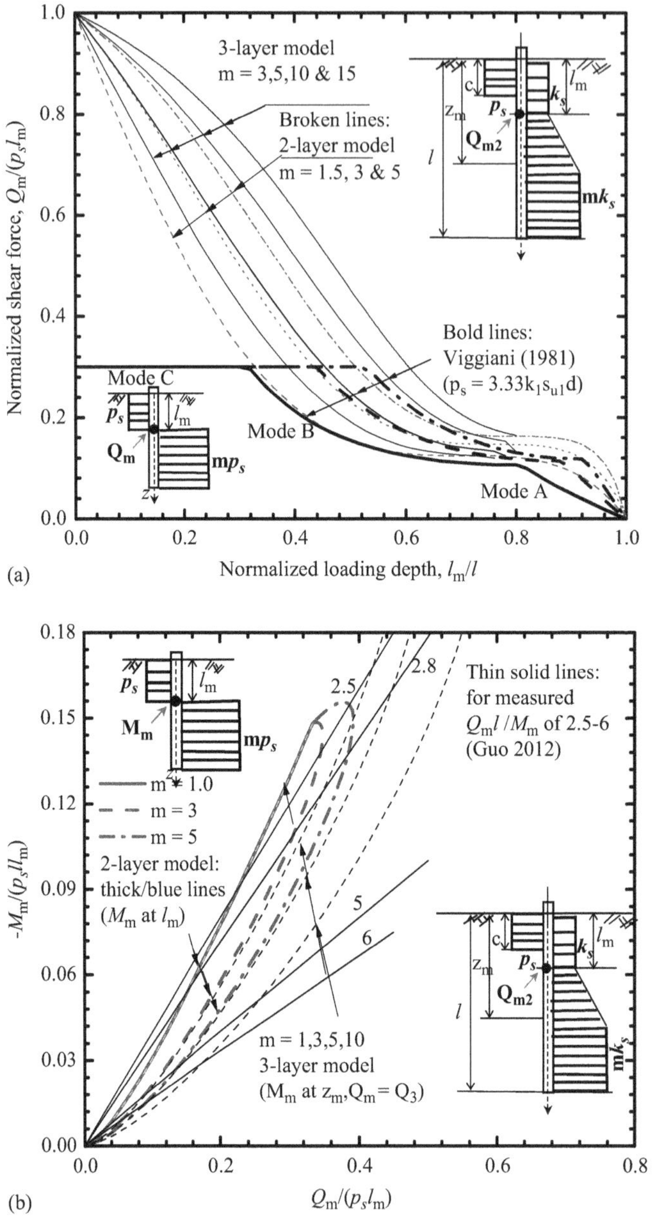

Figure 5.1 (a) $Q_m/(p_s l_m)$ versus normalized loading depth; (b) $M_m/(p_s l l_m)$ versus normalized shear force $Q_m/(p_s l_m)$ and typical test data (uniform p_s). [Adapted from Guo, W. D., *Int J Numer and Anal Meth in Geomech* **40**(14), 2016.]

values (see Figure 5.2a). The numerical ultimate-state results have a p_s value of 1.35 (= $1/\alpha$) times that of the 2-layer solutions. Likewise, the thrust (Q_m) from the 3-layer solutions matches well with the ultimate solutions (see Figure 5.2b), using $p_s = 0.3A_r l_m$ (= $0.5\alpha A_r l_m$, $\alpha = 0.6$). This comparison suggests that α is 0.6–0.74 for a *LIFPULD* p_s, compared to α of 1/3 for a uniform p_s. In other words, for active piles with A_r of $3\gamma_s' K_{p1} d$ (Muraro et al. 2014, Kourkoulis and Gelagoti 2011), once soil movement (passive loading) is inflicted, the A_r value should be reduced by the factor α to an A_r of $1.8\gamma_s' K_{p1} d$.

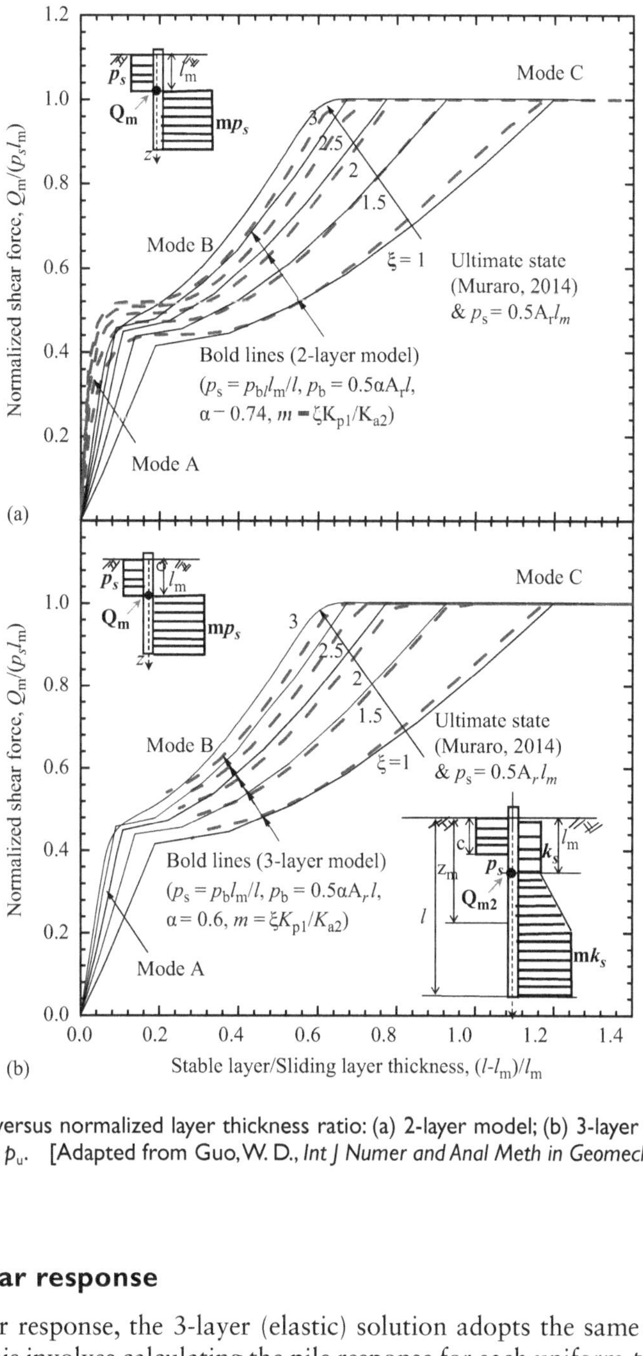

Figure 5.2 $Q_m/(p_s l_m)$ versus normalized layer thickness ratio: (a) 2-layer model; (b) 3-layer model for a linearly increasing p_u. [Adapted from Guo, W. D., *Int J Numer and Anal Meth in Geomech* **40**(14), 2016.]

5.2.4 Nonlinear response

To gain nonlinear response, the 3-layer (elastic) solution adopts the same approach as the 2-layer model. This involves calculating the pile response for each uniform p_s, and for a series of stepwise increased uniform p_s (= $\alpha p_{ub} l_m/l$, applied) using the following steps:

- Determine z_m using Equation (5.5) for an assumed l_m.
- Substitute z_m into Equations (5.3) and (5.4) to obtain the normalized displacement $\bar{w}_g$ and rotation $\bar{\omega}_r$, respectively. Also substitute z_m into the expressions for $M_2(z)$, $p_i(z)$, and $Q_i(z)$ (see Table 5.1) to gain the maximum bending moment M_m, net on-pile force per unit length $p(z)$, and shear force $Q(z)$, respectively.

- Estimate the pile response at a loading depth c by setting $l_m = c$ in the solution. This includes the magnitude and depth of M_m and the maximum in-pile shear force Q_m under the p_s.

Conversely, the factor α is deduced from the measured ratio of the movement w_s to the pile displacement w_g. The k_s, m, p_b and k_θ values are obtained from measured rotation angles, displacements, bending moments (at a typical sliding depth), as well as on-pile *FPUL* profile (if available). The loading depth c is deduced from measured 'ultimate' profile of bending moment.

The modelling uses effective soil movement $w_s^*(= p_s/k_s)$ and linearly increasing p_s. This induces an inverse trapezoidal shape of the on-pile *FPUL* $p(z)$ along an in situ slope stabilizing pile ($H \neq 0$ at pile head) (Frank and Pouget 2008); and a parabolic shape of $p(z)$ along the head-restrained ($k_\theta \neq 0$) piles subjected to lateral spreading (Dobry et al. 2003). A low limit w_s is observed for $\alpha < 1$ (with reduced p_s). Example calculations are provided later.

5.3 PARAMETRIC ANALYSIS ($H = 0$)

5.3.1 Normalized elastic solutions

The 3-layer model was utilized to examine the impact of rotational stiffness k_θ and modulus ratio m (dragging) on pile response, which was presented in normalized form of

- Mudline displacement over soil movement, $\alpha w_g/w_s$ (see Figure 5.3);
- Rotational movement of pile $\omega_r k_s l/p_s$ (Figure 5.4);
- Maximum bending moment $M_m/(p_s l_m l)$ at z_m and/or sliding depth l_m (Figure 5.4);
- Thrust $Q_m/(p_s l_m)$ at sliding depth, and shear force $Q_{m3}/(p_s l_m)$ in stable layer (Figure 5.5).

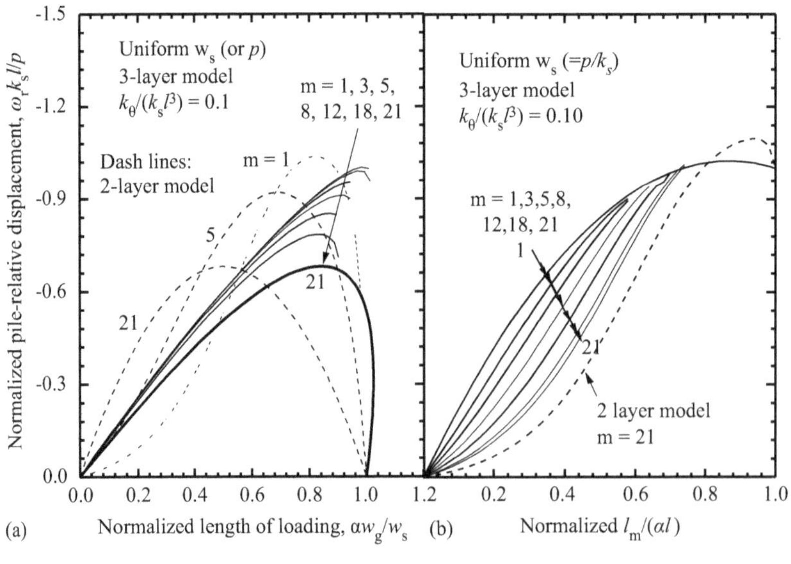

Figure 5.3 Normalized uniform soil movement w_g (constant k_s) with loading depth at (a) normalized rotational stiffness (m = 10); (b) modulus ratio $m\left(\bar{k}_\theta = 0.1\right)$. [Adapted from Guo, W. D., *Int J Numer and Anal Meth in Geomech* **40**(14), 2016.]

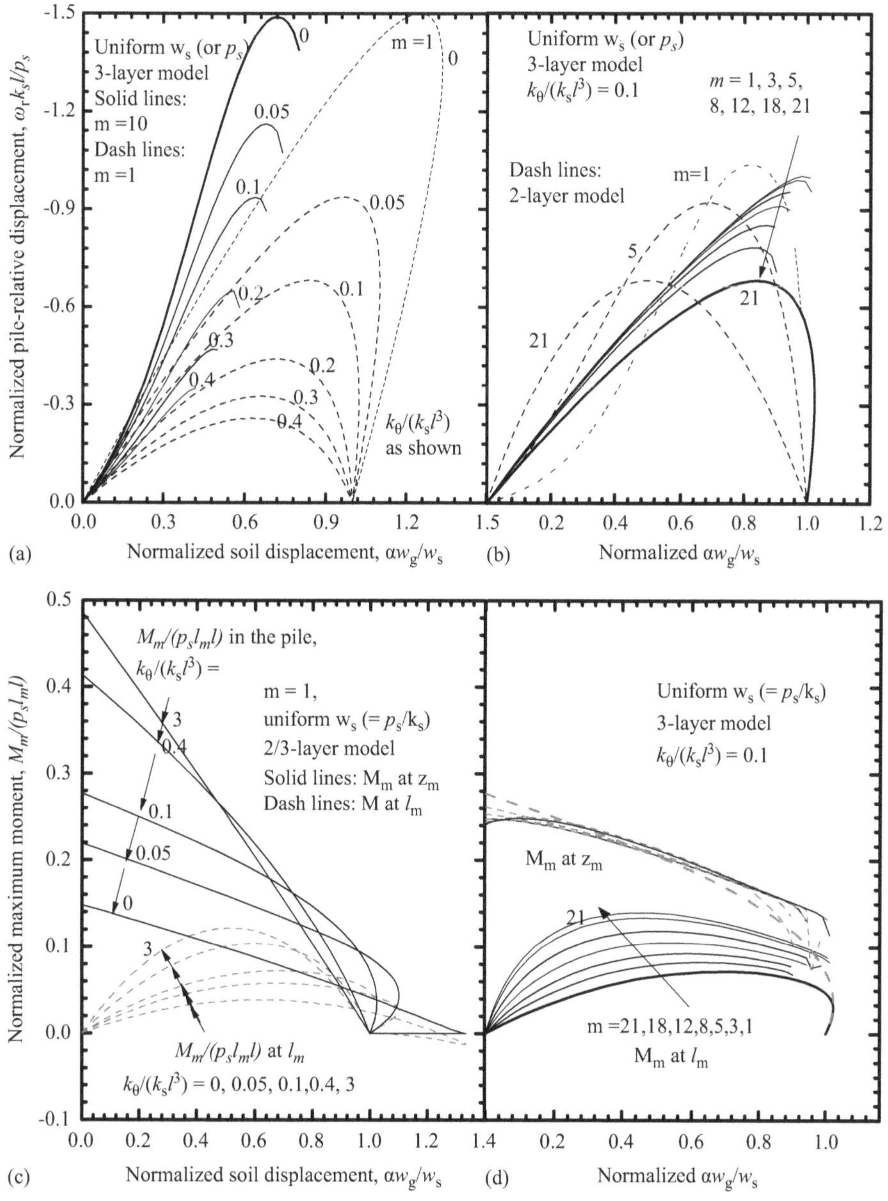

Figure 5.4 Pile-head displacement versus rotation (constant k_s) at (a) rotational stiffness (m = 10 or 1); (b) modulus ratio m $\left(\bar{k}_\theta = 0.1\right)$; and versus bending moment (for base restrained) ($k_B = k_\theta$) at z_m and l_m: (c) $\bar{k}_\theta = 0 - 3\bar{k}_\theta$ and m = 1; (d) $\bar{k}_\theta = 0.1$ and m = 1–21. [Adapted from Guo, W. D., *Int J Numer and Anal Meth in Geomech* **40**(14), 2016.]

The normalized displacement, $\bar{w}_g$ increases with the normalized sliding depth (see Figure 5.3) at the normalized stiffness $\bar{k}_\theta = 0 - 0.4$ and modulus ratio m = 10; as well as at $\bar{k}_\theta = 0.1$ and m = 1–21. It is worth noting that the 3-layer reduces to 2-layer at m = 1. Moreover, the normalized rotational displacement $\bar{\omega}_r \left(= \omega_r l\right)$ increases (see Figure 5.4) and may exceed unity for dragging. On the other hand, it reduces (until vanishes eventually) with an increase in $\bar{k}_\theta$

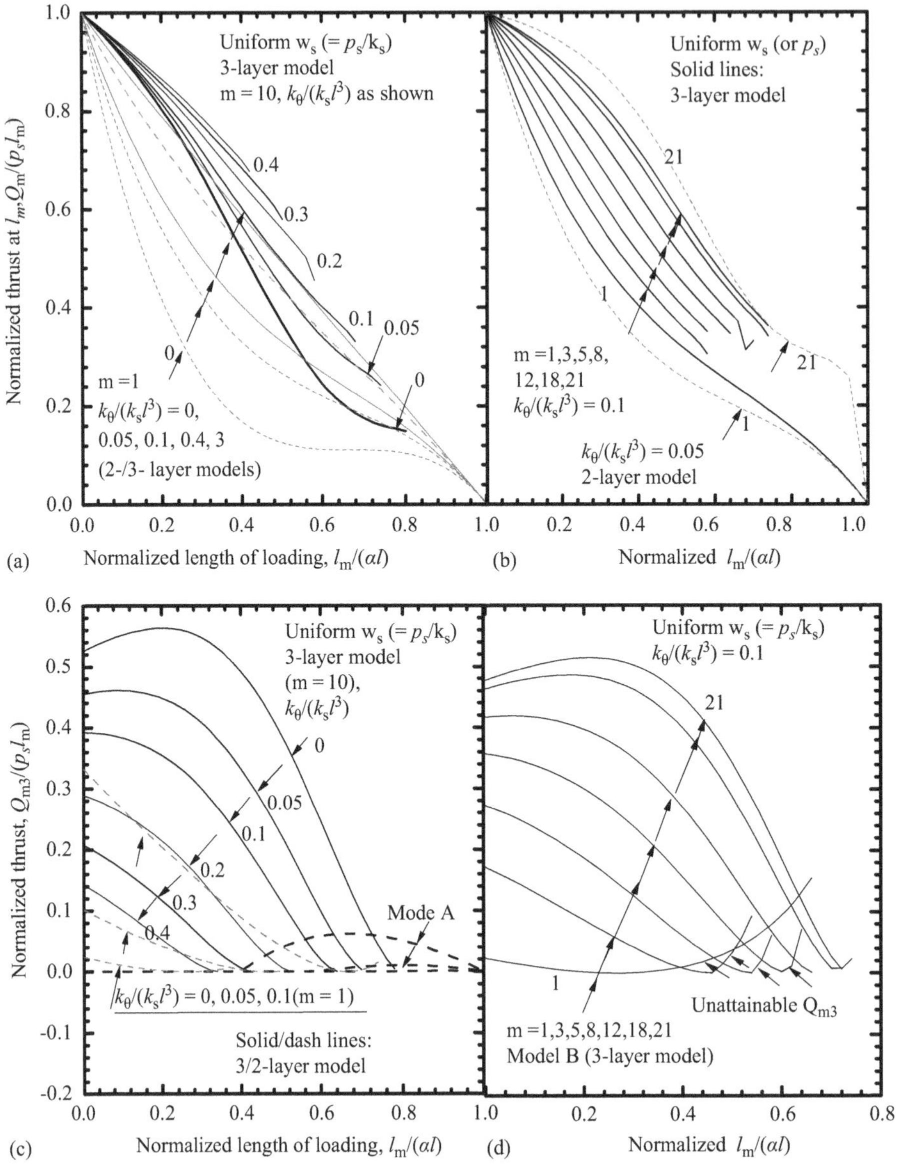

Figure 5.5 Normalized thrust Q_m at sliding depth at (a) normalized rotational stiffness (m = 10); (b) modulus ratio $m\left(\bar{k}_\theta = 0.1/0.05\right)$; (c) normalized rotational stiffness (m = 10); (d) modulus ratio $m\left(\bar{k}_\theta = 0.1\right)$. [Adapted from Guo, W. D., *Int J Numer and Anal Meth in Geomech* **40**(14), 2016.]

(e.g., $\bar{k}_\theta \geq 0.2$ at m = 1), or m (e.g., at $\bar{k}_\theta = 0.1$ and m = 21). Additionally, at a larger normalized displacement, the 3-layer model predicts a higher normalized rotation than the 2-layer model does.

The maximum bending moment M_m is affected by the location of rotational restraint, such as at the pile-head, base, or sliding depth. For a base restrained pile, the normalized moment, $M_m/(p_s l_m l)$, doubles at the sliding depth l_m (i.e., $\bar{k}_T = \bar{k}_\theta$) (see Figure 5.4) as the modulus ratio

m increases from 1 to 21 due to dragging, while it only varies by ~5% at the depth z_m (bottom of the transition layer). The normalized M_m at z_m, however, increases with the base rotational restraining stiffness to a maximum of 0.5 (see Figure 5.4c). At $\bar{w}_g > 1$ (in Figure 5.4a), two peaks of shear force Q_{m1} (sliding layer) and Q_{m3} (stable layer) occur. Otherwise (with $\bar{w}_g \leq 1$), only one peak force exists. The normalized thrust $\bar{Q}_m$ in the stable or sliding layer with dragging (at $m > 1$) differs between the 2-layer (i.e., $\bar{Q}_{m1}$ and $\bar{Q}_{m2}$) and the 3-layer models (i.e., $\bar{Q}_{m1} = \bar{Q}_{m2}$, and $\bar{Q}_{m3}$ for stable layer). The peak $\bar{Q}_{m3}$ and the stable layer vanish at $\bar{k}_\theta > 0.1$. When restraining pile-head only (i.e., $\bar{k}_G = \bar{k}_\theta$), the impact of dragging on the moment is illustrated in later example predictions.

5.3.2 $M_m/Q_m l$ ratios from rotation restraint and measured data

Figure 4.12a (Chapter 4) shows the measured $M_m/Q_m l$ ratios of test piles at various l_m/l ratios (Guo and Qin 2010) are grossly well predicted using the 2-layer model without normalized cap-rotational stiffness $\bar{k}_\theta$. The 3-layer model is adopted to obtain the $M_m/Q_m l$ ratios for m = 7.67 (design state), or 17.7 (ultimate state) and $\bar{k}_\theta$ [i.e., $k_\theta/(k_s l^3)$] of 0, 0.1, 0.2 and 0.3. The predictions are plotted in Figure 4.12b together with the measured data. The figure indicates that the centrifuge test piles (denoted as 'Leung et al') (Leung et al. 2000) were rotationally restrained (with $\bar{k}_\theta$ =~0.2) at a high normalized 'sliding' depth, due to the presence of a front retaining wall located 3 m away of the excavation. A restraining effect (with $\bar{k}_\theta = 0.1-0.3$) is also suggested for the Kamimoku-4 and Kamimoku-6 piles, although this needs to be confirmed.

5.4 MODEL TESTS OF PILES IN MOVING SAND

Model tests (Guo and Qin 2010) were conducted using the shear apparatus shown in Chapter 1. The tests are denoted by a second letter S or D for a sliding depth of 200 mm or 400 mm. Typical tests provide a standard k_s for Equation (1.2), Chapter 1, and insights elaborated next.

5.4.1 Five T-block tests on single piles

The T32-0 series (using the T-block) consists of five tests on free-head d_{32} piles, which were conducted to a maximum sliding depth of 125, 200, 250, 300 and 350 mm. These tests provide:

- The ultimate on-pile pressure profiles with depth (see Figure 5.6a);
- The maximum bending moment M_m, rotational angle ω_r and displacement w_g at GL (Figures 5.6b and c) respectively;
- The moment M_m versus shear force Q_{mi} induced in the piles (in stable and sliding layers) (Figure 5.6d);
- The evolutions of the moment M_m and shear force Q_{mi} (Figures 5.7a–d) with the frame (soil) movement w_f or the sliding depth ratio R_L (= l_m/l).

Note that the initial movement w_i (of 40–50 mm) causes negligible pile response. The M_m, Q_m, w_g and the ultimate $p(z)$ were ascertained from profiles of bending moment $M(z)$, shear force $Q(z)$, displacement $w(z)$ and on-pile *FPUL* $p(z)$, which are exemplified in Figure 5.8 for the test TS32-0 at a maximum l_m of 200 mm.

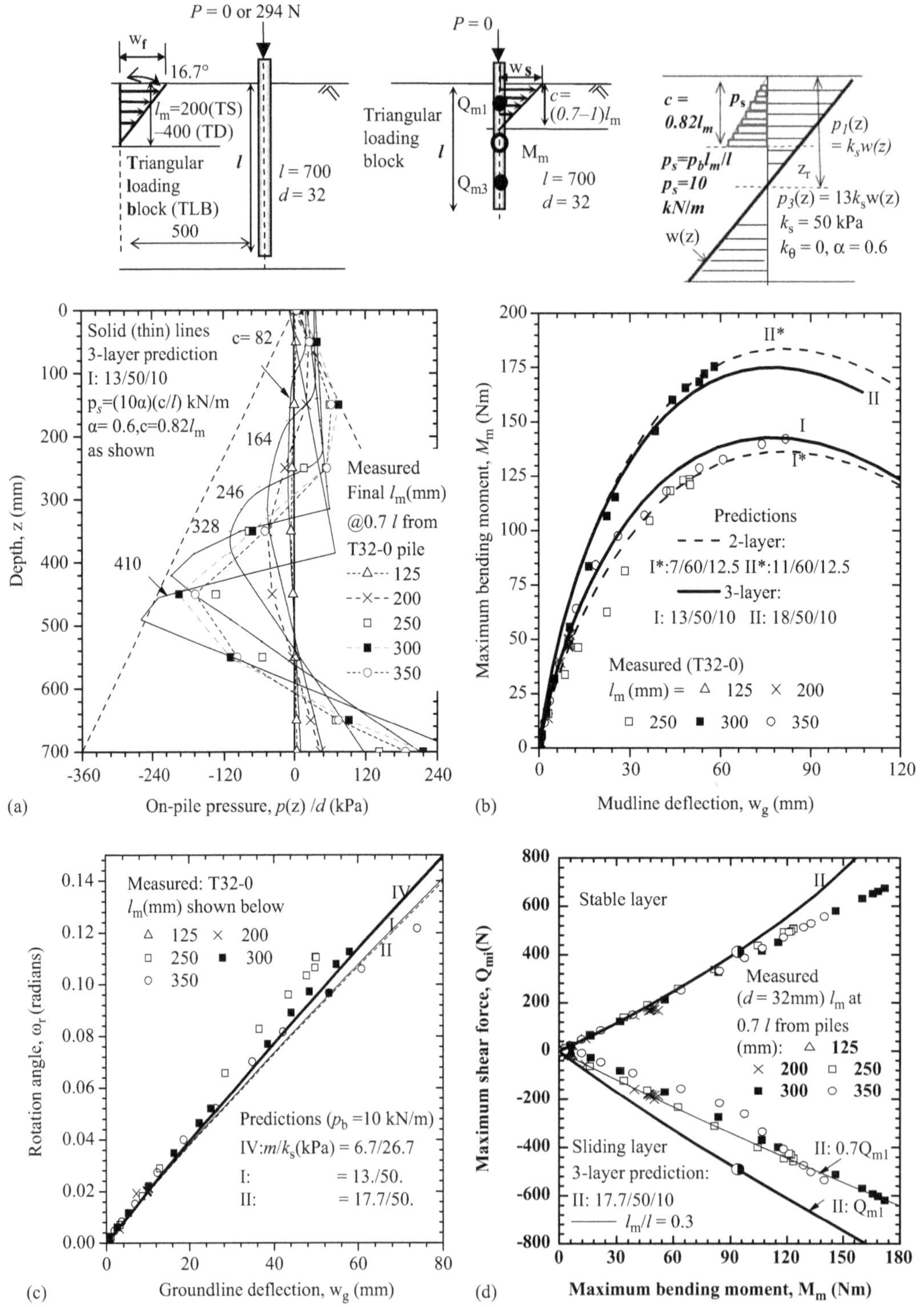

Figure 5.6 Predicted versus measured (a) on-pile pressure (Guo 2012); (b) $M_m \sim$ mudline displacement w_g curves; (c), (d) response of u_g–ω and M_m–Q_{mi}; respectively (Guo and Qin 2010). [Adapted from Guo, W. D., *Int J Numer and Anal Meth in Geomech* **40**(14), 2016.]

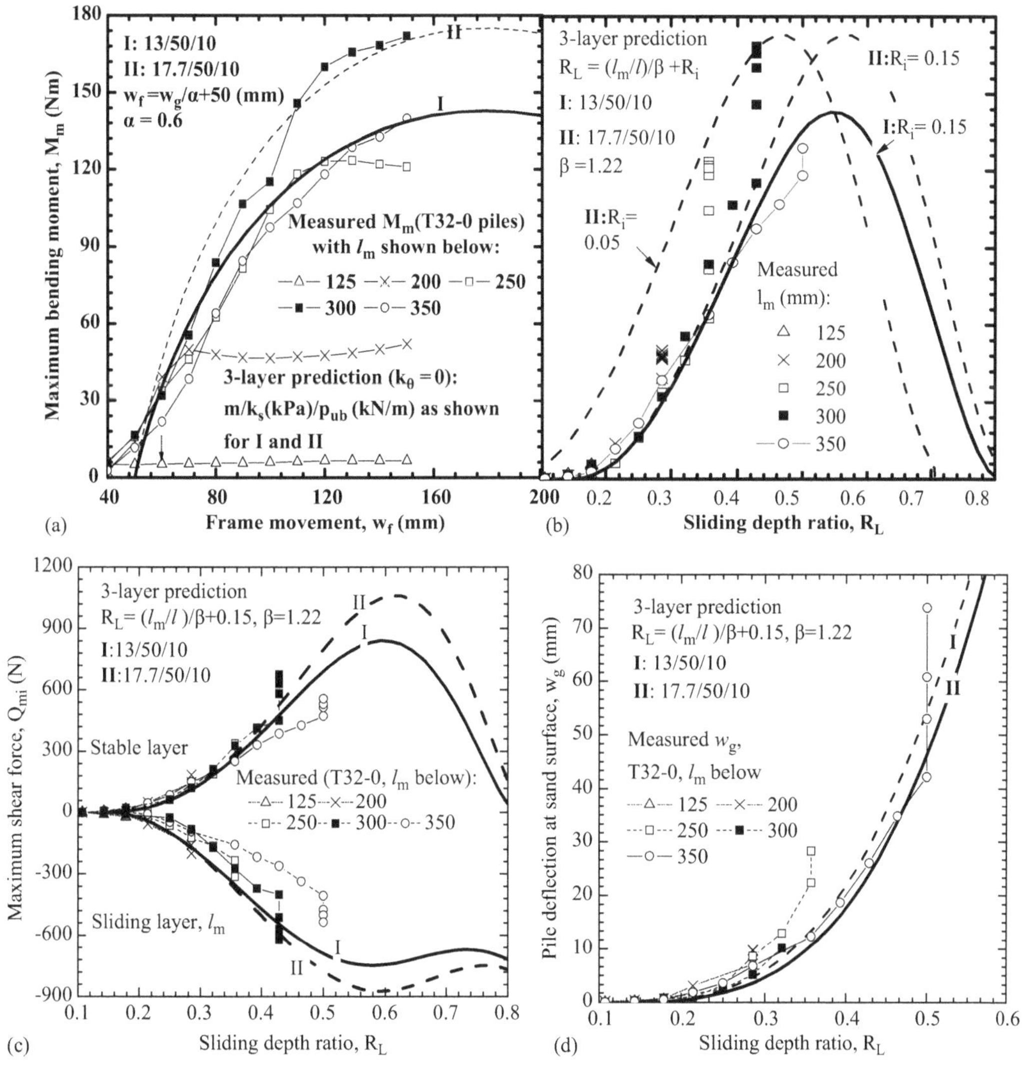

Figure 5.7 Predicted versus measured (a) bending moment M_m (proportional to shear force Q_{mi}) with soil movement w_f; (b) bending moment M_m; (c) shear force Q_{mi}; (d) pile displacement w_g with sliding depth ratio (l_m/l). [Adapted from Guo, W. D., *Int J Numer and Anal Meth in Geomech* **40**(14), 2016.]

5.4.2 Two U-block tests on 2-pile in-line groups

Tests US32-0 and US32-294 on capped-head 2-piles (d_{32}) in line, with a centre-to-centre spacing of $3d$, were conducted to a sliding depth of 200 mm using the uniform loading block, with $P = 0$ (no load), or $P = 294$ N per pile, respectively. The recorded response of pile A (close to loading block) includes:

- The maximum moment M_{mi} (i.e., in-pile M_{m2} and M_o (= M_{m1}) at GL) versus pile displacement w_g (see Figure 5.9a);
- The ultimate-state profiles of the on-pile *FPUL*, $p(z)$ (Figures 5.9b and 10e), pile deflection $w(z)$ (Figures 5.10c$_1$ and c$_2$), bending moment $M(z)$, and shear force $Q(z)$ (Figures 5.9c and d), respectively;

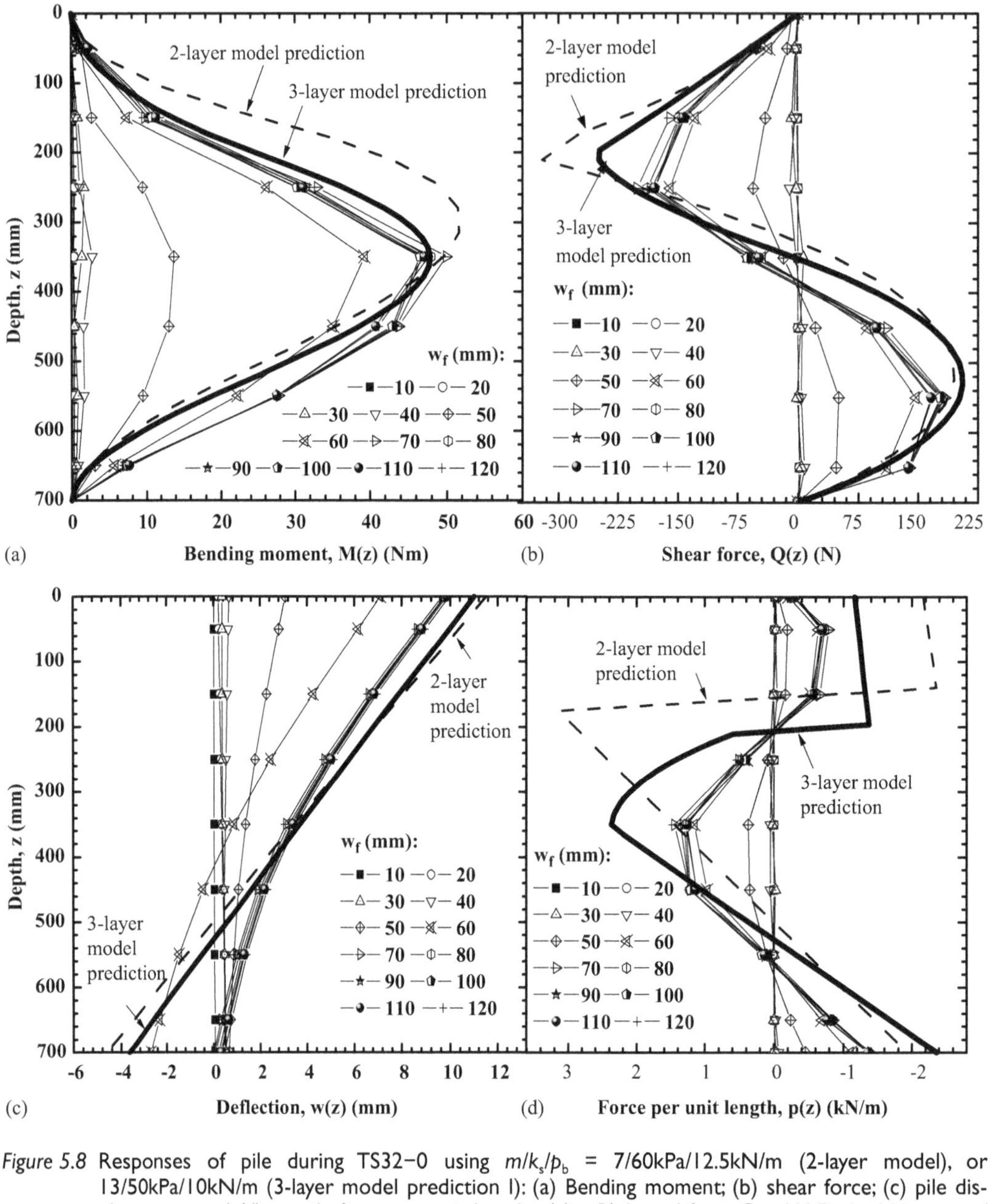

Figure 5.8 Responses of pile during TS32–0 using $m/k_s/p_b$ = 7/60kPa/12.5kN/m (2-layer model), or 13/50kPa/10kN/m (3-layer model prediction I): (a) Bending moment; (b) shear force; (c) pile displacement; and (d) on-pile force per unit length $p(z)$. [Adapted from Guo, W. D., *Int J Numer and Anal Meth in Geomech* **40**(14), 2016.]

- The variations of M_{mi} and shear force Q_{mi} with the frame movement w_f (Figures 5.10a, and d);
- The M_{mi}–w_g curves, and the $p(z)$ profiles for Pile A and Pile B (see the inset in Figure 5.9), as an example, in Figures 5.10b and e, respectively.

A linear *FPUL* p_s with the sliding depth (see the dash lines in Figure 5.6a) is observed for free-head piles up to a sufficiently high rotation, which is not evident along capped piles (see Figure 5.9b, Guo and Ghee 2005b). These tests provide valuable insight into the behaviour of capped-head piles under passive soil movement.

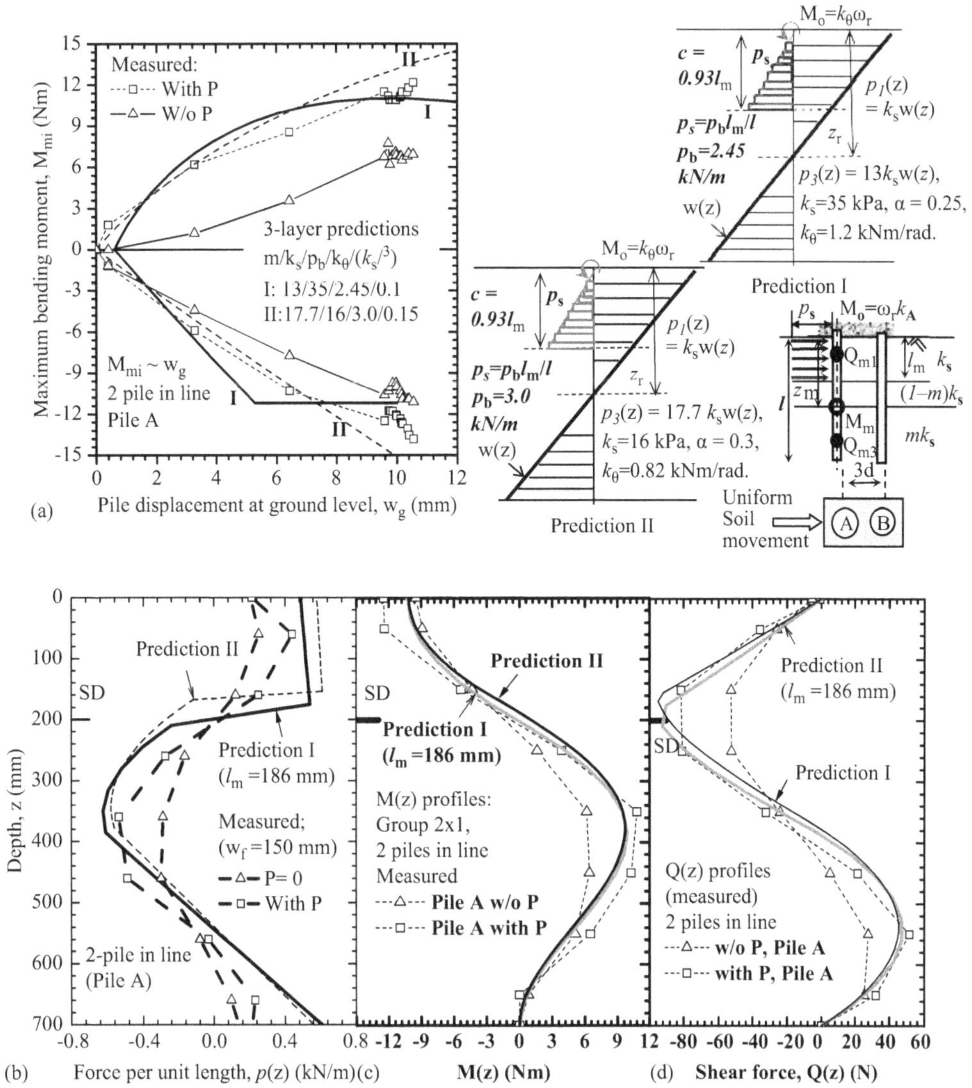

Figure 5.9 Predicted (3-layer model) versus measured (Guo and Ghee 2005) response of Pile A in 2-pile in-line group under a uniform soil movement: (a) Development of M_{mi} and w_{gi}; (b) on-pile force per unit length; (c) bending moment and (d) shear force profiles at ultimate state. [Adapted from Guo, W. D., *Int J Numer and Anal Meth in Geomech* **40**(14), 2016.]

5.4.3 Modelling free-head and capped piles

The 3-layer model was used to simulate the measured response of the single piles, and the capped 2-piles. The parameters are determined as follows:

- m = 17.7 (= K_{p1}/K_{a2}, for ϕ = 38°), and m = 13, reduced by 30–40% from 17.7 (G5) for limited mobilization of the resistance and no axial load P = 0, and/or without a dragging layer.
- k_s = 50 kPa using Equation (1.2), Chapter 1 for m = 17.7, d = 0.032 m, l_{2c}/l_m = 1 (rigid piles), p_{u2}/p_{u1} = 1 (i.e., identical sand over pile embedment), and λ_k = 1.73; k_s = 45–58 kPa gained using $k_s = 4.64p_{ub}$ (Guo 2014), the l/d ratio, and p_{ub} = 9.6–12.5 kN/m [= (300–390)d, see Figure 5.6a]. Additionally, the k_s is twice the secant value of 25–35 kPa deduced from bending moment (Guo and Qin 2010); and $k_s = 50[1 + (m - 1)(z - l_m)/(z_m - l_m)]$ (transition layer) and 885 kPa (= mk_s, stable layer).

- p_{ub} = 10.0–11.6 kN/m, given d = 0.032 m, l_m = 0.7 m, γ_s' = 16.5 kN/m³, and ϕ = 38° (sand layer); as well as s_g = 1.53–1.77 (to cater for jack-in installation of the piles, Guo 2012).

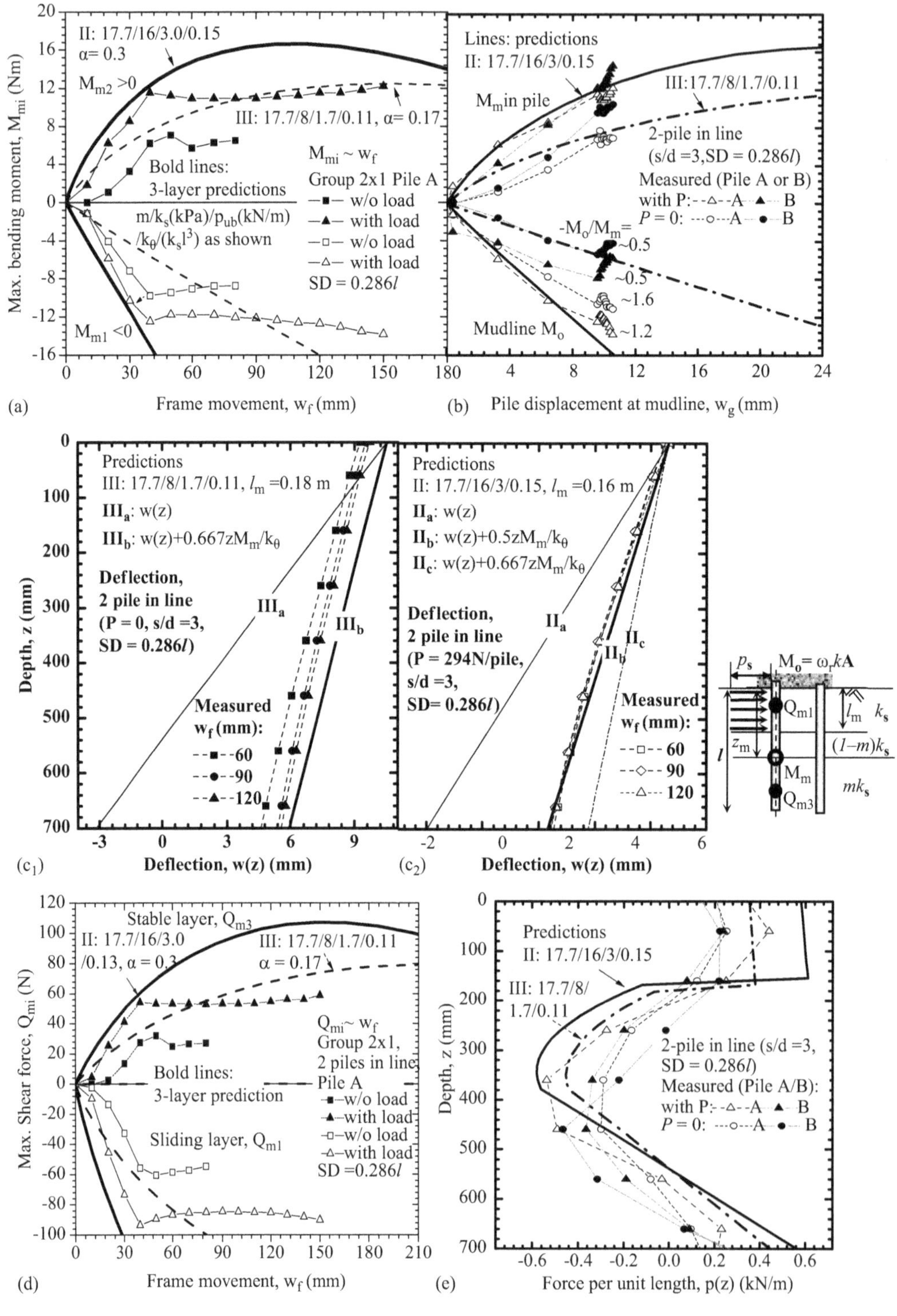

Figure 5.10 Predicted versus measured response of (a) M_{mi}–w_f; (b) M_{mi}–w_g; (c_1, c_2) $w(z)$; (d) Q_{mi}– w_f; (e) $p(z)$ (2-pile in-line). [Adapted from Guo, W. D., *Int J Numer and Anal Meth in Geomech* **40**(14), 2016.]

5.4.3.1 Free-head d_{32} piles (T tests, α = 3/5 for P = 0)

The free-head pile tests involve $k_G = k_T = k_\theta = 0$, and $H = 0$ (no 'lumped' constraints other than the sand resistance). Two sets of 3-layer predictions were made, using m/k_s(kPa)/p_b(kN/m) of 13/50/10 (Prediction I) or 17.7/50/10 (Prediction II), with $p_{ub} = p_b$ at the pile location. They are plotted respectively in Figures 5.6b and c for w_g–M_m, and w_g–ω_r (rotation) relationships (I and II); and in Figure 5.6d for the evolution of M_m and shear force Q_{m2} (stable layer) (II). The predictions generally agree with the measured data but overestimate the maximum shear force Q_{m1} (sliding layer) by 30%. As a comparison, the 2-layer prediction, using m/k_s(kPa)/p_b(kN/m) of 7–11/60/12.5 (Guo 2015b), also agrees well with the measured displacement w_g and moment M_m relationships (see Figure 5.6b).

Example 5.1 On-pile pressure profiles

The profiles of on-pile pressure $p(z)/d$ were predicted at a loading depth c of $0.82l_m$ on pile surface using p_s (= $6c/l$, α = 0.6). They are plotted in Figure 5.6a for c values of 82, 164, 246, 328 and 410 mm, respectively concerning l_m (at loading source) of 0.1, 0.2, 0.3, 0.4, and 0.5 m. Given $k_\theta = 0$, $H = 0$, $m = 13$, and p_b = 1.74 kN/m (= 0.6 × 10 × 0.286), taking $c = l_m$ = 0.2 m (i.e., l_m/l = 0.286) as an example, Equations (5.5) through (5.8) are resolved together to yield z_m = 0.352 m (which involves k_θ rather than p_s), w_g = 0.011 m [= 0.55/k_s (m)], and ω_r = -0.021 (= –1.046/k_s). The same calculations were conducted for all other c values.

Example 5.2 M_m and Q_{mi} with movement w_f or R_L

The evolution of the 'ultimate' M_m and Q_{mi} with the frame movement is closely tracked by Predictions I and II (see Figure 5.7a) using α = 0.6 (= w_g/w_f). The response of M_m, Q_{mi} and w_g for all tests with the sliding depth ratio R_L (= l_m/l) in Figures 5.7b–d, respectively, is also well replicated by β = 1.22 (with $c = 0.82l_m$), ignoring the negligible impact of the initial soil movement w_i (= 50 mm) and sliding depth ratio R_i (= 0.05–0.15). In particular, Q_{m2} (stable layer) is well predicted despite the overestimation of measured Q_{m1} (sliding layer) (see Figure 5.7c). The 3-layer predictions (with 13/50/10) indeed compare more favourably than the 2-layer model [with m/k_s(kPa)/p_b(kN/m) = 7/60/12.5] with the measured profiles, as shown in Figure 5.8 for a typical test TS32-0 at a sliding depth l_m of 200 mm.

Example 5.3 Stiffness $\bar{k}_\theta$ versus z_m

Previously Figures 5.5c and d presented the impact of normalized rotational stiffness $\bar{k}_\theta$ on the normalized shear force $Q_{m3}/(p_s l_m)$ in the stable layer. The effect is elaborated for pile TS32-0 (P = 0) at a sliding ratio l_m/l of 0.1 (see Table 5.3). The increase in $\bar{k}_\theta$ from 0 to 0.952 results in (i) a shift of the depth z_m towards the pile tip; (ii) a reduction in the M_m at z_m to zero (from 10.9 Nm); (iii) an increase in the ground-line bending moment M_m (M_o) from zero to –22.0 Nm; and (iv) a subtle rise in the Q_{m1} [= –(65.2–66.8) N] and the Q_{m2}, and a reduction in the Q_{m3} from 40.01 N to zero. At the $\bar{k}_\theta$ (= 0.952) and l_m/l (= 0.1), the lower layer effect fades, as M_m shits towards the pile-cap level ($z_m \to l$, or $\to$ 0), and Q_{m3} reduces to 0. The reduction in Q_{m3} is consistent with Figures 5.5c for m = 10. For instance, 'Q_{m3} = 0' occurs at $\bar{k}_\theta \approx 0.3$ at l_m/l = 0.4. Likewise, the same pile test with a vertical load P of 294 N (TS32-294) was simulated using the same parameters but for α = 0.9 and m = 18 (see Table 5.2). The prediction agrees well with the measured profiles (not shown herein).

5.4.3.2 Capped piles with limited p_b (α = 0.25–0.3)

The 3-layer solutions were employed to simulate the capped 2-piles (US32-0) in-line (see insert of Figure 5.9) under a uniform loading block. Characterized by $k_G = k_\theta$, $k_T = 0$, and H = 0, the predictions I and II adopted m/k_s (kPa)/p_b (kN/m) / $\bar{k}_\theta$ (see Table 5.2) of 13/35/2.45/0.1 (design-state), and 17.7/16/3.0/0.15 (ultimate state), respectively. The limiting p_b for the translating piles was estimated as 2.5–3.0 kN/m (= αp_{ub}, $\alpha \approx 0.25$–0.3, and p_{ub} = 10 kN/m). For $s/d = 3$, the interaction factor p_m was estimated as 0.83 (Pile A) and 0.35 (Pile B), which were reduced to 0.7 and 0.32 due to the loading distance of 0.7l.

Example 5.4 Design state

For design state, the 'prediction I' adopts m = 13, k_s = 35 kPa (= 50p_m, p_m = 0.7), and k_θ = 1.2 kNm/rad (= 0.1 × 35 × 0.7³). Figure 5.9a shows the predicted maximum moments in the pile and at the ground-line plotted against the displacement w_g. At l_m of 0.2m, the loading depth c was 0.186 m (= 0.93 × 0.2 m, β = 1.08), and p_s was 0.652 kN/m. The predicted $p(z)$, $M(z)$ and $Q(z)$ profiles for the l_m are plotted in Figures 5.9b through d. All the predictions compare well with the measured response, respectively. Notably, due to the limited mobilization of p_s, the maximum moments M_m peaked at 11.5 Nm that was taken as the M_o at large sand movement w_f (= ~150 mm). This is confirmed by the small pile-soil relative defection w_g of ~7 mm (see Figure 5.10c$_2$).

Table 5.2 Parameters for 3-layer model predictions of typical piles (Guo 2021b, 2022b)

Cases	*c/l$_m$*[a]	*m/p$_{ub}$ (kN/m)*[a]	*k$_s$ (kPa)/d (cm)*[a,b]	*$\bar{k}_\theta$*[a]	*α*	*ϕ*	*Reference*
TS32-294	0.82	18/12.5	50/3.2	0	0.9	38	Guo and Qin (2010)
TS32-0	0.82	13/12.5	50/3.2	0	0.6	38	
US32-0	0.93	13–17.7/10	35.0/3.2	0.1	0.25–0.3	38	Guo and Ghee (2005)
Test 9	0.77	17.7/46.4	20/43	0.007	0.5	38	Stewart et al. (1994)
Test 11	0.83	17.7/85	20/43	0.1	0.72	38	
BS test	0.97	13.6/610	881/127	0.15	0.72	35	Bransby and Springman (1996)
Kobe 1 × 6	1.0	5.2/471.1	26/72	0.0045	0.9	30	Armstrong et al. (2014)
Kobe 2 × 4	1.0	9/730.2	79.5/122	0.019	1.3	30	
SINE 2 × 4	1.0	9/730.2	70.5/122	0.0016	1.0	30	

Note:
[a] In the second through to fourth columns, the values of c/l_m, m, p_{ub}, k_s, $\bar{k}_\theta$ and α along with l, l_m, d are all input values.
[b] The reduced value of p_s is owing to limited depth of rotation, but largely translation of the pile.

Table 5.3 Impact of k_θ on d_{32} pile in sliding soil at l_m/l = 0.1[a] (Guo 2016)

$\bar{k}_\theta$	$\bar{z}_m / \bar{z}_{mt3}$	$\bar{w}_g$	$\bar{\omega}_r$	w_g *(mm)*	M_m *(Nm)*	ω_r *(rad)*	Q_{mi} *(N)*[c]
0	0.417/0.709	0.073	−0.148	1.8	10.9/0[b]	0.0037	−65.22/−65.81/40.01
0.2	0.513/0.756	0.063	−0.119	1.26	5.11/−8.16	0.0024	−65.88/−66.55/22.49
0.4	0.618/0.809	0.056	−0.099	1.12	2.06/−13.62	0.0020	−66.31/−67.07/11.55
0.6	0.736/0.868	0.052	−0.085	1.03	0.58/−17.50	0.0017	−66.59/−67.46/4.73
0.8	0.874/0.937	0.049	−0.074	0.97	0.06/−20.34	0.0015	−66.78/−67.78/0.94
0.952	1.0/1.02	0.047	−0.067	0.94	0/−22.00	0.0014	−66.86/−68.01/0.0

[a] m = 13, H = 0.
[b] $k_G = k_\theta$, value before and after sign '/' for M_m locates at depth z_m and cap-level, respectively.
[c] The first, the middle and the last values are for the sliding, transition, and stable layer, respectively.

Example 5.5 Ultimate state

For ultimate state, the 'prediction II' adopts m = 17.7, k_s = 16 kPa (= 50p_m, p_m = 0.32 ≈ α = 0.3), p_b = 3 kN/m, and $\bar{k}_\theta = 0.15$. The prediction also well captures the evolution of the M_{mi} (i.e., M_m and M_o) with w_g, the ultimate $p(z)$, $M(z)$, and $Q(z)$ in Figure 5.9; the evolution of the M_{mi} with head-displacement w_g (see Figure 5.10b); the ultimate $w(z)$ and the on-pile *FPUL* (see Figures 5.10c$_2$ and e), respectively. The prediction II offers a better agreement with the measured M_o and w_g curve, and a higher limit of M_m (see Figure 5.10b) than the prediction I, which should be capped as ~12.0 Nm as well.

Example 5.6 Pile B and k_θ

The response of Pile B (at P = 0) was captured using 'Prediction III' and 17.7/8/1.7/0.11 (see Figures 5.10b and e). To account for shadowing effect, the parameters for pile A were multiplied by a factor of 0.5–0.7 to result in k_s = 8 kPa (= 16p_m, p_m = 0.5–0.7), p_b = 1.7 (= 2.5p_m, p_m = 0.7), and $\bar{k}_\theta \left(= \bar{k}_A = 0.11\right)$. The predictions agree with the measured data in Figures 5.10b and e, respectively. Pile A had a head-rotation angle of 0.0123 radians and a moment M_o of 10.14 kNm at a stiffness k_θ of 0.823 kNm (via $\bar{k}_\theta = \bar{k}_A = 0.15$, and k_s = 16 kPa). On the other hand, Pile B head rotated by 0.0195 radians and a predicted M_o of 5.88 kNm for k_θ = 0.302 kNm (via $\bar{k}_\theta = \bar{k}_A = 0.11$, and k_s = 8 kPa). Despite having a rigid cap, Pile A and B experienced different rotation angles and discrepancies in their pile-cap connections. To account for the impact of cap-rotation on the increasing displacement with depth, a factor λ_d of 0.667 and 0.5 (see Figures 5.10c$_1$ and c$_2$) were adopted for tests under P = 0 and P = 294 N/pile. The factors are valid against the measured data.

5.5 PILES SUBJECTED TO LARGE EMBANKMENT MOVEMENT

Centrifuge tests were conducted on pile groups adjacent to embankments overlying a thick-clay layer (Test 9) or a clay-sand layer (Test 11 and BS). The measured response of the pile groups was studied next using the 3-layer model and input parameters tabulated in Table 5.2 (Guo 2016, 2021b).

5.5.1 UWA tests 9 and 11

Stewart et al. (1994) conducted Tests 9 and 11, where two rows of seven piles were held in a rigid cap 2 m (all at prototype scale) above the soil surface that deflects freely. The piles were 22.5 m in length and 0.43 m in diameter, and penetrated a soft clay layer before partially entering a dense sand stratum. Four piles in each group were instrumented to measure the pile response to nearby on-going construction of a sand embankment, which was built in six stages and reached to a height of ~8.5 m. The results of Tests 9 and 11 are simulated herein.

In Test 9, the clay layer was 18-m thick and s_u of 17 kPa. For increasing embankment loading (q), the measured pile-cap displacement w_g (front/back row) and maximum moment M_m (both rows) were measured. The q–w_g and q–M_m curves are plotted in Figures 5.11a and b, and the w_g–M_m (both front- and back-rows) in Figure 5.11c. The measured bending moment profiles with depth at the 'ultimate' state are depicted in Figure 5.11d for front- and back-row piles. The piles are simulated using p_{ub} = 51 kN/m [= $4s_u dl/(0.82l_m)$, with s_u = 17 kPa, l_m = 18 m, l = 22.0 m, d = 0.43 m], m = 17.7 (= K_p/K_a) at ultimate state, $k_\theta = k_G$ = 1.12 MN·m/radian (with $\bar{k}_\theta = 0.007$, at GL), and α = 0.5 (deep sliding). Given ϕ' = 23°, γ_s' = 16.5 kN/m^3, l_m = 11 m (= 0.5l), and q = 100 kPa, it follows an average p_s of 34.2 kN/m (= 51α), and $p_{ub}/(\sigma_v' d)$ = 0.72.

The $p_s/(dq)$ ratio for over-consolidated clay is 0.72, in between the short-term (0.75–0.792) and long-term (0.277–0.35) for embankment loading that were deduced from measured data (Jeong et al. 1995). The $p_s/(\sigma_v' dK_p)$ value is 0.286. The predicted q–w_g, q–M_m, and w_g–M_m

curves are shown in Figure 5.11, based on k_s = 15–20 kPa, and m = 12.3 (2-layer) or 17.7 (3-layer) (Guo 2016). The predicted 'limiting' bending moment profile are shown in Figure 5.11d for a loading depth c of 14.7 m (= $0.81l_m$, l_m = 18 m). The predictions are insensitive to the modulus k_s, and indicate that 50% (α = 0.5) of the surcharge loading q is transferred onto the on-pile pressure (p_s/d) during deep sliding. Alternatively, the k_s may be calculated as 18.1 kPa {= $G_{sec}/[2(1 + 0.5)]$, Poisson's ratio = 0.5} using G_{sec} = 54.4 kPa = $60s_u[1-0.985(c/l_m)^{0.2}]$ for c/l_m = 0.82 and s_u = 17 kPa.

In Test 11, the thickness of the clay layer was reduced to 8-m with s_u = 11 kPa. The piles were modelled using p_{ub} = 121 kN/m (= $3.1s_udl_m$, with s_u = 11 kPa, l_m = 8 m, l = 22.0 m and d = 0.43 m), k_s = 20 kPa, m = 17.7, k_θ = 26.6 MN·m/radian ($\bar{k}_\theta = \bar{k}_G$ = 0.15 for the pile-cap), and α = 0.72 (an inverse triangular movement profile). The average p_s was 86 kN/m (= 121α), with

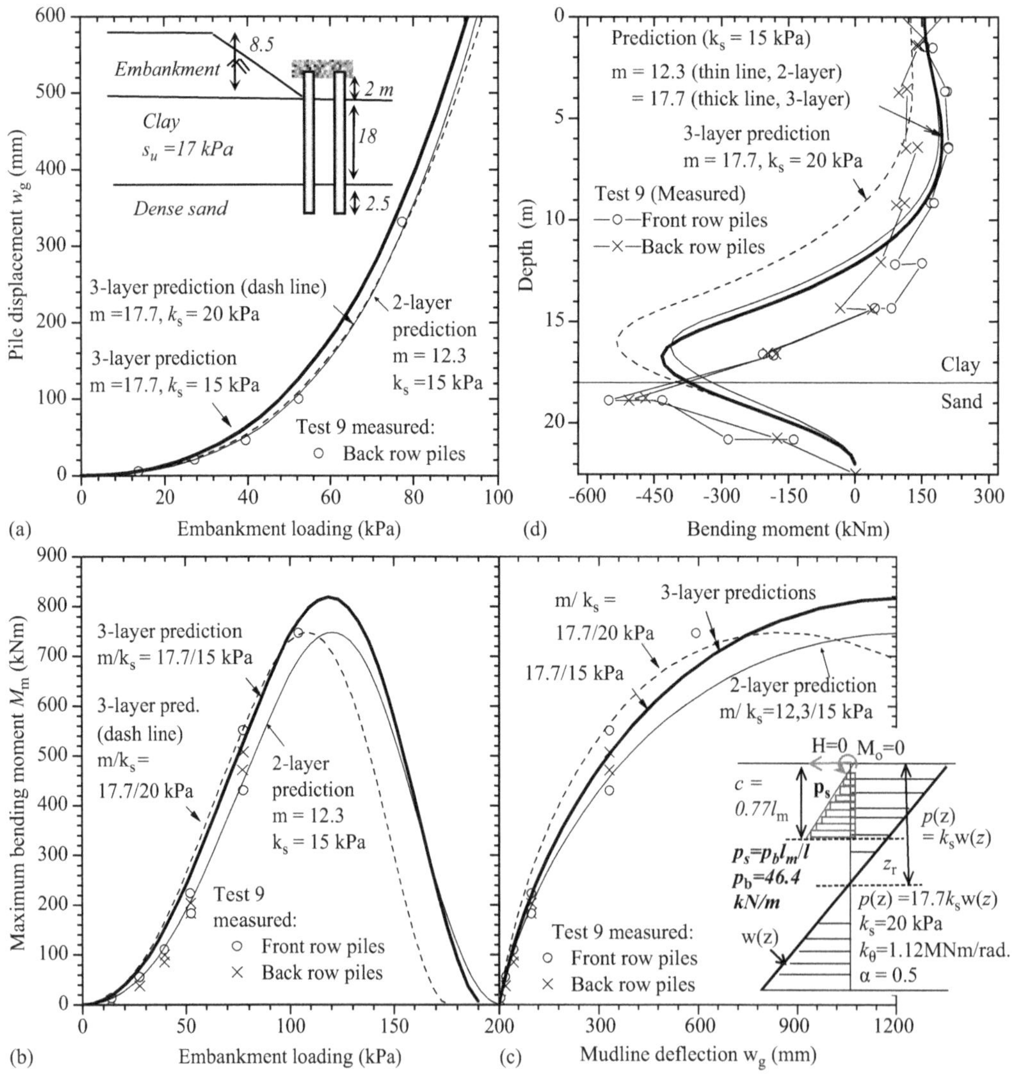

Figure 5.11 Predicted versus centrifuge Test 9 (Stewart et al. 1994): (a) embankment load (q) versus maximum bending moment (M_m); (b) q versus pile-head deflection (w_g); (c) w_g-M_m; (d) bending moment profiles. [Adapted from Guo, W. D., *Proc. NZGS Symposium, Dunedin, New Zealand*, 2021b.]

$p_{ub}/(d\sigma_v') = 1.5–2.1$ and $p_{ub}/(\sigma_v' d K_p) = 0.68–0.93$ (with $\phi' = 23°$, $\gamma_s' = 16.5$ kN/m^3, and l_m = 8–11 m). The normalized rotational stiffness $\underline{k}_\theta$ was raised to 0.15 to account for the impact of the underlying sand layer on the top clay layer. These parameters offered good prediction of the $q–w_g$, $q–M_m$, and $w_g–M_m$ curves (see Figure 5.12). For comparison, the 2-layer predictions were also made using the same parameters (except for a reduced m of 12.3), which also agree well with the measured data.

5.5.2 Test BS

Bransby and Springman (1996) conducted centrifuge tests (termed as Test BS) on a pile group consisting of two rows of infinitely long piles. Each pile (1.27 m in diameter, 19 m in length with E_p = 40 GPa, a Poisson's ratio of 0.33) was embedded in a 6 m-thick clay layer (s_u = 42.5 kPa)

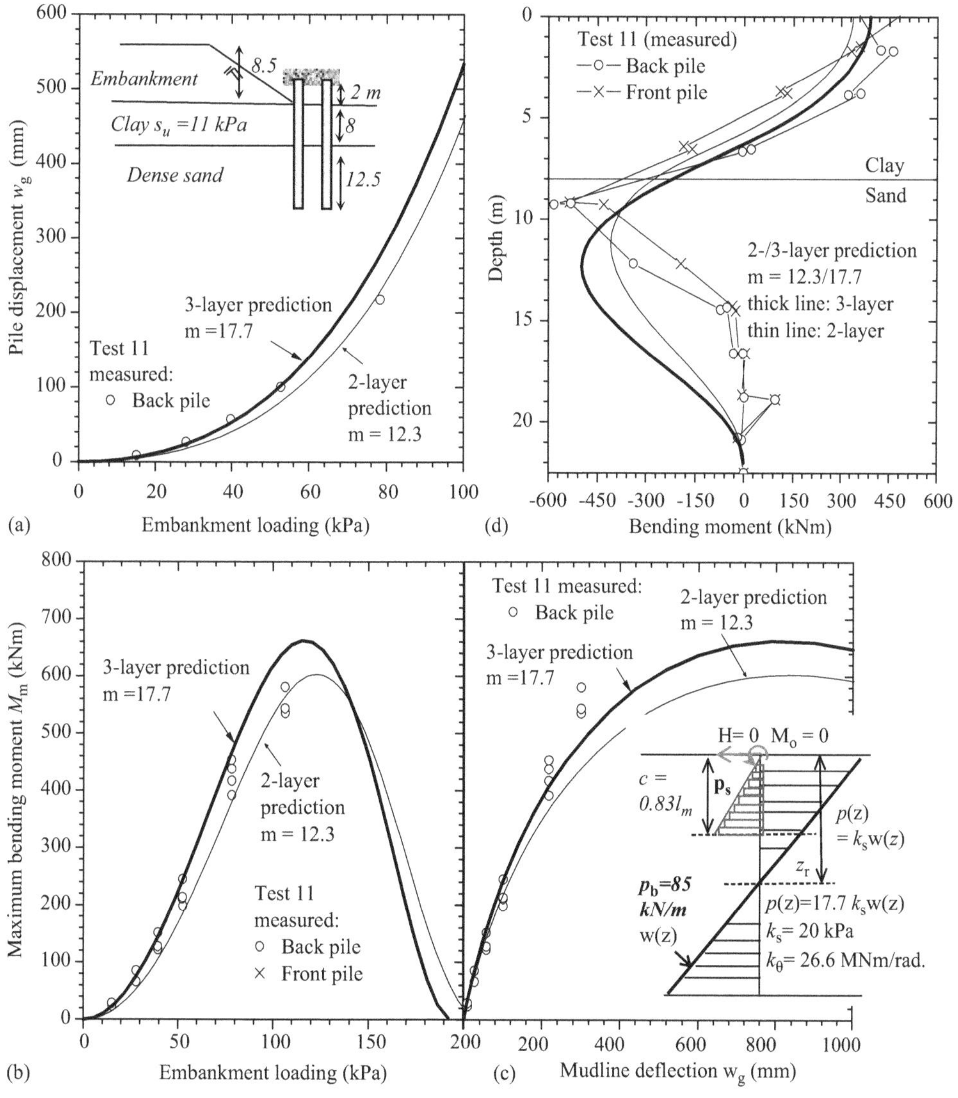

Figure 5.12 Predicted versus centrifuge Test 11 (Stewart et al. 1994): (a) embankment load (q) versus maximum bending moment (M_m); (b) q versus pile-head deflection (w_g); (c) $w_g–M_m$; (d) bending moment profiles.

overlying dense sand (ϕ = 35°). The piles were installed at a spacing of 6.67 m along the row and 5 m between the rows and were connected by a rigid pile-cap (9 m wide and 1 m thick). The shear modulus increased by 7.5 MPa/m downwards from 7.5 MPa at the clay-sand interface. The clay surface outside the pile cap was subjected to a uniform surcharge (sand cushion) of 17 kPa, while the top soil surface was subjected to a uniform surcharge pressure q applied to 1 m away from the nearest edge of the pile cap. Figures 5.13a and b show the displacement w_g, the on-pile lateral pressure p_s/d; the rotation angle of pile group, and the shear force mobilized along the interface between the pile-cap and the subsoil, as the pressure q increases. Figures 5.13c and d depict the profiles of bending moment $M(z)$ and on-pile lateral pressure (at q = 200 kPa), respectively. In addition, a 3-dimensional *FEM* analysis by Bransby and Springman (1996) is also presented in the figures.

5.5.2.1 3-layer model predictions

The 3-layer model was used to predict the behaviour of piles (l = 19.0 m, d = 1.27 m) in a sliding layer (l_m = 6.0 m) using m = 13.6, k_s = 881 kPa, p_b = 610 kN/m (p_{ub} = 845 kN/m, and α = 0.72 for an inverse triangular soil movement, Guo 2021b), and k_θ = 848 MNm/radian. These parameters were determined as follows: (i) $p_{ub} = 1.38 \times 11 s_u \times 1.27 l_m/c$ for $l_m/l < 0.5$, with s_u = 42.5 kPa; (ii) k_s = 51(1.27)(13.61/0.271)] kPa using Equation (1.2), Chapter 1 with K_{p1} = 3.69 and K_{a2} = 0.271 (ϕ = 35°); (iii) α = 0.72 for the clay layer, which is estimated as 0.4–1.0 [= 1.3 × (0.33–0.8) for 30 per cent increase from translation (α = 0.33) to rotation (α = 0.8) incurred by embankment surcharge]; and (iv) $k_\theta = 0.15 \times 881 \times 19^3$ using $\bar{k}_\theta = \bar{k}_G$ = 0.15 based on the current model piles.

The model predicted displacement w_g, rotation angle and on-pile pressure (p_s/d) for q of ~250 kPa to an accuracy of 95%, using the stipulated shear force on the pile-cap in Figure 5.13b. The predicted bending moment and on-pile pressure profiles at a loading depth c of 5.8 m ($0.97 l_m$) are plotted in Figures 5.13a and b, respectively, along with the measured data. The soaring bending moment at the sliding depth is owing to the cap shear force in the 3-layer model. The predictions (bold lines) generally concur with the measured data, but there is a discrepancy between the predicted and measured rotation angles, suggesting relative rotation between the pile and pile-cap of ~32% the pile rotation.

5.5.2.2 Justification of k_s Value

The modulus k_s may alternatively be estimated as $2.44 G_{s1}$ for l/d = 15 (Guo 2012) using the secant shear modulus G_{s1} where $G_{s1} = G_i[1-0.985(\text{SML})^{0.2}]$ (Stewart et al. 1994) with G_i = initial shear modulus, and SML = $(0.5/\alpha)c/l_m$, average stress mobilization level for the embankment. For an 'arc' profile and a linear profile of soil movement, the SML at c/l_m = 0.97 is estimated 0.97 (α = 0.5) and 0.674 (α = 0.72) (Guo 2021b). The G_i is 5.17 MPa for E_s = 15 MPa (Case I), or 7.5 MPa (at the clay and sand interface, Case II) for which the k_s is estimated as follows.

- Case I: 'G_i = 5.17 MPa' offers G_{s1} = 108.5 kPa {$= 5{,}170[1-0.985(0.97)^{0.2}]$} and k_s = 264 kPa at α = 0.5; and k_s = 1,134 kPa (with G_{s1} = 464.7 kPa) at α = 0.72. The average k_s for α = 0.5–0.72 is 699 kPa.
- Case II: 'G_i of 7.5 MPa' offers k_s of 384 kPa (α = 0.5)–1,644 kPa (α = 0.72), with an average of 1,014 kPa.

The average k_s of 699–1,014 kPa (Case I–II) is 856.7 kPa, which is only 3% less than 881 (kPa). It would lead to a similar prediction to what presented here.

Using the same parameters (but for m = 8.17), the 2-layer model prediction is presented in the figures. The model yields slightly smaller displacement and rotation, but a higher lateral

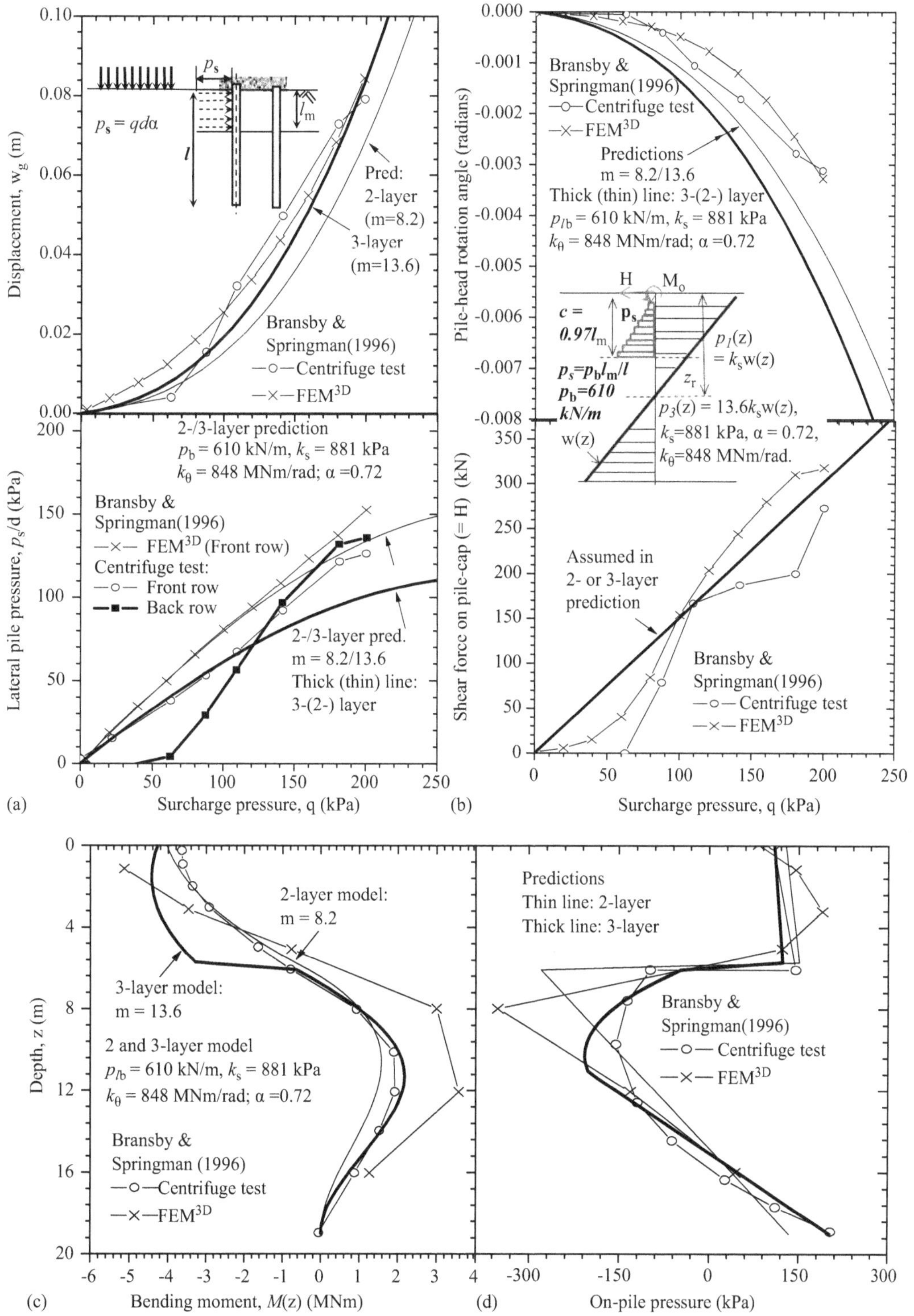

Figure 5.13 Predicted versus centrifuge test (Bransby and Springman 1996) of piles under typical surcharges: (a) pile-group displacement and (c) on-pile lateral pressure; (b) pile-rotation angle, and shear force mobilized along the pile cap, and under a surcharge of 200 kPa; (c) bending moment profiles; and (d) on-pile pressure profiles. [Adapted from Guo, W. D., *Int J Numer and Anal Meth in Geomech* **40**(14), 2016.]

on-pile pressure than the 3-layer model. The moment conforms to the measured data (without shear force), and the shear force profile is similar in shape to the 3-D *FEM* analysis. The inclusion of the transition layer allows for an excellent capture of the on-pile pressure (Figure 5.13d), which is not seen in the *FEM* analysis.

Considering the potential dragging on the piles, the 3-layer model is next adopted to analyse centrifuge model tests on pile groups subjected to 'lateral spreading' embankment loading (q).

5.6 PILES EMBEDDED IN LATERAL SPREADING EMBANKMENT

Armstrong et al. (2014) performed three centrifuge tests on an embankment-pile system subjected to lateral spreading. Each test comprised two identical approach embankments made of dry, dense Monterey sand (D_r = 100%), separated by a 12-m-wide channel. The embankment was 8 m to 11 m height between the crest and the model container wall with an average of 10 m, and had a crest width of 12 m and side slopes of 2:1 (horizontal to vertical) in all directions. The embankment was underlain by a 1.3-m-thick compacted, non-plastic silt layer, followed by a 5-m-thick loose sand layer (D_r = 30%), a second 0.7-m-thick silt layer, and a 17-m-thick, dense sand layer (D_r = 75%). In each test, one of the embankments (referred to as Kobe-1 × 6, Kobe-2 × 4, and Sine-2 × 4) included a pile group that extended into the dense sand layer. The Kobe-1 × 6 group consisted of a single row of six closed-ended, aluminium piles with an outer diameter of 0.72 m and a flexural stiffness (E_pI_p) of 174 MN·m². Both the Kobe-2 × 4 and Sine-2 × 4 groups had two rows of four closed-ended, aluminium piles, with an outside diameter d_o of 1.22 m, a center-to-center spacing of $3d_o$, and an E_pI_p of 1,876 MN·m². An aluminium-epoxy pile cap connected the piles together in each test. The model was subjected to longitudinal ground motions at the base. The Kobe-1 × 6 and Kobe-2 × 4 tests were exposed to a modified version of the ground motion recorded at a depth of 83 m at Port Island during the 1995 Kobe Earthquake, with a peak base acceleration of 0.8 g and 0.7 g, respectively. The Sine-2 × 4 test was initially subjected to a shaking event that contained a total of 20 sine wave cycles with packets at 0.2 g, 0.3 g and 0.5 g, respectively.

The centrifuge tests were simulated using an equivalent static analysis (*ESA*) by obtaining the embankment displacement and displacement profile for a range of pile/bridge restraining forces, determining the pile/bridge restraining forces for a range of imposed ground displacements, and identifying the point of compatibility in the forces and displacements between the two steps. However, this method tends to overestimates the embankments deformations (without piles) (Armstrong et al. 2014), and the bending moment of the piles (Figure 5.14). As a result, it is necessary to scrutinize the determination of input parameters using 3-layer model.

For the piles in the Kobe 2 × 4 test, the 3-layer model solutions adopted p_{ub} = 730.2 kN/m, m = 9 (ϕ' = 30°), k_s = 70.5 kPa [= $51d1.5m(l_m/l)^2 1.5$, d = 1.22 m], and k_θ = 17.6 MN·m/rad (see Figure 5.14). The surcharge q was calculated for the embankment height h_e of 8–11-m (above the loose sand). The lateral spreading *FPUL* p_s was estimated to be 162.26 kN/m (= $0.78\sigma_v'd$, with $\sigma_v' = \gamma_s' h_e$, γ_s' = 17 kN/m³, and h_e = 10 m) in light of $p_s/(\sigma_v' d K_p)$ = 0.26 (with ϕ' = 30°). The normalized on-pile pressure ratio $p_s/(dq)$ was estimated as 0.78 for short-term embankment loading (Jeong et al. 1995). With l = 22.5m, l_m = 5 m, and α = 1, the p_{ub} was obtained and $p_{ub}/(\sigma_v' d)$ = 0.6. The 3-layer model predictions were found to agree well the numerical predictions for the restraining force-embankment displacement relationship, and pile-head displacement-maximum bending moment curve, respectively in Figure 5.14. The prediction agrees with the measured profile of bending moment in Figure 5.14c.

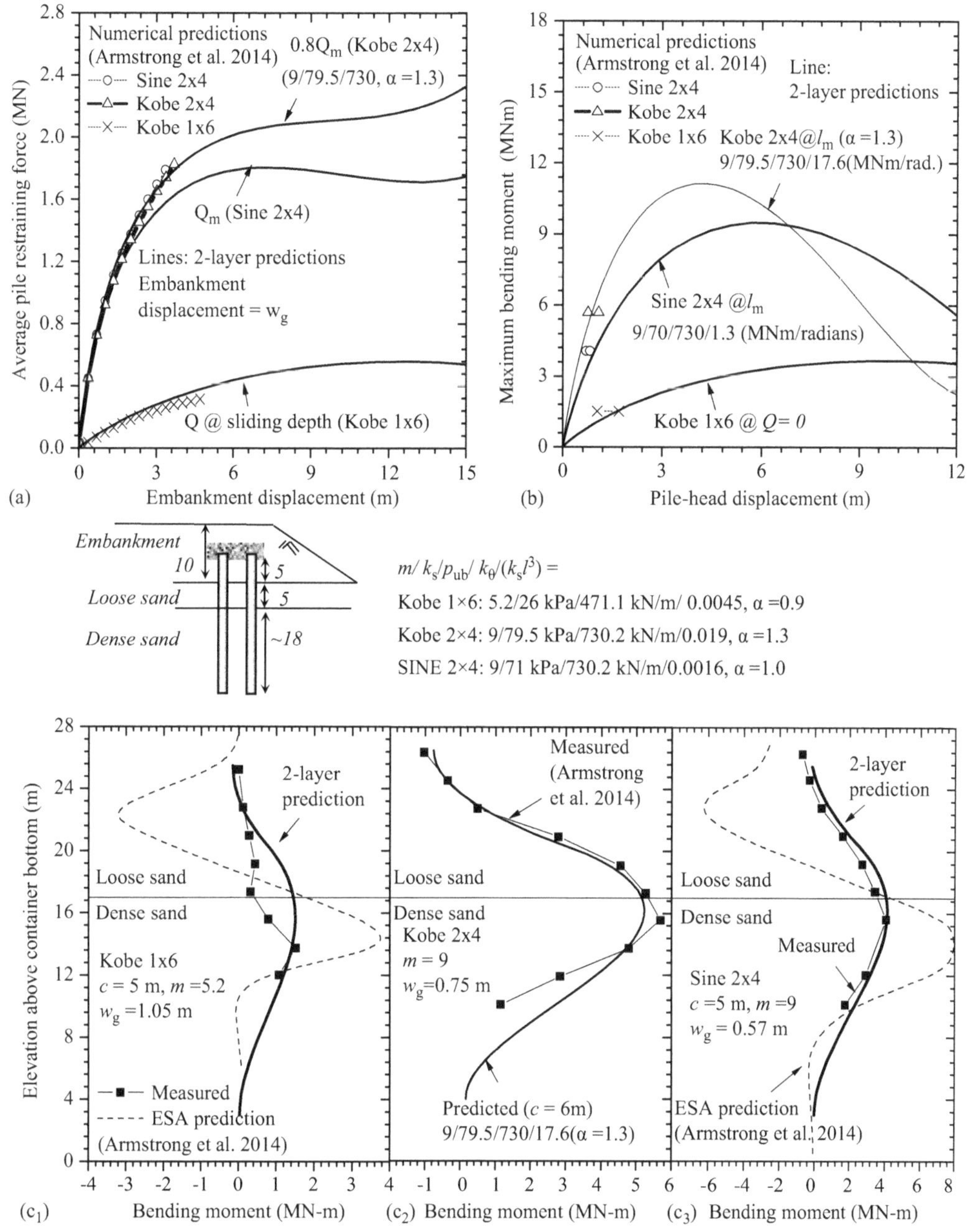

Figure 5.14 Predicted versus numerical solutions or measured (Armstrong et al. 2014) of (a) maximum shear force versus embankment displacement; (b) maximum bending moment versus pile-head deflection w_g; and bending moment profiles at w_g of (c_1) 1.03 m (Kobe 1 × 6); (c_2) 0.75 m (Kobe 2 × 4); and (c_3) 0.57 m (Sine 2 × 4); respectively. [Adapted from Guo, W. D., *Proc. NZGS Symposium, Dunedin, New Zealand*, 2021b.]

As with Kobe 2 × 4, the Kobe 1 × 6 and Sine 2 × 4 groups were simulated and shown in Figure 5.14 as well. To compensate the impact of flexibility (small diameter), a design level *m* of 5.2 (i.e., 60% ultimate value of 9) was adopted for Kobe 1 × 6. A similar ratio $p_s/(\sigma_v' d K_p)$ of 0.28 was adopted for both Kobe 1 × 6 and Sine 2 × 4, with p_s of 105.3 kN/m and 176.3

kN/m, respectively (based on $\sigma_v' = 170$ kPa and $\phi' = 30°$) [or $p_{ub}/(d\sigma_v') = 0.66$]. The 3-layer predictions for the pile-head displacement (at $c = l_m = 5$ m) were 1.05 m (Kobe 1 × 6), 0.75 m (Kobe 2 × 4) and 0.56 m (Sine 2 × 4), which agreed well with the measured values of 1.03 m, 0.75 m and 0.57 m, respectively. The profiles of bending moment also agreed well (Figure 5.14c). The 3-layer model solutions are simpler and more efficient to apply than other approaches, with good accuracy. However, it should be noted that the stipulated α value can only be verified with the evolution of pile response with soil movement, which is currently unavailable

5.7 CONCLUSIONS

This chapter introduces a 3-layer model and elastic solutions to capture the nonlinear response of rigid, passive piles in sliding soil. It includes a reduction factor (α) for on-pile force from active to passive loading; a method for deducing input parameters from test data, and example simulations of free-head and capped piles, as well as embankment pile interaction. Salient features of the model include the use of α to accommodate effect of rotational restraint (e.g., free to capped head and an underlying stiff layer), a subgrade modulus for large soil movement, a modulus ratio at the design and ultimate state, a normalized rotational stiffness (of 0.1–0.15) for capped piles, and an envelope response of pile at a series of sliding depths. Examples are presented to demonstrate the model ability to accurately predict the nonlinear response of piles under either static embankment loading or lateral spreading. The results indicate that:

- On-pile pressure p_s reduces by a factor α in rotationally restrained, sliding layer, similar to a p-multiplier for a laterally loaded, capped pile: $\alpha = 0.25$ and 0.6 (a shallow translating and rotating piles); $\alpha = 0.3$–0.5 (a uniform p_s) and 0.8–1.3 (a linearly increasing p_s) for a slide overlying a stiff layer; and $\alpha = 0.5$–0.72 for moving clay under embankment loading.
- $m = (0.7–1.0)K_{p1}/K_{a2}$ during a large, overlying soil movement; or $m = 2.1K_{p1}/K_{a2}$ for a sliding sand over soft rock. The subgrade modulus k_{s2} of underlying sand for a large soil movement (typically > 0.5 m) may be scaled from model tests using Equation (1.2), Chapter 1.
- $\bar{k}_\theta = 0.1 - 0.15$ for the model 2-piles and embankment piles, which reduces the pile displacement with depth by a magnitude of $(0.5–0.667)zM_m/k_\theta$.

Chapter 6

Passive piles

Model tests

6.1 INTRODUCTION

Model tests were conducted to gain pile response to rotational soil movement (Poulos et al. 1995; Chen et al. 1997), translational-rotational movement incurred by embankment (Ellis and Springman 2001), and translational movement resulting from slope and lateral spreading (Guo and Ghee 2004), respectively. Rotational tests offer a low-bound estimate of pile response, while translational tests provide more realistic soil movement profiles (Guo and Ghee 2005) and upper bound estimate of response of pile at typical sliding depths, axial load levels and pile diameters. Bending moment under translational movement is up to 3–5 times higher than under rotational movement (Guo and Qin 2010). This conclusion, nevertheless, veers from a limited impact of the movement profiles on pile response via numerical simulations (Tasiopoulou et al. 2013). Furthermore, soil movement profiles may alter during lateral spreading.

The 22 February Christchurch earthquake was caused by a 'hidden' fault lying along the southern edge of the city (GNS science). It may involve pulse-like, long-period, long-duration of ground motion (soil movement), which is more destructive than the ordinary ground shakings (Hayashi et al. 2018; Wang 2019). The earthquake incurred lateral spreading and hinges at the pile-cap level (reaching yielding bending moment M_o^y) and at the transition depth from liquefied to non-liquefied layers (with maximum bending moment M_m attaining yielding moment M_y of pile body), respectively (Cubrinovski et al. 2014). The in situ piles exhibit sliding, sway and back-rotation, as seen in South Brighton bridge. Interestingly, the Gayhurst bridge piles with a lesser tilt (of 1.2°) formed a M_o^y hinge, while the Anzac bridge piles with a large tilt (of 5°) did not.

The piles are modelled empirically as a fully fixed-head pile (with cap-level moment M_o of M_m) (Berrill et al. 2001) or experimentally using a scaled pile-soil system (Abdoun & Wang 2002; Dobry et al., 2003). Simple models (Haskell et al. 2013) offer $Q_m l/M_m$ of 1.5 (Q_m = in-pile maximum shear force), which is well below 2.5–6.6 gained from model and in situ test piles (Chapter 5). The yielding bending moments differ between ground-level (GL) (of M_o^y) and pile body (of M_m) (Brandenberg et al. 2005a, 2005b), as recognized for laterally loaded piles (Duncan et al. 1994; Mokwa & Duncan 2003; Guo 2015a), and adopted in design code (e.g., $-M_o^y/M_m$ = 0.4 in the Chinese design code JGJ 1994).

This chapter presents 1 g model tests on free-head piles, and capped pile groups (see Figure 6.1) and their response concerning the impact of pile diameter, axial load P and sliding-depth under pulse-like lateral soil (frame) movement of a uniform, an inverse triangular, or an arc frame (soil) movement, respectively. The 3-layer model and solutions (Guo 2015b, 2016) (see Figure 1.5, Chapter 1) are adopted to analyse the test results and to deduce the values

DOI: 10.1201/9781003315230-6

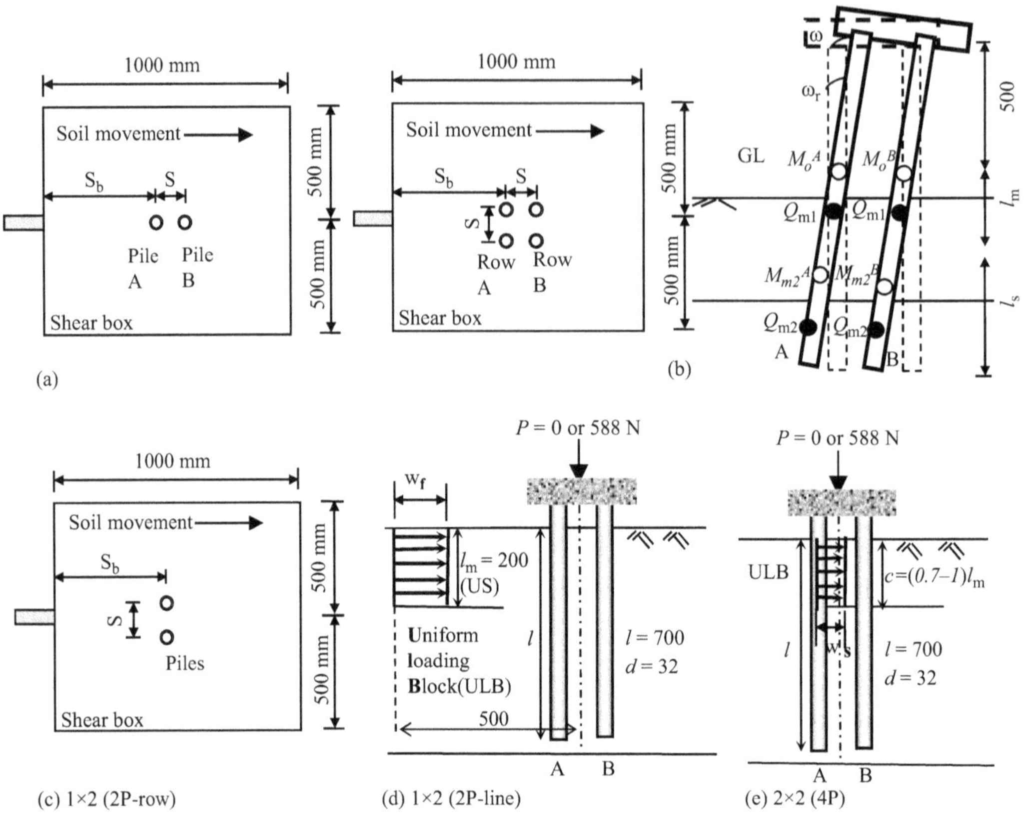

Figure 6.1 Modelling pile groups: (a) A-A Plan view of tests 2 × 1 and 2 × 2; (b) key response; (c) test 1 × 2; and schematic model of (d) test 2 × 1; (e) test 2 × 2.

of the five input parameters of modulus k_s, modulus ratio m, *LFPUL* at pile-tip level p_{ub}, cap rotational stiffness k_θ, and factor α of non-uniform soil movement on p_{ub} (Guo 2020a, b). The impact of sliding or sway is explored concerning alteration of bending moment δM_m, limiting bending moment M_m, maximum rotation angle ω_r^{max} and residual (sway) displacement w_g^{SY} (written together as $\delta M_m - M_m - \omega_r^{max} - w_g^{SY}$).

6.2 MODEL TESTS

The model tests were performed using the apparatus illustrated in Figure 1.8 (Chapter 1). During the 'shearing', the sand surface was free of loading, and the pile was only subjected to lateral pressure caused by the moving sand (at a sliding depth l_m of 200–400 mm), in addition to the overburden pressure resulting from self-weight and ~10% extra shear resistance from the single test-pile.

The U-block was used to induce uniform and simultaneous movement of the laminar aluminium frames at a constant sliding depth (l_m) of either 0.2 m (= 0.286l) or 0.4 m (0.57l). The inverse T-block enforced sand wedges that gradually extended towards the tested pile, while the A-block exerted a cone-face pressure. To reach a total lateral (frame) movement w_f of 110–150 mm (i.e., 11–15 steps, see Figure 6.2a), the lateral T block translated horizontally (see Figure 1.8a, Chapter 1) at an increment of 10 mm (measured on the top frame). This

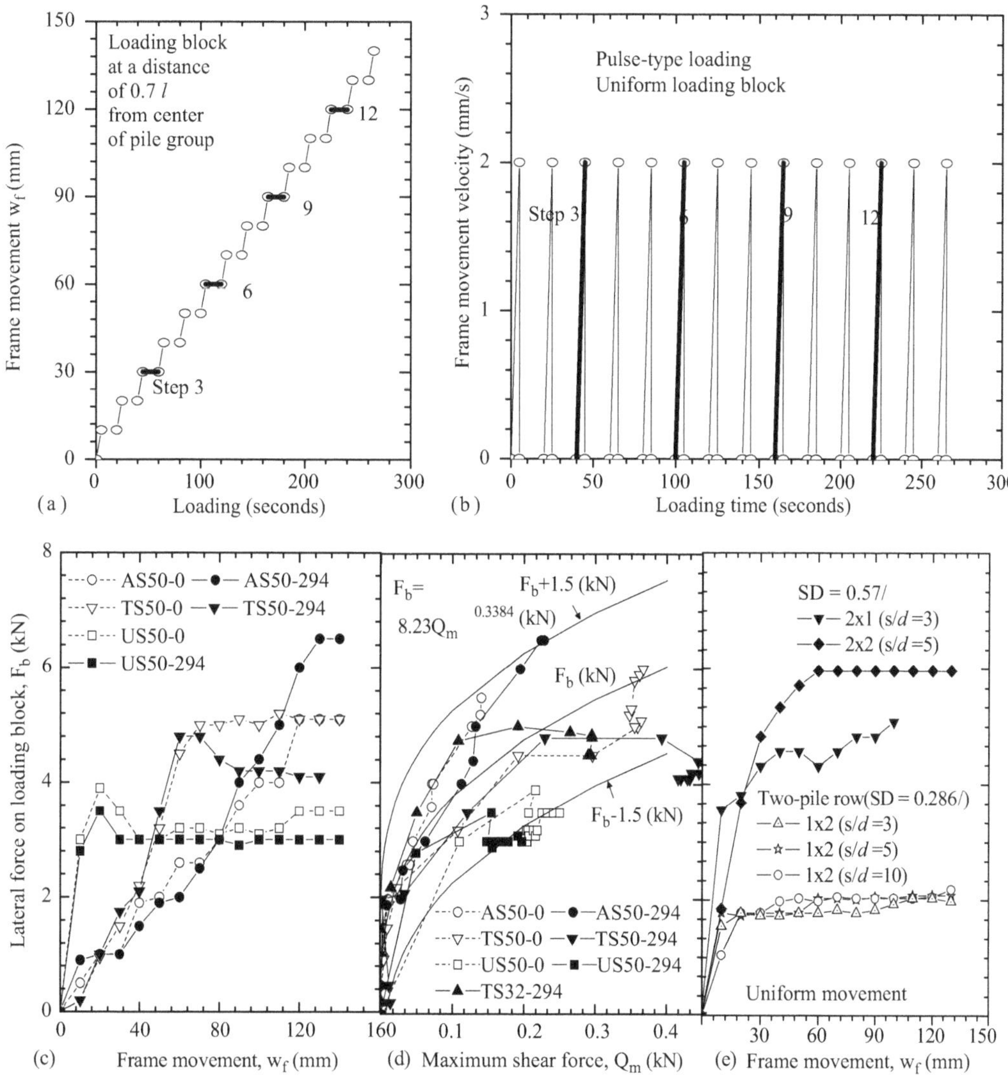

Figure 6.2 (a) Step loading versus soil (frame) movement; (b) Pulse-type soil (frame) movement; (c) Total loading on uniform loading block per pile (Guo & Ghee 2010); (c) Total loading F_b versus shear force Q_m. [Adapted from Guo, W. D., *Int J Physical Modelling in Geotechnics* **22**(4), 2022b; Guo, W. D., H. Y. Qin and E. H. Ghee, *Int J Geomechanics* **17**(4), 2017.]

induced a pulse-type loading at a speed of 2 mm/s (see Figure 6.2b), and a pulse duration of 5s, which effectively model long-period and long-duration ground motions (Hayashi et al. 2018). For the T- and the A-block tests, the sliding depths gradually increased until the final pre-set sliding depth l_m (SD) of 0.2 m (= 0.28l) or 0.4 m (= 0.57l). The shear modulus G_s of the sand was deduced as 13–35 kPa for single piles and 2-pile groups.

The lateral hydraulic jack (Figure 1.8a, Chapter 1) exerted a total loading force on the shear box, which was recorded to a total frame movement w_f of 120–140 mm (see Figures 6.2c and e, Ghee 2009). At the final SD of 0.28l, the minimum w_f required to mobilize the maximum loading force F_b are 20 mm for 3–4 kN, 60 mm for 4–5 kN, and 130 mm for 5–6 kN, respectively, using the U-, the T-, and the A-blocks (Figure 6.2c). This induces the maximum (in-pile) shear force Q_m as shown in Figure 6.2d. Interestingly, the T-block induced a Q_m (of 0.37–0.45 kN) that were twice as high as those induced by the A-block (of 0.15–0.25 kN), despite requiring only 70–80% the F_b on A-block, or F_b values lying between the A- and U-tests. Furthermore, the highest shear force Q_m (~$0.1F_b$) from the T-block was observed only at SD = $0.28l$ (not evident at SD = $0.57l$), which represents a worst-case scenario. This substantiates the pragmatic use of T-movement profile in real-world design (Poulos et al. 1995; Guo 2003, 2013c), especially for large movements (e.g., w_f > 90 mm in the current tests) at SD = $0.28l$. In contrast, at SD = $0.57l$, an A-profile should be considered as well.

6.2.1 Single piles

In the shear apparatus filled with medium-grained quartz (Queensland) sand, either a single flexible pile (d_{32}) or a rigid pile (d_{50}) was installed at the centre. Each pile was subjected to one of the three block loading: A-block, the inverse T-block or the U-block (see Figures 1.8b through d, Chapter 1). A total of 25 tests (see Table 6.1) were conducted, with 13 piles tested under an axial load P of 294 N/pile and the remaining tests without any load P (= 0).

Table 6.1 Details of pile tests using 3 types of loading blocks (Guo et al. 2017)

Test No.	*Test description*	*Pile diameter (mm)*	*Axial load (N)*	*Sliding depth (m)*	*Remarks or s_b (m)*
1/2	AS32-0/ AD32-0[a]	32	0	0.2/0.4	Arc block loading at a distance S_b = 0.5(m)
3/4	AS32-294/ AD32-294	32	294	0.2/0.4	
5/6	AS50-0/ AD50-0	50	0	0.2/0.4	
7/8	AS50-294/ AD50-294	50	294	0.2/0.4	
9/10	TS32-0/ TD32-0	32	0	0.2/0.4	Triangular block loading at a distance S_b = 0.5(m)
11/12	TS32-294/ TD32-294	32	294	0.2/0.4	
13/14	TS50-0/ TD50-0	50	0	0.2/0.4	
15/16	TS50-294/ TD50-294	50	294	0.2/0.4	
17/18	US32-0/ UD32-0	32	0	0.2/0.4	
19/20	US32-294/ UD32-294	32	294	0.2/0.4	Rectangular block loading at a distance S_b = 0.50(m) (piles at centre of the shear box)
21/22	US50-0/ UD50-0	50	0	0.2/0.4	
23/24	US50-294/ UD50-294	50	294	0.2/0.4	
25	US50-294(S)	50	294	0.2/0.4	

Notes:

[a] First letter 'A', 'T' or 'U' indicates test under an arc, inverse triangular or uniform loading; second letter 'S', or 'D' indicates an ultimate sliding depth (at loading position) of 0.2 m or 0.4 m; third number '32', or '50' indicates a pile diameter of 32 mm or 50 mm; and last number '0', or '294' indicates a vertical load (in N) per pile.

6.2.2 Capped piles

Capped (d_{32}) piles were installed in the centre of the shear box. These piles were socketed into a 50 mm-thick aluminium cap (with a Young's modulus of 70 GPa) to a depth of 0.8d. A small relative rotation angle w (= w_r-w_{cap}) was allowed between the pile (ω_r at GL) and pile-cap (ω_{cap}). The pile-pile cap connection had a bending capacity of M_o^y. The capped piles were jacked together to a depth of 0.7 m in the model sand at a distance s_b from the source of soil movement (Figures 1.8b and 6.1). Twenty tests were conducted on the capped piles either without any load P (= 0) or with a load of P = 294 N per pile, at an SD of 0.286l (see Table 6.2 for twelve tests), and 0.57l (see Table 6.3 for eight tests). For simplicity, the letter 'P' is used to denote 'pile', and a number in brackets to denote 's/d' (s = pile centre-to-centre spacing; d = 32 mm, outside diameter).

As explained in Chapter 1, every test records readings from strain gauges and LVDTs, as well as the force applied on the lateral jack for each frame movement. The $M(z)$ is obtained first. Numerical integration or differentiation of $M(z)$ then yields the rotation profile $\omega_r(z)$, the displacement $w(z)$, and the shear force $Q(z)$, from which the maximum shear force Q_m, maximum bending moment M_m and its depth x_m, rotation angle ω_r, as well as pile-head deflection w_g are determined.

Table 6.2 Model pile response (SD = 0.286l) (Guo 2022b)

Test No.	*Test layout*	*Load P (N)*	$k_s\,(kPa)\,/\,p_b\,(KN/m)/\bar{k}_\theta$ [a]	*Reference*
1, 2	Single pile	0,294	P = 0: 32/4.5/0 (I_c)	
3, 4	1 × 2 (row, s/d = 3)	0,588	P = 0: 16/2.5/0 (IV_b) −45/5/0(I_c)	2P(3)-row
7, 8	1 × 2 (s/d = 5)	0,588	P = 0: 45/5/0 (I_c)	2P(5)-row
9, 10	1 × 2 (s/d = 10)	0,588	P = 0: 16/2.5/0 (IV_b)	2P(10)-row
11, 12	2 × 1 (line, s/d = 3)	0,588	14/1.7/0.04–0.13 (II)	2P(3)-line
15, 16	2 × 2 (s/d = **3**)	0,1177	14/1.7/0.04 (II) and 16/4/0.17 (A) (IV)	4P(3)
19, 20	2 × 2 (s/d = **5**)	0,1177	14/1.7/0.04(B) (II) 14/1.7/0.13 (II)	4P(5)
Predictions I, II		I_c: 25–45/4.5–10/−0.04–0.2, II: 14/1.7/−0.04–0.13		
Predictions III, IV		III: 14–16/4–5/−0.04 to 0.17, IV: 16/4/0.17		
Predictions III_b, IV_b		III_b: 16/5/0, IV_b: 16/2.5/0		

Notes:
[a] m = 17.7, ratio of the subgrade modulus of the stable layer over sliding layer, and $\bar{k}_\theta = k_\theta / \left(k_s l^3\right)$ (see Guo 2016)

Table 6.3 Model pile response (SD = 0.57l) (Guo 2022b)

Test No.	*Test layout*	*Load P (N)*	$k_s\,(kPa)\,/\,p_b\,(KN/m)/\bar{k}_\theta$ [a]	*Reference*
5, 6	1 × 2 (row, s/d = 3)	0,588	P = 0: 16/5/0 (III_b)	2P(3)-row
13, 14	2 × 1 (line, s/d = 3)	0,588	P = 0: 16/2.5/0 (IV_b)	2P(3)-line
17, 18	2 × 2 (s/d = 3)	0,1177		4P(3)
21, 22	2 × 2 (s/d = 5)	0,1177		4P(5)

Note:
[a] m, p_b as explained in Table 6.2

6.3 RESPONSE OF SINGLE PILES

6.3.1 Envelope of M_m–w_f curves to simulate piles in spreading soil

The response of single piles to soil movement w_f at SD = 0.286*l* and 0.57*l* was thoroughly discussed by Guo et al. (2017). The U-, T- and A-block tests encountered with increasing thrust F_b (see Figure 6.2). Superposing the U- and the T-test results well replicated the evolving movement profiles (from uniform to triangular and eventually to trapezoidal) observed in an in-flight base shaking test (Sharp and Dobry 2002). The U-, T-, and A-tests were thereby simulated together. The response trajectory is generally similar (except where specified) across the two SDs; and only the larger response for SD = 0.57*l* is presented here. Figures 6.3a–d

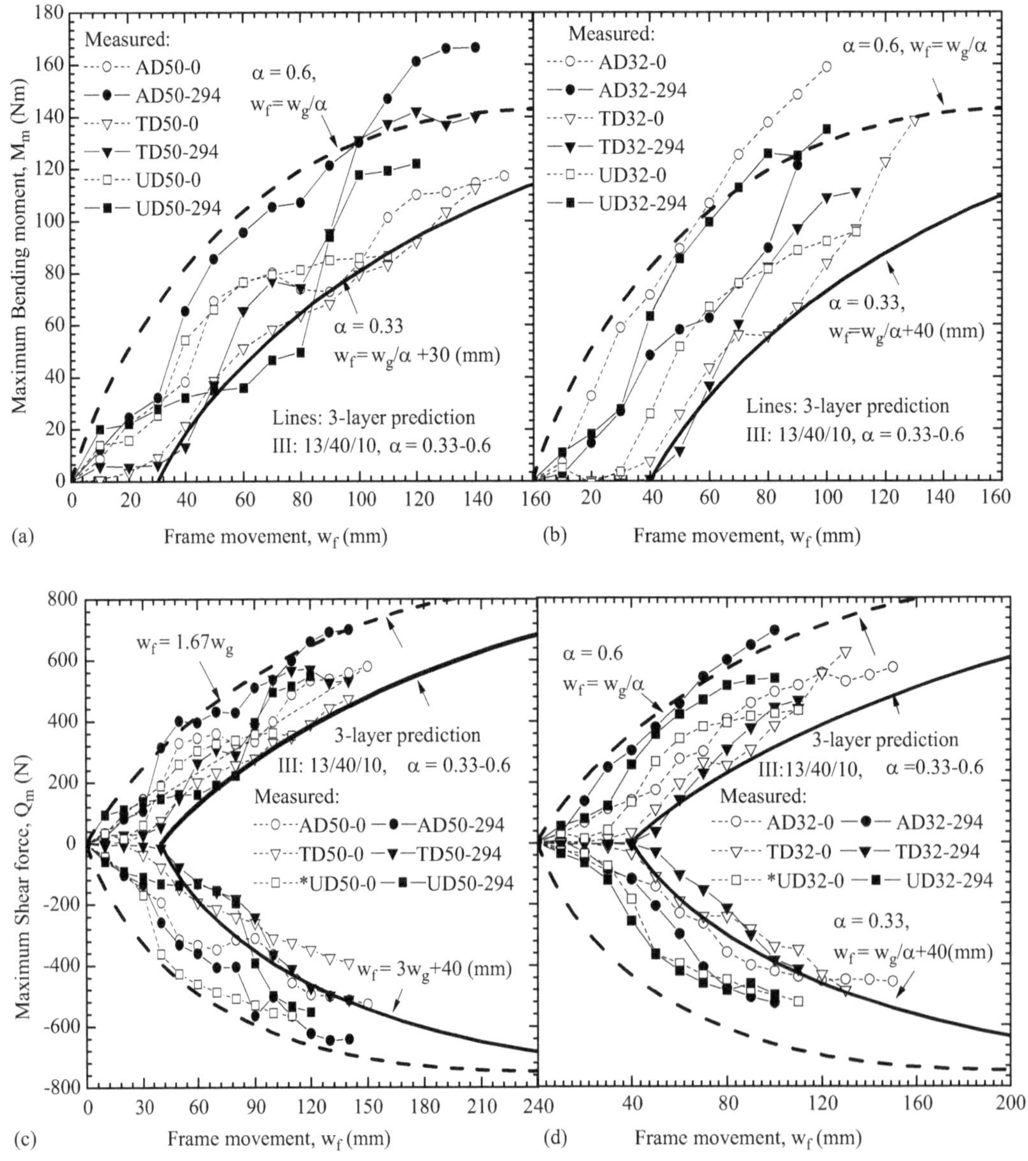

Figure 6.3 Evolution of (a, b) maximum bending moment; and (c, d) maximum shear force with frame movement (SD = 0.57*l*, (a, c) and (b, d) for d_{50} and d_{32} piles, respectively). [Adapted from Guo, W. D., H. Y. Qin and E. H. Ghee, *Int J Geomechanics* **17**(4), 2017.]

illustrate the evolution of the in-pile bending moment M_m and shear force Q_m with the movement w_f. The figures indicate that an initial movement w_f (rewritten as w_i) causes negligible pile response; an effective frame movement w_e (= w_f − w_i) is therefore used subsequently unless otherwise specified.

6.3.2 Response of single pile with sliding depth

Figure 6.4 provides the response of bending moment M_m, shear force Q_{mi}, and the pile-head deflection (at sand surface) w_g. Little increase is observed over an initial sliding depth ratio R_L (= l_m/l) of 0.1–0.17, but the response climbs rapidly thereafter, particularly for the tests conducted under the load P (e.g., AD50-294 in Figure 6.4b for M_m). The ratio of pile-head displacement over the soil movement (w_g/w_f) is employed to deduce the pressure reduction factor α.

Example 6.1 Deflection w_g, movements w_e and w_i

Figure 6.5 presents plots of the pile-head deflection w_g and its rotation angle ω_r as well as the deflection w_g under the movement w_f and their mathematical correlations. The math expressions attached allow estimation of the initial movement w_i (= – intercept/gradient) for each test. For instance, the initial movement w_i for the UD50-294 test is 4.6 mm (= −5.0/1.09) using the intercept and gradient of the No. 1 expression curve. The w_g/w_e ratio (i.e., the gradient of the curves) increases in the following order: AS (0.015–0.02), TS(0.17–0.18), AD(0.37–0.48), TD(0.94–0.97), and UD(1.09–1.41). The AS tests had the lowest gradient, which leads to a relatively low proportion of block force (F_b) being transferred to the in-pile shear force (Q_m), with Q_m/F_b = 0.025–0.05 (= 0.15–0.25/5–6) (see Figure 6.2d).

Example 6.2 FS of 2–3 at w_g = 10 mm and formation of rotational hinges

Figures 6.6a–d show the M_m–w_g and M_m–ω_r curves for d_{50} and d_{32} piles subjected to soil movement at SD = 0.57l, respectively. The UD50 test pile translated, otherwise with addition of the load P, the w_g reduces by ~40%. The AD and TD test piles represent worst-case scenario for M_m. The associated M_m at w_g = 10 mm is 2.5–4 and 2.3 times those induced in the UD tests, while the M_m at a 'failure' angle of ω_r = 5° is 1.67 and 1.27 times larger. A factor of safety (FS) of 2–3 for M_m at w_g = 10 mm would only warrant a FS of 1.0 at ω_r = 5° under lateral spreading loading. This overestimated FS using w_g needs to be addressed to avoid formation of rotational hinges at piles tilt reaching 5°–6°. Furthermore, Guo et al. (2017) shows that an increase in load P from 0 to 294 N on the d_{32} or d_{50} piles incurs 22–60% larger M_m (thus Q_m) under the U-, T-, or A-tests at SD = ~0.286l, as well as 22% larger bending moment M_m in TS50 test pile over that in US50 pile.

6.3.3 Rotational restraint on piles in AS tests

Guo et al. (2017) observed that the A-block induces flexible deformation 'jumps' at w_f = 50–60 mm, 80–90 mm, and 110–120 mm, respectively, which suggests the block confronts with a 'rotational restraint' during the sliding. In contrast, the TS tests exhibit a rigid (linear) deformation and does not show deflection leaps. The restraining effect is obliged to be considered in design as shown later in example.

6.3.4 Increasing on-pile pressure with sliding depth

Figures 6.7 and 6.8 provide the response (at ultimate state) profiles of bending moment, shear force, and deflection as well as on-pile pressure [i.e., $p(z)/d$] for all typical tests. The pressure in the sliding, transition and stable 'zones' is characterized by 30–50 kPa (TS, TD), –(70–130)

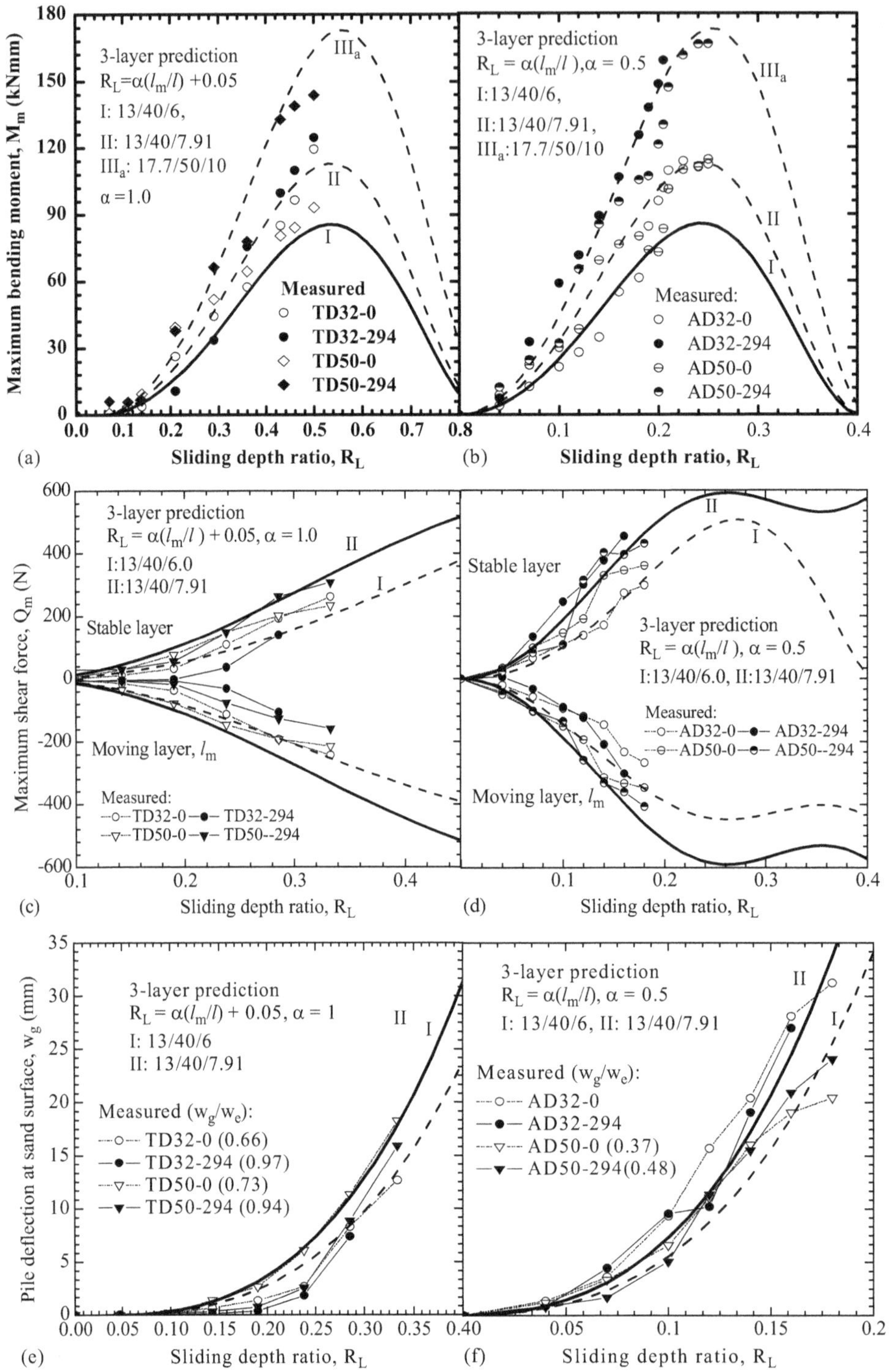

Figure 6.4 Variations of (a, b) bending moment M_m; (c, d) shear force Q_m; and (e, f) pile-deflection w_g with sliding depth ratio R_L (Guo 2016) (d_{32} and d_{50} piles, (a, c, e) and (b, d, f) for triangular, and arc tests, respectively). [Adapted from Guo, W. D., H. Y. Qin and E. H. Ghee, *Int J Geomechanics* **17**(4), 2017.]

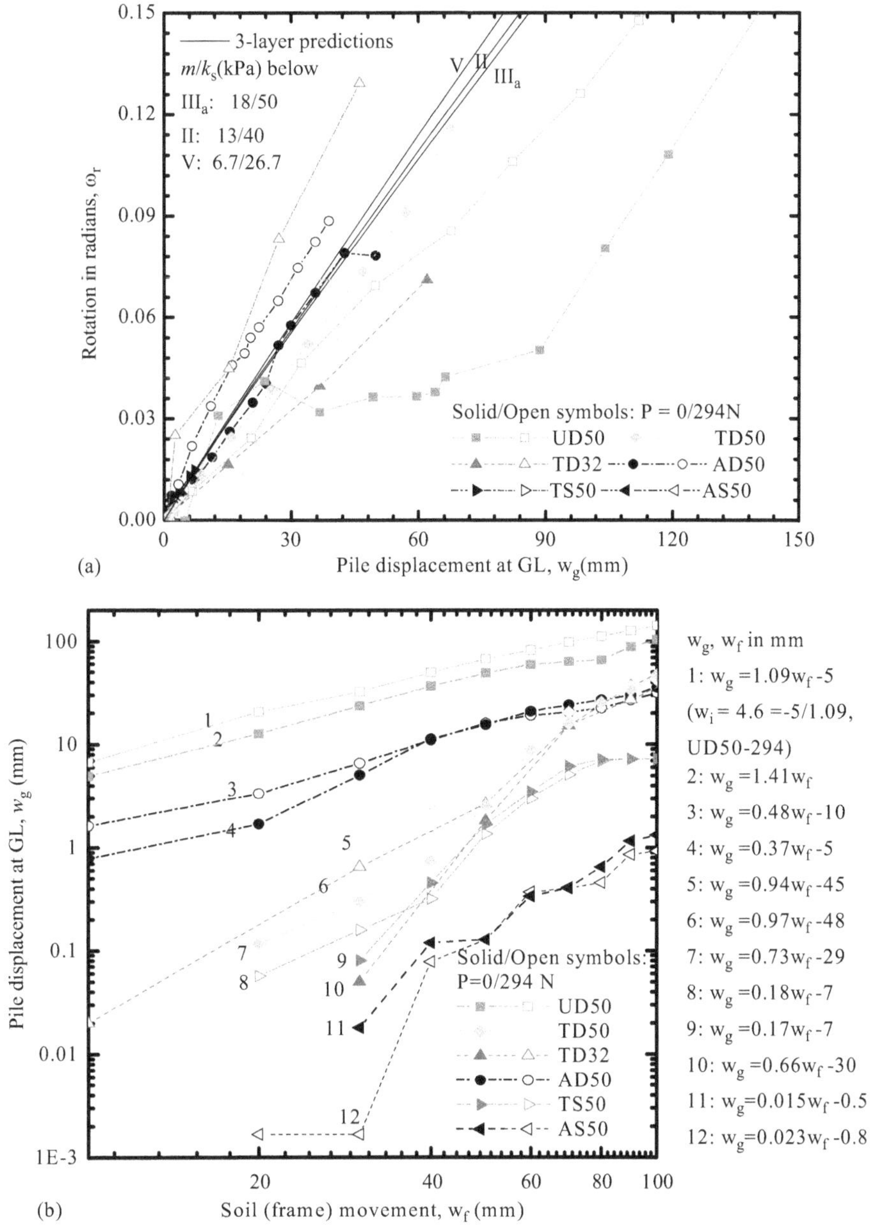

Figure 6.5 Measured response of (a) $w_g - \omega_r$; and (b) w_f (w_s) $- w_g$ for free-head single piles subjected to three types of soil movement profiles. [Adapted from Guo, W. D., H. Y. Qin and E. H. Ghee, *Int J Geomechanics* **17**(4), 2017.]

kPa, and 70–130 kPa [with $l_m = 0.286l$, and $l_m = (0.5–0.56)l$], respectively. In the sliding layer of $0.56l$, the on-pile pressure $p(z)$ is slightly higher under the T- and A-blocks than under the U-block. The 'driving' pressure from the dry-sand movement is over 10 times the overburden pressure of 3.3 kPa ($= \gamma l_m$) (JRA 2002). However, the pressure at pile-tip level is still close to 60% that on lateral piles, i.e., $0.6\gamma_s'(K_{p2})^2 l$ (= 122.46 with γ_s' = 16.5 kN/m$_3$, ϕ' = 38°, and l

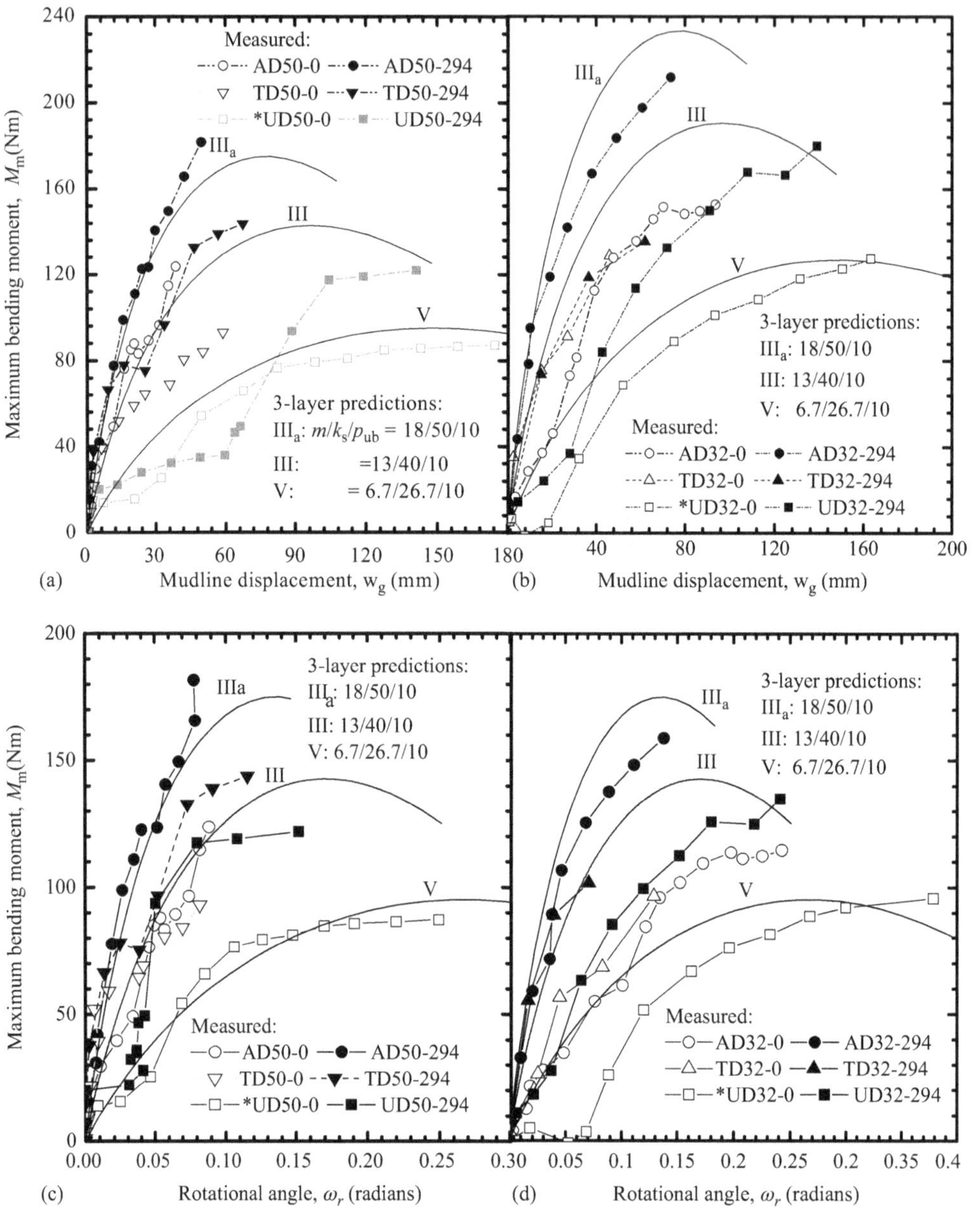

Figure 6.6 Predicted (3-layer model) versus measured response for free-head piles subjected to three types of soil movement profiles (SD = 0.57*l*): (a, b) $w_g - M_m$; (c, d) $\omega_r - M_m$. [Adapted from Guo, W. D., H. Y. Qin and E. H. Ghee, *Int J Geomechanics* **17**(4), 2017.]

= 0.7 m) (Chapter 1). In the transition zone, the on-pile pressure at SD = 0.57l on the small-diameter d_{32} piles (of 110–170 kPa) is 1.7–1.8 times that (of 60–100 kPa) on the d_{50} piles, which is not seen for SD = 0.285l. The on-pile pressures in the stable layer grossly double as the sliding depth rises from 0.286l to 0.57l. Interestingly, a similar pressure profile is prescribed in the Chinese 'cantilever beam' method, which applies limit equilibrium to design slope-stabilizing piles.

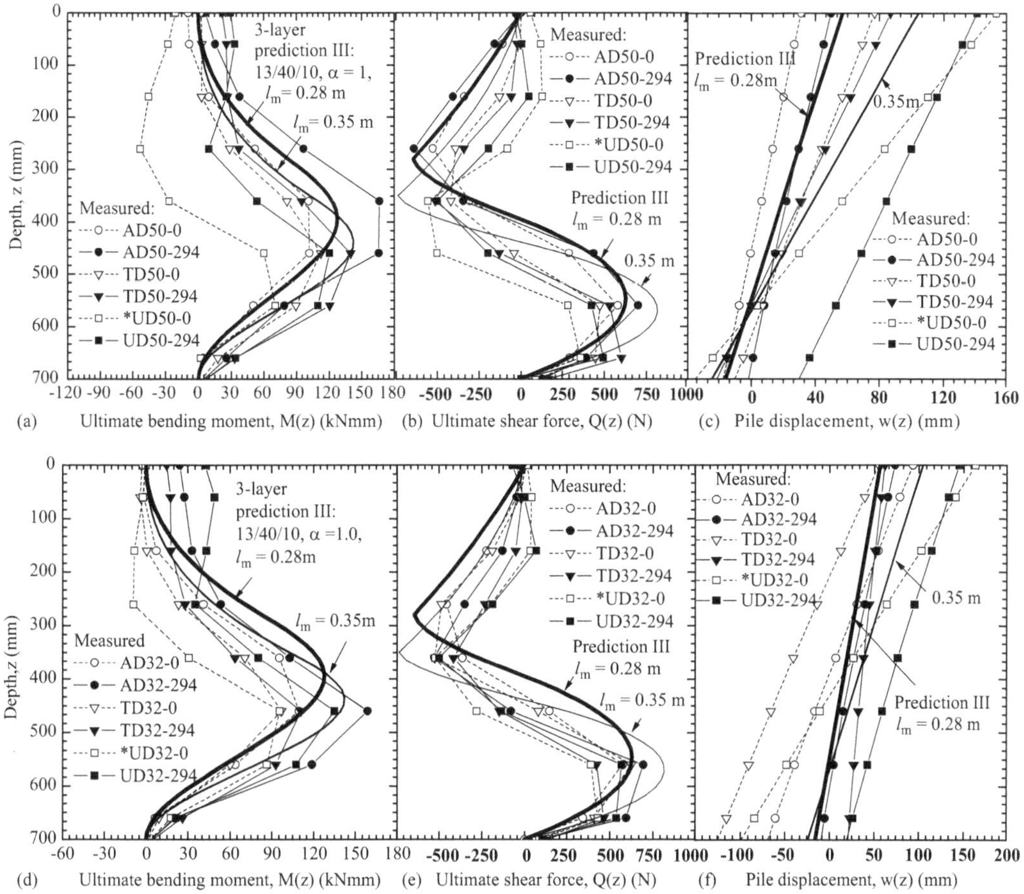

Figure 6.7 Pile response of $M(z)$, $Q(z)$, and $w(z)$ (SD = $0.57l$) due to arc, triangular, and uniform profiles (a, b, c); and (d, e, f) for d_{50} and d_{32} piles, respectively. [Adapted from Guo, W. D., H. Y. Qin and E. H. Ghee, *Int J Geomechanics* **17**(4), 2017.]

6.3.5 Thrust and M_m and k_s

Figure 6.9 provides the maximum bending moment M_m and shear force Q_{mi} (≈ sliding thrust) induced in d_{50} and d_{32} piles during the soil movement over an SD of $0.28l$ and $0.57l$. The M_m and Q_{mi} are linearly correlated, which resemble the correlation between the deflection w_g and the effective frame movement w_e (see Figure 6.5b). This finding supports elastic pile and soil interaction at a 'low' sliding modulus of subgrade reaction k_s (see Table 6.4). The k_s value is fraction of the subgrade modulus k of lateral piles (Guo 2012) calculated as $k_s \approx kw_g/w_e$.

6.3.6 Effect of loading block angle

To examine the impact of soil movement shape on pile response, the angle θ of inverse triangle loading blocks was varied to 22.5° and 30° (see insert of Figure 1.8, Chapter 1; Qin and Guo 2016b). This resulted in an increase in the frame movement w_a (i.e., w_f upon touching the final depth l_m of 200 mm) to 90 mm (θ = 22.5°), and 110 mm (θ = 30°) from the standard test of 60 mm (θ = 15°) (see Figure 6.10c). The initial frame movement w_i was also increased from 37 mm (θ = 15°, e.g., in TS32-0) to match the new loading angle (see Figure 6.10b). The maximum bending moment M_m increased appreciably, including the M_m for

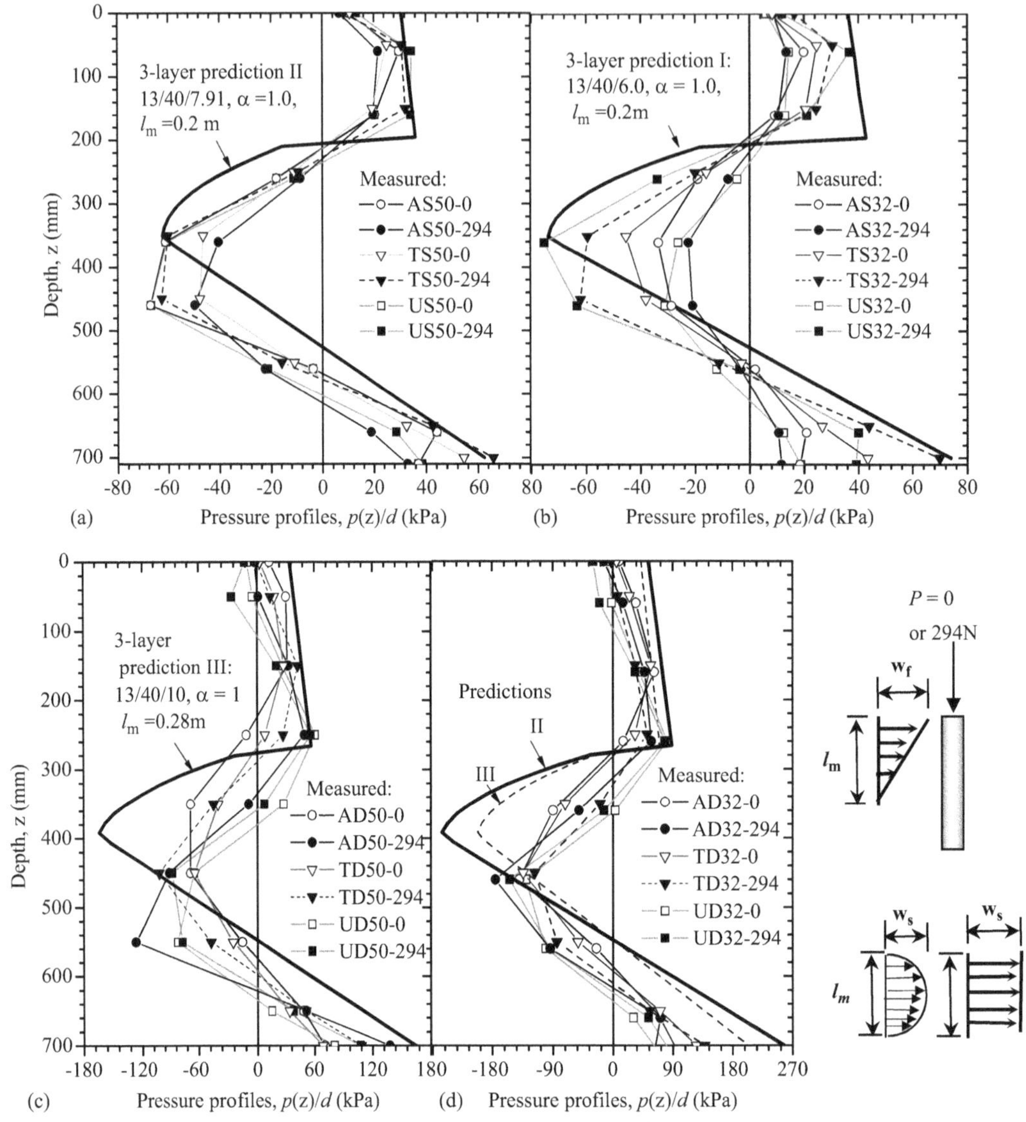

Figure 6.8 On-pile pressure profiles induced by arc, uniform and triangular blocks [(a, c) and (b, d) for d_{50} and d_{32} piles, respectively]: (a, b) SD = 0.286*l*; (c, d) SD = 0.57*l*. [Adapted from Guo, W. D., H.Y. Qin and E. H. Ghee, *Int J Geomechanics* **17**(4), 2017.]

$\theta = 0°$ of test RS32-0 [Guo and Ghee 2005a,b; Qin 2010, see Figure 6.10c (right)], but it did not from increasing w_a to w_p. Therefore, for a large soil movement, it is not critical to know the exact soil movement profile (with depth), but only the sliding depth. This explains why a good prediction was made of in situ pile behaviour using loading angle of the soil movement profile and the measured depth of sliding slope (e.g., Cai and Ugai 2003).

Furthermore, Chen et al. (1997) conducted tests that enforced rotation of the pile(s) about the depth L_s (i.e., the thickness of low stable layer, 325 mm, see inset in Figure 6.10c) at a constant sliding depth l_m (of 200–350 mm). These tests provide the moment M_m under a rotational angle θ [≈ $\arctan(w_f/l_m)$] about the depth L_s. Figure 6.10 indicates the moment M_m rises rapidly with the rotation angle θ (<12°), and thereafter only increases with sliding depth l_m (L_s = constant).

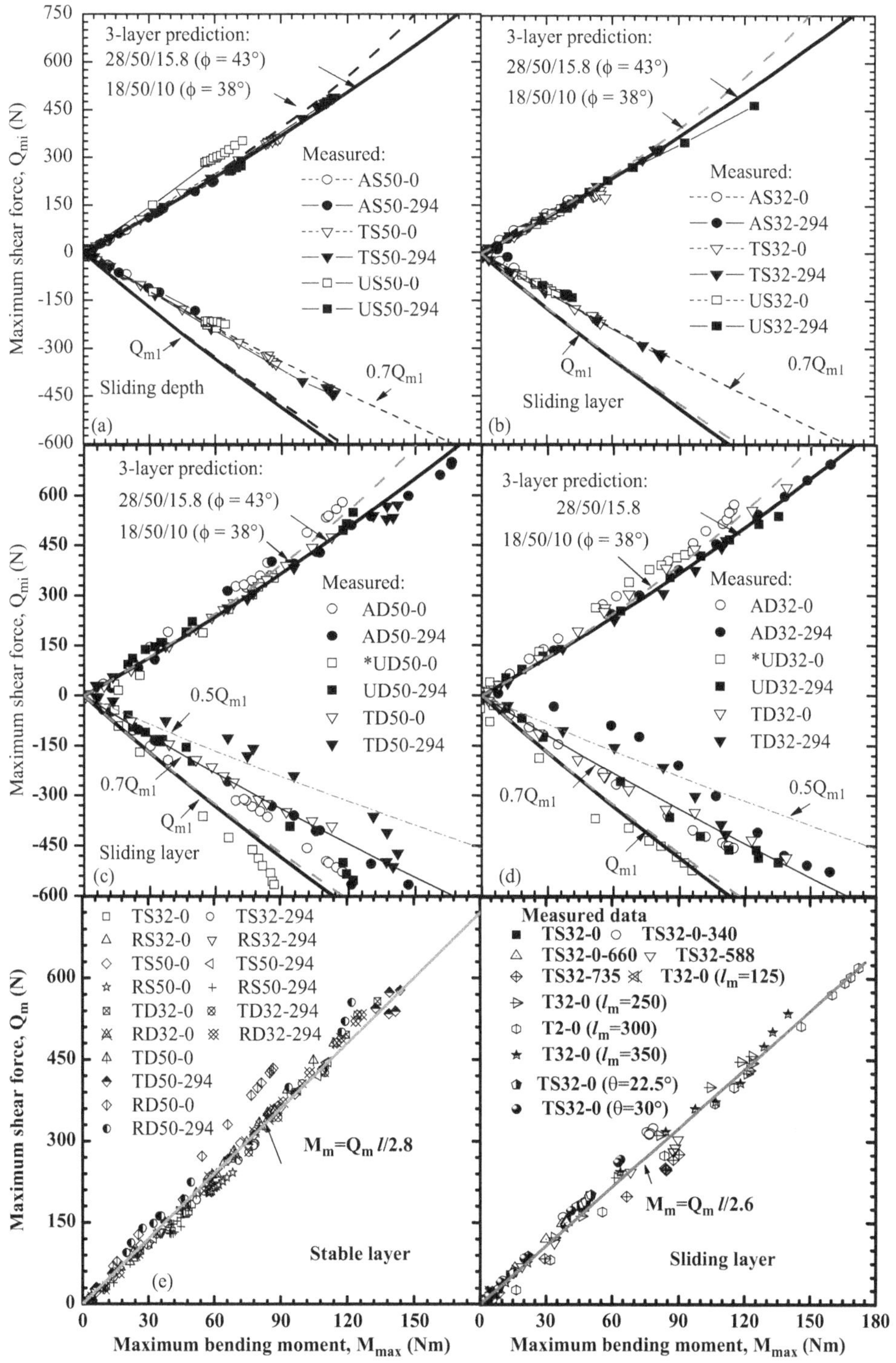

Figure 6.9 Maximum shear force versus maximum bending moment: (a) d_{50} piles; (b) d_{32} piles (SD = 0.286*l*); (c) d_{50} piles, and (d) d_{32} piles (SD = 0.57*l*); (e) stable and sliding layer.

Table 6.4 Input parameters for 3-layer predictions of single pile (Guo et al. 2017)

Prediction	*I*	*II*	*II_a*	*III*	*III_a*	*IV*	*IV_b*	*IV_c*	*V*
m	13	13	17.7	13	17.7	13	17.7	17.7	6.7
k_s (kPa)	40	40	98	40	50	170	97.5	125	26.7
p_{ub} (kN/m)	6.0	7.91	10	10.0	10.0	10.0	10.0	12.5	10.0
α (lateral movement)	0.4–0.6	0.6		0.33–0.6	0.5–1.0		0.77	0.9	
SD	0.286*l*			0.286*l*–0.57*l*		0.286*l*			0.57*l*
	d_{32}	d_{50}		d_{32}, d_{50}		d_{50}	d_{50} (T-profile)		U-tests

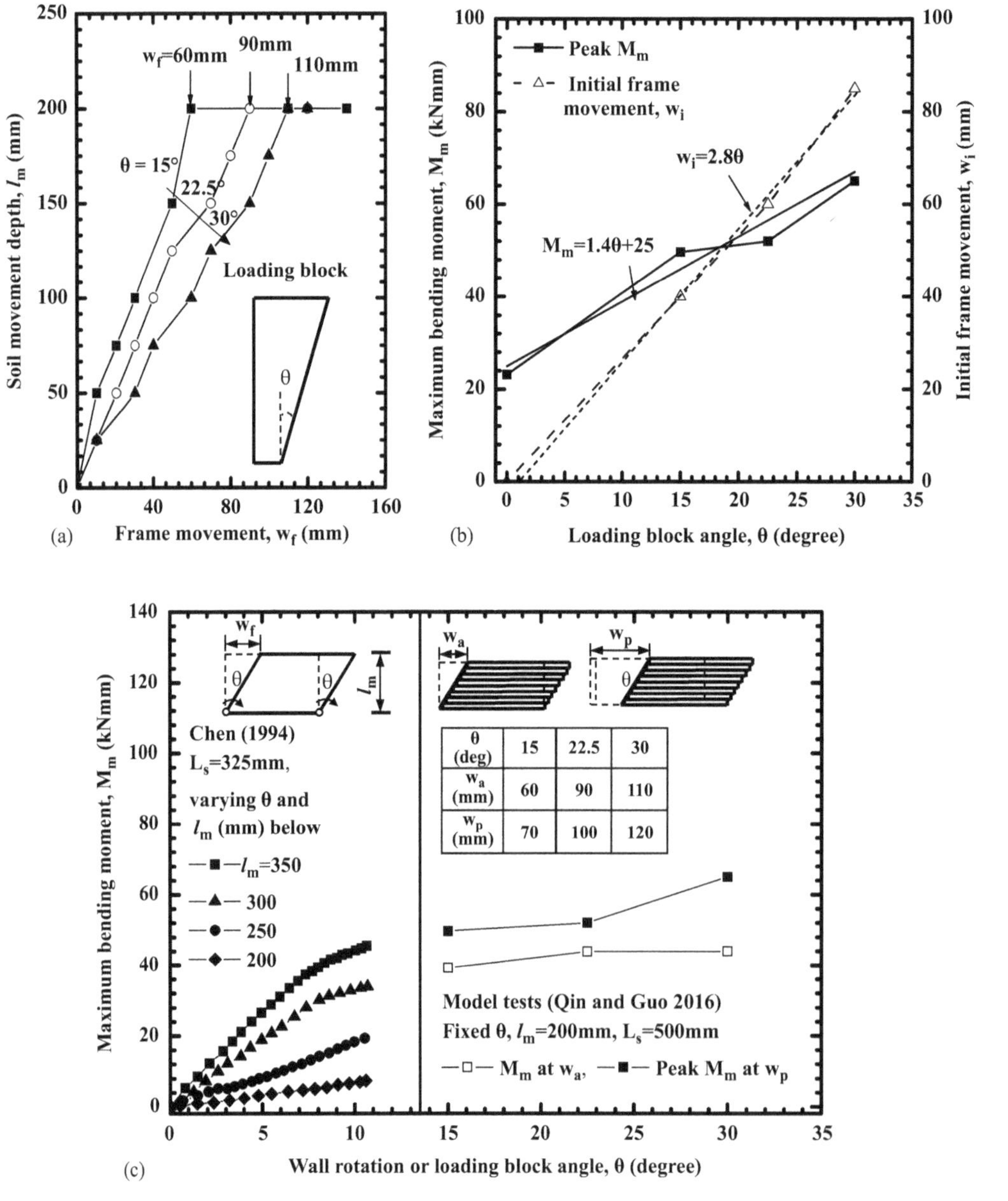

Figure 6.10 Soil movement depth, loading angle and maximum bending moment M_m. [Adapted from Guo, W. D., H. Y. Qin and E. H. Ghee, *Int J Geomechanics* **17**(4), 2017.]

The 1g tests allow experimental validation of the 2-/3-layer models in capturing nonlinear response of prototype piles subjected to lateral spreading (Guo 2015b, 2016, 2020a, b). The models utilize four input parameters, namely modulus k_s, modulus ratio m, limiting *FPUL* at pile base p_b (= αp_{ub}), and rotation stiffness of pile k_θ, to capture the impact of scale, pile rigidity, and soil properties on pile response, as elaborated next.

6.4 3-LAYER MODEL SIMULATION

The 3-layer model (via expressions in Table 5.1, Chapter 5) was matched with the-test results to assess the rational of the input parameters m, k_s, p_{ub} (α), and k_θ obtained using guideline outlined in Chapter 1 of Guo (2016). The input parameters were initially determined using the guideline outlined as follows:

- m = 17.7 (= K_{p1}/K_{a2}) for large soil movement, and m = 10–13 [= (0.6–0.7)K_{p1}/K_{a2}] for design-level movement, respectively, with ϕ' = 38°.
- k_s = 40–50 kPa for d_{32} piles in sliding layer, and k_s = 62–80 kPa under vertical load P.
- p_{ub} (passive d_{32} piles) = 6.0–7.9 kN/m, which is 60% that on lateral piles. p_{ub} (lateral d_{32} piles) = 10.0–13.3 kN/m with $p_{ub} = s_g\gamma_s'(K_{p2})^2 dl$, s_g = 1.5–2.0, γ_s' = 16.5 kN/m3, ϕ' = 38°, d = 0.032 m, and l_m = 0.7 m. p_{ub} (passive d_{50} piles) = 9.4–12.3 kN/m [= (6.0–7.9) × 50/32], or p_{ub} (lateral d_{50} piles) = 14.5–19.3 kN/m for ϕ' = 43°; and k_s = 97.5–125 kPa [= (62–80) × 50/32] kPa.

These parameters are grouped into nine sets (see Table 6.4), comprising I, II, III, and IV for design state (m = 13), II$_a$, III$_a$, IV$_b$ and IV$_c$ for ultimate state (m = 17.7), as well as V for constrained sliding state (m = 6.7). For instance, set I for d_{32} piles is defined as 13/40/6.0 (α = 0.4–0.6) and set II for d_{50} piles is defined as 13/40/7.91 (α = 0.6) (SD = 0.286l).

The 3-layer model predictions, using each of the nine sets of parameters, are compared with the experiment (at H = 0, k_θ = 0). The predictions I and II compare well with the measured response for SD = 0.286l (not shown herein, Guo et al. 2017). The predictions I, II, III and III$_a$ agree well with the measured response for SD = 0.57l, as is evident in $M_m(Q_m)$–w_f curves (see Figure 6.3), R_L-M_m, -Q_m, and -w_g curves (see Figure 6.4), w_g–M_m and ω_r–M_m curves (see Figure 6.6), $M(z)$, $Q(z)$ and $w(z)$ profiles (see Figure 6.7), and $p(z)/d$ profiles as well as the profiles for SD = 0.286l (see Figures 6.8a and b), respectively.

Figure 6.6 reveals interesting features of the vertically loaded (P > 0) UD tests, which remain intact despite a large translation of about 60 mm (d_{50}) or 15 mm (d_{32}). These tests involve m_u/k_{su} = 6.5/26.7 kPa (subscript 'u' denotes U-tests), $m_t = 2m_u$, and $k_{st} = 1.5k_{su}$ ('t' for T-tests), and $m_a = 3m_u$, and $k_{sa} = 2k_{su}$ ('a' for A-tests). A shallow slide (SD = 0.286l) involves m = (2–3)m_u, and k_{st} = (~6.5)k_{su} for m = 6.5 (US), 13 (TS), and 17.7 (AS), and k_{st} = 173 kPa. Regardless of pile diameters, the modelling requires α values of 0.4–0.6 (lateral movement) and 0.6 (depth movement) at SD = 0.286l; and α values of 0.33–0.6 (lateral movement) and 0.5 (AD)–1.0 (TD) (depth movement) at SD = 0.57l.

The prediction III aligns with all the AD, UD and TD tests (see Figure 6.3) when considering the initial movement w_i of 0 (A-profile) to 30–40 mm (T-profile), and the increasing on-pile sliding depth l_m from 0.28 m (A-profile) to 0.35 m (T-profile) (see Figure 6.7). The w_f in each M_m-w_f curve was scaled from the predicted M_m–w_g curve using the factor α ($\approx w_g/w_e$, see Figure 6.3). For evolving profiles of soil movement, it is recommended to adopt the 'envelope' predictions I and II (SD = 0.286l, Guo et al. 2017), and III and III$_a$ (SD = 0.57l). The predicted sliding force at SD = 0.57l is obtained as Q_{m1} (see Figure 6.9) assuming a uniform k_s, otherwise the force reduces to 0.5Q_{m1} if the modulus linearly increase with depth (linear k_s). The predictions of 0.5Q_{m1} (linear k_s) and Q_{m1} (uniform k_s) well bracket the measured values. The rotational restraining effect needs to be considered, as is elaborated next.

Example 6.3 Unknown soil movement profile for in situ tests

White et al. (2008) conducted large-scale direct shear tests on single composite grouted piles which extended through compacted soil into the underlying subsoils within the shear box. The two test piles had nominals d of 115 mm, or 178 mm and length l of 2.1 m. The soil was a glacial till with a peak undrained shear strength s_u of 53 kPa. The shear boxes were pushed to translate the compacted soil laterally over a sliding thickness of 0.6 m. The profiles of bending moment (at a soil movement w_s of 13 mm), and shear box load-displacement relationships were measured and are replotted in Figures 6.11a and b, respectively.

The modelling of 115-mm-diameter pile involves input parameters (Guo 2016) of (i) m = 5 assuming ϕ' = 25°; (ii) k_s = 500 kPa (= $25s_u/2.6$); (iii) p_u = 81.3 kN/m (= $3.05s_u dl_e/l_m$) in which l_e = 1.6 m (effective embedment), and d = 0.115 m; and (iv) k_θ = 116.6 kNm/radian $\left(\bar{k}_\theta = 0.04\right)$ to cater for restraining effect. In estimating the l_e, the length z_c [with $w(z) > w_s$] was subtracted from the pile embedment. The modelling of 178-mm-diameter pile is associated with m = 5, k_s = 700 kPa, p_u = 115.4 kN/m, and k_θ = 40.8 kNm/radian $\left(\bar{k}_\theta = 0.01\right)$. With these parameters, the 3-layer predictions compare well with the measured moment profiles. The predicted (solid) lines of Q_{m1} (maximum in-pile shear force) versus w_g (pile-head displacement) are indeed softer than the measured (applied) load-(box) displacement curve for either pile, as the load encompasses both soil and pile resistance. Adding the measured soil resistance (without pile, not shown herein) into the calculated Q_{m1}, the 'revised' predictions (dash lines) match well in trend with the loads. The overestimation (at a displacement > 15 mm) may originate from a gradually reduced soil resistance on the tests with increasing shear-box displacement w_f (e.g., at w_f = 140 mm for the 178-mm-diameter pile). The 3-layer model is thus sufficiently accurate to model the shear tests using the modulus k_s, sliding depth l_m, effective embedment length l_e and without knowledge of the soil movement profile.

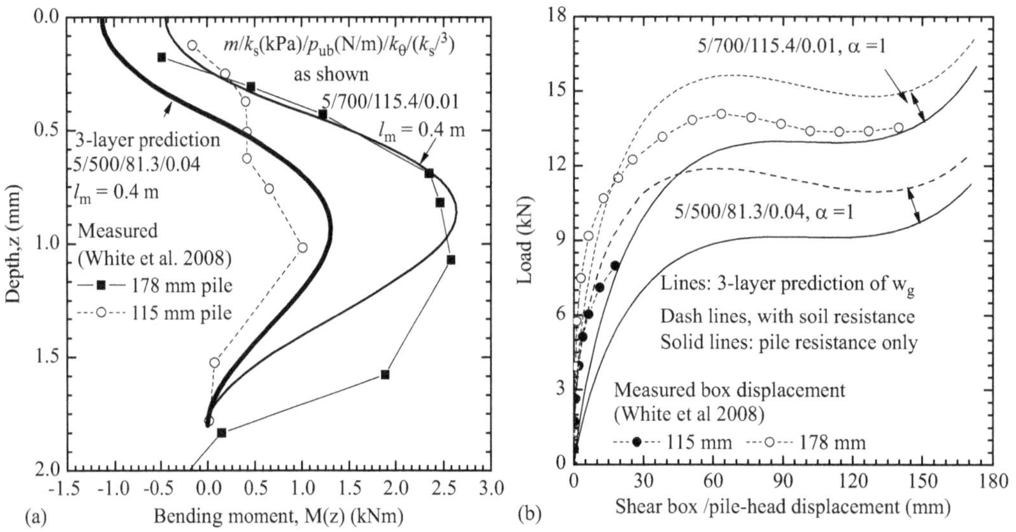

Figure 6.11 Predicted versus measured (White et al. 2008) response of piles in Glacial till: (a) $M(z)$ at w_s = 13 mm, (b) Load versus shear box displacement. [Adapted from Guo, W. D., H.Y. Qin and E. H. Ghee, *Int J Geomechanics* **17**(4), 2017.]

Example 6.4 Progressive increase in soil movement on passive piles

Smethurst and Powrie (2007) studied the behaviour of bored concrete piles in stabilizing a railway embankment. The piles (l = 10 m, d = 0.6 m, and E_pI_p = 10^5 kNm2) were installed at a centre-to-centre spacing of 2.4 m, in a four-layered subsoil consisting of rockfill (with ϕ' = 35° over a depth z of 0–2 m), embankment fill (ϕ' = 25° over z = 2–3.5 m), weathered Weald clay (ϕ' = 25° over z = 3.5–4.5 m), and intact Weald clay (ϕ' = 30° over z = 4.5–10 m), respectively. The lateral movement of subsoil around the piles was measured, and it was 5.5 mm on day 42, and 26 mm on day 1,345 over z of 0–3.5 m, which linearly reduced to zero from z of 3.5 to 7.5 m. The response of Pile C was measured, including the pile-head movement w_g, maximum moment M_m (at a depth z_m of 5.6 m), and head rotation angle ω_r. They were 68 mm, 81.8 kNm, and 0.67 × 10^{-4}, respectively by day 42; and 35–38 mm, 170.7 kNm (at z_m = 5.8 m); and 48.7 × 10^{-4} by day 1,345. The response is presented in Figures 6.12a–f, including $w_g(w_s)$–M_m curves, w_g–ω curve, and profiles of soil movement, pile deflection $w(z)$, bending moment, and on-pile force per unit length. Note the 'R_3' is the estimated thrust in stable layer.

The 3-layer prediction was made using m = 6 (with ϕ' = 25°); k_s = 789 kPa (= G_s/2.6, G_s = 2.052 MPa); p_u = 138.9 kN/m (= 127.7α, α = 1.05), k_θ = 7.5 MNm/rad (Guo 2012), and l_e = 3.9 m (= $l - z_c$, l = 5.9 m, equivalent length of rigid pile, and z_c = 2 m). The p_{ub} was obtained as 116.1 kN/m, using $p_{ub} = 0.6\gamma_s'(K_{p1})^2 dl_e$ in which γ_s' = 9 kN/m³, $K_{p1} = K_{p2}$ = 2.464 (ϕ' = 25°), d = 0.6 m, and l = 5.9 m. The p_{ub} was raised by 10% to 127.7 kN/m to account for the impact of the top, stiff layer of rock fill.

At w_s = 5.5 mm and 26 mm, the predicted $w(z)$, $M(z)$ and $p(z)$ profiles (day 42 and day 1,345) agree well with the measured data (see Figures 6.12c–f), respectively, but not for the net on-pile force $p(z)$ (see Figure 6.12e) at w_s = 5.5 mm. The $p(z)$ is dominated by pile flexibility (Guo 2012). With the l_e and k_θ, the prediction, based on l_m = 1.8 m (at w_s = 5.5 mm) or l_m = 3.5 m (w_s = 26 mm), well captures the impact of movement profiles on the response.

6.5 RESPONSE OF CAPPED PILES

The impact of evolving soil movement profiles such as those encountered during lateral spreading can be effectively encapsulated as the 'envelope' of pile response induced by the T-, U- and A-block tests. Although the response from each loading block may differ in terms of initial movement or sliding, the trajectory of the pile response is similar. The tests on capped piles are therefore confined to using uniform loading (U-) block. Test evidence will be provided to illustrate modes of sliding, back-rotation, sway, as well as their evolution concerning bending moments (M_o^y and M_m), angle of pile rotation (ω_r), sway displacement (w_g^{SY}), and moment raise (δM_m). It was thoroughly analysed to gain the δM_m - M_m - ω_r^{max} - w_g^{SY} values and soil movement at ultimate state. These results provide insight into the observed behaviour of the piles underpinning the Christchurch bridges.

6.5.1 Sliding, back-rotation, sway and relative rotation angle ω

First, 2-pile in-line and 4-pile groups tested at an SD of 0.286l were scrutinized to trace modes of sliding, sway and back-rotation. As with free-head piles, Guo (2022b) demonstrate that all tests exhibited linear correlation between pile-displacement w_g (at GL) and the frame movement w_f (<w_f^* of 0.04l or ~30 mm). The gradients of the w_f and w_g lines offer k_s of 20–35 kPa (regardless of $M_o \neq 0$), and α of 0.25 (≈w_g/w_e Guo 2015b, 2016). Figure 6.13a shows typical w_g–ω_r curves, which indicate

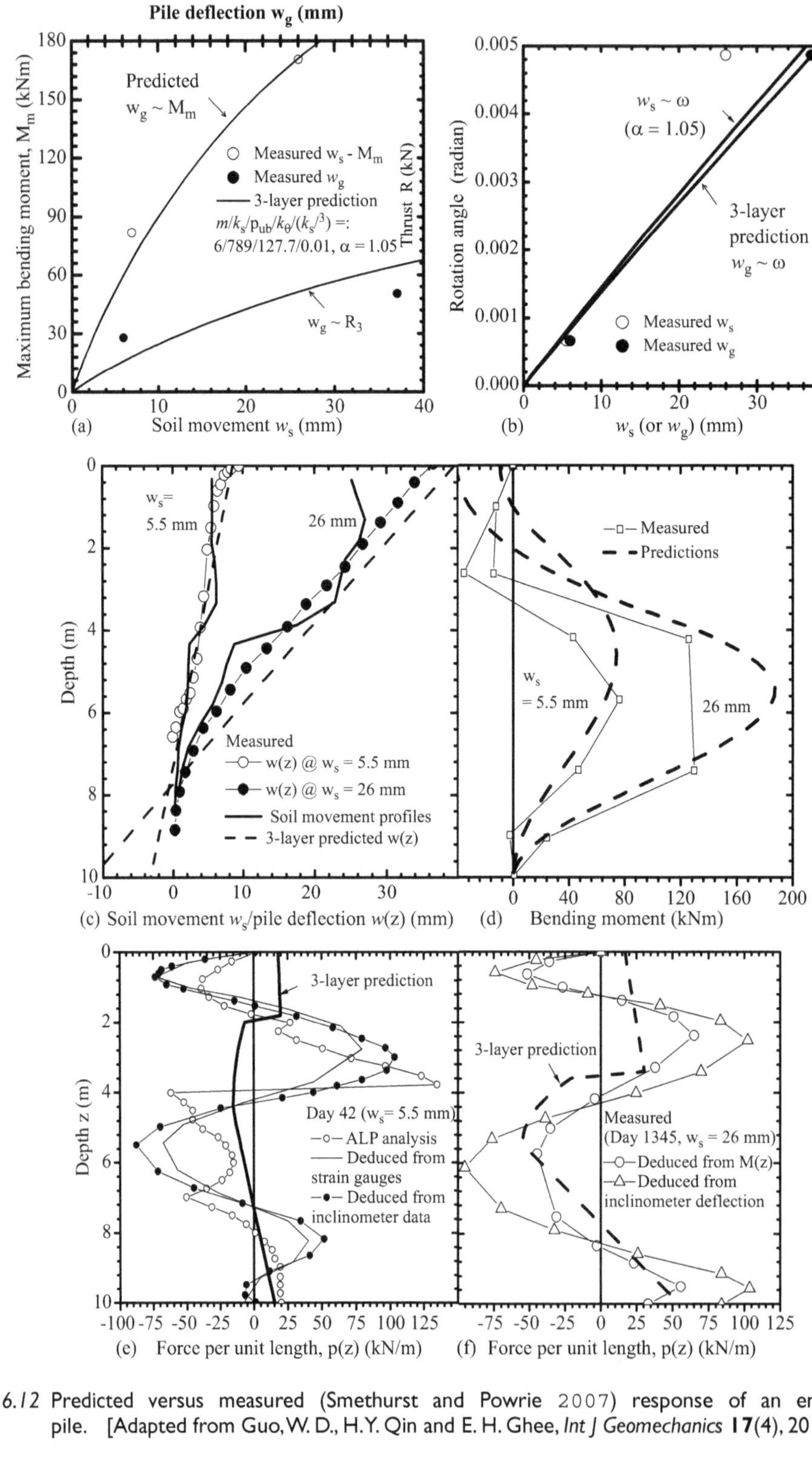

Figure 6.12 Predicted versus measured (Smethurst and Powrie 2007) response of an embankment pile. [Adapted from Guo, W. D., H. Y. Qin and E. H. Ghee, *Int J Geomechanics* **17**(4), 2017.]

- sliding at which the maximum bending moment M_m is nearly constant over an irreversible displacement or rotation angle;
- rotational sway ω_r^{SY} (see magenta lines) defined as the differential angle of ω_r^{m} measured (denoted by superscript 'm') at GL between starting and ending levels of a sway cycle;
- displacement sway (w_g^{SY}) calculated as the differential displacements (w_g^{m}) in the cycle.

Figures 6.13c–f explain back-rotation with a reduced angle ω between the pile-cap and piles in the loading direction. Sway is presented as ω_r^{SY}/ω_r and w_g^{SY}/w_g, and the angle ω is presented as a fraction of the angle of pile ω_r which is negative in 2P(3) and 4P(3) and positive at

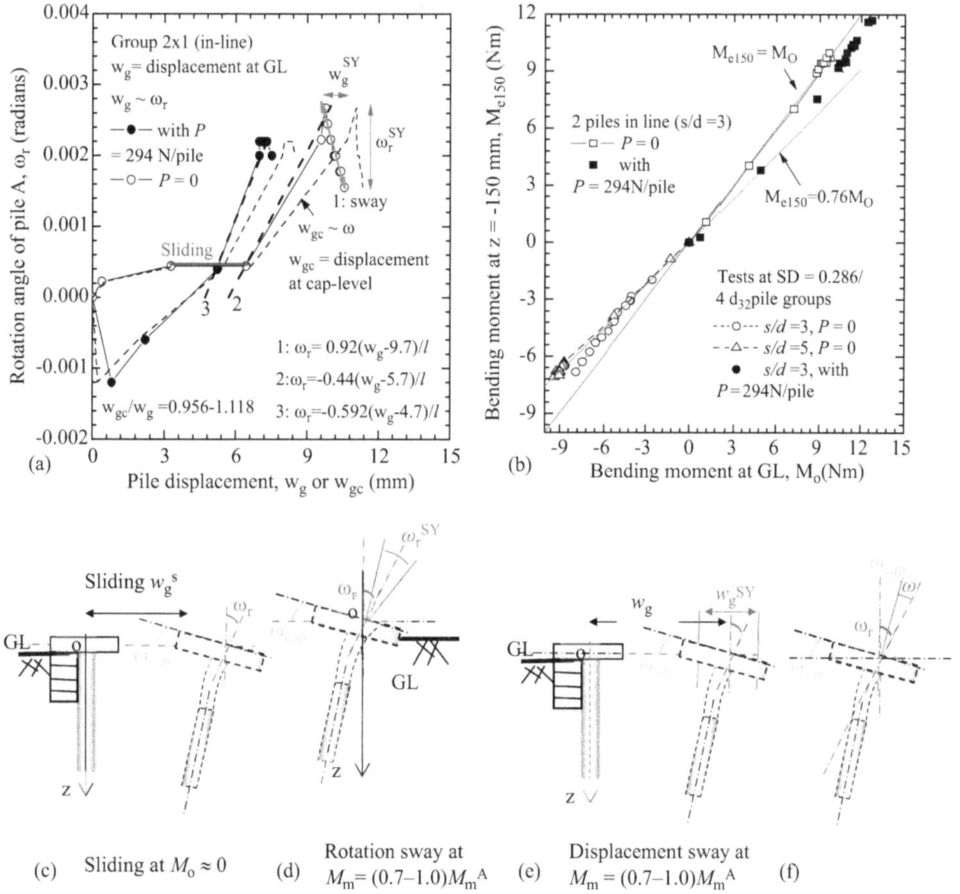

GL = Ground level; ω_{cap}= Angle of rotaion at pile cap level (unknown); ω_r = angle of rotation of pile body (measued); $\omega = \omega_r - \omega_{cap}$, relative angle of rotation between the pile and the cap at GL.
The impact of the free-length is converted into an eqivalent rotational stiffness k_θ (= M_o/ω).
The k_θ and ω values are uniquely back-estimated using the measured M_o–w_g and M_o–ω_r curves, and $\omega = \omega_r - \omega_{cap}$, and $M_o = -k_\theta\omega$. The ω for current model pile and cap connections are deduced as $\omega = (1\sim2)\%\omega_r$
(i) $\omega_{cap} > \omega_r$, and $\omega < 0$, bacward rotation resulting positive M_o
(ii) $\omega_{cap} < \omega_r$ and $\omega > 0$, and forward rotation associated with negative M_o

Figure 6.13 Cap conditions on pile displacement w_g and w_{gc} and bending moment M_o and M_{e150} (SD = 0.286*l*): (a) variations of w_g or w_{gc} with rotation ω_r; (b) M_{e150}/M_o = 0.76–1.0; (c) sliding at initial loading stage; (d) rotational sway; (e) displacement sway; (f) equivalent k_θ and ω values. [Adapted from Guo, W. D., *Int J Physical Modelling in Geotechnics* **22**(4), 2022b.]

SD = 0.57l. Importantly, the sway occurs only at $w_f > w_f^*$, and is not observed at SD of 0.57l. Relevant features are discussed concerning groups 4P(3) and 4P(5) at ultimate state (except where specified).

The pile-cap of the free-standing groups located at 0.5 m above GL (see Figure 6.1). The displacement at cap-level w_{gc} is within 12% difference of the w_g at GL (see Figure 6.13a). The bending moment at GL (M_o) is within 24% difference of the measured moment (M_{e150}) at an 'eccentricity' e of 150 mm above GL (Figure 6.13b). This value is used to estimate rotational stiffness k_θ^A (= M_o/ω_r) for pile A (denoted by superscript 'A'), and pile B, respectively. This stiffness is a 'fictitious' value of the 'free-standing capped' pile, as the actual stiffness is equal to M_{oc}/ω (i.e., bending moment M_{oc} at the pile and cap connection over ω of (0.05–0.1)ω_r, see Figure 6.14). This treatment is sufficiently accurate for practice purpose (Guo 2016, 2020a, b)

For the 2-pile in-row, the rotation angle ω_r and displacement w_g^m (at GL) observes $\omega_r = 1.5w_g^m/l$ ($\omega_r = \omega_r^{FH}$) for a free-head (FH) pile (Guo 2012, 2014). The equivalent ω_r^{FH} is calculated using the measured pile displacement w_g^m at GL regardless of cap restraint. The displacement and rotation at starting and ending levels of each sway (including w_g^{SY} and ω_r^{SY}) were provided previously (Guo 2022b) for all pile groups at SD = 0.286l. For instance, normalized $\bar{\omega}_r$ $(= \omega_r^m / \omega_r^{FH})$ are tabulated in Table 6.5. The translating sand together with the angle-change ω incur the sway at a moment M_o of (0.7–1.0)M_m^A (M_m^A = limiting M_m of pile A, e.g., Figure 6.13b), or (0.7–1.0)M_o^{yA} for pile A and a similar magnitude for pile B. Back-rotation incurs a positive moment M_o, subsequent to yielding at negative moment M_o^{yA} (pile A) from forward rotation (see Figure 6.13f).

Figure 6.14 provides the measured pile-displacement w_g and rotation angle ω_r at GL versus bending moments M_m and M_o curves. Guo (2022b) demonstrates that (i) the positive gradient of rotation $\omega_r(z)$ with depth z shifts to negative beyond the SD; (ii) the gradients along piles A and B are markedly different (T_1) for a 'mono-pile' (Figure 6.14a_1), different (T_2) in a general pile group (Figure 6.14b_1), or of opposite sign within SD (T_3) for a 'raked-pile' group, respectively; (iii) sliding reduces the sway compared to the non-sliding case; (iv) the load P may enforce non-uniform, rotational restraint on the base between pile A and pile B; otherwise without the load, a large translation occurs over entire depth. Finally (v) sway induces back-rotation of pile-B within SD (Figure 6.14c_1).

Table 6.5 Measured normalized angle $\bar{\omega}_r$ and moment M_o (SD = 0.286l) (Guo 2022b)

(s/d)	*P = 294 N/pile*			*P = 0*		
	2P(3)-line	*4P(3)*	*4P(5)*	*2P(3)-line*	*4P(3)*	*4P(5)*[c]
$\bar{\omega}_r$	0.06897	0.0416	0.103159	0.06897	−0.03147	0.114519
M_o^A(N·m)	−13.96	−12.66	−12.94	−11.13	−9.04	−10.15
M_o^B(N·m)	−4.17	0.22	−2.05	−5.35	4.31	−1.19
M_o^{yA}/M_m^A	1.2	1.0	0.8	1.2	0.5	0.6
M_o^{yB}/M_m^B	0.5	0	0.3	0.5	0.25	0.25
			Yielding moment M_o^y			
M_o^{yA}(N·m)	−13.96	−12.66	−12.94	−11.13	−9.04	−10.15
M_o^{yB}(N·m)	−7.90	−1.76	−2.83	−5.35	4.61	−3.94
M_o^{yA}/M_o^{yB}	1.77	7.19[a]	4.57[b]	2.08	1.96	2.58

Notes:
[a] Pile B failed.
[b] large sway.
[c] no cap sway.

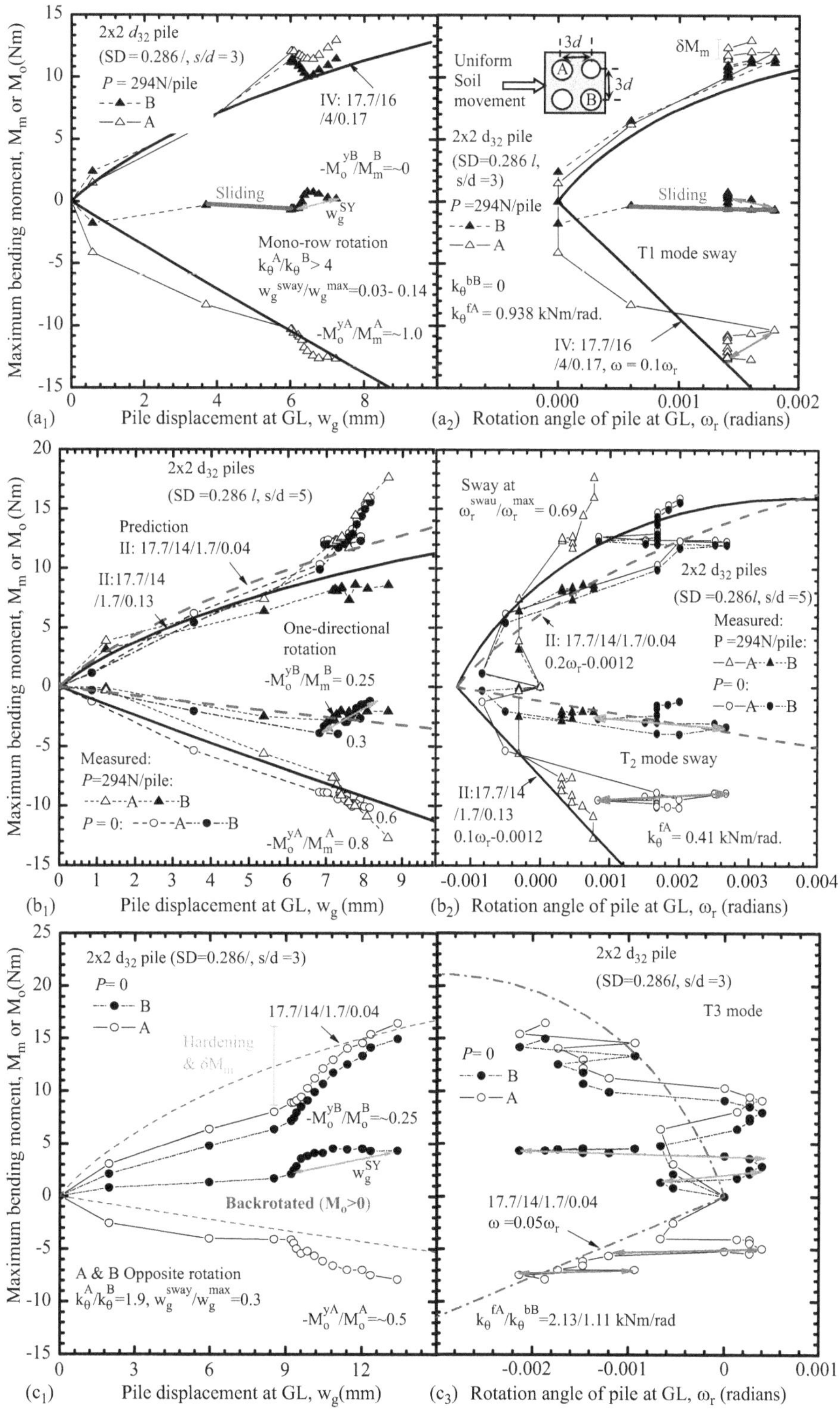

Figure 6.14 Response of 4-pile groups (s/d = 3, 5) of (a$_1$) (b$_1$)(c$_1$) w_g – M_m or M_o; (a$_2$) (b$_2$)(c$_2$) ω_r – M_m or M_o. [Adapted from Guo, W. D., *Int J Physical Modelling in Geotechnics* **22**(4), 2022b.]

6.5.2 Sway at SD of 0.286*l* and δM_m–M_m–ω_r^{max}–w_g^{SY}

The measured w_g (ω_r) – M_m (or M_o) curves are fitted with 3-layer model predictions (Guo 2016) using a modulus ratio m of 17.7 and the sets of I, II, III, IV parameters k_s (kPa, modulus), p_b (kN/m), and $\bar{k}_\theta\left[=k_\theta/\left(k_s l^3\right)\right]$ (see Table 6.2). The parameters and the angle-change ω were deduced and are shown in pertinent figures. The M_m, M_o^y, M_o and $\bar{\omega}_r$ values (for piles A and B) are obtained and tabulated in Table 6.5 for ultimate state, and Table 6.6 for middle stress level and ultimate state (note that $\bar{\omega}_r^{max}=\omega_r^{max}/\omega_r^{FH}$) at SD = 0.57*l*. The response, for instance, reveals T_1, T_2, and T_3 modes of sway (see red and magenta lines) (Guo 2022b).

Example 6.5 Mono sway (T_1)

Under the load P, Figures 6.14a$_1$ and a$_2$ show mono sway of the pile A in 4P(3) group: (i) the bending moment in pile-A row is much larger than that of pile-B ($k_\theta^{fB}\approx 0$), and $M_o^{yA}/M_o^{yB}=7.2$; (ii) the 2-way sways at M_m^A incurs a small w_g^{SY}/w_g ratio (of 0.03–0.14) and a moment raise δM_m (of $0.127M_m$) for the free-head condition of $\bar{\omega}_r^{max}$ $(=0.97)$. The sliding alters the shape of rotation $\omega_r(z)$ profile and raises the $M(z)$, which barely varies during the sway. Non-homogeneity of k_s needs to be considered to estimate pile response profiles.

Example 6.6 General rotational sway (T_2)

At $P=0$, the 4P(5) group back-rotates (Figure 6.14b$_2$), then rotates forward at $k_\theta^{fA}/k_\theta^{fB}=3$. At a moment of $0.7M_m^A$, it back-rotates again to incur a large (T_2) sway of $0.69\omega_r^{max}$ within the soil (with negligible pile-cap movement, Figure 6.15a$_1$ and a$_2$), and δM_m of $0.326M_m$ at k_θ^{SY} (of 2 kN·m/radians). At the same $\bar{\omega}_r^{max}$ (of 0.97) as T_1 mode, the group yields at $M_o^{yA}/M_o^{yB}=2.6$ and $0.7M_m^A$. During the T_2 sway (of pile B, Figure 6.14b$_2$, at w_f = 50–90 mm), the shape change in the rotation profile and the on-pile FPUL $p(z)$ is limited, similar to T_1 mode (see Figure 6.15).

Table 6.6 Measured normalized angle $\bar{\omega}_r$ and moments (SD = 0.57*l*) (Guo 2022b)

Groups	*P = 294 N/pile*				*P = 0*			
	2P(3)-row	*2P(3)-line*	*4P(5)*	*4P(3)*	*2P(3)-row*	*2P(3)-line*	*4P(5)*	*4P(3)*
$\bar{\omega}_r\left(\omega_r^m\right)$	1.0 (<0.11)	0.4 (<0.03)	0.23 (<0.01)	0.51 (<0.02)	0.93 (<0.12)	0.13 (0.01)	0.235 (<0.01)	0.42 (<0.01)
M_o^{yA}(N·m)	-5.59	-11.0	-21.1	-2.61	-4.42	-1.45	-5.91	-7.44
M_o^{yB}(N·m)	-7.98	-20.84	-15.13	-11.25	6.09	-1.07	4.91	-5.45
M_m(N·m)	92	76	72	52[a]	82	40	0	48
$\bar{\omega}_r^{max}$ [c]	1.36 (1.10)	1.15 (0.80)	1.05 (0.90)	1.4 (1.26)	1.47 (0.92)	0.49 (0.42)	1.05 (0.97)	1.33[b] (1.12[d])
M_m^A–M_m^B(N·m)	111–90	78–143	72–100	78–80[b]	82–79	39–36	44~	63
M_o^A(N·m)	35.3	6.4	12.1	39.2	4.42	1.69	5.58	5.3
M_o^B(N·m)	19.48	5.59	13.08	17.25	1.67	-4.31	4.94	-8.41
M_o^{yA}/M_o^{yB}	1/1.43[e]	1/1.9	1.39[f]	1/4.3[g]	0.73	1.36[f]	1.20[f]	1.37[f]
M_o^A/M_o^B	1.81	1.14	0.93	2.27	2.65	0.39	1.13	0.63

Notes:
[a] As with tests for $P=0$, owing to 'sliding'.
[b] The middle phase is $\bar{\omega}_r=0.82$, M_m^A–M_m^B = 63–47 (N·m).
[c] gained from gradients of w_g–ω_r plots.
[d] in brackets from measured ω_r^m, and w_g^m at the M_m.
[e] $M_o^{yA}/M_o^{yB}=0.7$ or $M_o^{yB}/M_o^{yA}=1.43$.
[f] Pile A slid initially.
[g] piles failed.

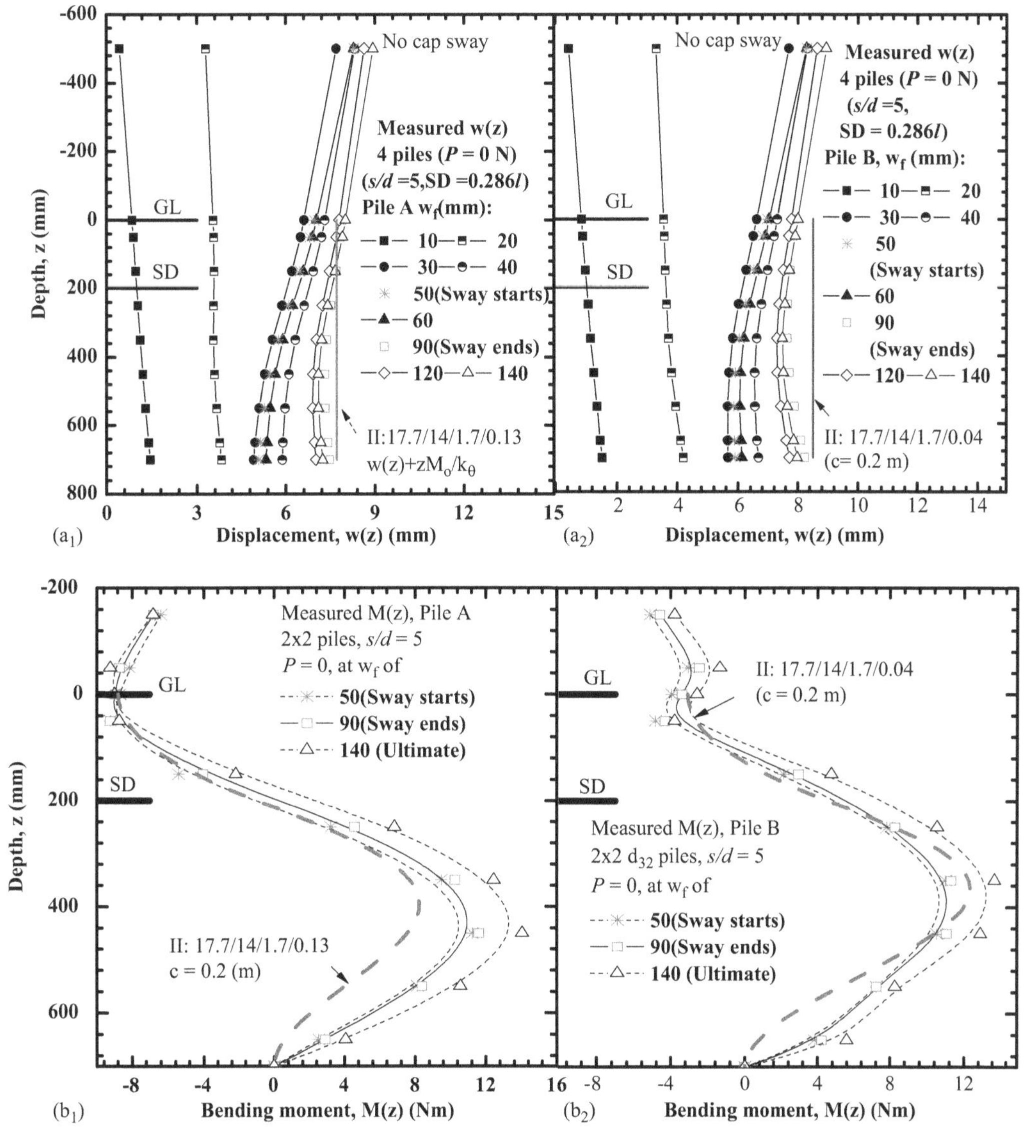

Figure 6.15 Response profiles of 4-pile groups from sway (T_2): (a_i) $w(z)$; (b_i) $M(z)$; and (c_i) $p(z)/d$ (i = 1, 2 for pile A and B, SD = 0.286l and s/d = 5). [Adapted from Guo, W. D., *Int J Physical Modelling in Geotechnics* **22**(4), 2022b.]

Example 6.7 Raked, inward rotational sway (T_3)

Heads of Piles A and B in the 4P(3) group (P = 0) rotate towards each other (countering rotations, T_3) within SD at k_θ^{fA} and k_θ^{fB} (Figure 6.14c$_1$), and yield at M_o^{yA}/M_o^{yB} = 2. An unequal 2-way sway occurs twice at a moment of (0.5–0.7)M_m^A, as is evident in the $w(z)$ profiles (Figure 6.16). The group shows cap-sway movement (see Figure 6.16). This reduces the normalized rotation angle $\bar{\omega}_r^{max}$ to 0.47, and the sway-displacement w_g^{SY} to 0.3w_g (Figure 6.14c$_1$), but raises δM_m to .84M_m at k_θ^{SY} (of 3.28 kN×m/radians).

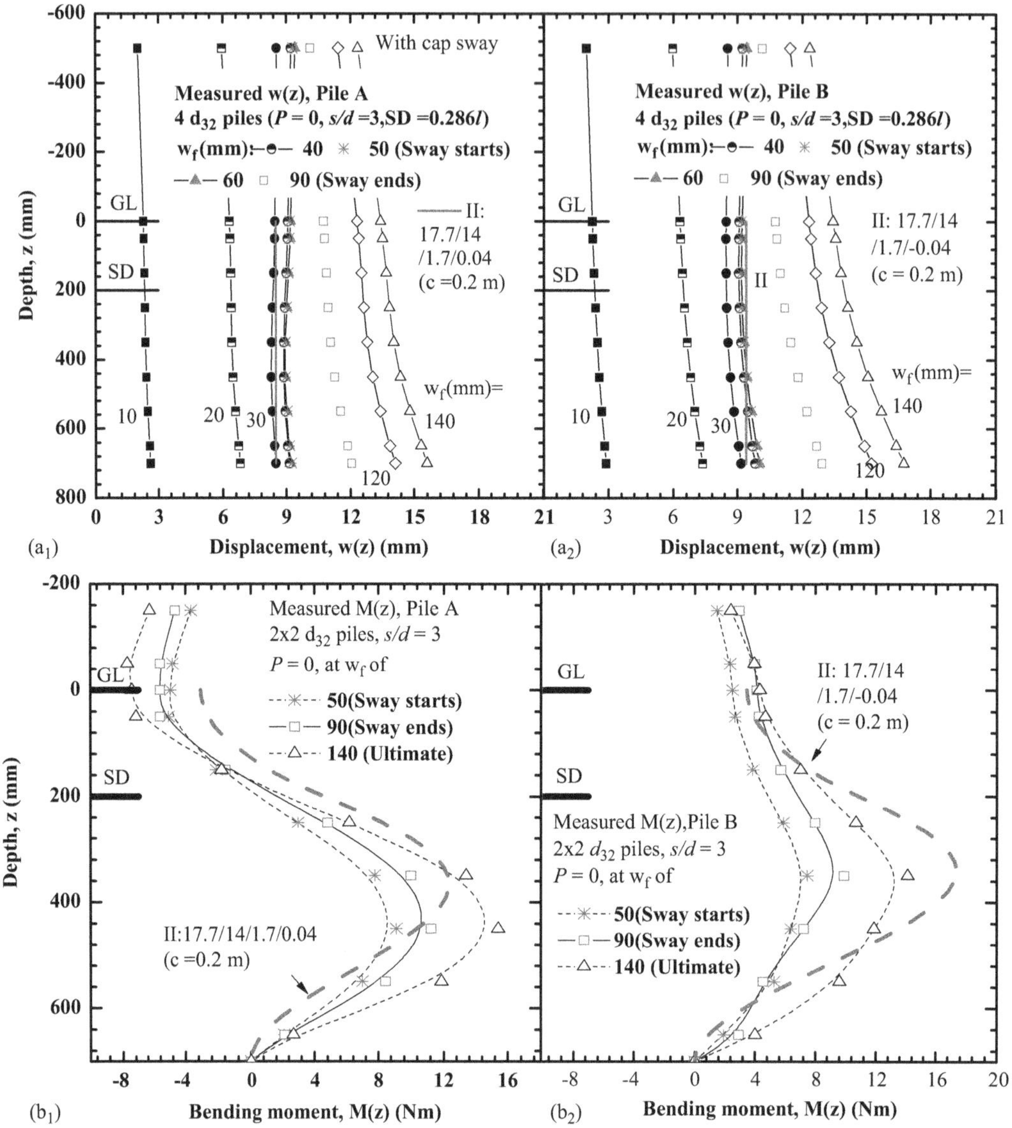

Figure 6.16 Response profiles with back-rotated piles (T_3): (a_i) $w(z)$; and (b_i) $M(z)$ (i = 1, 2 for pile A and B at SD = 0.286l and s/d = 3). [Adapted from Guo, W. D., *Int J Physical Modelling in Geotechnics* **22**(4), 2022b.]

The T_2 and T_3 modes of sway (for 'capped' piles) induce a limiting M_m that is about 20% larger than that of the 'mono' (T_1) pile. The sway incurs moment redistributions to M_o^{yA}/M_m = 1.0 (T_1) and 0.6–0.8(T_2), and M_o^{yA}/M_o^{yB} = 7.6(T_1), and 2.6(T_2) at near free-head rotation $\bar{\omega}_r^{max}$ of 0.97 (T_1 and T_2 modes). 'Raked' (T_3) piles involves a low rotation $\left(\bar{\omega}_r^{max} = 0.47 - 0.61\right)$, and relatively evenly distributed moment (M_o^{yA}/M_m = 0.5, M_o^{yA}/M_o^{yB} = 2.0), but a larger moment raise. The ratios are different from those for lateral piles, and need to be considered in pile design.

Example 6.8 Normalized sway displacement and rotation

The ratio of w_g^{SY}/w_g^{max} reaches 0.03–0.14 ('mono' and general piles), and 0.3 ('raked' piles), with an average of $w_g^{SY} = 0.2w_g^{max}$ and $\omega_r^{SY} = 0.4\omega_r^{max}$. These ratios are scalable. At $w_g^{max} = 1$ m and $\omega_r^{max} = 12.5°$, for instance, a prototype pile may incur a residual displacement of 0.2 m for a sway of 0.2 m to 5° rotation. These scaled values roughly agree with the measured (residual) displacement of 0.15–0.26 m, and 5° tilt of the piles underpinning the Anzac bridge experiencing a soil movement ($\approx 1.2w_g^{max}$, assuming $\alpha = 0.833$) of 0.9–1.2 m.

Example 6.9 Normalized rotation angle

The maximum $\bar{\omega}_r^{max}$ values were estimated at the 'sway' stage (see Table 6.7) using the measured displacement and rotation angle of a few project piles. The $\bar{\omega}_r^{max}$ values of 0.491–0.77 indicate 'stable' piles but not for cap-sway piles. In contrast, the high values of 1.05–1.4 indicate a degeneration of capped-head restraint for exceeding unity of free-head piles, as observed in the 1 g model tests at SD = 0.286l, and possibly in the Anzac and South Brighton bridges. Finally, neglecting the 0.5 m- free-length (see Figure 6.13), the 3-layer model overestimates the on-pile pressure $p(z)/d$ (see Figure 6.17), but it is still sufficiently accurate to capture other response of the piles. Notably, the predicted on-pile pressure (of 5–25 kPa) matches well with that induced by lateral spreading (Dobry et al. 2003; Guo 2015b).

Table 6.7 Normalized angle $\bar{\omega}_r^{max}$ estimated using measured (real projects) (Guo 2022b)

References	*Length (m)*	*w_g (m)*	*ω_r^m (×10−^{3}) radians*	*ω_r^{FH} (×10−^{3})*[a]	$\bar{\omega}_r$
In situ test piles					
Bransby and Springman (1996)	19	0.08	3.1	6.32	0.491
UCSD pile 1	8.5	0.32	38.4	56.47	0.68
UCSD pile 2	8.5	0.6	64.58	105.88	0.61
WU pile	8.8	0.42	48.87	71.59	0.683
9-pile group	9.0	0.18	6.11	30	0.204
Abdoun and Wang (2002)	8.0	0.85–0.93			0.667
Chow (1996)	6.5	0.0027	0.48	0.623	0.77
Anzac	15–22	0.15–0.26[b]	87	20–25	3.5–4.4[c]
South Brighton	18.7	0.22–0.26[b]	122	18–21	5.9–6.9[c]
Model test piles					
Ghee (2009)		SD = 0.286l, $\bar{\omega}_r^{max}$		SD = 0.571l, $\bar{\omega}_r^{max}$	
2-pile (line, P = 0/294 N/pile)	0.7	0.29–0.6[d]/0.39[d]		0.13[e]–0.49/0.4[e]–1.15	
4-pile (s/d = 3, P = 0/294N/pile)		0.06–0.47[d]/0.12–0.97[d]		0.82[e]–1.33/0.51[e]–1.4	
4-pile (s/d = 5, P = 0/294N/pile)				0.1[e]–1.05/0.23[e]–1.05	

Notes:
[a] $\omega_r^{FH} = 1.5w_g/l$
[b] residual displacement.
[c] drop by 80% if using pile-displacement (< w_f).
[d] sway level.
[e] at M_o^Y point.

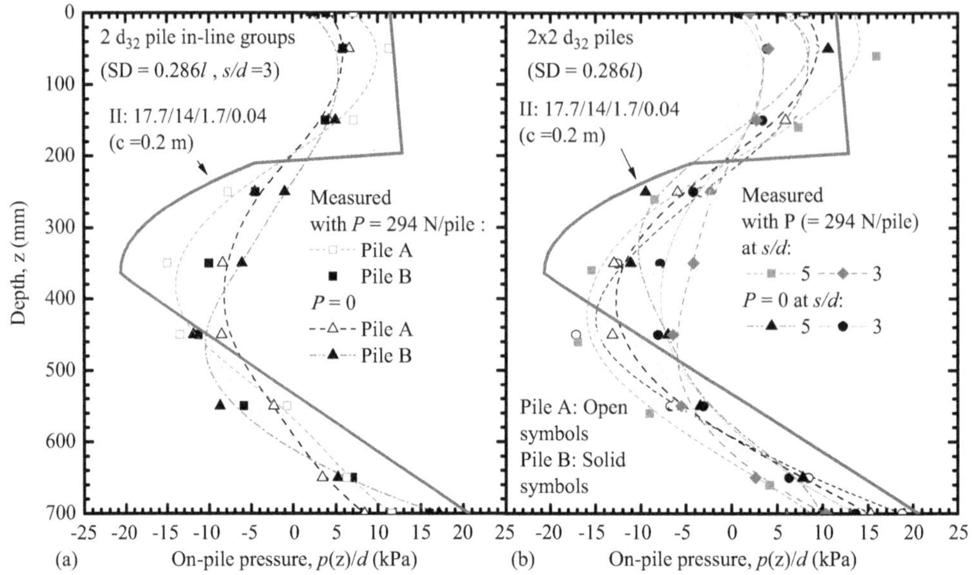

Figure 6.17 Ultimate on-pile pressure $p(z)/d$: (a) 2-pile in-line groups; (b) 4 pile groups. [Adapted from Guo, W. D., *Int J Physical Modelling in Geotechnics* **22**(4), 2022b.]

6.6 SLIDING, BACK-ROTATION AND LIMITING STATE AT SD = 0.57*L*

At a sliding depth of $0.57l$, model piles slide and back-rotate to an angle of 8°–16.5° (see Figure 6.18), which conform to the pile performance during Christchurch quake. Guo (2022b) analysed these tests [especially on 4P(3) and 4P(5) groups] and offered simple expressions to estimate the limiting movement w_f^*, angle of pile rotation $\bar{\omega}_r^{max}$ and bending moment M_m (along four failure paths). The main findings are presented in this section.

6.6.1 Moment M_o^y, limiting soil movement w_f^*, no sway

At SD = $0.57l$, the observations are as follows:

- The displacement w_g and angle ω_r increase with soil movement w_f (see Figure 6.18) without sway in the w_g–ω_r curves.
- $\alpha = 0.5$–1.1, and in line with $\alpha = 0.58$–0.67 for lateral spreading (Guo 2015b).
- The piles may slide (at $M_o \approx 0$) to a large rotation angle; and/or attain the GL (hinge) moments $-M_o^y$ at a small angle (similar to SD = $0.286l$); and subsequently, back-rotate to raise values of w_g, ω_r, M_m and M_o (positive) continuously until they are restrained by the maximum angle ω_r^{max}, the critical w_f^* - ω_r^{max} line in Figure 6.18b, or the M_m hinges.
- The maximum back-rotation M_o increases with $\bar{\omega}_r$ at a rate up to 50% less than that of forward rotation M_o^y.
- The sliding of piles (at $P = 0$) renders $M_o^{yA}/M_o^{yB} > 1$; otherwise without sliding, $M_o^{yA}/M_o^{yB} < 1$ (e.g., the 2-pile in-line groups in Figure 6.19).

Compared to SD = $0.286l$ (see Table 6.5), the $-M_o^{yA}$ value at SD = $0.57l$ (see Table 6.6) is largely small but for 4P(3). The $-M_o^y/M_m$ ratio for all tests (see Table 6.6) reduces by a

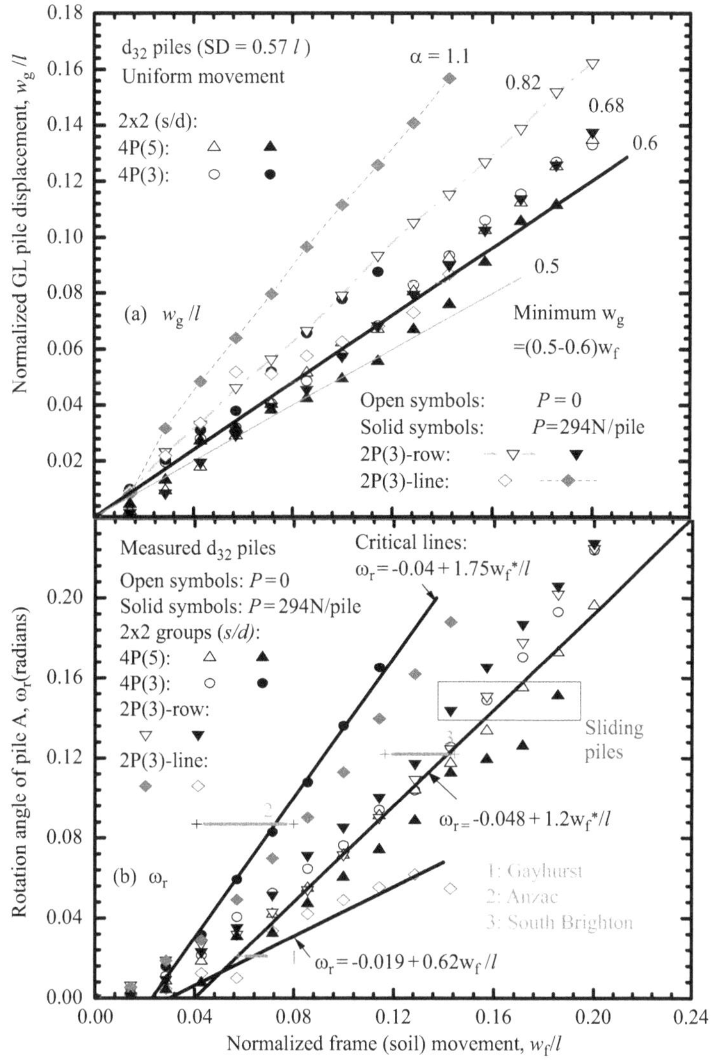

Figure 6.18 Development of ground-level pile displacement w_g and rotation ω_r (all tests at SD = 0.57*l*): (*a*) $w_f/l - w_g/l$; (**b**) $w_f/l - \omega_r$. [Adapted from Guo, W. D., *Int J Physical Modelling in Geotechnics* **22**(4), 2022b.]

factor of 25–60, from 0.5–1.2 (SD = 0.286*l*, see Figure 6.14) to 0.02 (SD = 0.57*l*). The ratio at SD = 0.286*l* largely exceeds 0.4 that is adopted for lateral piles (JGJ 1994). The pile-cap connection M_o^y should take a maximum of $1.2M_m$ during passive loading. Indeed, restraints at the pile-cap and base (as seen at SD = 0.286*l*) will promote sway of piles during ground movement.

6.6.2 Normalized angle $\bar{\omega}_r^*$ at critical state

Each measured pair of ω_r^{max} and w_g^{max} are used to estimate the $\bar{\omega}_r^{max}$. The M_m values are measured for the normalized angle $\bar{\omega}_r$ and $\bar{\omega}_r^{max}$ for stable and ultimate state. This offers the critical lines (see Figure 6.18b) of $\omega_r = -0.04 + 1.75(w_f^*/l)$ (without sliding), or $\omega_r = -0.048 + 1.2(w_f^*/l)$ (with sliding). Piles may fail along one of the 'four' lines (Figure 6.20) as discussed later.

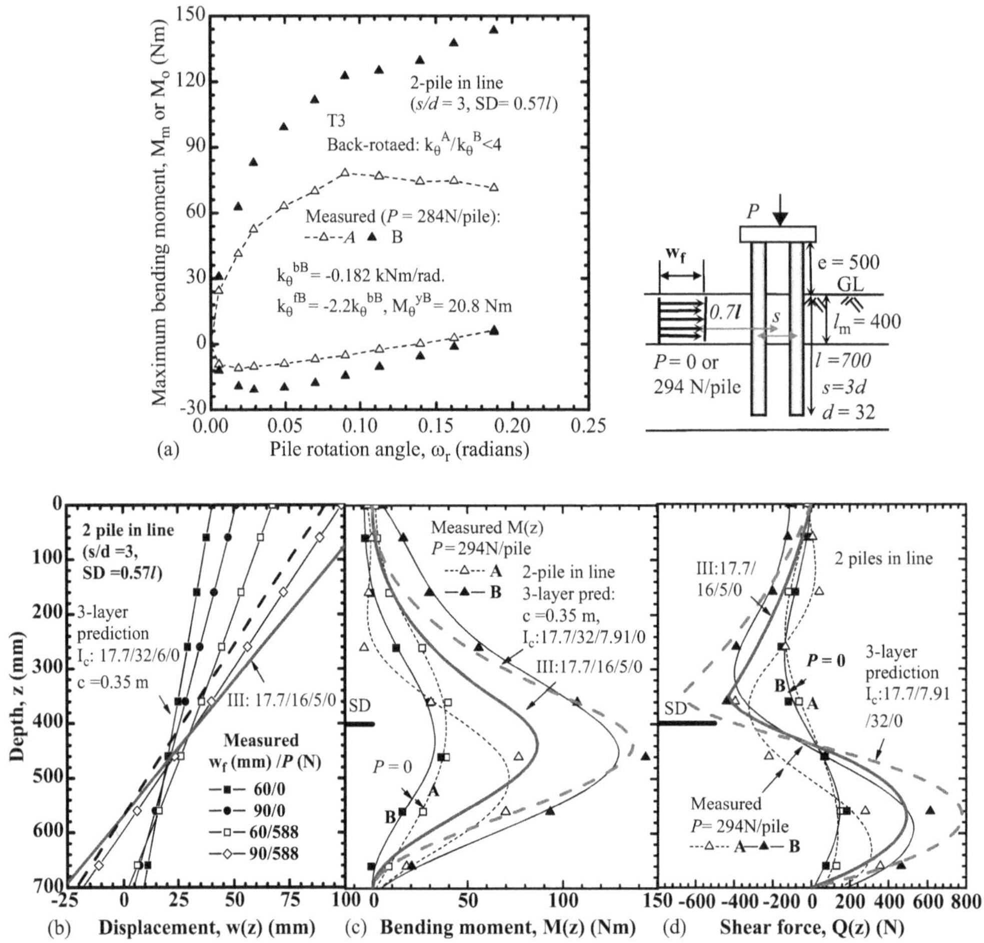

Figure 6.19 Predicted versus measured profiles of piles in 2-pile in-line group: (a) M_o (M_m) − ω_r; (b) $w(z)$; (c) $M(z)$; (d) $Q(z)$ (SD = 0.57l and s/d = 3). [Adapted from Guo, W. D., *Int J Physical Modelling in Geotechnics* **22**(4), 2022b.]

6.7 M_M REDUCTION/INCREASE FROM SLIDING/BACK-ROTATION

Bending moment increases by δM_m without sliding at $P > 0$, or remains negligible up to a sliding angle $\delta\theta^S$ at $P = 0$. The variation of M_m (i.e., δM_m) due to sliding and back-rotation is quantified through a sliding stiffness k_θ^S (= $\delta M_m/\delta\theta^S$) (Guo 2022b).

Example 6.10 Stiffness and bending moment raise

For pile B in 4P(5) (see Figure 7.16b, Chapter 7), the k_θ^S is deduced as 742.9 N·m/radians for δM_m of 48.3 N·m over $\delta\theta^S$ of 0.065 radians. The k_θ^S value is similar to forward rotation k_θ^{fA} (of 0.67–2.13 kN·m/radians).

The maximum moment M_m (i.e., M_m^A, M_m^B for pile A or B) encompasses standard M_m (of 100 N·m) and back-rotating moment (of 13.1 N·m), a sliding component (at k_θ^S = 742.9 N·m/radians) over the angle $\delta\theta^S$, and back-rotating moment M_o measured (or estimated

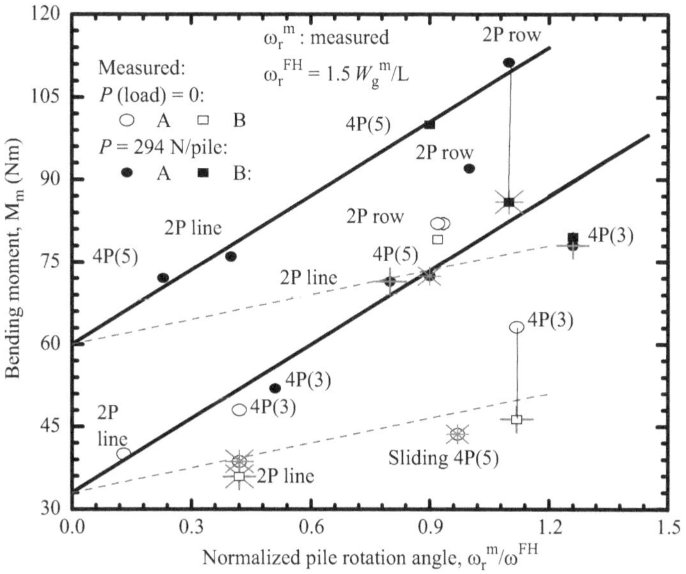

Figure 6.20 Measured moment M_m versus normalized ω_r^m/ω_r^{FH}. [Adapted from Guo, W. D., *Int J Physical Modelling in Geotechnics* **22**(4), 2022b.]

using 3-layer model, Guo 2020b). Assuming hinge forms at a depth l_h of 0.7l, the maximum moment M_m for a prototype pile is given by:

$$M_m = \left(86.9 - k_\theta^s \delta\theta^s + M_o\right)\left(l_h / l_o\right)^3 \tag{6.1}$$

where M_m in N·m, l_o = 0.7 m, embedment of model pile, l_h = length (m) of rigid pile or length between hinges.

Example 6.11 Estimation of maximum bending moment

For pile A (with P) of 4P(5) (see Table 6.8, Figure 7.16b, Chapter 7), the δM_m is estimated as 29.7 N·m (= 0.04 radians × 742.9 N·m/radians) for a sliding $\delta\theta^s$ of 0.04 radians. Using the measured M_o^A (of 12.1 N·m), the M_m^A is estimated as 69.3 N·m (= 86.9–29.7 + 12.1) given $l_h = l$. This value agrees with the measured 72 N·m. Likewise, the M_m^A and M_m^B values for other piles were estimated using the measured $\delta\theta^S$. The estimations are largely within 10% difference from the measured M_m values (see Table 6.8), as overestimation (P = 0) and underestimation (P > 0) for the piles in 2-pile-line groups are owing to a higher and a lower sliding k_θ^S than the adopted 742.9 N·m/radians.

6.8 FOUR $M_m - \bar{\omega}_r$ FAILURE PATHS

Figure 4.20 provides the four failure lines, as follows.

Lower bold-line for piles with sliding: The measured moment M_m is plotted against angle $\bar{\omega}_r$ (at P = 0). The 4P(3) group failed below the 'bold-line' with 1.33 $\left(\bar{\omega}_r^{max}\right)$ and 63 N·m (M_m).

Upper bold-line (for piles without sliding at P = 294 N/pile): Compared to the sliding piles, the moment (δM_m) rises by 27 N·m over a sliding angle $\delta\theta^S$ of 0.0363 radians (see Figure 7.16b). Piles A and B in 2P(3)-line group stretch to 1.15 $\left(\bar{\omega}_r^{max}\right)$, which lifts the M_m transiently and causes nonhomogeneous-rotation.

Table 6.8 Estimation of M_o and M_m considering sliding reduction (Guo 2022b)

Groups	*P = 294 N/pile*				*P = 0*			
	2P(3)-row	*2P(3)-line*	*4P(5)*	*4P(3)*	*2P(3)-row*	*2P(3)-line*	*4P(5)*	*4P(3)*
$\delta\theta^{Sa}$	0.01	0.0	0.04	0.05	0.02	0.02	0.065	0.04
δM_m	7.4	0	29.7	37.1	14.9	14.9	48.3	29.7
M_o^A(N·m)	35.3	6.4	12.1	39.2	4.42	1.69	5.58	5.3
M_m^A(N·m)	115	78.4	69.3	89	76.5	73.7	40	62.5
M_m^{Am}(N·m)	111	93.3	72	78	82	39[c]	44	63
M_m^A/M_m^{Am}	1.03	0.83	0.96	0.88	0.93	0.53	1.0	1.0
$\delta\theta^{Sa}$	0.02	0	0.	0–0.04[b]	0.02	0.02	0.065	0.04[b]
δM_m	14.9	0	0	0–29.7	14.9	14.9	48.3	29.7
M_o^B(N·m)	19.48	5.59	13.08	17.25	1.67	−4.31	4.94	−8.41
M_m^B(N·m)	91.5	92.5	100	105–74.4	73.7	67.7	43.55	48.8
M_m^{Bm}(N·m)	90	143	100	80	79	36 [c]	44	48
M_m^B/M_m^{Bm}	0.98	1.55	1.0	0.8–1.1[b]	0.93	0.53	1.01	0.98

[a] measured for (equivalent) sliding of negligible moment;
[b] Equivalent sliding for M_o = constant.
[c] not reaching the upper limit, which cause overestimation.

Lower dash-line for yielding piles at $P = 0$, including 4P(5), 4P(3). The pile-rotation jumps inward at a stable displacement prior to yielding at 0.42–0.49 $\left(\bar{\omega}_r^{max}\right)$, comparable to the 'raked' 4P(3) piles at SD = 0.286l, Figure 6.14c$_1$.

Upper dash-line for yielding piles with back-rotation (at P = 294 N/pile): A large back-rotation (of 0.15 radians) raises the moments by 36 N·m at a stiffness of 0.24 kN·m / radians, which shifts the lower dash-line upper.

The 2P(3) and 4P(3) groups slide to over 10 times larger angle (of 0.01–0.03 radians) than those at SD = 0.286l (see Figure 6.14). The 4P(3) group constitutes the equation without sliding (see Figure 6.18b) and residual displacement (w_g^{SY}). The M_m and ω_r^{max} values along the four paths may be adopted to assess impact of sliding and back-rotation along with the 3-layer model (Guo 2020a). The features (of sway, sliding and back-rotation) embedded in model piles are consistent with the behaviour discussed next on piles underpinning Christchurch bridges.

Example 6.12 Estimation of critical soil movement

Formation of hinges at $w_f > w_f^*$(SD = 0.57l): The 'pile-heads' underpinning Gayhurst, Anzac, and South Brighton bridges record a permanent rotation angle ω_r of 1.2°, 5° and 7° (= 0.021, 0.087, and 0.122 radians), respectively (Cubrinovski et al. 2014). The limiting w_f^*, based on the upper critical line (see Figure 6.18), is estimated as 0.0349l, 0.0726l and 0.0926l, respectively, and w_f^* = 0.36 m (l = 10.4 m) for Gayhurst, 1.09–1.6 m (l = 15–22 m) for Anzac, and 1.72 m (l = 18.7 m) for South Brighton bridges. Figure 6.18b demonstrates that the measured w_f (of 0.6–0.75 m) around Gayhurst bridge reaches (1.7–2.1)w_f^* (w_f^* = 0.36 m); and the measured w_f (of 2.18–2.7 m) around South Brighton bridge reaches (1.27–1.57)w_f^* (w_f^* = 1.72 m). The measured w_f (of 0.9–1.2 m) around the Anzac bridge was largely less than its limiting w_f^* (of 1.1–1.6 m). It did not induce sufficient sway and/or hinges. South Brighton Bridge, underpinned by raked piles, fell into the 'sliding' zone, and indeed slid by 0.2 m, With $w_f > w_f^*$, sway may have incurred a limiting pile-soil interaction or even GL hinges.

Table 6.9 Calculation of w_g and M^* of Christchurch piles (Guo 2022b)

Name	w_f^*/l (m)	Total w_g^*(m)[a]	Residual w_g (m)[b]	M^* (kN·m)[c]	Measured			
					Tilt [d]	w_g(m)	w_f(m)	M_o^y (kN·m)
Gayhurst	0.3/10.4	0.18	0.04–0.05	112	1.2°	~	0.6–0.75	128.6
Anzac	1.1–1.6/ 15–22	0.66–0.96	0.15–0.26	386–1,217	5.0°	0.2–0.25	0.9–1.2	689.0
South Brighton	1.6/18.7	0.96	0.22–0.26	747	7.0°	0.2	2.18–2.7	468.0

Notes:
[a] $w_g^* = 0.6w_f^*$(m).
[b] $w_g = (0.23–0.27)w_g^*$(m).
[c] Maximum moment for back-rotation M^* (kN·m) = $0.0392(l/0.7)^3$.
[d] Backward tilt(angle).

Example 6.13 Estimation of permanent displacement using w_g^{SY} (SD = 0.57l)

The sway displacement w_g^{SY} reaches $(0.14–0.3)w_g^{max}$ at limit state of back-rotated pile B, and SD = 0.286l. It may attain $(0.23–0.27)w_g^{max}$, as induced in the yielding, 4-pile group [4P(3) at P = 294 N/pile] assuming rotational restraints at pile-cap and base levels at SD = 0.57l (Guo 2022b). On the other hand, at critical state, the w_g^{max} is, on average, equal to $0.6w_f^*$ (see Figure 6.18a, Table 6.9), which offer $w_g^{SY} = (0.138–0.162)w_f^*$. The w_g^{SY} values for the movement w_f^* are estimated as 0.15–0.22 m [= 0.6(0.23 × 1.1 to 0.27 × 1.6), w_f^* = 1.1–1.6 m] for the straight piles of Anzac bridges; and 0.237 m (w_f^* = 1.72 m) for the raked piles of South Brighton bridges. The estimated w_g^{SY} values (regarded as residual displacement) agree with the (reported) pile movement of 0.2–0.25 m and 0.2 m, respectively (see Table 6.9). This success implies a sufficient accuracy of the model tests on socketed pile-cap connection to 'replicate' field pile behaviour.

Example 6.14 Estimation of yielding bending moment

Bending moment increase from sway may be offset by the decrease from sliding. The back-rotating moment M_o^A [4P(3), Figure 7.13, Chapter 7] of 39.2 N·m (see Table 6.6) is then scaled to M_o^A at GL of 128.6 kN·m [= $(l/0.7)^3$ × 39.2, $l_h = l$ = 10.4 m] (Gayhurst), 385.7–1216.9 kN·m (Anzac), and 743.3 kN·m (South Brighton bridges), respectively. Taking a typical M_m^A of 80 N·m [non-sliding, 4P(3) at l = 0.7 m], the 'hinge' moment at l_h of 0.7l is up-scaled to 90.0 kN·m (= $80l_3$, l = 10.4 m) (Gayhurst), and 270.0–851.8 kN·m (Anzac), using Equation (6.1). It is scaled to 458 kN·m (South Brighton bridges) using M_m^A of 70 N·m [sliding, 4P(5)]. The estimated M_o^A – M_m^A values are 90–128.6, 385.7–851.8, and 458–743.3 kN·m, which are consistent with the reported yielding moments of 112.0, 689.0 and 468.0 kN·m, respectively (Haskell et al. 2013) despite 'unknown' pile-cap connections.

6.9 CONCLUSIONS

This chapter presents model tests and 3-layer modelling of single and capped piles subjected to pulse-like soil movement. It provides:

- a simple way to simulate evolution of soil movement;
- conditions to induce sliding, sway and back-rotation of piles;
- methods to estimate limiting soil movement, pile displacement, and hinge bending moment.

The chapter also quantifies the factor of safety under passive loading; and reveals the rotational restraint caused by non-uniform soil movement. It also provides simple estimations of limiting soil movement and permanent displacement of the piles underpinning Christchurch bridges. The methods complement the 3-layer modelling, allowing for a realistic assessment of piles exhibiting sliding, sway and back-rotation.

ACKNOWLEDGEMENTS

The model tests were supported by an Australian Research Council Discovery Grant (DP0209027). The financial assistance is gratefully acknowledged. The tests were conducted in Griffith University during 2003–2008 by Enghow Ghee, together with 14 Master's and PhD students. Gilles Ravanelli, Malcolm Duncan and Geoff Turner offered the technical support.

Chapter 7

Lateral spreading and instability of back-rotated piles

7.1 INTRODUCTION

The response of piles subjected to lateral spreading has been extensively investigated using shaking table tests, centrifuge tests and numerical simulations (e.g., Boulanger et al. 2003; Brandenberg et al. 2005a, b; Dobry et al. 2003; He et al. 2009; Bhattacharya et al. 2005). These studies to date provide insight into response of 'forward rotating' piles, but it is important to note that working piles may also back-rotate (Haskell et al. 2013) due to earthquake, tsunami, hurricane or erosion (Guo 2020a, b). To address these problems, a 2-layer model (Figure 1.4 and Chapter 4) has been developed to encapsulate the rigid pile (of length l) embedded in a sliding layer (with thickness l_m and modulus of subgrade reaction k_s) over a stable layer (subgrade modulus of mk_s). The movement of sliding soil (w_s) exerts a linearly increasing *FPUL* with depth up to p_b (= αp_{ub}, p_{ub} = limiting *FPUL* for free-head piles, α = modification factor) at pile-tip level. This *FPUL* forces the pile to rotate about a depth (z_r) to an angle (ω_r) and a ground level (GL) deflection (w_g). The rotational stiffness of the pile encompasses stiffness k_G at GL, stiffness k_T at tip and a total stiffness k_θ over its length.

To account for 'dragging effect', a 'fictitious' layer is added between the two layers to form the 3-layer model (Figure 1.5 and Chapter 1). Explicit solutions have been developed for both the 2-layer and 3-layer models (Guo 2015b, 2016), which are highlighted in non-dimensional charts, and exemplified for slope-stabilizing piles (Chapter 4), embankment piles (Chapter 5) and piles subjected to lateral spreading (Guo 2020a, b).

Initially, piles may rotate forward (see Figure 7.1b) at a stiffness k_θ^f (superscripts 'f' and 'b' denote forward- and back-rotation, respectively) until they yield at the moment M_o^y. Subsequently, the piles may rotate backward (see Figure 7.1c) at a stiffness of $-k_\theta$. Response of the piles, including displacement w_g and maximum bending moment M_o (at GL) and M_m (in-pile) (see Figure 7.1d), can be obtained in two steps (Guo 2020a, b): (i) by considering k_θ as negative to gain the response [e.g., $M_o(M_m)$–w_g curves] of a pure back-rotating pile (e.g., $M_o^{fA} = 0$ at $\omega_r = 0$, for pile A); and (ii) by translating the obtained M_o ($-w_g$) curves down through adding M_o^{fA} to M_o (see Figure 7.1d). It is critical to note that Guo (2020a,b) discovered that a negative k_θ incurs amplification of the response, especially around singularity stiffness of the 2- and 3-layer models. Amplified response from back-rotating piles caused collapse of Showa Bridge (Guo 2020b), which are characterized by m = 1.5, k_s = 30–50 kPa and p_{ub} = 30 kN/m, and $-\bar{k}_\theta\left[= k_\theta / \left(k_s l^3\right)\right]$ of 0.08–0.12 [written together as M/k_S(kPa)/p_{ub}(kN/m)/k_θ later].

The simple 2-layer model, in particular, effectively captured the pile response subjected to lateral spreading (for limited dragging effect) (Guo 2015b); and the response from back-rotation (e.g., induced by P–Δ_o effect where P is the vertical load on piles; and Δ_o is the initial lateral displacement) (Guo 2022a).

DOI: 10.1201/9781003315230-7

Relationship among moments and angle of rotations (Pile A)

$$M_m = -M_o^{fA} + k_\theta \omega_r, \quad k_\theta^f = M_o^y/\omega_r^f \quad M_o^{fA} = M_o^y + w_g^f \tan(\omega_r)$$

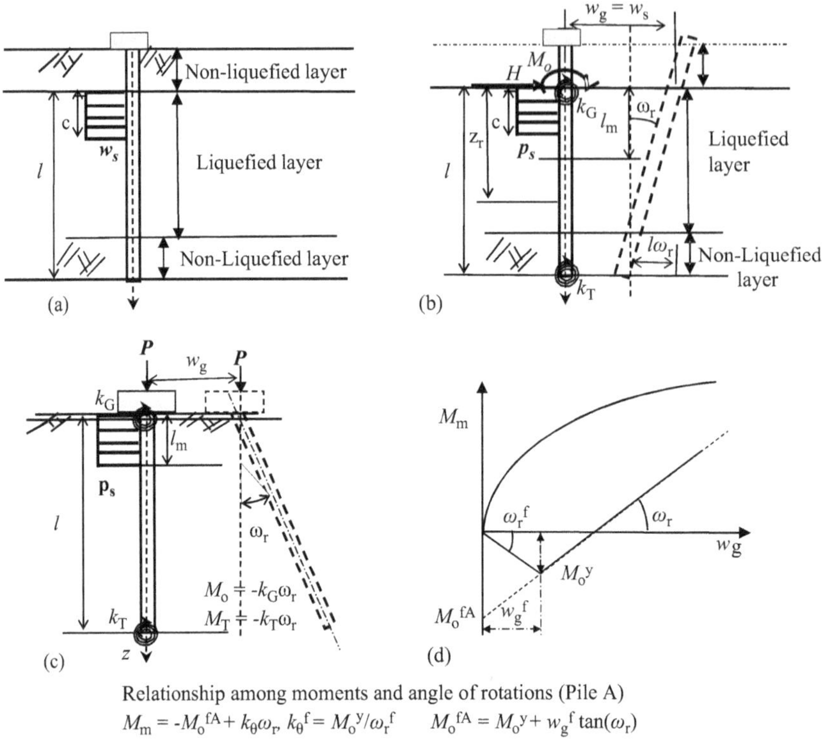

Figure 7.1 Schematic model of a capped pile subjected to lateral spreading: (a) a 3-layer lateral spreading pile-soil system; (b) forward-rotating piles ($k_\theta > 0$); (c) back-rotated pile ($k_\theta < 0$); (d) moments and angle of rotations (Pile A).   [Adapted from Guo, W. D., *Can Geotech J* **52**(7), 2015b.]

7.2 2-LAYER MODEL PREDICTIONS

During lateral spreading, each on-pile p_s induces a set of pile deflection $w(z)$, net *FPUL* $p_i(z)$, shear force $Q_i(z)$, and bending moment $M_i(z)$ for the sliding layer ($i = 1$), and the stable layer ($i = 2$), respectively. Chapter 4 provides expressions [e.g., Equations (4.1) through to (4.4)] for estimating the response. To predict nonlinear response of the pile, a linearly increasing force per unit length p_s (= $\alpha p_{ub} l_m/l$) with the normalized sliding depth (SD) (l_m/l) is used in the elastic solution. This simulation is also provided in this chapter for response of model piles with $k_G \neq 0$, and base rotationally constrained ($k_T \neq 0$) ($H = 0$), as well as piles subjected to full-length lateral spreading ($k_G = H = M_o = 0$, $k_T \neq 0$).

7.3 RESPONSE OF FORWARD-ROTATING PILES

Figure 7.2 shows the impact of base-rotational stiffness (k_T) (see Figure 7.1) on the distribution profiles along a free-head pile subjected to lateral spreading. The profiles shift from dashed lines for a floating-base pile to 'uniform' net *FPUL* $p_i(z)$ in each layer and displacement $w(z)$ with depth z for a fixed-base pile with sufficiently large stiffness. A 'triangular' p_s profile over depth $0\sim l_m$ is observed for a lightly head-restrained pile (e.g., $k_\theta = 0.1$), as stipulated previously (Dobry et al. 2003; He et al. 2009). The figure highlights the need to consider base-rotational stiffness (k_T) on the response of piles subjected to lateral spreading, as it significantly affects the pile response.

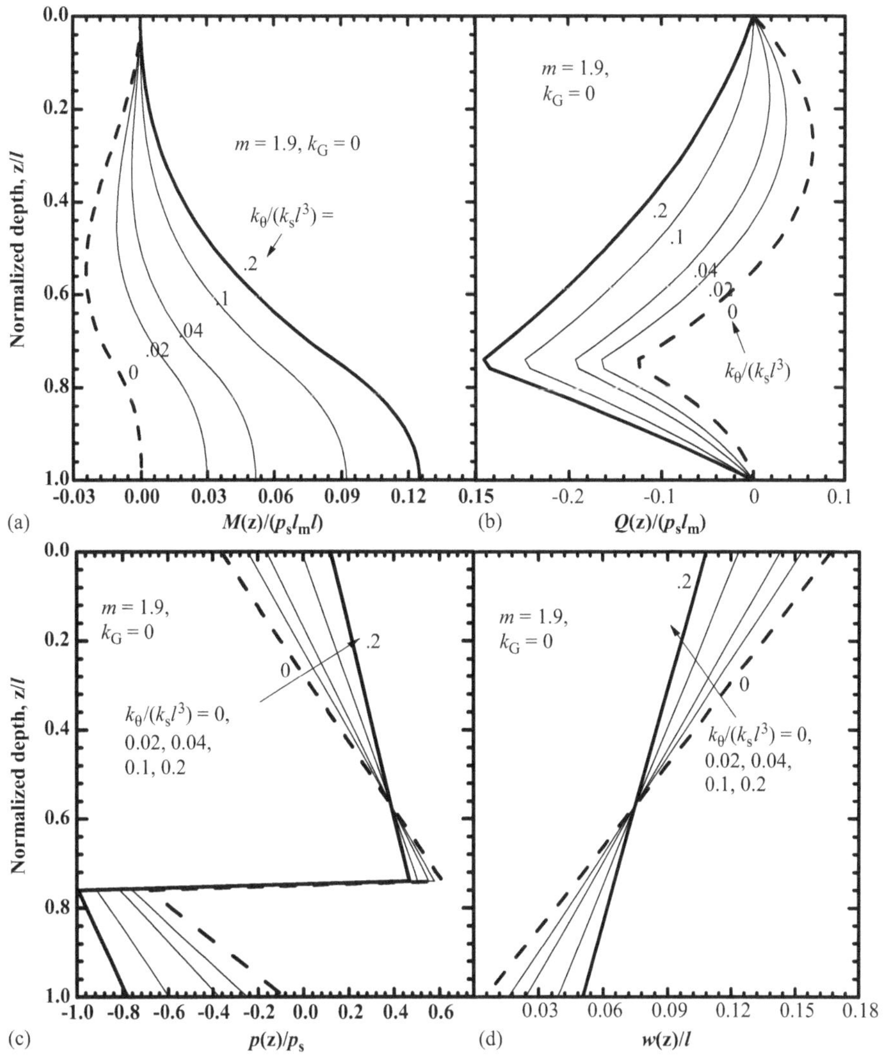

Figure 7.2 Normalized (a) $M(z)$; (b) $Q(z)$; (c) $w(z)$; (d) $p(z)$ for a normalized cap stiffness of 0–0.2 (m = 1.9, $\bar{l}_m$ = 0.75). [Adapted from Guo, W. D., *Can Geotech J* **52**(7), 2015b.]

7.3.1 Modelling centrifuge tests

Abdoun et al. (2003) conducted a series of eight centrifuge tests on nine models of single piles and pile groups. The models were excited in flight at an acceleration of 50 times gravity (50 g) and 40 cycles of base acceleration at a frequency of 2 Hz, with a uniform amplitude of 0.3 g (prototype). Accelerometers and pore pressure transducers were embedded to measure lateral accelerations and excess pore pressures in the soil, respectively; LVDTs were mounted on walls of the laminar box to monitor the free-field soil lateral deformations; and strain gauges were installed along the piles to gain bending moments.

The Model 3 test (renamed as C1-M3, see Table 7.1) on a single pile in a two-layer soil profile is simulated herein. The pile (l = 8 m) was embedded in a liquefiable sand layer (6 meters thick) with a relative density D_r of 40% overlying a slightly cemented sand layer (2

Table 7.1 Model predictions for 10 piles (Guo 2015b)

Cases	*l or c^a/l_m (m)*	*$m/p_b/p_s$ (kN/m)*	*k_s (kPa)/d (cm)*	*k_θ(MN m/rad)/$\bar{k}_{\theta*}$*	*Mea/Pre w_s (cm)*	*Mea/Pre M_m (kNm)*	*Mea/Pre. w_g (cm)*	*Reference*
			2-layer model for piles subjected to lateral spreading					
C1-M3[c]	8/6	1.9/22.5/4.5–9.7	23.5/47.5	3.82/0.317	78/76.8	113/114	27/23.9	Dobry et al. (2003)
C2-M5a[c]	8/6	5.3/30/6.7–14.4	23.5/47.5	3.93/0.327	77/67.2	170/170.3	35/35.2	
C3-M1[c]	3.6[a]/4.8	7/50/10.5–14.4	40/31.8	4.90/1.108	105/98.7	132/131	9.8[b]/9.8	He et al. (2009)
C4-M2[c]	1.28[a]/1.6	8/7.3/1.3–2.8	50/25.4	0.10/0.407	12.5/12.0	2.65/2.59	3.4[b]/3.0	
C5-M3[c]	1.2[a]/1.56	8/9.9/1.5–3.0	120/25.4	0.200/0.439	7.8/7.1	3.0/2.2	1.2[b]/1.4	
C6-M6[c]	1.2[a]/1.54	8/8.8/1.7–2.3	50/15.2	0.300/ 0.913	—/14.1	1.86/2.26	1.4/1.3	
			2-layer and 3-layer models for model piles in sliding soil					
C7-TS32-294	0.7/–0.7	11/12.5/2–7.0	60/3.2	0/0	0–14/—	0–0.175/0–0.175	0–6./0–6.	Guo and Qin (2010)
C8-TS32-0	0.7/~0.7	7/12.5/2–6.0	60/3.2	0/0	0–14/—	0–0.14/0–0.14	0–8./0–8.	
C9-TS50-0	0.7/~0.7	5.5/12.5/1.5–3.	280/5.0	0/0	0–14/—	0–0.08/0–0.08	0–0.8/0–0.8	Guo et al. (2006)
C10-TS50-294	0.7/~0.7	7.5/12.5/2.0–3.5	280/5.0	0/0	0–14/—	0–0.11/0–0.11	0–0.8/0–0.8	

Note: Note in the second through to fifth columns, the values of m, p_b, and k_s, along with the given values of l, l_m, d, and k_θ are all input values. Values of p_s are all calculated from p_1, and p_2 (Table 4.2, Chapter 4).

[a] The loading depth c as (−0.70–0.9)l_m for the single-layer, fixed-bead piles.
[b] At soil surface level.
[c] Letters 'Mi' and 'M5a' denotes original test name of 'Model i' and 'Model 5a'.

meters thick) with a cohesion of 5.1 kPa, and a friction angle ϕ of 34.5°. Figure 7.3 provides the measured ground movement (w_s), pile-soil relative displacement ($l\omega_r$), and maximum bending moments (M_m), which encompass both cyclic and permanent components. In particular, the moment M_m (located at a depth of 5.75 m in the liquefied layer) increased to 113 kNm as the pile-head deflection approaches 350 mm (?), and subsequently decreased with

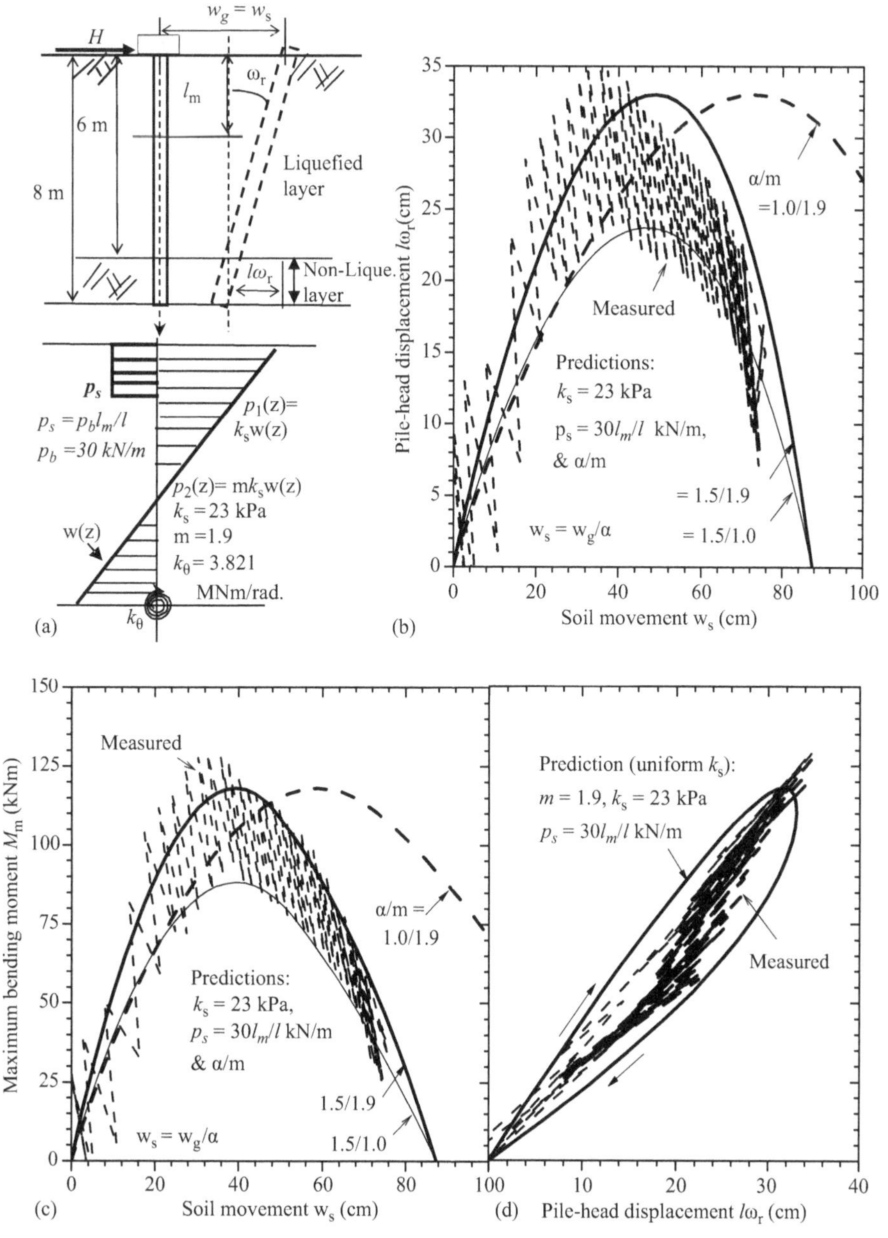

Figure 7.3 Prediction versus measured (Abdoun et al. 2003) response of Model 3 subjected to lateral spreading: (a) pile-soil interaction model; (b) w_s-pile-soil relative movement; (c) w_s–M_m; (d) pile-soil relative movement versus M_m. [Adapted from Guo, W. D., *Can Geotech J* **52**(7), 2015b.]

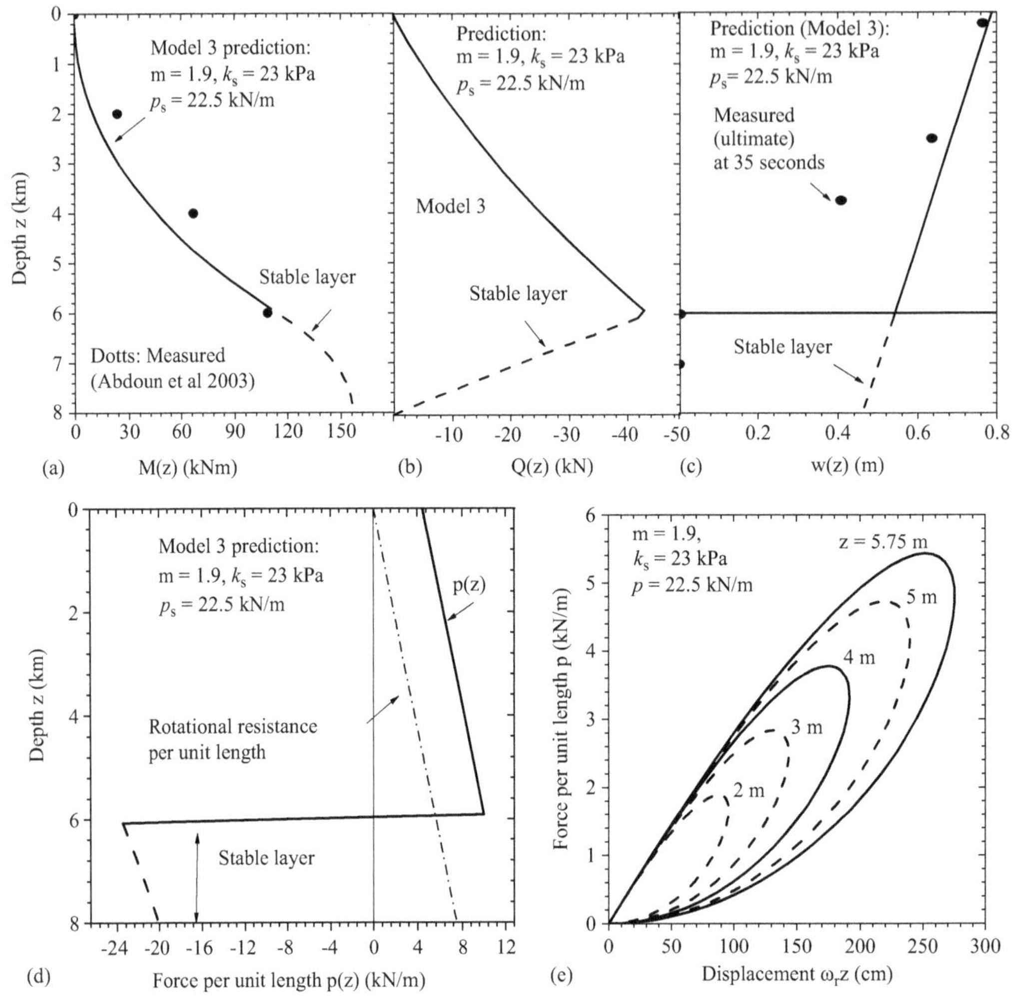

Figure 7.4 Predicted profiles of Model 3 pile ($p_s = 30l_m/l$ kN/m, l_m = 6 m, l = 8 m, k_s = 22.9 kPa, $k\theta$ = 3821 kNm/radian): (a) $M(z)$; (b) $Q(z)$; (c) $w(z)$; (d) $p(z)$; (e) p–y curves. [Adapted from Guo, W. D., *Can Geotech J* **52**(7), 2015b.]

further increase in the deflection and lateral spreading. The measured profiles of bending moments and soil movements at ultimate state are plotted in Figures 7.4a and c, respectively.

The 2-layer model of Model 3 test was based on m = 1.9, k_s = 23 kPa, p_b = 30 kN/m, l = 8 m, H = 0, and $k_T = k_\theta$ = 3.821 MNm/radian ($\bar{k}_\theta$ = 0.317). The k_θ value for k_s = 23 kPa is two-thirds of 5.738 MNm/radian for k_s = 0 (Dobry et al. 2003). The calculation includes (in sequence): (i) normalized rotation $\bar{\omega}_r$ and displacement $\bar{w}_g$ using Equations (4.2) and (4.3) (Chapter 4) for a given SD l_m (= c = 0.1l < final SD); (ii) the moment M_m (at a depth of 5.75 m), shear force Q_m and on-pile *FPUL* p_s using the expressions in Table 4.2. The calculations (i) and (ii) are repeated for new l_m of 0.2l, 0.4l, 0.6l, 0.8l and l. The results for each l_m/l are presented in Table 7.2. They are plotted as bold, solid lines in Figures 7.3b, c and d for the w_s–$l\omega_r$, w_s–M_m and $l\omega_r$–M_m curves, respectively.

It is worthy to stress that (i) the measured pile movement is the relative displacement between pile-head and toe; (ii) the effective soil movement w_s at pile location is equal to

Table 7.2 Response of Model 3 (C1-M3) pile subjected to lateral spreading[a] (Guo 2015b)

l_m/l	$\bar{w}_g$	$\bar{\omega}_r$	$\omega_r l$ (cm)[b]	w_g/w_s (cm)[c]	M_m (kNm)[d]	Q_m (kN)[d]	p_s (kN/m)[d]
0.2	0.222	−0.024	5.1	5.8/3.9	23.18	−1.39	0.84
0.4	0.452	−0.042	17.7	23.7/15.8	75.88	−8.33	2.90
0.5	0.561	−0.047	24.5	36.8/24.5	101.0	−15.36	4.03
0.6	0.663	−0.048	30.1	52.2/34.8	116.4	−25.71	4.95
0.8	0.844	−0.037	31.1	88.6/59.1	90.0	−36.19	5.11
1.0	1.0	0	0	131.1/87.5	0	0	0

[a] $m = 1.9, k_G = 0, k_T = k_\theta = 3.821$ MNm/radian, and $H = 0$.
[b] Relative pile displacement between head and toe;
[c] w_g estimated from Equation (4.3), Chapter 4, $w_s = \alpha\omega_r l$ and $\alpha = 1.5$.
[d] Bending moment M_m, shear force Q_m and the on-pile force per unit length p_s (= $\omega_r z_c p_b l_m/l$) at a depth of $z_c = 5.75$ m.

$0.667w_g$ ($\alpha = 1.5$); and (iii) stepwise increase of the on-pile *FPUL* p_s (via $l_m = c$) allows non-linear response. The predicted bending moment M_m and pile-soil relative displacement $l\omega_r$ agree well with measured data (including increase-decrease cycles, see Figure 7.3) for the GL free-field displacement w_s. The bending moment M_m eventually stops at 27 kN-m (?) for a stable layer ($k_T > 0$), as is evident in other tests (Motamed and Towhata 2010). A lower bound response is obtained using $m = 1$, and shown in the figures as well.

Given the set of parameters ($H = 0$, $c = l_m = 6$ m, $l = 8$ m, $\lambda = 0.333$, $k_s = 23$ kPa, $p_b = 30$ kN/m, and $k_\theta = 3.821$ MNm/radian), the expressions in Table 4.2 (Chapter 4) are used to gain the profiles of bending moment $M(z)$, shear force $Q(z)$, pile displacement $w(z)$, the net *FPUL* $p_1(z)$ at ultimate state; and the p–$y(w)$ curves at depths of 2 m, 3 m, and 4 m, and 5.75 m. They are plotted in Figures 7.4a through e, respectively. The $M(z)$ and $p_i(z)$ are well predicted against the measured data and similar centrifuge tests (González et al. 2009). The associated on-pile pressure is 9.47–20.4 kPa [i.e., $p_1(z)$ = 4.46 to 10 kN/m over the 6-m], which agrees well with the previous suggestions. The impact of $\bar{k}_\theta$, (= 0.326) is evident in subsequent examples.

Example 7.1 Modelling centrifuge tests of Case C2-M5a

Given identical conditions to the Model 3 (C1-M3) test, Abdoun et al. (2003) conducted Model 5a test (i.e., C2-M5a in Table 7.1), to investigate the effect of an extra rectangular pile cap [2 × 2.5 × 0.5 m (in thickness)] rigidly connected to the top of the pile on the response. The side area of 2.5 × 0.5 m was subjected to lateral spreading. The prototype M_{max} was measured as 170 kNm at a pile-head deflection of 350 mm. The test results allow parameters k_s, m, k_θ and p_b to be deduced (see Table 7.1). The deduced p_b value was 33% higher than 22.5 kN/m for the C2-M3 test. Note that p_b was estimated using $p_b = 0.8\gamma_s' K_p^2 dz$, given $z = 6$ m, γ_s'(effective unit weight) = 9.0 kN/m^3, $\phi = 0°$, and $d = 0.475$ m. Other response is not detailed herein.

Example 7.2 Deducing parameters for test piles (with known k_G) in single layer

He et al. (2009) conducted Models 1, 2, 3 and 6 tests (renamed as C3-M1 through to C6-M6 in Table 7.1) to examine response of single piles subjected to liquefaction flow of sloping ground (up to 6 degrees). The piles were 'fixed' to the base before constructing the soil stratum (D_r = 40–50%, and γ_{sat} = 19 kN/m^3). Each pile was instrumented with a displacement transducer at the pile head and strain gauges along the shaft, and sand stratum with accelerometers and pore pressure sensors. The measured maximum bending moment and ground-line pile-deflection at an 'ultimate' soil movement for each pile are tabulated in Table 7.1; and the response profiles are plotted in Figure 7.5.

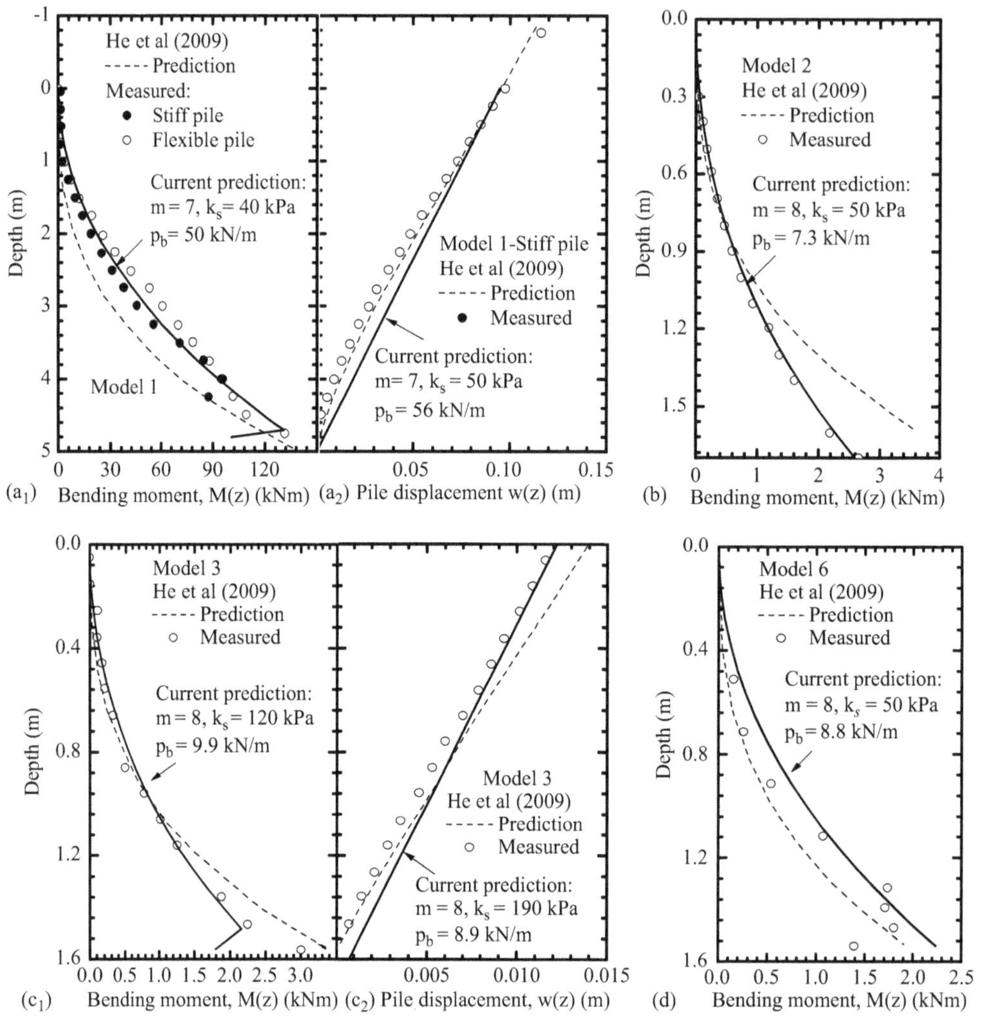

Figure 7.5 Prediction versus measured (He et al. 2009) response for piles subjected to lateral spreading. [Adapted from Guo, W. D., *Can Geotech J* **52**(7), 2015b.]

As with the abovementioned calculation, the 2-layer model was used to match with the measured data to deduce the parameters k_s, m, k_θ and p_b (see Table 7.1) for the single, base rotationally restrained piles C3-M1 through to C6-M6. The model stipulates (i) a full-length liquefied soil ($l_m = l$) and (ii) a loading depth c of $(0.75–0.9)l$, for a reduced bending moment at a distance of $(0.1–0.25)l$ above the base (e.g., in Figure 7.5d), as observed along retaining walls. The exact loading depth c for each case was deduced by fitting model solutions to measured bending moment profile at the given rotational stiffness k_T (see following examples).

Example 7.3 Deducing parameters for Case C3-M1

He et al. (2009) conducted his Model 1 test on a flexible pile (with k_T of 185.0 MNm/rad), and on a rigid pile (with k_T of 8.5 MNm/rad) in Kasumigaura saturated sand (5.0 m in thickness) using a large laminar soil container [12 × 3.5 × 6 m (high)]. The sand (Kagawa et al. 2004) has D_{50} = 0.31 mm, fines content F_c = 3%, and uniformity coefficient C_u = 3. Displacement transducers were mounted on the exterior wall of the laminar container to measure free-field lateral displacement.

Example 7.4 Deducing parameters for Cases C4-M2–C6-M6

The Model 2, 3, and 6 tests (He et al. 2009) adopted silica sand (sourced from a San Diego, CA quarry) with D_{50} = 0.32 mm, F_c < 2%, and C_u = 1.5. The sand was saturated in a medium laminar test container [4 m × 1.8 m × 2 m (high)] (Jakrapiyanun 2002). The model C4-M2 and C5-M3 pile involve a single pile with a tip stiffness k_T of 0.11 MNm/rad, and 0.2 MNm/rad, respectively. The pile was installed in the container (with a 2° inclined to the GL). In contrast, Model 6 (C6-M6) was conducted on a single, concrete pile with k_T = 0.3 MNm/rad, using a levelled, rigid-wall container [4 m × 1.8 m × 2 m (high)] with the soil surface inclined at 6%.

Prior to liquefaction, the pile-head undergoes similar movement as the surrounding soil. Subsequently, the lateral displacement renders the pile displacement peak and then slightly decreases, which resembles the variation in bending moment. The maximum measured pile-head displacements and moments are provided in Table 7.1. The moment M_m (of $k_\theta \omega_r$) largely occurs around the pile-tip level, as evidenced by the moment and displacement profiles in Figure 7.5.

Using the parameters in Table 7.1, the expressions from Table 4.2 (the 2-layer model, Chapter 4) were adopted to predict (a_1) the bending moment profile $M(z)$ and (a_2) the pile displacement profile $w(z)$ for test C3-M1; (b) the $M(z)$ for test C4-M2; (c) the $M(z)$ and $w(z)$ for test C5-M3, and (d) the $M(z)$ for test C6-M6. The predicted $M(z)$ profiles agree well with the measured data in Figures 7.5a_1, b, c_1 and d, respectively, and the predicted $w(z)$ profiles match the available data in Figures 7.5a_2 and c_2.

7.3.2 Back-estimating input parameters

Measured response is matched with the 2-layer model by adjusting parameters k_s, m, k_θ and p_b. The modulus k_s is adjusted to match evolution of soil movement w_s, while the m and k_θ values are intended to match the maximum bending moment, rotational angle and displacement of a pile. The p_b is adjusted to fit on-pile pressure and distribution of bending moment with depth. The back-estimation of the parameters is rigorous. Nevertheless, the deduced parameters for full-length lateral spreading are provided for reference only, as they are yet to be confirmed using M_m–w_s curve which are not available.

Table 7.1 shows that the deduced normalized stiffness $\bar{k}_\theta$ values are consistent between C1-M3 and C2-M5a tests, and between C4 and C5 piles (He et al. 2009). Additionally, the k_T values for C5–C6 tests are also in good agreement with reported data. The stiffness k_T of C4 is lower than the reported value of 18.5 MN-m/rad, indicating other rotational constraints along the pile. The piles with $\bar{k}_\theta$ of 0.32–1.1 may exhibit the features of fixed-head piles (k_θ = 10). For instance, the ratio $\alpha w_g/w_s$ may increase linearly with the sliding depth.

The estimated p_b for C6 pile is 4.33 kN/m (= $\gamma_s dz$), which is about 50% the deduced p_b of 8.8 kN/m. In contrast, the p_b (= $\gamma_s dz$) is estimated as 7.72 kN/m, and 7.52 kN/m for C4 and C5 piles, which agree well with the deduced values of 7.3, and 9.9 kN/m, respectively. The associated on-pile pressures (≈ p/d) range from 9.5 to 30 kPa for C1–C2 tests, and 2.7 to 4.7 kPa for C4–C6 tests, which agree with reported values (He et al. 2009). In particular, the estimated p_b for C3 pile is 19.48 kN/m using $p_b = \gamma_s' K_p^2 dz$, (with z = 4.8 m, γ_s' = 9.0 kN/m^3, ϕ = 5°, and d = 0.318 m) or 29 kN/m using $p_b = \gamma_s dz$ (He et al. 2009). Both are lower than the deduced 50 kN/m. This is due to a large k_T value and twice the pressure on C2 in the C3 test. The average on-pile pressure (over pile embedment) and pile-tip level pressure roughly increase with the tip rotational stiffness k_T (see Figure 7.6a) for C1–C6 tests; while the pile-head level pressure increases with the pile diameter (see Figure 7.6b).

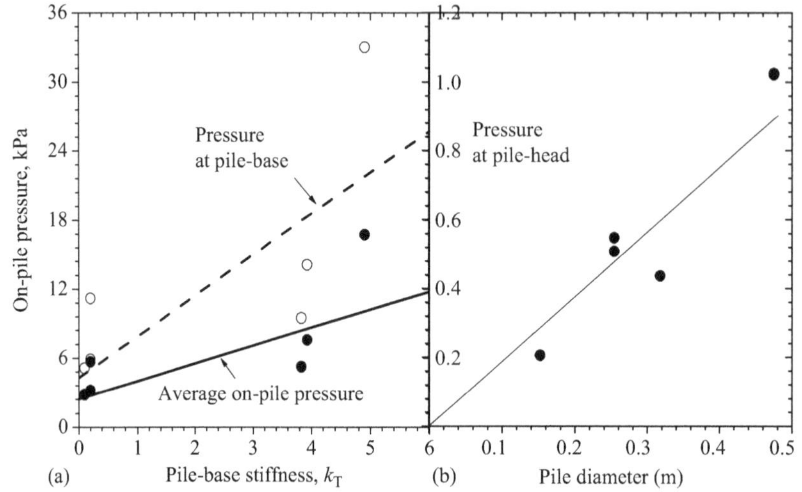

Figure 7.6 On-pile pressure versus (a) pile-base stiffness k_T; (b) pile diameters. [Adapted from Guo, W. D., *Can Geotech J* **52**(7), 2015b.]

Finally, the measured response of the model piles C7–C10 is well predicted using 2-layer model (Guo 2014) and the parameters (see Table 7.1), which includes normalized rotational displacement versus normalized displacement, and bending moment versus displacement relationship. The parameters may be determined using low-cost 1 g model tests as an alternative to expensive shaking table experiments.

7.3.3 Summary for forward rotation

The 2-layer model and closed-form solutions well capture nonlinear response of rotationally restrained, rigid piles subjected to lateral spreading. At a movement w_s ($\approx w_g/\alpha$) of lateral spreading, the pile-head displacement is measured as relative displacement $\omega_r l$ rather than the w_g. The impact of the spreading w_s on nine model piles is effectively incorporated by using a modified SD of l_m/α and movement w_g/α, respectively (e.g., α = 1.39–1.5). The on-pile pressure increases with tip-rotational stiffness, and the pressure distribution gradually shifts from triangular towards uniform with increasing head-restrained stiffness.

7.4 MODEL FOR RESPONSE AMPLIFICATION

To date, model tests, analytical and numerical simulations have largely been confined to response of forward-rotating piles. However, research by Guo (2020a, b) indicates that back-rotation of pile-head is responsible for the majority of pile failure observed in practice (Knappett & Madabhushi 2009; Haskell et al. 2013). This phenomenon inflicts amplified pile response around a specific rotational stiffness, which can be accurately estimated using 2-layer and 3-layer models.

7.4.1 2-layer model

In Equation (4.3) of Chapter 4, setting the denominator to zero yields a normalized singularity stiffness (NSS) $\bar{k}_\theta$ denoted as $\bar{k}_\theta^*$

$$\bar{k}_\theta^* = \frac{\left(1+4m\lambda+6m\lambda^2+4m\lambda^3+m^2\lambda^4\right)\bar{l}_m^3}{-12\left(1+m\lambda\right)} \tag{7.1}$$

where $\lambda = \left(1-\bar{l}_m\right)/\bar{l}_m$. Note that typo errors are identified in the original expressions of λ and $\bar{k}_\theta^*$ (Guo 2020a). Using Equations (4.2) through to (4.4) (Chapter 4) for $\bar{l}_m = 0-1$, the response of $\bar{w}_g, \bar{\omega}_r$ and $\bar{M}_{m2}$ was estimated for m = 1.5 and 4, as shown in Figure 7.7. The figure indicates (i) an amplified response for $\bar{k}_\theta$ adjacent to the NSS $\bar{k}_\theta^*$ estimated using Equation (7.1); (ii) proportional $\bar{M}_m/\left(p_s l_m l\right)\left[=\bar{M}_{m2}/\left(l_m/l\right)\right]$ and $\bar{\omega}_r$ ratios to $\bar{w}_g$; and (iii) instability over $\bar{k}_\theta$ = -(0.09 – 0.13) at m = 1.5 (e.g., dilative sand), and across $\bar{k}_\theta$ = -(0.10 – 0.32) at m = 4. Here 'unstable' denotes that the response amplification (e.g., $\bar{w}_g^b/\bar{w}_g^f$) ratio from forward to backward rotation exceeds factor of safety. Using Equation (4.3) (Chapter 4), the $\bar{w}_g^b$ and $\bar{w}_g^f$ are estimated with $\bar{k}_\theta$ and $-\bar{k}_\theta$, respectively (Guo 2020a, b). As with lateral loading, capped piles exhibit a lower mobilized p_b than free-head piles (Guo 2012), as illustrated late using 1 g model tests.

7.4.2 3-layer model

The subsoil around the pile is also simplified as 3-layer, which consists of an upper, sliding layer with a subgrade modulus k_s (at depth z of 0 ~ l_m), a transition layer with a modulus that linearly increases from k_s to mk_s (at $z = l_m$ ~ z_m, z_m = depth of maximum bending moment), and a lower, stable layer with a modulus of mk_s (at $z = z_m$ ~ l), respectively. The model, described in Chapter 5, is underpinned by five input parameters k_s, m, p_{ub}, α and k_θ, under a lateral load H and bending moment M_o exerted at ground level. As with the 2-layer model, given a normalized depth $\bar{l}_m\left(=l_m/l\right)$ and p_s, the displacement w_g and rotation ω_r at GL, as well as the depth z_m and maximum moment M_m are obtained, which allow for estimating response profiles of shear force $Q(z)$, bending moment $M(z)$, net *FPUL* $p(z)$, deflection $w(z)$ with depth z, and the maximum M_m and Q_m (Guo 2016). By repeating the calculation for a series of $p_s\left(=\alpha p_{ub}\bar{l}_m\right)$, via raising $\bar{l}_m$), the model provides accurate predictions of the nonlinear response of the piles. The approach has been validated against measured data and finite element analysis for piles under forward rotation (Guo 2016; Guo et al. 2017).

Given a specified external loading p_s (related to normalized SD$\bar{l}_m$), and a normalized head-level shear force $\bar{H} = H/\left(p_s l\right)$, the normalized pile-head displacement $\bar{w}_g\left(=w_g k_s/p_s\right)$ and rotational angle $\bar{\omega}_r\left(=\omega_r k_s l/p_s\right)$ are given by

$$\bar{w}_g = \frac{w_g k_s}{p_s} = \frac{-6\left(1+\bar{z}_m\right)\left(\bar{H}+\bar{l}_m\right)}{(m-1)\bar{z}_m^2+\left[-3(1+m)+(m-1)\bar{l}_m\right]\bar{z}_m+\bar{l}_m\left(-2\bar{l}_m+3\right)(m-1)} \tag{7.2}$$

$$\bar{\omega}_r = \frac{\omega_r k_s l}{p_s} = \frac{-2\bar{w}_g}{1+\bar{z}_m} \tag{7.3}$$

The maximum bending moment at depth z_m and the moment at the SD l_m, are obtained using $M_{zm} = M_3(z_m)$ and $M_{lm} = M_2(l_m)$, respectively. They are normalized as $M_{zm}/(p_s l^2)$ and $M_{lm}/(p_s l^2)$:

$$\frac{M_{zm}}{p_s l^2} = \frac{-\bar{\omega}_r}{12}\left\{\bar{z}_m^3+\left[2+m+(1-m)\bar{l}_m\right]\bar{z}_m^2+(m-1)\bar{l}_m\left(\bar{l}_m-1\right)\left(2\bar{z}_m-\bar{l}_m\right)\right\}$$
$$-\bar{H}\bar{z}_m-\left(\bar{z}_m-0.5\bar{l}_m\right)\bar{l}_m+\bar{k}_\theta\lambda\left(\bar{z}_m\right)\bar{\omega}_r \tag{7.4}$$

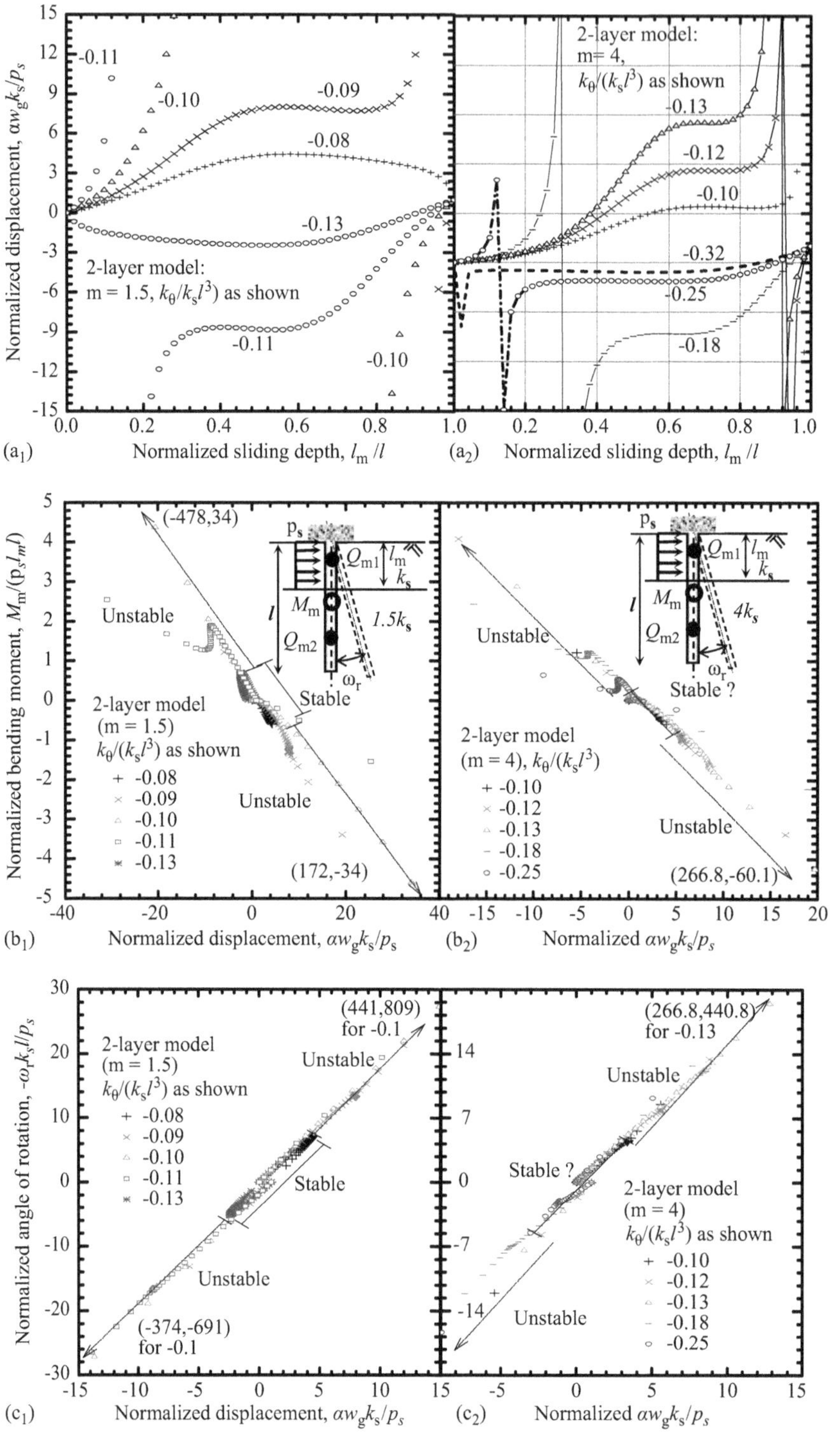

Figure 7.7 Amplified response at range of rotational stiffness (a_1–c_1) m = 1.5; (a_2–c_2) m = 4. [Adapted from Guo, W. D., *J of Engrg Mechanics* **146**(9), 2020a.]

$$\frac{M_{lm}}{p_s l^2} = \left[-1.5\left(1+\bar{z}_m\right)+\bar{l}_m\right]\frac{\bar{\omega}_r \bar{l}_m^2}{6} - \left[0.5\bar{l}_m^2 - \bar{l}_m\bar{H} + \bar{k}_\theta\lambda\left(\bar{l}_m\right)\bar{\omega}_r\right] \tag{7.5}$$

where $\bar{k}_\theta = k_\theta / \left(k_s l^3\right)$. The maximum shear force $Q_{3m}\left(= Q_3\left(z_t\right), z_t = -\bar{w}_g / \bar{\omega}_g\right)$ occurs in stable layer, and is given by

$$\frac{Q_{3m}}{p_s l} = \left[\frac{1-m}{2}\left(\bar{l}_m + \bar{z}_m\right) + m\left(-\frac{\bar{w}_g}{\bar{\omega}_r}\right)\right]\bar{w}_g + \left[\frac{1-m}{6}\left(\bar{z}_m^2 + \bar{z}_m\bar{l}_m + \bar{l}_m^2\right) + \frac{m}{2}\left(-\frac{\bar{w}_g}{\bar{\omega}_r}\right)^2\right]\bar{\omega}_r - \bar{H} - \bar{l}_m \tag{7.6}$$

The moments at the pile-head (of $k_G\omega_r$) and base (of $k_T\omega_r$) are transferred to a depth z by a factor $\lambda(z)$. This factor reaches 1 for rotation restrained (fixed)-head and floating-base piles. The normalized depth of maximum bending moment $\bar{z}_m\left(= z_m/l\right)$ can be determined using the expression provided in Table 5.1 (Chapter 5).

The response of the pile (including displacement w_g, rotation ω_r, maximum moment M_m and shear force Q_m) is amplified (as shown in Figure 7.8) to varying degrees with increasing SD (l_m). The enlarged ratio of the displacement from forward to backward rotation also depends on the normalized depths $\bar{z}_{mb}$ and $\bar{z}_{mf}$:

$$\frac{\bar{w}_{gb}}{\bar{w}_{gf}} = \frac{\left(1+\bar{z}_{mf}\right)\left[\left(m-1\right)\bar{z}_{mb}^2 + \left[-3\left(1+m\right)+\left(m-1\right)\bar{l}_m\right]\bar{z}_{mb} + \bar{l}_m\left(-2\bar{l}_m+3\right)\left(m-1\right)\right]}{\left[\left(m-1\right)\bar{z}_{mf}^2 + \left[-3\left(1+m\right)+\left(m-1\right)\bar{l}_m\right]\bar{z}_{mf} + \bar{l}_m\left(-2\bar{l}_m+3\right)\left(m-1\right)\right]\left(1+\bar{z}_{mb}\right)} \tag{7.7}$$

The enlarged ratios of rotation and specified bending moments are expressed as follows:

$$\frac{\bar{\omega}_{gb}}{\bar{\omega}_{gf}} = \frac{\left(m-1\right)\bar{z}_{mb}^2 + \left[-3\left(1+m\right)+\left(m-1\right)\bar{l}_m\right]\bar{z}_{mb} + \bar{l}_m\left(-2\bar{l}_m+3\right)\left(m-1\right)}{\left(m-1\right)\bar{z}_{mf}^2 + \left[-3\left(1+m\right)+\left(m-1\right)\bar{l}_m\right]\bar{z}_{mf} + \bar{l}_m\left(-2\bar{l}_m+3\right)\left(m-1\right)} \tag{7.8}$$

$$\frac{M_{zmb}}{M_{zmf}} = \frac{\begin{array}{l}\frac{-\bar{\omega}_r}{12}\left\{\bar{z}_{mb}^3 + \left[2+m+\left(1-m\right)\bar{l}_m\right]\bar{z}_{mb}^2 + \left(m-1\right)\bar{l}_m\left(\bar{l}_m-1\right)\left(2\bar{z}_{mb}-\bar{l}_m\right)\right\}\\ -\bar{H}\bar{z}_{mb} - \left(\bar{z}_{mb}-0.5\bar{l}_m\right)\bar{l}_m + \bar{k}_\theta\lambda\left(\bar{z}_{mb}\right)\bar{\omega}_r\end{array}}{\begin{array}{l}\frac{-\bar{\omega}_r}{12}\left\{\bar{z}_{mf}^3 + \left[2+m+\left(1-m\right)\bar{l}_m\right]\bar{z}_{mf}^2 + \left(m-1\right)\bar{l}_m\left(\bar{l}_m-1\right)\left(2\bar{z}_m-\bar{l}_m\right)\right\}\\ -\bar{H}\bar{z}_{mf} - \left(\bar{z}_{mf}-0.5\bar{l}_m\right)\bar{l}_m + \bar{k}_\theta\lambda\left(\bar{z}_{mf}\right)\bar{\omega}_r\end{array}} \tag{7.9}$$

$$\frac{M_{lmb}}{M_{lmf}} = \frac{\left[-1.5\left(1+\bar{z}_m\right)+\bar{l}_m\right]\frac{\bar{\omega}_{rb}\bar{l}_m^2}{6} - \left(0.5\bar{l}_m^2 - \bar{l}_m\bar{H} + \bar{k}_{\theta b}\lambda\left(\bar{l}_m\right)\bar{\omega}_{rb}\right)}{\left[-1.5\left(1+\bar{z}_{mf}\right)+\bar{l}_m\right]\frac{\bar{\omega}_{rf}\bar{l}_m^2}{6} - \left(0.5\bar{l}_m^2 - \bar{l}_m\bar{H} + \bar{k}_{\theta f}\lambda\left(\bar{l}_m\right)\bar{\omega}_{rf}\right)} \tag{7.10}$$

The degree of amplification for the shear force Q_m is estimated using Equation (7.6), or simply taking the ratio of M_{zmb}/M_{zmf}, for 'linear' correlation between the two (Guo 2016). Equation (7.2) has a singularity at a normalized depth $\bar{z}_m^*\left(= z_m^*/l\right)$ of

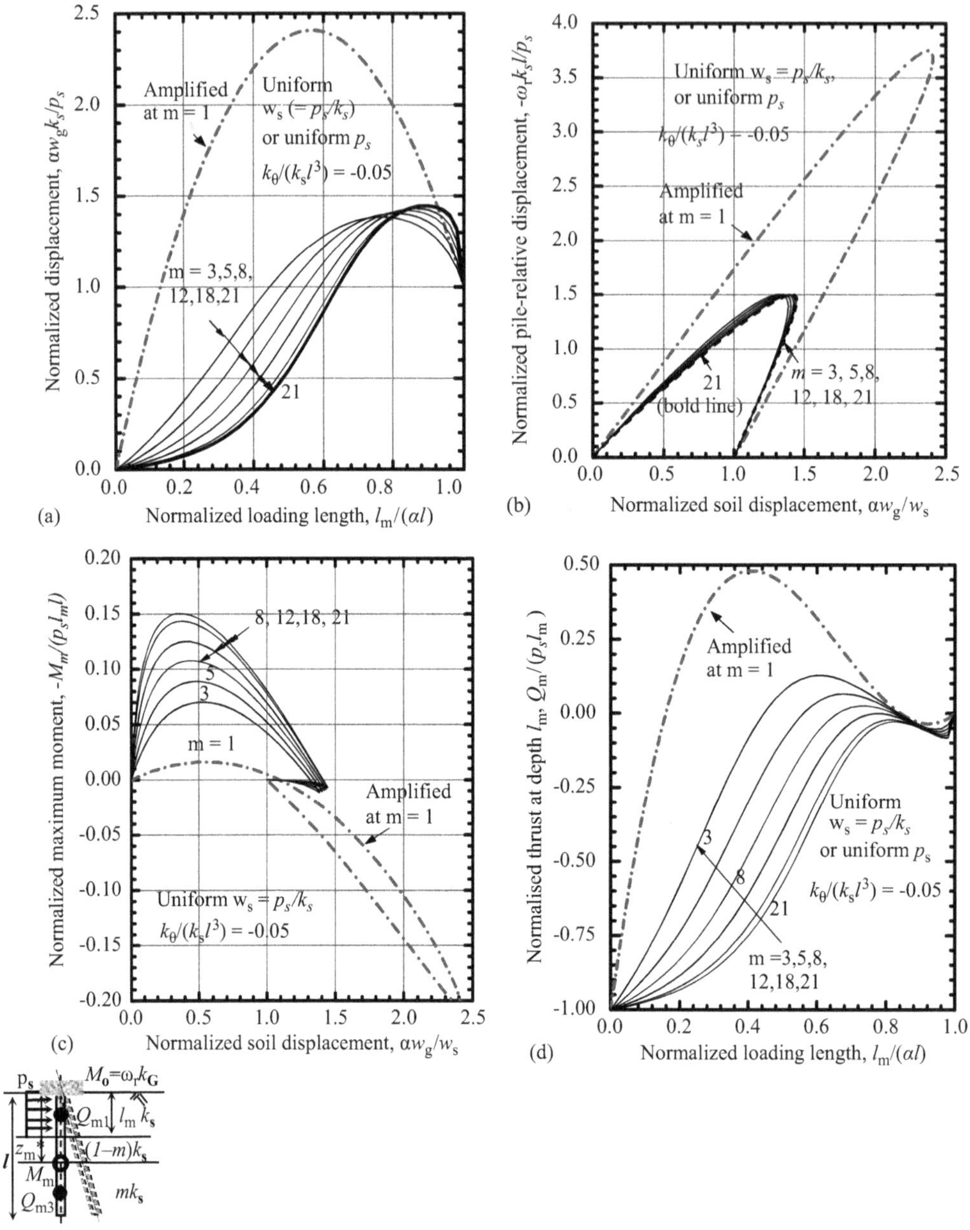

Figure 7.8 Normalized response of (a) displacement w_g; (b) angle of rotation; (c) maximum bending moment; (d) thrust for piles with $\bar{k}_\theta = -0.05$ at $m = 1-21$; without transition layer. [Adapted from Guo, W. D., *Can Geotech J* **52**(7), 2015b.]

$$\bar{z}_m^* = \frac{1}{2}\left\{-\bar{l}_m + 3\frac{1+m}{m-1} - \left[9\bar{l}_m^2 - 6\frac{3m-1}{m-1}\bar{l}_m + 9\left(\frac{1+m}{m-1}\right)^2\right]^{0.5}\right\} \qquad (7.11)$$

Table 7.3 Amplified times of response from forward to backward rotation (Guo 2020b)

l_m/l	$\bar{k}_{b\theta}{}^a / \bar{k}_\theta^*$	$\bar{w}_{gb} / \bar{w}_{gf}$	$\bar{\omega}_{gb} / \bar{\omega}_{gf}$	M_{mb}/M_{mf}	$-\bar{k}_\theta^*$	m
0.2, 0.4	0.8	5.9, 5.5	9.4, 9.7	12.6, 12.5	0.113, 0.107	1.5
0.2, 0.4	1.1	−11.5,−10.6	−22.2,−23.3	−27.6,−28.6		1.5
0.2, 0.4	0.8	6.2, 6.3	9.5, 9.9	12.7, 14.3	0.239, 0.187	4.0
0.2, 0.4	1.1	−12.5, −12.8	−22.7, −24.4	−28.1, −32.6		4.0
0.2, 0.4	0.8	6.3, 6.6	9.4, 9.6	12.5, 15.0	0.375, 0.257	7.0
0.2, 0.4	1.1	−12.7, −13.7	−22.3, −23.3	−27.7, −32.5		7.0
0.2, 0.4	0.8	6.4, 7.1	9.2, 9.2	12.4, 17.2	0.981, 0.529	21
0.2, 0.4	1.1	−12.9, −15.0	−21.6, −21.7	−26.9, −34.1		21

Note:
[a] $\bar{k}_{b\theta} = -\bar{k}_{f\theta}$.

At $\bar{z}_m^*$, the normalized singularity stiffness (NSS) $\bar{k}_\theta^*$ is recast into

$$\bar{k}_\theta^* = \frac{1}{12}\left[(m-1)\left(\bar{l}_m - 1\right)^3 - 1 - 3\bar{z}_m^*\right] \tag{7.12}$$

The displacement (w_g) and rotation (ω_r) are amplified to different degrees at given rotational stiffness (RS) $k_{\theta f}$ and $k_{\theta b}$ (= $-k_{\theta f}$ for Table 7.3). For instance, at m = 1.5 and $\bar{k}_\theta = 0.8\bar{k}_\theta^*$, over $\bar{l}_m$ of 0.2–0.4, the normalized displacement and rotation angle increase by 5.5–5.9 times and 9.4–9.7 times, while the moment increases by 12.5–12.6 times. In practice, piles initially rotate forward, and then back-rotate owing to pull-back and/or non-uniform sliding restraints from adjacent soil.

Setting $\bar{k}_\theta = \bar{k}_\theta^*$, Equation (7.12) can be used to estimate the SD l_m that incurs the singularity. Should the transitional layer be neglected, the NSS $\bar{k}_\theta^*$ is then given by

$$\bar{k}_\theta^* = \frac{\left[\bar{l}_m^4 + 4m\left(1-\bar{l}_m\right)\bar{l}_m^3 + 6m\left(1-\bar{l}_m\right)^2\bar{l}_m^2 + 4m\left(1-\bar{l}_m\right)^3\bar{l}_m + m^2\left(1-\bar{l}_m\right)^4\right]}{-12\left[\bar{l}_m + m\left(1-\bar{l}_m\right)\right]} \quad \text{(2-layer model)} \tag{7.13}$$

An example response of the 2-layer model for a pile discussed by Guo (2015b) is illustrated in Figure 7.8.

7.4.3 Singularity stiffness

The singularity stiffness $\bar{k}_\theta^*$ for back-rotated piles ($\bar{k}_\theta < 0$) at typical soil modulus ratio is obtained using Equations (7.12) and (7.13). The resulting stiffness is plotted against normalized thickness of passive loading $\left(\bar{l}_m\right)$ in Figure 7.9a, or against the normalized depth of maximum bending moment $\left(\bar{z}_m\right)$ in Figure 7.9b. A pile with $-\bar{k}_\theta \le 0.5$ is expected to have response amplification to a spreading depth of 0.6l in a spreading soil with m = 1–6. This is a concern, as a majority of piles have a normalized $\bar{k}_\theta$ of 0.317–1.108 (at a SD of 0.286l)

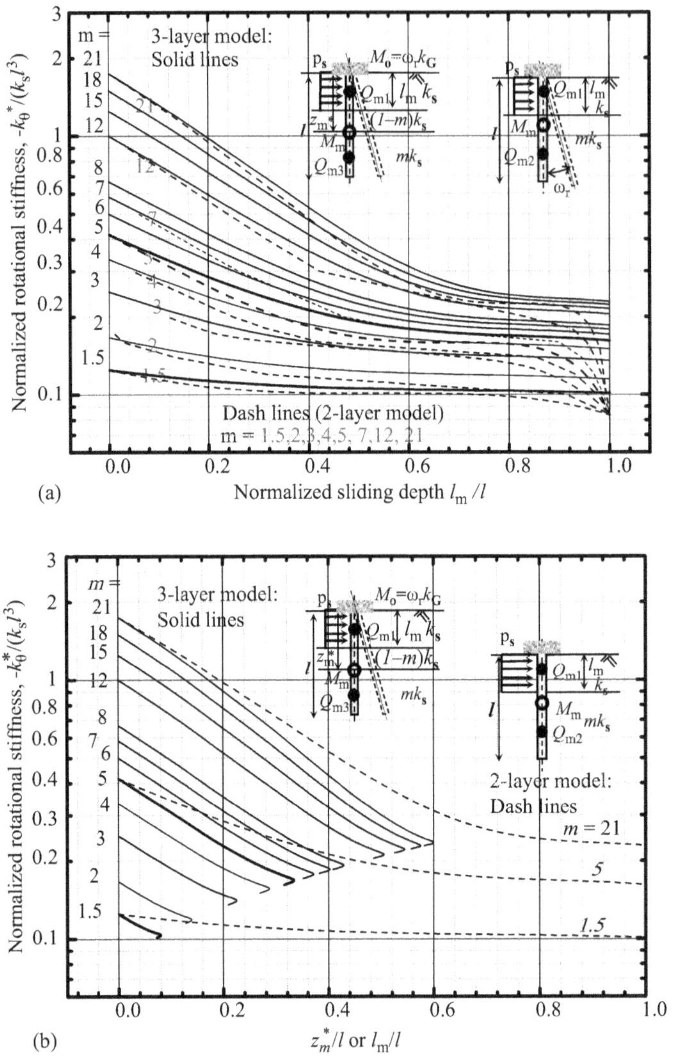

Figure 7.9 Normalized singularity stiffness against: (a) normalized sliding depth; (b) normalized depth of maximum bending moment or l_m/l. [Adapted from Guo, W. D., *Can Geotech J* **52**(7), 2015b.]

and 0.0463 (SD = 0.57l) when subjected to lateral spreading or embankment loading, or in a sliding slope (Guo 2015b). In comparison with Equation (7.8), the back-rotating $-\bar{k}_\theta$, for instance, raises the angle from ω_{rf} to ω_{rb}

$$\frac{\omega_{rb}}{\omega_{rf}} = \left(1 + \frac{\bar{k}_\theta}{\bar{k}_\theta^*}\right) \bigg/ \left(1 - \frac{\bar{k}_\theta}{\bar{k}_\theta^*}\right) \text{(2-layer model)} \qquad (7.14)$$

The amplification from back-rotation was first discovered by Guo (2020a,b) despite a high rate of pile failure. To ensure a safe and economic design, Equations (7.2)–(7.14) can be used to estimate the amplified displacement, rotation angle, maximum bending moment

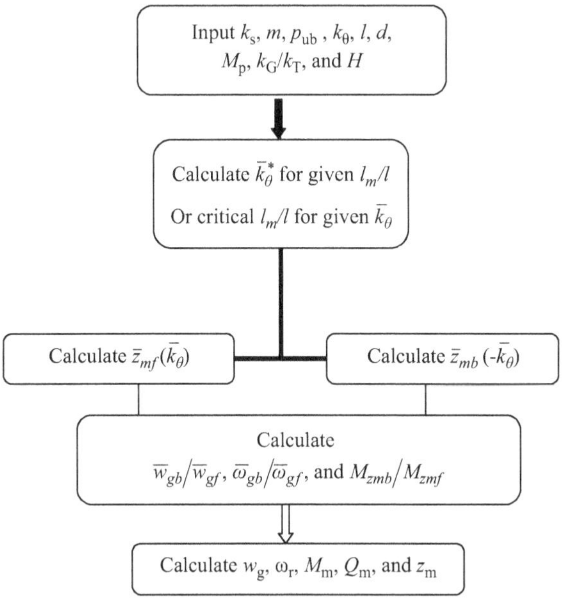

Figure 7.10 Estimation of response of back-rotated piles. [Adapted from Guo, W. D., *Can Geotech J* **52**(7), 2015b.]

and maximum shear force in the procedure outlined in Figure 7.10. This is illustrated next concerning slope-stabilizing piles and piles subjected to lateral spreading.

7.5 SLOPE-STABILIZING PILES

Chapter 4 of this book discusses slope-stabilizing piles, and Figure 4.12 provides the measured ratio of $M_m/(Q_m l)$ (i.e., moment M_m over shear force Q_m and equivalent rigid pile embedment l, Guo 2013c) as well as the predicted ratios (using 3-layer model) for m = 7.67 and 17.7 for forward-rotating piles. This figure now includes new predictions assuming backward rotation. The figure indicates some field piles align with predictions of 'back-rotation' ($\bar{k}_\theta < 0$). Back-rotation of pile-head may occur owing to pull-back of the pile-head (Frank and Pouget 2008); restraint of moving soil in the inverse triangular profile (ITP) or the uniform profile (*UP*) (see Figure 1.8, Chapter 1) in the 1 g model tests, or restraint of the top soil (above transition layer) of piles named Esu and D'Elia, Katamachai-B and Carrubba, respectively. The 'rigid pile' prediction assumes no pile-head load (H = 0); otherwise, Chapter 4 should be consulted. The singularity amplification is explained next concerning two typical cases.

Example 7.5 Amplification capped by hinges

The Katamachai-B pile (see Figure 4.12, Chapter 4) had a rigid length l of 6 m and was pulled back by the top 2.8-m-thick soil (Guo 2013c) with a normalized sliding ratio $\bar{l}_m$ of 0.53. The pile rotated to an angle ω_r of 0.039 radians and yielded at M_o^y of 40 kNm. The pile-soil system had a subgrade modulus k_s of 60 kPa (= 1% static case, Guo 2015b) and $\bar{k}_\theta$ of −0.08. The $\bar{k}_\theta^*$ is estimated as −0.1638 using Equation (7.13) for $\bar{l}_m$ = 0.53 and m = 5.

Following the procedure outlined in Figure 7.10, the amplification degrees were estimated using expressions provided in Table 5.1 (Chapter 5): (i) Without lateral load (H = 0), the values of β and $\overline{B}$ were estimated as 0.265, and 2.95, respectively; (ii) The forward rotation was described by $\bar{k}_\theta = 0.08$, $\bar{C} = 1.836$, and $\bar{z}_{mf} = 0.893$; (iii) The pull-back was quantified by $\bar{k}_\theta = -0.08$, and $\bar{C} = 1.183$ as well as a reduced, normalized depth $\bar{z}_{mb}$ of 0.479; and (iv) finally, Equations (7.7) and (7.8) offer w_{gb}/w_{gf} of 2.1, and ω_{rb}/ω_{rf} of 2.7 (see Table 7.4).

The amplification of displacement and rotation reaches 2.1 and 2.7 times, which is supported by the 2–2.5 times underestimation of the measured displacement and angle of rotation (Guo 2013c) without amplification. The amplification exerts limited impact on the bending moment and shear force, which are capped by plastic hinge. This allows a good estimation of the moment and force using the *P-EP* solutions (Guo 2013c).

Example 7.6 Amplification restrained by dragging layer

Compared to the Katamachai-B pile, the Carrubba pile exhibits similar features such as (i) $-\bar{k}_\theta$ of 0.077 (estimated using M_o^y = 0.8 MNm, ω_r = 0.007 radians, l = 21.5 m and k_s = 150 kPa, Guo 2015b); (ii) formation of plastic hinge, and (iii) a restraining force applied by the top 2-m-thick soil to pull back the pile-head (Guo 2013c). The calculation results are detailed in Table 7.4. The displacement and rotation of the pile should have enlarged by 1.7 and 1.9 times given m = 7.67 and $-\bar{k}_\theta^*$ (of 0.254) at $\bar{l}_m$ of 0.442, but for the restraint of the 2-m transition layer. The good prediction of the pile response using the *P-EP* solution suggests that the pull-back and the restrain amplification may offset by the top layer in this circumstance.

The process of back-rotation is impacted by the *P*-Δ effect. However, it is the stiffness of the back-rotation that ultimately leads to significant amplification, rather than the *P*-Δ effect. This conclusion is supported by the following analysis on centrifuge tests.

Table 7.4 Calculations of amplification degrees (Guo 2020b)

Name	*Katamachai-B*[a]	*Carrubba*[a]		*Showa*[a]	
M_p (kNm)	40	800		1,000–1,620	
l_m/l	0.53	0.442	0.08	0.10	0.146
k_s (kPa)	60	150	30	30	30
m	5	7.67	1.5	1.5	1.5
β, Table 5.1	0.265	0.221	0.04	0.05	0.073
$\overline{B}$, Table 5.1	2.95	−2.88	−8.67	−8.584	−8.382
$\bar{k}_\theta^*$	−0.188	−0.254	−0.12	−0.118	−0.116
			Forward rotation		
$\bar{k}_\theta$	0.08	0.077	0.119	0.119	0.078
$\bar{C}$, Table 5.1	1.836	1.529	6.069	6.159	5.244
$\bar{z}_{mf}$, Table 5.1	0.893	0.702	0.772	0.79	0.681
			Back-rotation		
$\bar{k}_\theta$	−0.08	−0.077	−0.119	−0.119	−0.078
$\bar{C}$, Table 5.1	1.183	1.173	0.147	0.146	1.205
$\bar{z}_{mb}$, Table 5.1	0.479	0.491	0.017	0.017	0.146
$\bar{w}_{gb}/\bar{w}_{gf}$	2.087	1.697	238.2	−237.4	5.268
$\bar{\omega}_{gb}/\bar{\omega}_{gf}$	2.671	1.916	415.0	−417.9	3.592
M_{zmb}/M_{zmf}			−8.76	10.76	

Note:

[a] $\overline{H} = 0$ for all piles.

7.6 BACK-ROTATED PILES WITH *P*-Δ EFFECT

7.6.1 Constrained back-rotation with *P*-Δ effect (centrifuge tests)

Knappett and Madabhushi (2009) conducted centrifuges tests to study the response of vertically loaded pile groups ($P > 0$, and with an initial lateral deflection Δ_o) subjected to lateral spreading of silica sand (frictional angle of 32°). The piles ($d = 0.496$ m, $t = 82$ mm, $E_pI_p = 164$ MNm^2) were arranged as 2 × 2 piles (with $l = 14.4$ m) in groups I2, I4, I5 and I6 and as a line of 2 piles ($l = 8$ m) in groups I9 and I14. The normalized spacing s/d (s = pile centre-to-centre spacing) was 5.6 for groups I2, I4, I9 and I14, and 2.6 for groups I5 and I6. The piles were allowed to translate laterally at head-level but are socketed (fully fixed) at tip level. The test results indicate (i) a low $P\Delta_o$ value (of I4 piles) incurred large distortion after shaking, while a large $P\Delta_o$ (of I6 piles) did not; and (ii) the distorted I14 piles (during shaking) exhibited large back-rotation. Amplification from P–Δ_o effect is incompatible with the observed behaviour. To gain further insight into the incompatibility, the 2-layer model [via Equations (7.2) through (7.3)] was adopted using parameters of $c = l_m$, $H = 0$, $M_o = 0$, $k_s = 30$–120 kPa, $m = 2$–8, $p_{ub} = 20$–75 kN/m and $\alpha = 1$ (Guo 2015b). All piles yielded at M_y of 4,204 kNm and ω_r of 5° (i.e., 0.087 radians, as observed in I6) and were subjected to a head stiffness k_θ of $(M_y - P\Delta_o)/0.087$. Given the imposed P and Δ_o values, the $\bar{k}_\theta$ values for each test were obtained (see Table 7.5).

The 2-layer model provides the maximum capacity P ($= M_o/w_g$, $M_o \approx M_m$) and displacement w_g curves of the piles with fully rotation-restrained head $\left(\bar{k}_\theta = 10\right)$, at the specified p_{ub} ($= p_b$) of 20, 40 and 75 kN/m, and k_s of 30, 45 and 60 kPa for $m = 5$. The predicted curves are plotted in Figure 7.11, along with the measured P and Δ ($\approx w_g$ at failure) of each pile. The comparison indicates forward-rotating I2 piles is stable at $p_{ub} > 20$ kN/m, and I5 and I4 piles are stable at $p_{ub} \geq 75$ kN/m (in early stage). As the strength ($\approx p_{ub}/d$) reduces at later stage from subsequent shaking, the I4 piles back-rotated. Response amplification at NSS (see Table 7.5) comes into play as examined next.

Table 7.5 Possibility of amplification for six typical tests (Guo 2020a)

	P/pile (MN)	Δ_o *(mm)*	$P\Delta_o$ *(MNm)*	$\bar{k}_\theta\ k_s$ [c]	$-\bar{k}_\theta$ [d]	*S*[e]	*Amplification (features, conditions)*
I2[a]	6.38	56	0.357	14.81	0.123–0.494	√	No (no back-rotation)
I4[a]	7.98	120	0.9576	12.50	0.104–0.417	×	Yes (Front row back-rotated against back-row piles)
I5[b]	5.58	117	0.6529	13.67	0.114–0.456	×	No (no back-rotation)
I6[b]	7.98	560	4.4688	−1.02	−0.0085	×	No ($\bar{k}_\theta$ too small)
I9[a]	14.35	56	0.8036	76.34	0.636–2.54	×	No ($k_s > 120$ kPa, $m > 9$)
I14[a]	19.15	114	2.8131	45.37	0.378–1.51	×	Yes ($k_s > 120$ kPa, $m > 5$)

[a] $s = 5.6d$.
[b] $s = 2.6d$.
[c] $\bar{k}_\theta = \left(M_y - P\Delta_o\right)/\left(0.087k_sL^3\right)$, assuming a bending moment of $M_y - M_o$ at pile-head, and M_y = 4.204 MNm at angle of rotation of 5°.
[d] At k_s = 30–120 kPa.
[e] Stable (√) and unstable (×) during forward rotation.

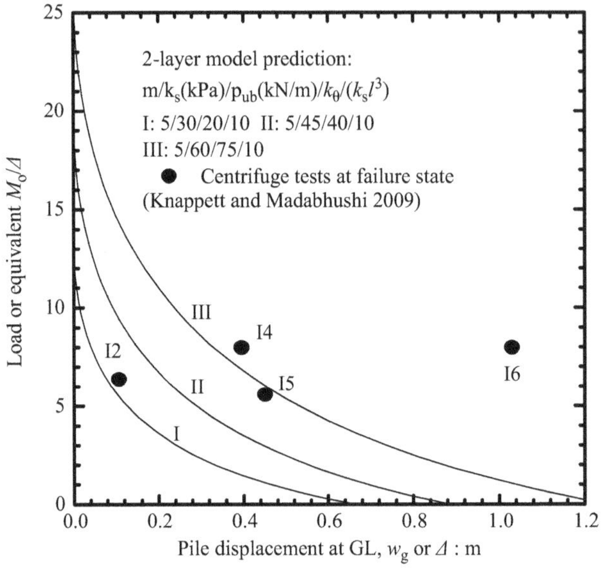

Figure 7.11 Equivalent load (M_o/Δ) and pile displacement at GL w_g or Δ. [Adapted from Guo, W. D., *J of Engrg Mechanics* **146**(9), 2020a.]

In late stage (p_{ub} = 20 kN/m), the long I4 piles back-rotated at $-\bar{k}_\theta$ of 0.104–0.417 (at k_s = 30–120 kPa). Considering $\bar{k}_\theta^* = -0.108$ and m = 1.5, for instance, the 'singularity' occurs at $\bar{l}_m = 0.191$. Using 1.5/30/20/–0.108, as l_m rises from 2.55 to 3.05 m (see Figure 7.12), the displacement w_g sways between 5.75 m and -6.1 m, the angle ω_r from –43.4° to 46.1°, and the bending moment M_m from -7.17 to 7.58 MNm, respectively. This prediction grossly agrees with the ω_r of –(27°–40°) and w_g of –4.5 m measured from the photo (see Figure 7.12) of the front piles, considering that the restraint from the 'test chamber' had limited the magnitude of the distortion. Likewise, analysis of other tests is provided in Examples 7.7 to 7.8.

Example 7.7 None amplification of I9 and I6

The I9 piles involve a calculated $\bar{k}_\theta$ of 0.636–2.54 (at M_o = 0.8036 MNm), which does not inflict amplification for $m \le 9$ (see Figure 7.9) and $k_s \le 120$ kPa. The front-row, long I6 piles yield at pile-cap connection and at a depth of 7.2 m, as the $P\Delta_o$ (of 4.469 MNm imposed) exceeds M_y. The highest $P\Delta_o$, nevertheless, results in a $-\bar{k}_\theta$ of 0.0085, which is less than $0.1\bar{k}_\theta^*$. It did not trigger *NSS* amplification. These predictions are confirmed by the test results.

Example 7.8 Amplification of I14

The back-rotated, short I14 piles yield at a head moment of 1.391 MNm, and in the pile-body of M_y. This results in $-\bar{k}_\theta$ of 0.38–1.51 [= (4,204–2,813)/(0.087k_s8^3)], At $\bar{k}_\theta = -0.38$, for instance, a slow movement (m = 5–8) will amplify the displacement w_g at early stage $\left(\bar{l}_m = -0.2\right)$ (see Figure 7.12a), as the $\bar{l}_m$ (at singularity) is estimated as 0.0384 using Equation (7.12), $\bar{k}_\theta^* = -0.38$ and m = 5. With 5/120/51/–0.38, Equations (7.2), (7.3) and (7.4) offer w_g, ω_r and M_m values of 4.3 m, –59.2° and –24.12 MNm at l_m of 0.307. As l_m rises to 0.308 m, these values swayed to opposite sign of –2.1 m, 29.4°, and 11.97 MNm, respectively. This sway distorted the front pile (as is evident in the rotation and deformation). The restraint from the 'test chamber' again had limited the magnitude of the distortion (see photos in Figure 7.12). The prediction is valid, although based on less steps of p_s (see Figure 1.4a, Chapter 1), some amplified points were filtered out.

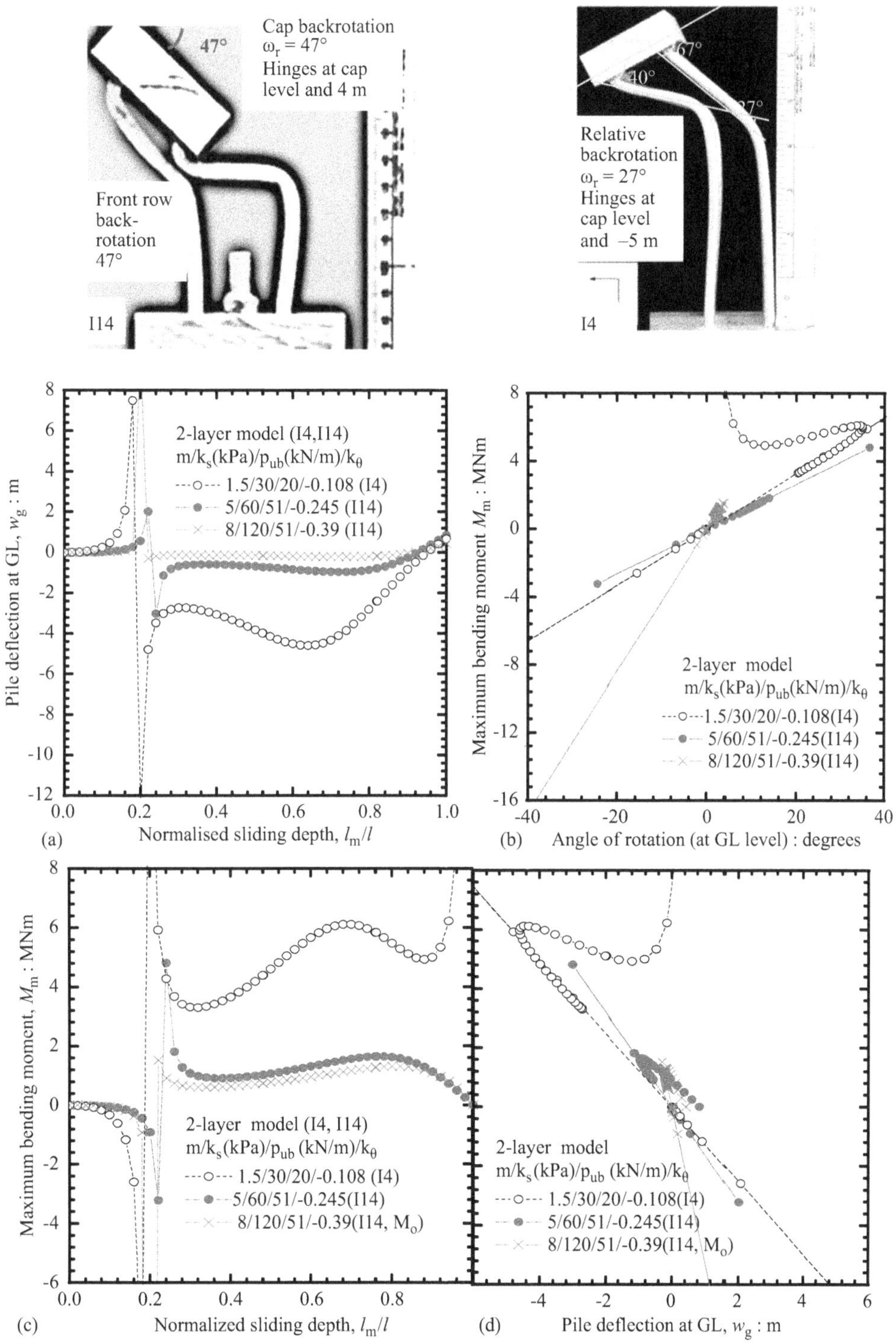

Figure 7.12 Collapsed piles (I4 and I14) in centrifuge tests (Knappett and Madabhushi 2009) and 2-layer model predictions of (a) w_g–l_m/l; (b) M_m–ω_r; (c) M_m–l_m/l; (d) M_m–w_g. [Adapted from Guo, W. D., *J of Engrg Mechanics* **146**(9), 2020a.]

7.6.2 P–Δ_o effect and lateral spreading

The 2-layer model proves effective in capturing the response of distorted and back-rotated piles, subjected to lateral spreading and set with P–Δ_o effect, by using the modulus (k_s), modulus ratio (m), rotational stiffness (k_θ) and limiting force per unit length (p_b). The study shows that regardless of the group configurations, the imposed $P\Delta_o$ values cause a reduction in the normalized stiffness $\bar{k}_\theta$ values to −(0.14–2.54) or (0.114–0.494). This reduced $\bar{k}_\theta$ together with p_b value at m = 1.5–8 incur the diverse pile response, such as (i) stable I2 and I5 piles with sufficient forward-rotating capacity; (ii) stable I9 piles for a high $\bar{k}_\theta$; (iii) distorted I14 piles at *NSS*; (iv) initially stable to later distorted (at *NSS*) I4 piles at reduced p_b and (v) hinged I6 piles despite its largest $P\Delta_o$ value and back-rotation (but no amplification). These findings highlight the flexibility of 2-layer model in simulating restrained piles affected by lateral spreading.

To determine amplification degree, obtaining the $\bar{k}_\theta$ value is critical. The next section discusses this at length, based on 1 g model tests under two soil movement profiles.

7.7 RESPONSE OF BACK-ROTATED PILES (1 G MODEL TESTS)

Chapter 1 of this book provides the detail of 1 g model tests to investigate the behaviour of the d_{32} piles socketed into an aluminium cap (50 mm in thickness) under vertical loading (P) and translating sand. The pile and the cap connection has a bending capacity (M_o^y). The vertical loading P was 0 or 294 N/pile, and shear force at head-level was negligible (H = 0). The tests provide 'measured' profiles of bending moment, shear force, soil reaction, and deflection, the maximum bending moment M_m, maximum shear force Q_m, pile-rotation angle ω_r, pile-head deflection w_g, as well as a k_s value of 20–35 kPa.

The focus herein is to deduce the k_θ by comparing the results of 1 g model test with those of the 2-layer model using the parameters from Guo (2016) of k_s = 25 kPa (= 2.1G_s, and G_s = 12 kPa), m = 17.7 (= K_p/K_a, ultimate state), p_{ub} = 10 kN/m (= $s_g\gamma' K_p^2 dl$, using γ' = 16.5 kN/m³, ϕ = 38°, d = 32 mm, l = 0.7 m, and s_g = 1.53). For translating piles, these p_{ub} and k_s values are reduced to 5.0–5.6 kN/m and 14–16 kPa, respectively. These values are identical to those used for forward-rotating piles, except for a negative value of $\bar{k}_\theta$. The parameters are presented together as $m / k_S(\text{kPa}) / p_b(\text{kN/m}) / \bar{k}_\theta$ to facilitate presentation of the subsequent analysis

Example 7.9 Yielding bending moment M_o^y of pile and cap connection

The yielding moment of the aluminium pile body, M_y is estimated as 0.4 kNm [= 350 × 10³ × 1.28/(70 × 10⁶)/(0.5d) using a yielding stress of 350 MPa and Young's modulus of 70 GPa. The M_o^y at the pile-to-cap connection is estimated as 20–60 Nm, as $M_o^y \le sP$ of 28.6–47.0 Nm, rotating about the other row with a pile-centre-centre spacing s of (3–5)d, and P = 294/pile (Juirnarongrit and Ashford 2006; Mokwa and Duncan 2003). This estimation is confirmed by M_o^y (= $M_y\omega/\omega_r$) using ω/ω_r = 0.05–0.2. The M_o^y/M_y ratio (of 0.05–0.01) is about 20% of 0.25–0.6 gained from steel pile-to pile-cap connection due to stress concentration (Xiao et al. 2006; Xiao et al. 2006; Richards et al. 2020).

The accuracy of the obtained $\bar{k}_\theta$ is checked against the yielding bending moment of pile body M_y and the M_o^y at the pile and cap connection. In a group of yielding piles subjected to a vertical load P = 294 N/pile and soil movement to SD of 0.57l, the moment M_o^{yA} and maximum back-rotated M_o^A (of pile A) were measured −21.0 Nm and 39.7 Nm, respectively, (Guo 2022b). It is worth noting that the pile-cap connection, especially the trailing pile A, is more susceptible to failure under back-rotation than lateral piles. The connection inflicts a large stress concentration between the piles and the pile-cap, as it involves a relative rotation of about (5–20)% of the pile body ω_r.

Example 7.10 Back-rotated piles without amplification

This example discusses the tests on a typical 4-pile group (s/d = 3). Figure 7.13 shows the evolution of displacement w_g, rotational angle ω_r, maximum moment M_m (stable layer) or M_o (at GL), and shear force Q_{mi} with the frame movement w_f. The piles initially rotate to M_o^{yA} (= −10 Nm), and then back-rotate to a maximum M_o (of 39.7 Nm) (see Figures 7.13e and f).

The 2-layer model prediction adopts m = 9.3, k_s = 16 kPa, p_{ub} = 4.5 kN/m ($\alpha = w_g/w_f$ = 0.71–0.82), and $\bar{k}_\theta = -0.05$ (see Figure 7.13b). The predictions for Pile A match well with the measured w_g (or ω_r)–M_m(M_o) curves, w_f–M_m(M_o) or w_f–Q_{m2}(Q_{m1}) curves (see Figure 7.13), as well as M_{m2}–Q_{m2} and M_{m1}–Q_{m1} curves (see Figure 7.14a), respectively. After deducting the M_o^{yA} of 10 Nm, the moment M_o-10 (Nm) agrees well with the measured data across the maximum shear force Q_{m1}. Additionally, the on-pile loading depth reached 0.32 m (see Figures 7.14b–e) to match with the measured profiles of $p(z)$, $w(z)$, $M(z)$ and $Q(z)$, respectively. Similar good predictions are observed for 2-pile in row and 4-pile groups (with s/d = 5) using the parameters (except for m = 17.7 for the 3-layer model, Guo 2016). The model is validated for the 1 g tests without amplification ($\bar{k}_\theta \le 0.08\bar{k}_\theta^*$ and $\bar{k}_\theta^* = -0.6$, Figure 7.9b).

7.8 DETERMINING ROTATIONAL STIFFNESS

The rotational stiffness of the model pile is determined using the shear apparatus described in Chapter 1. The sand is translated in an increment of 10 mm until a total movement w_f of 140 mm. The sliding gradually approaches a maximum l_m of 0.286l under the *ITP* movement (T-block). The response of the 2-pile in line groups (with s/d = 3, 5 and 7), and the 4-pile groups (s/d = 3, 5) is discussed previously (Guo 2020b), which includes the evolution of maximum moment M_{mi}, and the displacement w_g (see Figure 7.15). Under a *UP* movement (U-black) at an l_m of 0.286l or 0.57l, the response is depicted in Figures 7.16 and 7.17. It is worth noting that the piles may slide initially before rotating forward (if any), and subsequently back-rotate to yield at a positive moment M_{m1} (M_o).

7.8.1 Rotational restraints under ITP movement (SD = 0.286l)

Guo (2020b) shows that the three 2-pile in-line groups behave quite consistently in the M_m–w_g curves under the *ITP* movement, but for the slightly stiff reaction at s/d = 7 (see Figure 7.15). Pile B back-rotated to inflict positive bending (at GL), while Pile A experienced negative bending. The behaviour is also noticed in Pile B of the 4-pile groups. The on-pile pressure (and p_{ub} or p_b) reduces by ~50% from the 2-pile to 4-groups, which is compatible with the reduction in lateral pile-capacity from fixed-head to free-head by a maximum factor of 4 (Guo 2012). The response of the translating and back-rotated piles is captured using parameters $m / k_S(\text{kPa}) / p_b(\text{kN/m}) / \bar{k}_\theta$ of 17.7–9.3/64/2.5/−0.02 (IT_f and IT_g, see Table 7.6).

7.8.2 Restraints under uniform movement (SD = 0.286l–0.57l)

Under the *UP* movement at SD = 0.286l, the Pile B back-rotated (k_θ < 0) in only one inward-rotated 4-pile group without sliding (Guo and Ghee 2016) (Figure 7.17a). In contrast, all vertically loaded piles back-rotated at SD = 0.57l after some initial sliding (Figure 7.16). The k_θ, p_{ub} and k_s values are deduced by matching the measured pile response [of the ω_r–M_m(M_o)

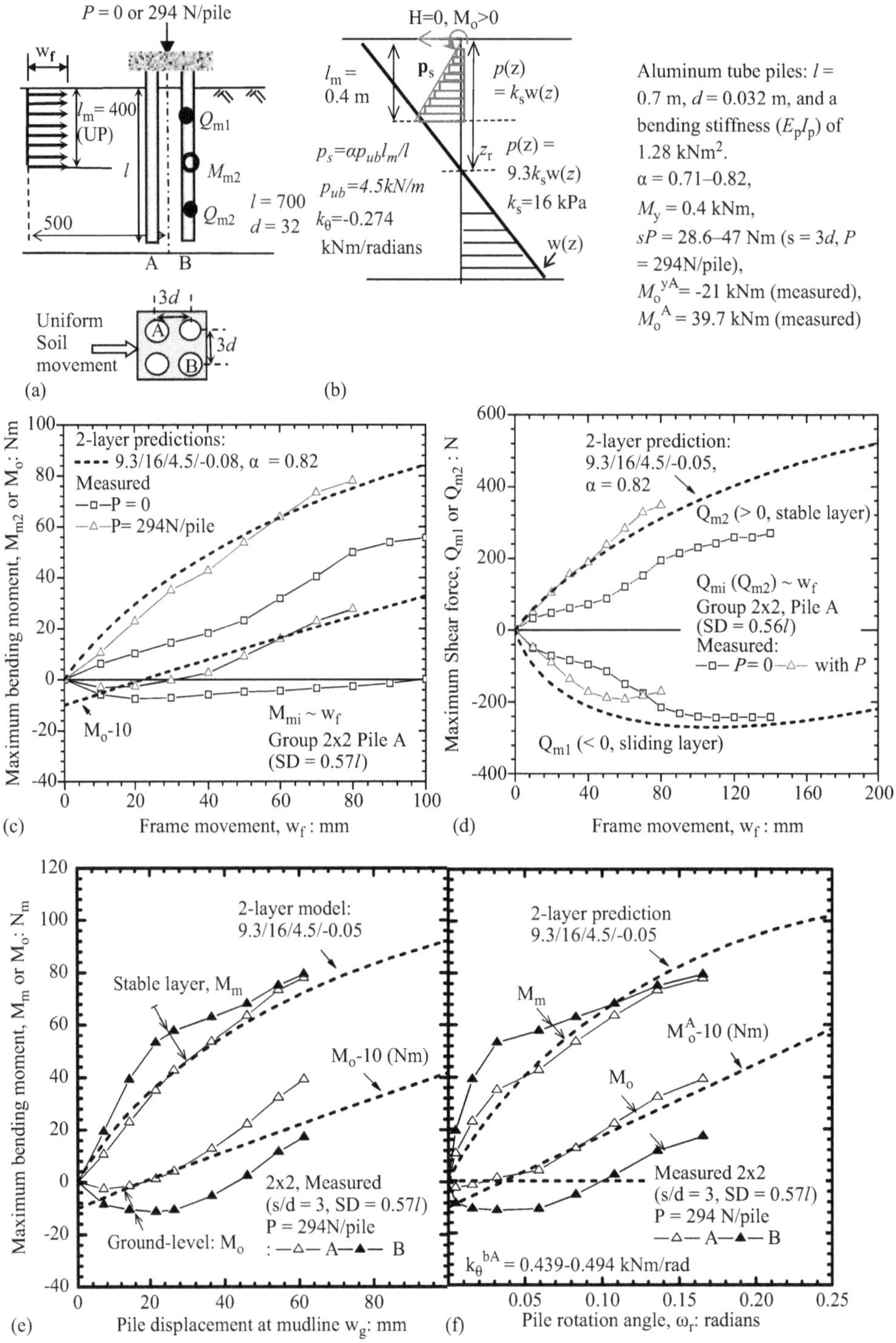

Figure 7.13 Yielding and sliding piles – Predicted versus measured development of 4-pile groups: (a) test schematic; (b) model parameters; (c) M_{mi}–w_f; (d) Q_{mi}–w_f; (e) M_{mi}–w_g; (f) M_{mi}–ω_r. [Adapted from Guo, W. D., *J of Engrg Mechanics* **146**(9), 2020a.]

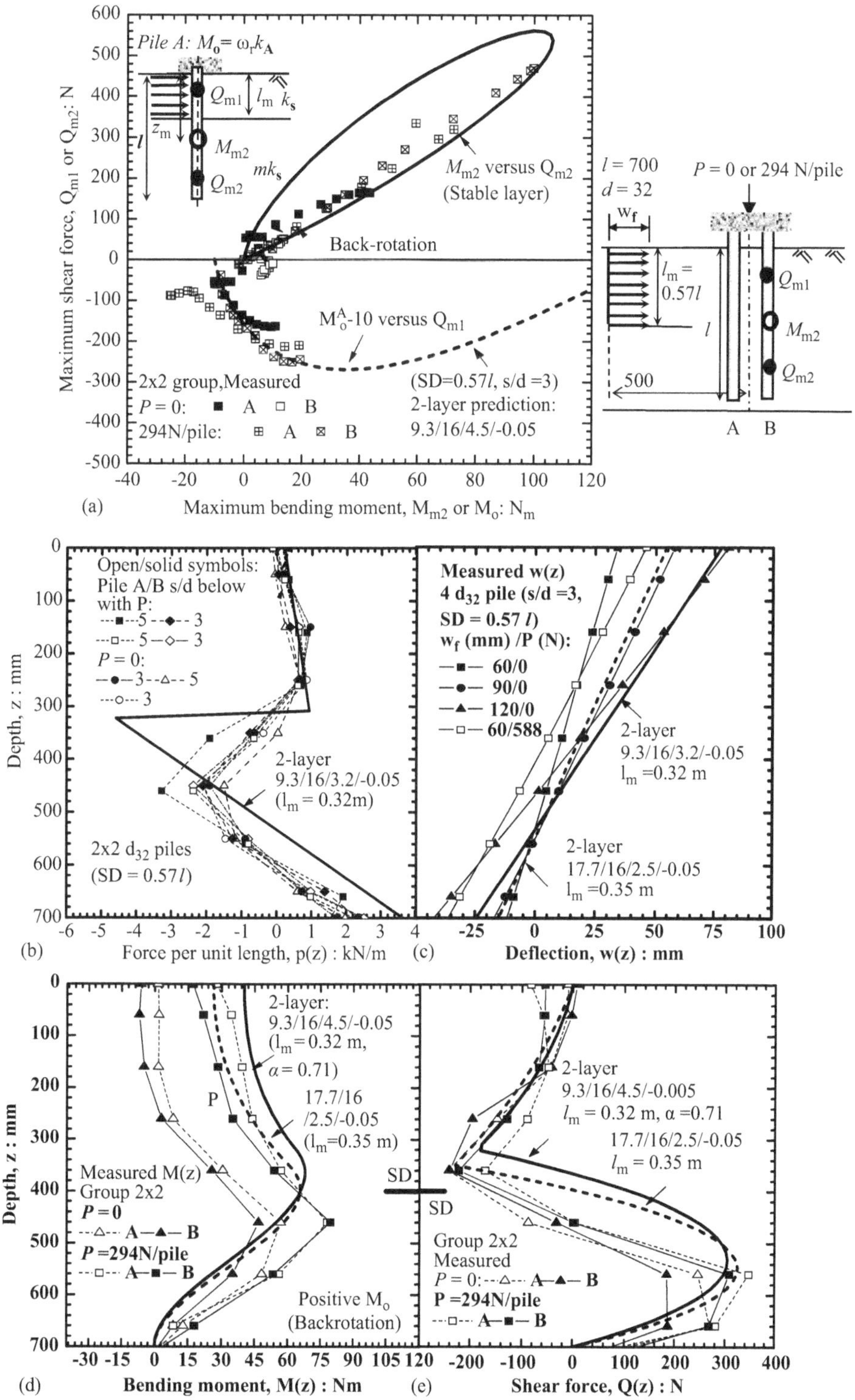

Figure 7.14 Predicted (2-layer model) versus measured (a) maximum shear force Q_{mi} and bending moment M_{mi}; (b) on-pile pressure; (c) $w(z)$; (d) $M(z)$; (e) $Q(z)$ (4-pile group, s/d = 3). [Adapted from Guo, W. D., *J of Engrg Mechanics* **146**(9), 2020a.]

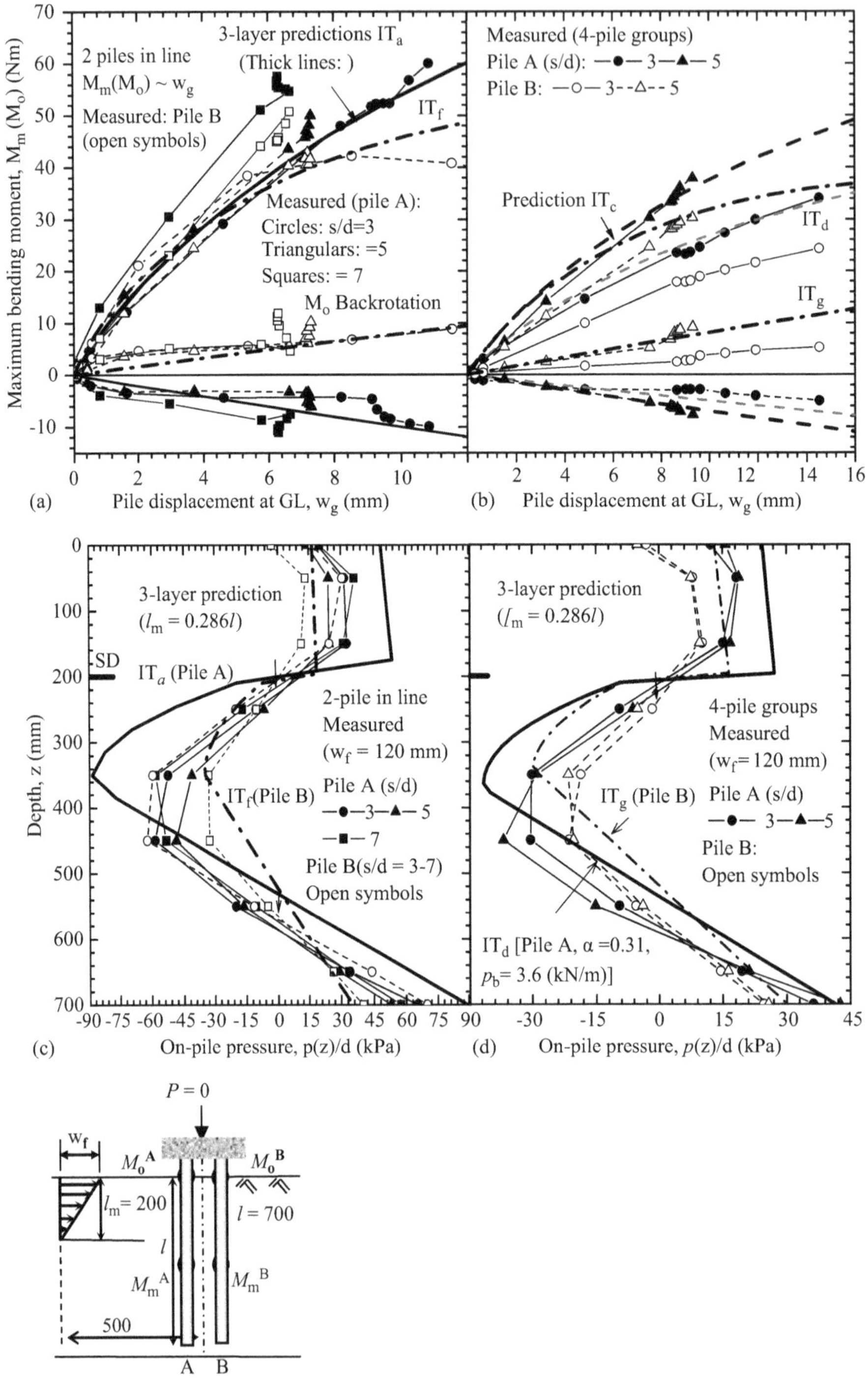

Figure 7.15 Predicted versus measured (Qin and Guo 2014) response of a pile in 2-pile-in-line (s/d = 3–7) and 4-pile groups (s/d = 3–5) under ITP soil movement: (a) (b) $M_m - w_g$; (c) (d) ultimate on-pile pressure. [Adapted from Guo, W. D., *Can Geotech J* **52**(7), 2015b.]

Table 7.6 Parameters for 3-layer model (ITP, SD = 0.286*l*) (Guo 2020b)

Prediction	*I*	IT_a	IT_f	IT_c	IT_d	IT_g	IT_e
α	0.62	0.62	0.22	0.43	0.31	0.22	0.43
k_s (kPa)	50	50	64	35	25	64	35
p_b (kN/m)	7.2	7.2	2.5	5.0	3.6	2.5	5.0
k_θ (kNm/ radian)	0	0.18	−0.44	0.12	0.09	−0.44	0
Row	A and B	A	B	A		B	
	Single	2-pile line		4-pile			row
Input parameters (generic)	m = 17.7 except m = 9.3 for pile B in 4-pile group; $p_b = \alpha p_{ub}$ (kN/m), p_{ub} = 11.6 kN/m, $\bar{k}_\theta = 0 - 0.031$ (A) and −(0.04–0.09) (B), and $k_\theta = \bar{k}_\theta \cdot k_s l^3$						

curves and the w_g–M_o lines of Pile A and B, Figure 7.16] with the 3-layer model. As shown in Tables 6.2 and 6.3 Chapter 6, the k_s (of 14–16 kPa) and $\bar{k}_\theta$ (of −0.04) remain unchanged from SD of 0.286*l* to 0.57*l*, whereas the p_b increases from 1.7–2.5 kN/m to 2.5–5.6 kN/m.

Example 7.11 Impact of group interaction and sliding on k_θ and p_b

The following describes the impact of group interaction and sliding on k_θ and p_b.

- The group effect in the 2-pile in-row (see Figure 7.16a) suggests a high p_b of 5.6 kN/m (via I_b). The calculated w_g–M_o curve is shifted downwards by M_o^{fA} = 0 (pile A), and M_o^{fB} = −7 Nm (pile B), respectively in the 2-pile in-row, and by M_o^{fA} = −7 Nm and M_o^{fB} = −20 Nm in Figure 7.16b (4-pile groups).
- The two measured curves for the sliding, 2-pile in-line or 4-pile ($\bar{P}$ = 0) groups (Guo 2022b) allow for the deduction of p_b (of 2.5 kN/m) and $\bar{k}_\theta\ (= 0, \mathrm{IV_b})$ (see Figure 7.16d).
- The parameters $\bar{k}_\theta\ /\ p_b\ (\mathrm{kN\ /\ m})$ are −0.04/5.0–5.6, 0/2.5, and −0.09/2.5 for the rotating, sliding and 'yielding' piles, respectively. The high $\bar{k}_\theta$ value is intended for vertically loaded pile. The p_b is over 2 times higher on rotating piles than on sliding and 'yielding' piles.
- The piles A and B rotate together (k_θ < 0). In no-sliding Pile B, the moment M_{m2}^B exceeds M_{m2}^A of the sliding Pile A ($M_{m2}^B > M_{m2}^A$) with a high M_o^{yB} of Pile B (> M_o^{yA}), as observed in Figure 7.16b. Conversely, $M_{m2}^A > M_{m2}^B$ is observed in the 2-pile in-row (see Figure 7.16a) and the back-rotated 4-pile group (*s*/*d* = 3) at SD = 0.286*l*.

(see Figure 7.15b)

7.8.3 Q_m–M_m for back-rotated, capped piles

Figure 4.12 (Chapter 4) indicates the relationship between rotational stiffness and the ratio of ultimate bending capacity over sliding thrust (thus shear force Q_m). The evolution of the ratio also allows the stiffness to be estimated against the model tests under the UP movement.

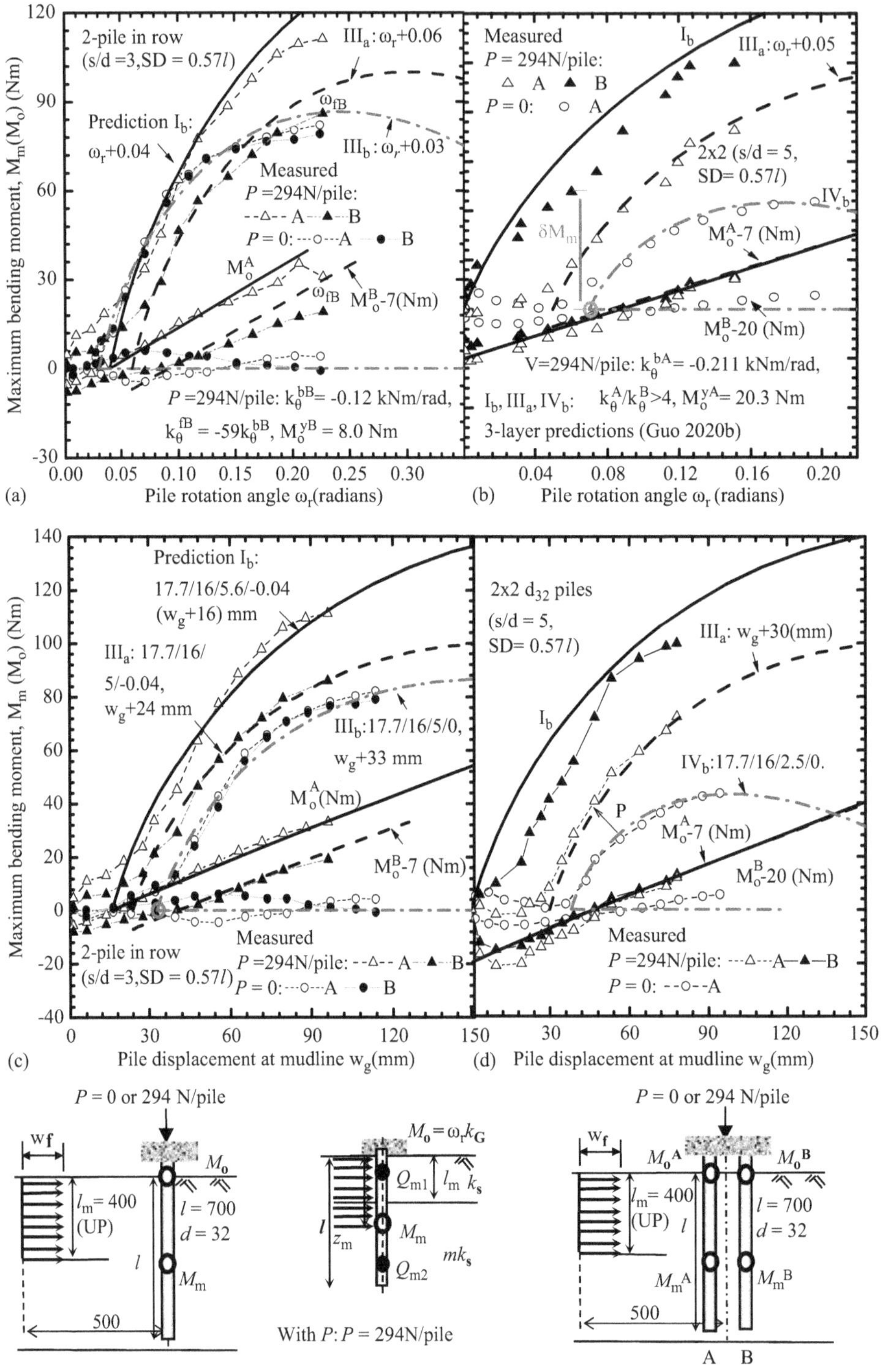

Figure 7.16 Predicted versus measured (Ghee 2009) displacement (w_g), rotation (ω_r) and bending moment M_m of piles (SD = 0.57l): (a) (c) 2-pile in-row; and (b) (d) 4-pile group. [Adapted from Guo, W. D., *Can Geotech J* **52**(7), 2015b.]

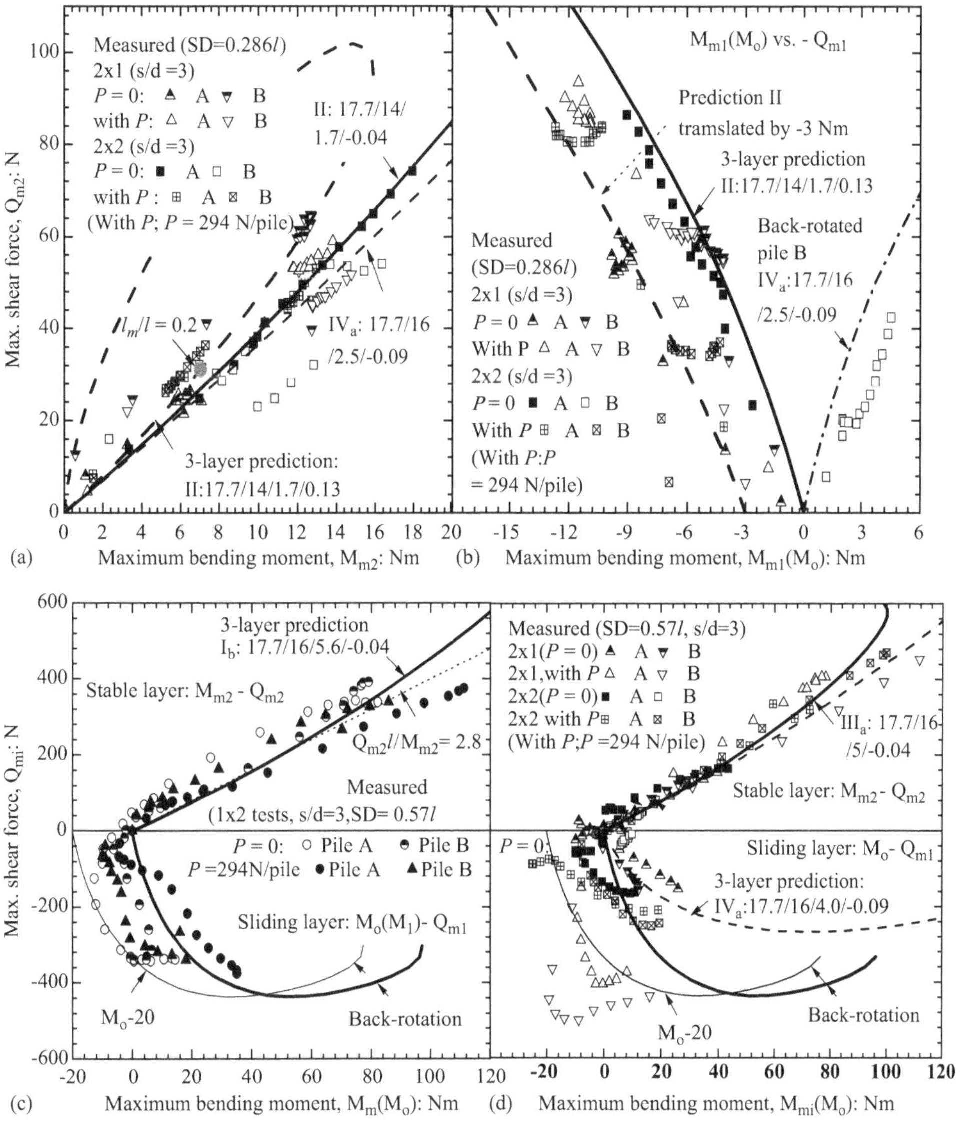

Figure 7.17 Predicted versus measured maximum shear force (Q_{mi}) and bending moment (M_{mi}) relationships: (a and b) $Q_{mi} - M_{mi}$ (SD = 0.286*l*); (c and d) back-rotated piles in 2-pile in-row, 2-pile in-line and 4-pile groups (SD = 0.57*l*). [Adapted from Guo, W. D., *Can Geotech J* **52**(7), 2015b.]

Example 7.12 Measured M_{mi}–Q_{mi} and k_θ

Figures 7.17a and b provide the measured M_{m2}–Q_{m2} (stable layer) and M_{m1} (M_o)–Q_{m1} (sliding layer) curves, for 2-pile in line (2 × 1) and 4-pile (2 × 2) groups at SD = 0.286*l* (*s*/*d* = 3). For SD = 0.57*l*, the curves are plotted together in Figure 7.17c for pile A or B in the capped 2-pile in-row (1 × 2); and in Figure 7.17d for the 2-pile in-line, and the 4-pile groups (*s*/*d* = 3). At SD = 0.286*l*, the measured moment M_o (at Q_{m1} = 0) is –(1.0–3.0) Nm (forward rotation) and 1.0–3.0 Nm (backward rotation), respectively. The latter increases to –(8–25)Nm at SD = 0.57*l* (see Figure 7.17c). As with single piles (Guo 2016), the non-homogeneous movement of the sand imposes a reduced shear force of (0.5–0.7)Q_{m1}. By matching with the measured M_{m1} and Q_{m1} curves, the 3-layer model allows an estimated $\bar{k}_\theta \left(= \bar{k}_A, \text{pile A}\right)$ of 0.11–0.13 (forward rotation for M_o = –3.0 Nm, see Figures 7.17a and

b) and $\bar{k}_\theta\left(=\bar{k}_B\right)$ of -0.04 (backward rotation of one pile) at SD = 0.286l, and $\bar{k}_\theta$ of -(0.04–0.09) (for M_o = -20 Nm, see Figures 7.17c and d) at SD = 0.57l. The stiffness k_θ deduced differs by 2–3 times between pile A $\left(\bar{k}_A\right)$ and pile B $\left(\bar{k}_B\right)$.

The deduced $\bar{k}_\theta$ value [of –(0.04–0.13)] roughly agrees with those deduced in Figure 4.12 (Chapter 4) using the measured ratios of $M_m/(Q_m l)$. The k_θ is estimated as –(0.192–0.624) kNm/radians, which is close to –0.44 kNm/radians deduced from the *ITP* tests. Nevertheless, the $\bar{k}_\theta$ is largely lower than those obtained for forward-rotating piles as shown in Table 7.1.

Interestingly, the k_θ value of appears to be insensitive to the uniform movement profile or the *ITP* movement, and for both SD of 0.286l and 0.57l. This suggests a reliable means to deduce the rotational stiffness k_θ by matching between measured data and 3-layer model. The back-rotated piles in Figures 7.13–7.17 involve a modulus ratio m of 9.3–17.7 (Guo 2016), which are not subjected to amplification. At those $\bar{k}_\theta$ values, nevertheless, lateral spreading (with a lower m of 1.5–5.0) can inflict amplification, as is illustrated next concerning the Showa Bridge.

7.9 AMPLIFICATION OF P5 AND COLLAPSE OF SHOWA BRIDGE

The Showa Bridge collapsed one month after construction due to an earthquake strike in 1964. Figure 7.18 shows the position of the girders and piles aftermath. Each span (over a width of 24 m) consisted of 12 composite girders, and was supported by a line of 9 steel piles (600 mm diameter, 9–16 mm in wall thickness). The girders rested on movable (left) and fixed (right) joints on each of P1–P5 piers, and on movable (right) and fixed (left) joints on each of P7-P10 piers. They also rested on two movable joints of P6 to allow for thermal expansion of bridge. The piles (l = 16 m and d = 0.6 m) had yielding moment M_y of 1.0–1.62 MNm.

The earthquake induces a lateral spreading of liquefied sand (k_s = 30–50 kPa) to incur forward rotation of the pile-heads (to the right) by 0.12 m (pier 1), 0.34 m (pier 2), 0.36 m (pier 3), and 0.43 m (pier 4), respectively (Fukuoka 1966; Hamada and O'Rourke 1992) at an estimated angle of 1.77°, 5.5° and 6.61° [= tan(δ/l), Figure 7.18]. The estimated angle of the hinged P4 piles, for instance, roughly agrees with the collapsed angle of 6° (= 0.105 radians), which suggests a normalized $\bar{k}_\theta\left[=M_y/\left(0.105k_s l^3\right)\right]$ of 0.078–0.126 that are consistent with previous estimations.

The lateral spreading caused the upper portions of piers P1–P5 and thus Girder D to move forward to the right until the Girder was halted by the fixed joints of P5 for a maximum sliding range of 0.43 m. Subsequently, the P5-head had to rotate backwards (to the left). At $\bar{k}_\theta$ = –0.078 and m = 1.5, for instance, a thin, moving layer of 1.2–2.3 m deep (i.e., l_m/l = 0.076–0.146, Figure 7.18) would amplify the pile-head displacements and rotation by 3.4–3.6, and 4.8–5.3 times. A low $-\bar{k}_\theta^*$ value is associated with a high l_m/l value, and the singularity of $\bar{k}_\theta^*$ = –0.119 occurs at a low l_m/l of 0.09 as shown in Table 7.7.

Example 7.13 Opposite directions of response at critical l_m/l

At the adjacent l_m/l values of 0.08 and 0.10, the amplification factor of w_g reaches 238.2 and -237.45 times, respectively (for k_s = 50 kPa, see Table 7.4). At l_m/l = 0.08, the normalized displacement $\bar{w}_g$, rotational angle $\bar{\omega}_r$ and maximum bending moment $M_m/(p_s l^2)$ are estimated to be 37.892, -74.52, and -8.756, respectively. With p_b = 30 kN/m, p_s = 2.4 kN/m (= $p_b l_m/l$) and l = 16 m, the displacement w_g, the rotational angle ω_r and the moment M_m are estimated to be 1.819 m, -0.224 radians (–12.8°), and –5.38 MNm, respectively. At l_m/l = 0.10, the response is obtained as $\bar{w}_g$ = –46.926, $\bar{\omega}_r$ = 92.278, and $M_m/(p_s l^2)$ = 10.757, leading to w_g = -2.816 m, ω_r = 19.83°, and M_m = 8.261 MNm (for p_s = 3.0 kN/m).

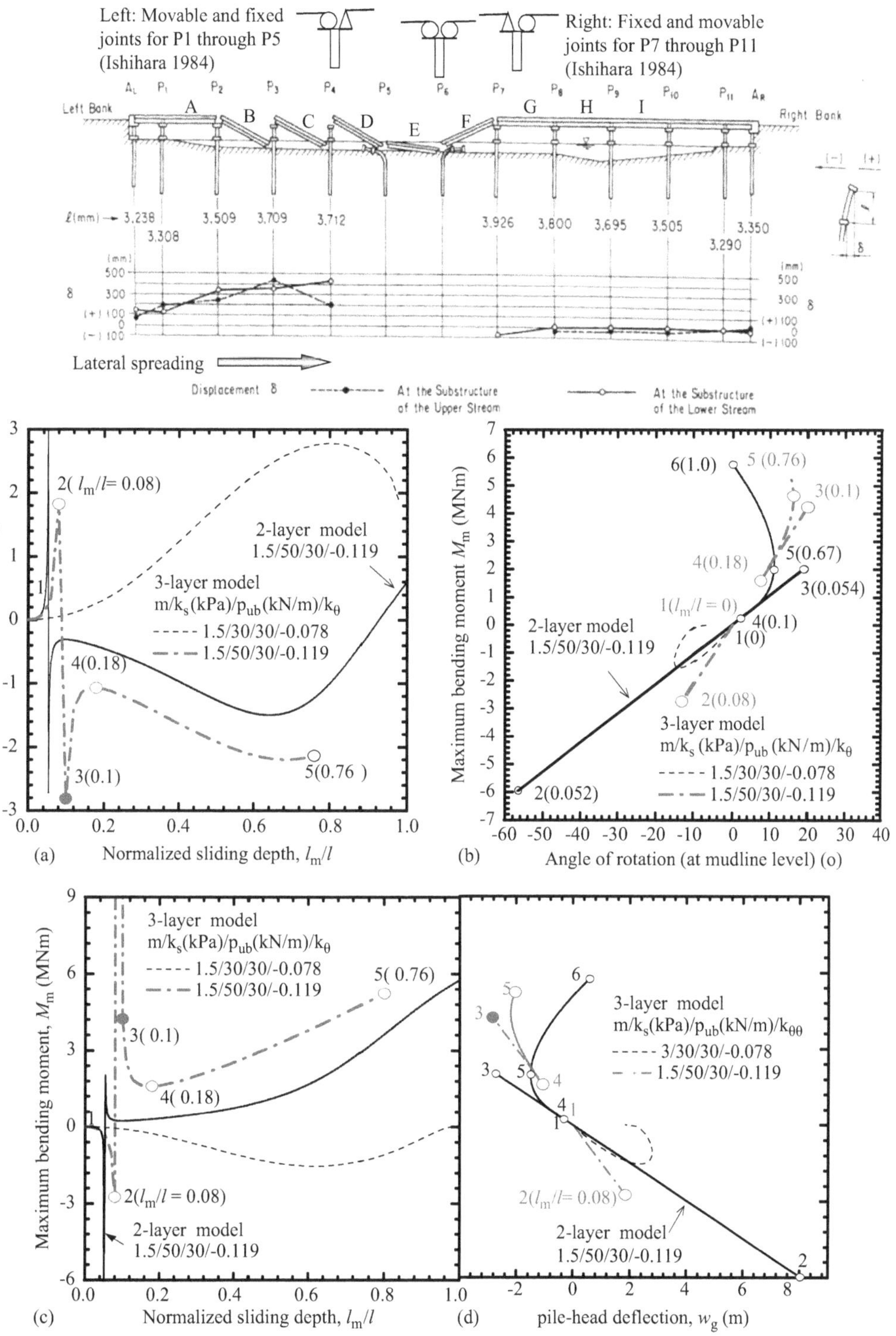

Figure 7.18 Showa Bridge with schematic joints and pile-cap deflections (Iwasaki 1986) and 2-/3-layer predictions of: (a) w_g–l_m/l; (b) M_m–ω_r; (c) M_m–l_m/l; (d) M_m–w_g. [Adapted from Guo, W. D., *Can Geotech J* **52**(7), 2015b.]

Table 7.7 Normalized singularity stiffness $\left(-\bar{k}_\theta^*\right)$ for Showa Bridge (Guo 2020b)

m	1.3	1.5	2	3
l_m/l = 0.146	0.103	0.116	0.147	0.205
l_m/l = 0.25	0.101	0.112	0.136	0.181
l_m/l = 0.50	0.098	0.106	0.124	0.151

As the l_m/l rises from 0.08 to 0.10, the pile is predicted to undergo significant oscillation, with shift of displacement w_g from 1.8 m to −2.8 m, rotation ω_r from −13° to 20° and bending moment M_m from −5.4 to 8.3 MNm, respectively (see Figure 7.18). The predictions agree with the measured displacement of P4 piles (Fukuoka 1966), and the collapse rotation angle of 18.4° of Guider D [= $\tan^{-1}(9/27)$, 9 m in free-length, and 27 m in span].

More specifically, the predicted response (using the 3-layer model) evolves from point 1 through to point 5 as the l_m/l increases from 0, 0.08, 0.1, 0.18, to ultimately 0.76. Note that at l_m/l = 0.09, the amplified response is estimated as w_g = −181.6 m, ω_r = 1278.6° and M_m = 270.68 MNm, which is not practically achievable (represented by gaps in the figure). As shown in Figure 7.18b, the piles rotate forward to point 2, back-rotate to point 3 and rotate forward again to point 4 (l_m/l = 0.18). The first forward rotation (with $l_m < 0.09l$) should have failed the piles, and the subsequent stages never occurred. Without considering the dragging layer (i.e., the 2-layer model), the singularity for $\bar{k}_\theta$ of −0.119 occurs at l_m/l of 0.05247 using Equation (7.13). The 2-layer model (see Figure 7.18) predicts a similar forward-backward-forward rotation (and failure) mechanism to the 3-layer model, albeit with an early failure at $l_m = 0.052l$.

7.9.1 Displacement amplification of P5

To the left of the bridge, the enlarged displacement of P5 pulled the Girder E (over P5–P6) to slide over P6, and forced Girder D to move towards C (P3–P4), and C towards B (P2–P3), until it pulled the Girder E off P6 first. This process may take 1–2 minutes to overcome sliding resistance between the girders and all the piers (P2, P3, P4, P5 and P6) after the cease of the strongest ground motion (Horii 1968; Hamada and O'Rourke 1992). The back-rotation of P5 (7.8°) and its displacement caused Girder D to fall, generating a forward rotation of 6° [= $\tan^{-1}(2/19)$] at P4 before being fully transferred to Girder C. The forward rotation at P4 in-turn caused the fall of movable joint at P4 and Girder C. This behaviour was replicated at P3 (e.g., with a slide of 0.31–0.43 m) but not at P2, where the slide was limited.

7.9.2 Movable joints eliminating displacement amplification

To the right of the bridge, as the Girder E was pulled off P6, the weight of Girder F on P6 was then shifted entirely to the right movable point, resulting in eccentric loading to engender forward (to right) rotation that caused fall off of the Girder F at P6. Simultaneously the fixed point of P7 back-rotated towards the left. The movable-head of piles at P6 has a negligible small stiffness $\bar{k}_\theta\left(\approx 0\right)$, and was distant to the amplification. Likewise, the movements of P7 (100 mm as measured), and P8 (50 mm) were not large enough to cause joints fall off the piers. In brief, the collapse of Showa Bridge in 1964 was due to the P5 piles rotating backward and undergone singularity amplification. Fortunately, the free-head of P6 effectively stopped the amplification and saved P7, P8 and P9. Had the spreading started from the right bank, then a mirrored imaging of the failure would have occurred.

Example 7.14 Limited amplification of piles underpinned Christchurch bridges

The region of Christchurch, New Zealand, experienced a series of strong earthquakes between September 2010 and December 2011, resulting in widespread liquefaction and lateral spreading along the Avon River. The bridges in the area, ranging from short to medium spans, were not immune to the effects of the seismic activity. The abutments of these bridges suffered from back-rotation, compression and deck damage (Haskell et al. 2013). Five typical bridges (see Figure 7.19) are examined here by using Equations (7.7) and (7.8) to gain enlargement of pile rotation and displacement resulting from back-rotation. Table 7.8 provides the pile length l, yielding moment M_y and the measured ultimate angle of back-rotation ω_r for each of the five bridge. The values of $-\bar{k}_\theta$ are obtained for each bridge using $-\bar{k}_\theta = M_o^y / \left(k_s \omega_r l^3\right)$ and $M_o^y = M_y$. Assuming k_s = 30 kPa, the NSS $\bar{k}_\theta^*$ was obtained for m = 1.5–4. By assigning an identical rotational stiffness from forward to backward rotation, the rotation (thus bending moment) increase by 1.6–1.4 times (South or S) for the Fitzgerald bridge; 1.9–2.7 times (North or N), and 1.3–1.5 times (S) for the Avondale bridge; 1 times (N, S) for the Gayhurst Bridge; 1.2–1.4 times (East) and 1.3–1.5 times (West) for the South Brighton bridge; and 1.4–1.7 times (N) and 1.5–2.0 times (S) for the Anzac Bridge. Gayhurst Bridge piles would be sensitive to amplification if there were a back-rotation between the abutment and the piers. The use of an integral abutment and piers prevents the damage from back-rotation.

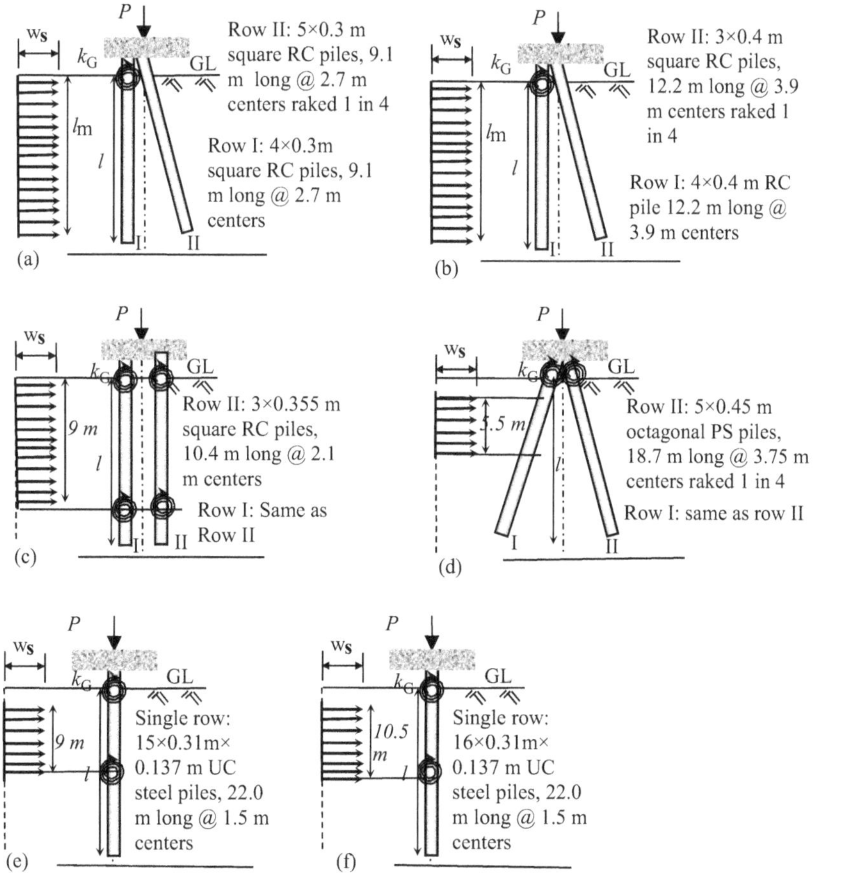

Figure 7.19 Piles in Christchurch subjected to lateral spreading: (a) Fitzgerald Avenue Bridge; (b) Avondale Bridge; (c) Gayhurst Bridge; (d) South Brighton Bridge; (e) Anzac Bridge (South); (f) Anzac Bridge (North). [Adapted from Guo, W. D., *Can Geotech J* **52**(7), 2015b.]

Table 7.8 Piles underpinned Christchurch bridges (Guo 2020b)

Bridge	*Fitzgerald*	*Avondale*	*Gayhurst*	*South Brighton*	*Anzac*
M_p (kNm)	69.4	132.8	111.6	175.2	689.1
l_m (m)	9.14	12.2	9.0	5.5	10.5
l (m)	9.14	12.2	10.4	18.7 and 13.3	22
ω_r[a](o)	7.5°	3.2°(N)–7.9°(S)	2.2°(N)–0.5°(S)	7°(E)–8°(W)	5°(N)–6°(S)
$\bar{k}_\theta \cdot k_s$	0.694	0.53–1.31	2.584–4.737	0.533–0.61	0.823–0.987
$\bar{k}_\theta\left(k_s = 30\text{KPa}\right)$	0.023	0.044–0.018	0.086–0.158	0.018–0.02	0.027–0.033
w_s (m)		0.87	0.65–0.75	2.18–2.70	0.94–1.12
$\bar{k}_\theta^*\left(m = 1.5\right)$	0.101	0.101	0[b]	0.107	0.106
ω_r raise (times)	1.6	2.7(N)–1.5(S)	1	1.4(E)–1.5(W)	1.7(N)–2.0(S)
w_g raise (times)	1.4	2.0(N)–1.3(S)	1	1.3(E)–1.3(W)	1.4(N)–1.6(S)
$-\bar{k}_\theta^*\left(m = 4\right)$	0.15	0.15	0[b]	0.185	0.176
ω_r raise (times)	1.4	1.9(N)–1.3(S)	1	1.2(E)–1.3(W)	1.4(N)–1.5(S)
w_g raise (times)	1.3	1.6(N)–1.2(S)	1	1.2(E)–1.2(W)	1.3(N)–1.3(S)
Amplification	No	No	No[b]	No	No

Notes:
[a] Measured angle of abutment (Haskell et al. 2013).
[b] The integral design does not allow back-rotation between abutment and piers.

7.10 CONCLUSIONS

Capped rigid piles may rotate forward and/or backward subjected to lateral spreading. The 2-layer model and solutions well capture the nonlinear response of the piles. The associated parameters [i.e., a movement factor α (= 1.39–1.5) and rotational stiffness of pile-cap connections] are deduced from ten model tests with diverse movement profiles under forward rotation. As for back-rotated piles, their response (deflection, rotation and bending moment) is amplified around a singularity stiffness. The 2-/3-layer models provide the expressions to estimate the stiffness, the amplification degree and the sliding depth at which the singularity occurs. The back-rotation stiffness of typical pile-cap connections is determined extensively using 1 g model tests. The models and expressions well capture the response of the 1 g model piles, and the centrifuge test piles with P-Δ effect, and reveal new failure mechanism of Showa Bridge and safety state of the piles underpinning Christchurch bridges.

Chapter 8

Excavation on adjacent piles

8.1 INTRODUCTION

Basement excavation work normally incurs soil movements behind retaining structure, and even catastrophic damage (Whittle and Davis 2006; Chai et al. 2014). The safety of nearby piles was assessed using centrifuge tests (Leung et al. 2000; Ong et al. 2003; Bourne-Webb et al. 2011), and numerical analysis. Guo (2021a) integrated parameters for modelling the piles with other cases, to gain the impact of mixed modes of soil movement on pile safety.

In modelling piles subjected to lateral spreading, Dobry et al. (2003) developed a simple (limit equilibrium) method underpinned by a rotational stiffness and a constant limiting force per unit length (*FPUL*) with depth. Guo (2014, 2015b, 2016) developed 2-/3-layer models and the associated closed-form solutions, which are capitalized on a set of (at most) five input parameters (see Chapters 4 and 5). These solutions well capture nonlinear response of rigid piles in sliding soil displaying diverse movement profiles as incurred by lateral spreading, sliding sand and sliding slope. The models, for instance, well predict 3–5 times higher bending moment under translational (Guo and Ghee 2004) than under rotational soil movement (Guo 2016).

In light of the 2-layer model (Guo 2014), BiP-k_θ and BiP-η models are developed to simulate piles subjected to mixed (sliding-rotational) modes of soil movement incurred by excavation loading (Guo 2021a). A flexible pile is modelled as upper and lower (**bi**-) portions, which join together by a rotational spring (of stiffness k_θ) at the depth of maximum bending moment z_m (see Figure 8.1). The BiP-k_θ model assumes a constant stiffness k_θ of the upper portion about the depth z_m (which includes elastic restraint from a lower elastic portion of BiP, termed as R-E model, see Chapter 4). The BiP-η model stipulates a constant rotation ratio η between the bi-portions. The models are tailored to capture response arising from elastic rebound, brittle failure and stiffness singularity, in comparison with 2-layer model.

8.2 UPPER PORTION INTERACTION UNDER SOIL MOVEMENT

In previous chapters, we introduced the 2-layer model [e.g., Figure 1.4, Chapter 1, Guo 2015b), which simulate behaviour of a rigid pile under external loading p_s (of $w_s k_s$) induced by a sliding layer (depth $0 \sim l_m$) of a modulus of subgrade reaction k_s, while embedded in a stable layer (depth $z_m \sim l$) with a modulus of mk_s. Under a lateral force H (at depth a below ground-level, or GL), a sliding of soil to a depth l_s forces the pile (with a diameter d) to rotate rigidly about the depth z_r (= $-w_g/\omega_r$) to an angle ω_r and a GL deflection w_g (Figure 1.4c) for a linear displacement $w(z)$ (= $\omega_r z + w_g$) with depth z (see Figures 1.4c and e). The net on-pile *FPUL* $p(z)$ is given by $k_s w(z) - p_s$ (sliding layer) and $mk_s w(z)$ (stable layer). The pile (see Figure 1.4c) has a total rotational restraining stiffness k_θ along its embedment length l,

DOI: 10.1201/9781003315230-8

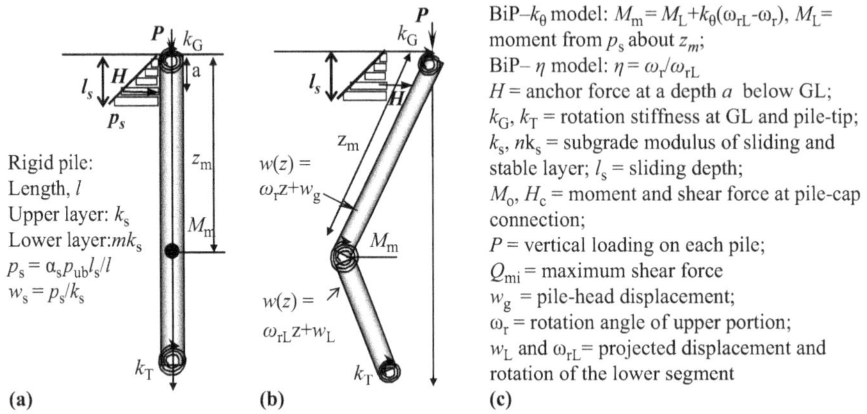

Figure 8.1 BiP and R-E models: (a) 2-layer model; (b) BiP/R-E models; (c) salient features of new models. [Adapted from Guo, W. D., *Int J Geomechanics*, **21**(1), 2021.]

including the stiffnesses k_G and k_T to impose constraining moments M_G (= $k_G\omega_r$) and M_T (= $k_T\omega_r$). In the framework of the 2-layer model, explicit expressions were developed to estimate the displacement $w(z)$, the shear force $Q(z)$ and the bending moment $M(z)$ at depth z; as well as the maximum bending moment M_m and its depth z_m; and the maximum shear force Q_{m1} in sliding layer and Q_{m2} in stable layer, respectively. To gain nonlinear response of the pile, these estimations are repeated for a series of $\bar{l}_s$ values ($l_s/l < 0.8$, e.g., by a step of 0.1) to form a stepwise, uniform *FPUL* p_s (= $\alpha p_{ub} l_s/l$) over the depth l_s.

The 2-layer model is underpinned by five parameters of the modulus k_s, the modulus ratio m, the limiting *FPUL* at pile-base level p_b (= αp_{ub}, p_{ub} = standard value of free-head piles), the stiffness k_θ, and a soil movement factor α. The four parameters (k_s, m, p_{ub} and k_θ) remain constant throughout the entire loading process (see Chapters 1 and 5). The soil movement factor α is also a constant for a consistent mode of pile-soil interaction as incurred by lateral spreading, sliding slope, and embankment loading (Guo 2015b, 2016), as well as in vertically loaded piles ($P \neq 0$) (Guo et al. 2017). For modelling flexible piles in centrifuge tests, the 2-layer model (rigid-pile based) was adapted by reducing values of k_s and increasing α values to estimate response profiles that reflect shift of soil movement modes from stable to collapse walls. However, a constant α of 1 is still used to model the overall response.

Furthermore, a 'hinge' spring is incorporated into the 2-layer model (see Figure 8.1, Guo 2021a). This renders the BiP models to inherit the five input parameters of the 2-layer model, namely k_s, m, p_{ub}, α and k_θ (see Figure 8.1b). Notably, the k_θ value now represents the total rotational stiffness k_θ along the upper segment, including k_G, rather than the entire pile length. The bi-portion (BiP) models and solutions are developed individually (Guo 2021a), which will be recapitulated next.

8.3 BIP MODELS AND SOLUTIONS

8.3.1 Non-dimensional analysis (H = 0, M_o = 0)

A flexible pile may be characterized by the displacement (w_g) and rotation (ω_r) of its upper portion and the projected displacement (w_L) and rotation (ω_{rL}) along the low segment at GL (z = 0). The pile displacement $w(z)$ (at depth z) is described by $w(z)$ (= $\omega_r z + w_g$) above the depth of maximum bending moment (z_m) and $w(z)$ (= $\omega_{rL} z + w_L$) below z_m (see Figure 8.1b).

Table 8.1 Solutions for 2-layer BiP models (Guo 2015b, 2021a)

The horizontal force and bending moment equilibrium of upper segment of the rigid pile (see Figure 8.1) are given by Equation (T8.1a) and (T8.1b), respectively

$$\int_0^{z_m}\left(\bar{\omega}_r s+\bar{w}_g\right)p_s ds-\int_0^{l_s} p_s ds = H \quad \text{(T8.1a)}$$

$$\int_0^{z_m}\left(\bar{\omega}_r s+\bar{w}_g\right)p_s\left(z_m-s\right)ds-\int_0^{l_s} p_s\left(z_m-s\right)ds = p_s l^2 \bar{k}_\theta\left(\bar{\omega}_{rL}-\omega_r\right)+M_L \quad \text{(T8.1b)}$$

The horizontal force and bending moment equilibrium of lower segment of the rigid pile (see Figure 8.1) are given by Equation (T8.1a) and (T8.1b), respectively

$$\int_{z_m}^{l}\left(\bar{\omega}_{rL} s+\bar{w}_L\right)p_s m ds = 0 \quad \text{(T8.1c)}$$

$$\int_{z_m}^{l}\left(\bar{\omega}_{rL} s+\bar{w}_L\right)p_s m\left(s-z_m\right)ds = p_s l^2\left[\bar{k}_\theta\left(\bar{\omega}_r-\bar{\omega}_{rL}\right)-\bar{k}_T\bar{\omega}_{rL}\right] \quad \text{(T8.1d)}$$

$$\bar{\omega}_r = \eta\bar{\omega}_{rL} \quad \text{(T8.1e)}$$

Given a constant stiffness $\bar{k}_\theta$ (BiP-k_θ model), first, Equations (T8.1a) and (T8.1b) are resolved to obtain solutions of $\bar{\omega}_r$ and $\bar{w}_g$ as functions of $\bar{\omega}_{rL}$ and $\bar{w}_L$; second, Equations (T8.1c) and (T8.1d) are resolved to obtain $\bar{\omega}_{rL}$ and $\bar{w}_L$. The $\bar{w}_g$ and $\bar{\omega}_r$ expressions are subsequently simplified into expressions in Table 8.3, respectively; along with the $\bar{w}_L$ expression. Given a constant rotation ratio η (BiP-η model), the condition of Eq. (T8.1e) is substituted into the solutions for BiP-k_θ model as explained in the text of this paper.

Regardless of pile rigidity, the length l and modulus k_s are adopted in rigid-pile based solutions to predict overall pile response. The modulus reduces from 'static' loading to lateral spreading or creep loading by 90% (Cubrinovski et al. 2006; Guo 2020a). The 'static loading' is associated with $k_s^{profile}$ (as deduced from response profiles) and the 'lateral spreading' involves a minimum rotation depth > 0.5l. The modulus is simply correlated with l_e by $k_s = [2 - (l/l_e)]k_s^{profile}$ (l_e > 0.5l) to gain response profiles of flexible piles. The 'reduction' is designed to maintain the pile-head displacement and to offset the resistance and moment gain (from rigid-pile solution on the embedment l) for the non-effective portion below l_e. The l_e value may be taken as z_m for a thick sliding layer and justified against measured data.

At the depth z_m, the pile-displacement w_h satisfies the compatibility condition of $w_h = \omega_r z_m + w_g = \omega_{rL} z_m + w_L$ (but not the rotation). The displacement is recast into following

$$w(z) = \left(\bar{\omega}_r \bar{z}+\bar{w}_g\right)p_s / k_s \tag{8.1a}$$

$$w(z) = \left(\bar{\omega}_{rL} \bar{z}+\bar{w}_L\right)p_s / k_s \tag{8.1b}$$

where $\bar{w}_g\left(=w_g k_s / p_s\right), \bar{\omega}_r\left(=\omega_r k_s l / p_s\right), \bar{w}_L\left(=w_L k_s / p_s\right)$ and $\bar{\omega}_{rL}\left(=\omega_{rL} k_s l / p_s\right)$ are normalized displacement and rotation for the upper and lower portions. As with the development of 2-layer model (see Table 4.1), the solutions for the BiP models are obtained by satisfying the force and bending moment equilibrium of each portion, while ensuring the displacement compatibility at the joint depth z_m (as highlighted in Table 8.1). It is worthwhile to stress that:

- The lateral force H (at a depth a below GL) and the *FPUL* p_s (up to a sliding depth l_s) incur a bending moment of $M_L = H(z_m - a) + 0.5p_s l_s(2z_m - l_s)$ about depth z_m;
- The lower segment is assumed to have a smaller rotation than the upper segment (termed as passive loading).
- The total moment about the depth z_m due to passive loading amounts to $M_L + k_\theta(\omega_{rL} - \omega_r)$.

The solutions are presented as explicit expressions in Table 8.2, irrespective of the BiP models (see Figure 8.1c). They offer the response profiles (with depth z) of net *FPUL* $p(z)$, displacement $w(z)$, shear force $Q(z)$, and bending moment $M(z)$. The maximum bending moment M_m [= $M_2(z_m)$], for instance, is given by

$$\frac{M_m}{p_s l^2} = \frac{1}{6}\bar{\omega}_r \bar{z}_m^3 + \frac{1}{2}\bar{w}_g \bar{z}_m^2 - \bar{M}_L \tag{8.2}$$

where $\bar{z}_m = z_m / l$, $\bar{M}_L = M_L / (p_s l^2) = \bar{H}(\bar{z}_m - \bar{a}) + 0.5\bar{l}_s(2\bar{z}_m - \bar{l}_s)$, $\bar{H} = H / (p_b l_s) = H / (p_s l)$, $\bar{a} = a / l$, and $\bar{l}_s = l_s / l$. In the BiP model, the rotational stiffness is lumped into the depth z_m, and thereby it is not included in Equation (8.2). In addition, the moment M_m is normalized by $p_s l_s l$ [i.e., $M_m/(p_s l_s l)$] for the 'nonlinear' case. The shear force attains its peak values Q_{m1} in the sliding layer and Q_{m2} in the stable layer, and has a value of $Q(l_s)$ at the sliding depth.

$$\frac{Q_{m1}}{p_s l} = \frac{-1}{2}\frac{\bar{w}_g^2}{\bar{\omega}_r} - \bar{P}_h \text{ and } \frac{Q_{m2}}{p_s l} = -0.5m\frac{(\bar{w}_g + \bar{z}_m\bar{\omega}_r)^2}{\bar{\omega}_r} \tag{8.3}$$

where $\bar{P}_h = \bar{H} + \bar{l}$

$$\frac{Q(l_s)}{p_s l} = 0.5\bar{\omega}_r \bar{l}_s^2 + \bar{w}_g \bar{l}_s - \bar{P}_h \tag{8.4}$$

The normalized displacement $(\bar{w}_g, \bar{w}_L)$ and rotation $(\bar{\omega}_r, \bar{\omega}_{rL})$ alter with the depth z_m and the pile deflection modes, which are discussed in more detail next.

Table 8.2 BiP models for passive piles (Guo 2021a)

Depth z	*Sliding layer with k_s*
$(z \le z_m)$	Expressions for $w(z)$, $p(z)$, shear force $Q(z)$, and bending moment $M(z)$ are identical to those presented in Table 4.5 for R-E model (Chapter 4)
$z_m \le z \le l$	Layer with mk_s $p_2(z) = (\bar{\omega}_{rL}\bar{z} + \bar{w}_L)mp_s$; $w(z) = (\bar{\omega}_{rL}\bar{z} + \bar{w}_L)p_s / k_s$
$Q(z)$	$Q_2(z)/(p_s l) = m\bar{w}_g(\bar{z} - \bar{z}_m) + 0.5m\bar{\omega}_r(\bar{z}^2 - \bar{z}_m^2)$
$M(z)$	$\frac{M_2(z)}{p_s l^2} = \frac{m\bar{\omega}_{rL}}{6}(\bar{z} - \bar{z}_m)^2(\bar{z} + 2\bar{z}_m) + \frac{m\bar{w}_g}{2}(\bar{z} - \bar{z}_m)^2 - \bar{k}_T(1 - \beta)\bar{\omega}_{rL} - \bar{M}_p$

Table 8.3 BIP-k_θ model (constant rotational stiffness) (Guo 2021a)

Normalized $\bar{z}_{me}\left(= Z_{me}/l\right)$	$\bar{z}_m^6 - 3(1+H_c)\bar{z}_m^5 + 3(1+3H_c)\bar{z}_m^4 - \left[\frac{12\bar{k}_T - 6(m+1)\bar{k}_\theta}{m} + 1 + 9H_c\right]\bar{z}_m^3 + \left[\frac{3(12\bar{k}_T + m)H_c - 18(m+1)\bar{k}_\theta}{m}\right]\bar{z}_m^2 + \frac{18(m+2H_c)\bar{k}_\theta}{m}\bar{z}_m - \frac{6(12\bar{k}_T + m)\bar{k}_\theta}{m} = 0$
Normalized w_g and ω_r	$\bar{w}_g = 2\frac{\left[\bar{z}_m^3 + 6\bar{k}_\theta\right]\bar{P}_h - 3\bar{z}_m^2\bar{M}_L - 3\bar{z}_m^2\vec{k}_\theta\bar{\omega}_{rL}}{-\left(\bar{z}_m^3 - 12\bar{k}_\theta\right)\bar{z}_m}$ $\bar{\omega}_r = \frac{6\bar{B}_w\left(\bar{z}_m\bar{P}_h - 2\bar{M}_L\right)}{\bar{B}_w\left(\bar{z}_m^3 - 12\bar{k}_\theta\right) + 144\bar{k}_\theta^2}$
Normalized w_L and w_h	$\bar{w}_L = \frac{-36\bar{k}_\theta\left(1+\bar{z}_m\right)\left(\bar{z}_m\bar{P}_h - 2\bar{M}_L\right)}{\bar{B}_w\left(\bar{z}_m^3 - 12\bar{k}_\theta\right) + 144\bar{k}_\theta^2}$ $\bar{w}_h = \frac{w_h k_s}{p_s} = \frac{-6\bar{k}_\theta\bar{\omega}_r\left(1-\bar{z}_m\right)}{\bar{B}_w}$
Normalized ω_{rL} and $k\theta_e$	$\bar{\omega}_{rL} = \frac{12\bar{k}_\theta\bar{\omega}_r}{\bar{B}_w},\ \bar{k}_T = k_T/\left(k_s l^3\right)$ $\bar{k}_{\theta e} = \frac{k_{\theta e}}{k_s l^3},\ k_{\theta e} = -\frac{1}{l'_c}$

Note: $\bar{B}_w = m\left(1-\bar{z}_m\right)^3 + 12\left(\bar{k}_T + \bar{k}_\theta\right), H_c = \left(\bar{H}\bar{a} + 0.5\bar{l}_m^{\,2}\right)/\left(\bar{H} + \bar{l}_m\right), \bar{z}_m = z_m/l, \bar{H} = H/\left(p_s l\right), \bar{k}_\theta = k_\theta/\left(k_s l^3\right),$ $\bar{k}_T = k_T/\left(k_s l^3\right), \bar{M}_L = \bar{H}\left(\bar{z}_m - \bar{a}\right) + 0.5\bar{l}_s\left(2\bar{z}_m - \bar{l}_s\right).$

8.3.2 BiP-k_θ model (a constant rotational stiffness)

The rotational stiffness k_θ is stipulated as a constant (Dobry et al. 2003; Guo 2015b), from which the BiP-k_θ model was developed (see Table 8.3). The maximum moment M_m at depth z_m is determined as $M_L + k_\theta(\omega_{rL} - \omega_r)$ for a sliding depth l_s, while the net moment from the lower segment is $-k_\theta(\omega_{rL} - \omega_r) - k_T\omega_{rL}$ where k_T denotes the rotational restraining stiffness along the lower segment. The force equilibrium and the bending moment equilibrium of each segment are used to determine the normalized displacement and rotation $\left(\bar{w}_g, \bar{\omega}_r\right)$ and the projected ones $\left(\bar{w}_L, \bar{\omega}_{rL}\right)$ from both the upper and lower segments for each $\bar{z}_m$. The displacement compatibility allows the joint depth z_m to be estimated by

$$\bar{z}_m^6 - 3(1+H_c)\bar{z}_m^5 + 3(1+3H_c)\bar{z}_m^4 - \left[\frac{12\bar{k}_T - 6(m+1)\bar{k}_\theta}{m} + 1 + 9H_c\right]\bar{z}_m^3 + \left[\frac{3(12\bar{k}_T + m)H_c - 18(m+1)\bar{k}_\theta}{m}\right]\bar{z}_m^2 + \frac{18(m+2H_c)\bar{k}_\theta}{m}\bar{z}_m - \frac{6(12\bar{k}_T + m)\bar{k}_\theta}{m} = 0 \tag{8.5}$$

where $H_c = \left(\bar{H}\bar{a} + 0.5\bar{l}_s^2\right)/\left(\bar{H} + \bar{l}_s\right)$, $\bar{k}_T = k_T/\left(k_s l^3\right)$, and $\bar{k}_\theta = k_\theta/\left(k_s l^3\right)$. The expressions in Table 8.3 offer (i) The H_c for a specific sliding depth l_s; and (ii) The normalized displacement $\bar{w}_g$ and rotation angle $\bar{\omega}_r$ and the projected $\bar{w}_L$ and $\bar{\omega}_{rL}$. Nonlinear response of the pile is obtained by repeating the calculation for a series of l_s.

8.3.3 BiP-η model (a constant rotation ratio)

The BiP-η model stipulates a constant ratio $\eta\left(= \bar{\omega}_r/\bar{\omega}_{rL}\right)$ between the rotation angles of the upper segment (ω_r) and the lower one (ω_{rL}) during the soil sliding. The displacement compatibility allows the normalized depth $\bar{z}_m$ to be gained from

Table 8.4 BiP-η model for passive loading (Guo 2021a)

$\bar{z}_m^3 + \frac{-3(m+1)+6(\eta-1)H_c}{3+m-2\eta}\bar{z}_m^2 + \frac{3m+6H_c}{3+m-2\eta}\bar{z}_m - \frac{m+12\bar{k}_T}{3+m-2\eta}\bar{z}_m = 0$	
$\bar{w}_g = \frac{\left[m(1-\bar{z}_m)^3 + 2\eta\bar{z}_m^3 + 12\bar{k}_T\right]\bar{P}_h - 6\eta\bar{M}_L\bar{z}_m^2}{\bar{z}_m\left[m(1-\bar{z}_m)^3 - \eta\bar{z}_m^3 + 12\bar{k}_T\right]}$	$\bar{\omega}_r = \frac{-6\eta(\bar{z}_m\bar{P}_h - 2\bar{M}_L)}{m(1-\bar{z}_m)^3 - \eta\bar{z}_m^3 + 12\bar{k}_T}$
$\bar{w}_L = \frac{-(1+\bar{z}_m)}{2\eta}\bar{\omega}_r$ $\quad \bar{w}_h = \frac{-(1-\bar{z}_m)}{2\eta}\bar{\omega}_r$ $\quad \bar{\omega}_{rL} = \bar{\omega}_r/\eta$	$\bar{k}_\theta = \frac{m(1-\bar{z}_m)^3 + 12\bar{k}_T}{12(\eta-1)}$
Passive moment about z_m: $M_L - k_\theta(\omega_r - \omega_{rL})$	

$$\bar{z}_m^3 + \frac{-3(m+1)+6(\eta-1)H_c}{3+m-2\eta}\bar{z}_m^2 + \frac{3m+6H_c}{3+m-2\eta}\bar{z}_m - \frac{m+12\bar{k}_T}{3+m-2\eta}\bar{z}_m = 0 \tag{8.6}$$

where $\eta \neq 0.5(m + 3)$. The associated z_m should exceed $0.5l$, unless there is a localized bending weakness. A constant $\eta\left(=\bar{\omega}_r / \bar{\omega}_{rL}\right)$ warrants the following condition.

$$\bar{B}_w = 12\eta\bar{k}_\theta \tag{8.7}$$

This relationship allows the expressions of $\bar{\omega}_r$, $\bar{w}_g$, and $\bar{w}_L$ to be uniquely determined (see Table 8.4), as with the BiP-k_θ model for each $\bar{z}_m$.

8.3.4 Remarks about using BiP models

The normalized stiffness $\bar{k}_\theta$ (see Table 8.4) is given by

$$\bar{k}_\theta = \frac{m(1-\bar{z}_m)^3 + 12\bar{k}_T}{12(\eta-1)} \tag{8.8}$$

where $\eta \neq 1$, otherwise the 2-segments would then behave as a single segment and the 2-layer model should be used instead. The stiffness $\bar{k}_\theta$ and the rotation ratio η are interchangeable using Equation (8.7). The two models generally have identical values of modulus ratio m, normalized depth $\bar{z}_m$, and normalized stiffness $\bar{k}_T$, and can be easily compared by interchanging the values of $\bar{k}_\theta$ and η. As with the 2-/3-models (Guo 2020a, b), the BiP models have singularities at rotational stiffness k_θ* and k_T*, around which the pile response is amplified.

$$\bar{k}_\theta^* = \frac{\eta\bar{z}_m^3}{12(\eta-1)} \tag{8.9a}$$

and

$$\bar{k}_T^* = \frac{-m(1-\bar{z}_m)^3 + \eta\bar{z}_m^3}{12} \tag{8.9b}$$

To prevent potential response amplification, the conditions of $\bar{k}_\theta < 0.5\bar{k}_\theta^*$ and $\bar{k}_T < 0.5\bar{k}_T^*$ need to be satisfied, with $\bar{k}_T(> 10\bar{k}_\theta)$ in initial calculations. Singularity also occurs at $\eta = 0.5(3 + n)$ [see Equation (8.6)]. Large η values result in reduced rotation of the lower segment. The normalized depth $\bar{z}_m$ largely stays around 0.75–0.8 (except for weak locations) after the hinge formation. The $\bar{z}_m$ is capped at 1.0, and $\bar{z}_m > 1$ (i.e., without base restraint) implies 'brittle' failure. The lower segment of the BiP models observe '$\bar{w}_L / \bar{\omega}_{rL} = -0.5(1+\bar{z}_m)$'. A '$\bar{k}_\theta > 0$' is noticed for amplification case, which is opposite to '$\bar{k}_\theta < 0$' defined in the 2-/3-layer models (Guo 2016, 2020a, b). And, finally the tensile force H is taken as negative compared to positive on-pile *FPUL* p_s (>0) induced by excavation.

The BiP-k_θ and -η models tend to underestimate the rotation of the lower segment. To account for elastic segment-soil interaction, the R-E model is elaborated next against Chapter 4.

8.4 BIP RIGID-ELASTIC MODEL AND SOLUTIONS

The moving soil engenders a bending moment M_m (= M_L – $k_\theta\omega_r$) about the rotation depth of the pile ($z_r \approx z_m$). The lower segment of the pile (with a bending stiffness E_pI_p) may exhibit elastic interaction with the surrounding 'stiff' layer (a subgrade modulus mk_s, $m > 1$). Assuming a free rotation at the depth z_m and floating base, the displacement (w_h) and rotation (ω_r) at the depth z_m are given by

$$\bar{w}_h = M_m I_c k / p_s \tag{8.10a}$$

$$\bar{\omega}_r = M_m I'_c kl / p_s \tag{8.10b}$$

where I_c, I_c' k and N are estimated using the expressions tabulated in Tables 4.6 and 4.7 (Chapter 4) for the subgrade modulus of mk_s. The interaction is simulated by R-E model, which meet five conditions: First, the normalized rotation $\bar{\omega}_r$ is identical to that of lower segment at z_m. Second, the M_m (= w_h/I_c) of Equation (8.10) is recast into the following using Equation (8.2)

$$\frac{\bar{w}_h}{I_c}\frac{p_s}{k} = \left(\frac{1}{6}\bar{\omega}_r\bar{z}_m^3 + \frac{1}{2}\bar{w}_g\bar{z}_m^2 - \bar{M}_L\right)p_s l^2 \tag{8.11}$$

Third and fourth, the equilibrium of shear force along the upper segment, and bending moment (of M_L – $k_\theta\omega_r$) about the depth z_m offer the normalized $\bar{w}_g$, $\bar{\omega}_r$ and $\bar{w}_h$ which can be estimated using expressions in Table 4.6 (Chapter 4) with $\bar{z}_m = \bar{z}_{me}$, and $\bar{k}_\theta = \bar{k}_{\theta e}$. The relationship of M_m (= $k_\theta\omega_r$) = w_h/I_c together with Equation (8.11) are recast into

$$\frac{\left(-4\bar{z}_m^3 + 12\bar{k}_\theta\right)\bar{P}_h + 6\bar{z}_m^2\bar{M}_L}{-6\left(\bar{z}_m\bar{P}_h - 2\bar{M}_L\right)\bar{z}_m} = \frac{\bar{w}_h}{\bar{\omega}_r} = \frac{k_\theta}{l}I_c \tag{8.12}$$

Finally, the bending moment compatibility of $w_h/I_c = \omega_r/I_c'$ allows the z_m to be gained from

$$\left(-4\bar{z}_m^3 + 12\bar{k}_\theta\right)\bar{P}_h + 6\bar{z}_m^2\bar{M}_L = -6\bar{z}_m\left(\bar{z}_m\bar{P}_h - 2\bar{M}_L\right)\frac{I_c}{I'_c l} \tag{8.13}$$

Importantly, '$k_\theta = -1/I_c$'' is successfully deduced from Equations (8.12) and (8.13).

The response profiles of the pile can be obtained using the expressions in Table 4.5 for $z \le z_m$, and in Table 4.7 (Chapter 4) for $z_m \le z \le l$. These solutions are obtained for the passive moment of $M_L - k_\theta\omega_r$. The elastic k_θ alters with length of elastic portion l_e (= $l - z_m$), and offers a 'semi' circle of M_m–ω_r curve. It is the upper limit of the rotational stiffness (k_θ or k_T) of the pile-soil system. The $\bar{k}_\theta$ has been previously deduced as 0.3–1.18 for 'full length' piles embedded in dense sand and subjected to lateral spreading (Dobry et al. 2003; Guo 2015b); while the $|\bar{k}_\theta|$ is equal to 0.04–0.15 for capped piles subjected to moving sand, embankment loading (Guo 2016) or lateral spreading (Guo 2020a).

To obtain nonlinear response for the BiP-k_θ (or -η) models, the same procedure as the 2-layer model is adopted. A series of sliding depths (l_s) is stipulated to gain a set of $\bar{z}_m$ values, which allow estimation of the $\bar{\omega}_r$, $\bar{w}_g$, $\bar{\omega}_{rL}$ and $\bar{w}_L$ using the expressions in Tables 8.3 or 8.4, as well as the normalized moment $M_m/(p_s l_s l)$ using Equation (8.2), respectively. The loading above the depth z_m is identical between the BiP models and the 2-layer model, with the response being similar among the models, but for the effect of restraining lower segment.

As an example, Figure 8.2 provides the normalized response of translating-rotating piles predicted using the BiP-η model (with η = 12 or −12.5 and $\bar{k}_T = -0.17$), and BiP-k_θ model $\left(\text{with}\, \bar{k}_\theta = -0.013, -0.0145\right)$, respectively. Figure 8.3 shows the normalized response profiles (at $l_s = 0.5l$) of $M(z)/(p_s l_s l)$, $Q(z)/(p_s l)$, $w(z)k_s/p_s$ and $p(z)/p_{ub}$ for bending moment, shear force, displacement and *FPUL*, respectively. The impact of the lower-segment restraint underpinning the three BiP models is evident in Figure 8.4, showing the normalized relationships of moment $M_m/(p_s l_s l)$ versus shear force $Q(l_s)/(p_s l)$, and shear force $Q(l_s)/(p_s l)$ versus rotation $\omega_r k_s l/p_s$, respectively. Compared to the 2-layer model, the BiP models (with η = −12.5–12.0) offer a 2–5 times higher shear force at a given rotation (Figures 8.4b and d), and a similar degree of increase in the M_m values at a specified sliding (shear) force. The magnification can be detrimental, as explored next using available centrifuge tests.

8.5 PILES ADJACENT TO EXCAVATION

Ground movement during excavation is customarily monitored in situ, to ensure safety of piles located at a distance l_d from the excavation face (see the inset of Figure 8.5). Five centrifuge tests were conducted to examine the impact of excavation (within five days) on piles (at l_d = 3 m unless otherwise specified) behind a retaining wall (Ong et al. 2003, 2006; Leung et al. 2006).

The tests were analysed using the 2-layer model, and the BiP models (Guo 2021a). The piles (at prototype scale) are characterized by l of 12.5 m, d of 0.63 m, and E_pI_p of 2.2 × 10^5 kNm^2. They were embedded in either a single sand-layer (of 12.5 m in depth) or a clay-sand layer (i.e., 6.5-m thick clay layer underlain by 6.0-m sand layer). The Toyoura sand has an effective angle of internal friction (ϕ') of 43°, unit weight (γ_s') of 15.28 kN/m^3, angle of dilatancy (ϕ_d') of 5°, and $\sigma_v' dK_p$ of 318.23 kN/m (with an overburden stress σ_v' of 95.5 kPa at 0.5l of 6.25 m, and K_p = 5.289). The clay has an average undrained shear strength (s_u) of 10 kPa, an angle ϕ' of 23°, and $\sigma_v' dK_p$ of 148.288 kN/m (with σ_v' = 103.1 kPa at 0.5l and K_p = 2.283).

To assess pile response in the clay-sand layer, excavation was conducted to a maximum depth l_{exc} of 1.2 m (Test 2 or T2, stable wall), 1.8 m (Test 5 or T5, marginally stable wall) and 2.8 m (Test 7 or T7, collapsed wall), respectively. The excavation induced a linearly reduced

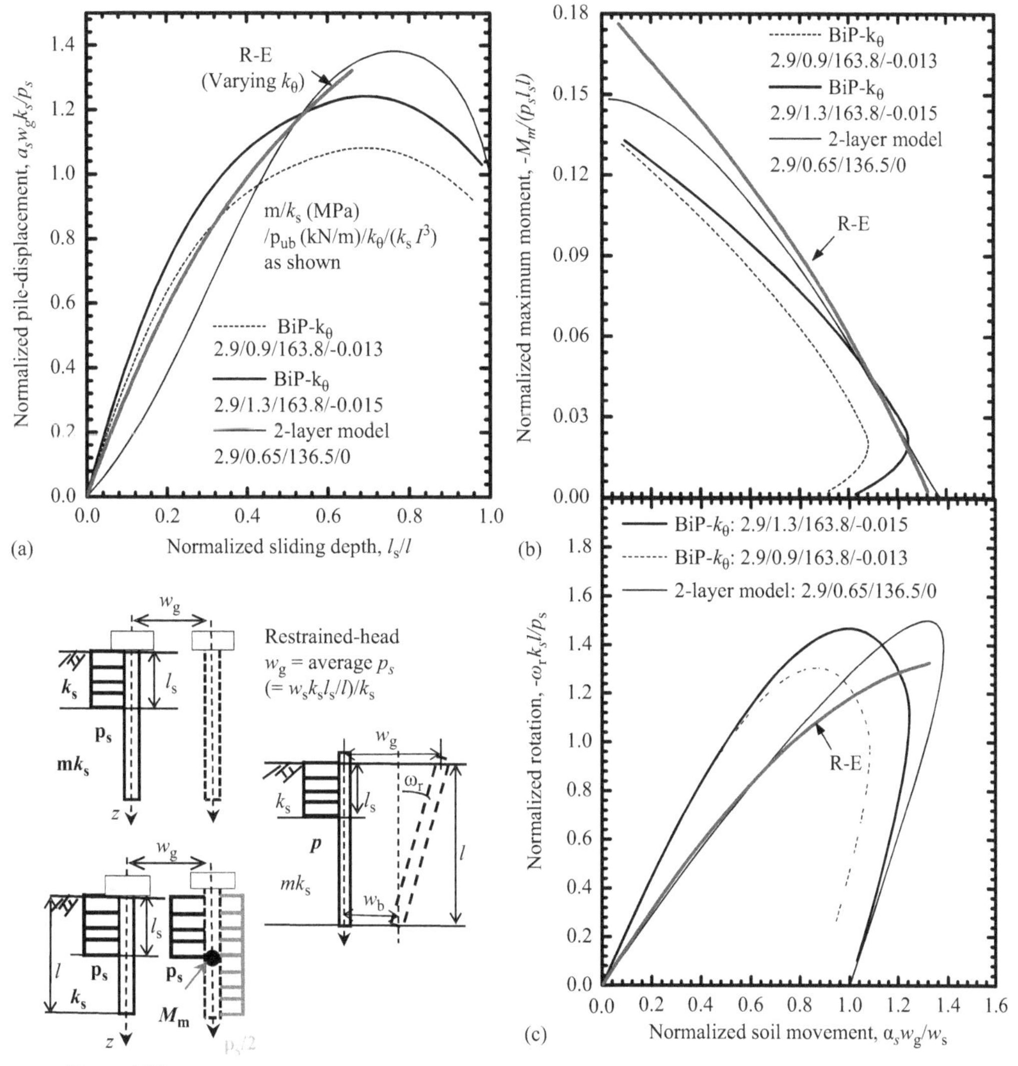

Figure 8.2 Normalized response from BiP-k_θ and R-E models: (a) pile displacement; (b) soil movement versus maximum bending moment; (c) soil movement versus rotation of pile [m = 2.9, k_s = 0.9–1.5 MPa, and $\bar{k}_\theta = -(0 - 0.032)$ (R-E model)]. [Adapted from Guo, W. D., *Int J Geomechanics*, **21**(1), 2021.]

profile (at a gradient of rotation ω_s) of soil-movement (w_s) with depth; as well as a rotation angle (ω_r), and a displacement (w_g) of the pile(s) rotating about the depth z_m at a stiffness $k_{e\theta}$ (of M_m/ω_r), respectively (Table 8.5). The soil movement extends to a (measured) depth of $\beta_s l_{exc}$ (see Figures 8.4e and f). The β_s increases rapidly from 0.7–1.2 to 2.6–3.3 before collapse of the walls, which suggest a mixed mode of rotating and sliding soil. The measured M_m–w_g and M_m–ω_r (angle of rotation) curves (see Figure 8.6) follow a consistent trend until failure for all the T2, T5, and T7 piles. In contrast, embedded in 10-m thick clay layer over 2.5 m-thick sand layer, the T6 piles collapsed at a low, final l_{exc} of 1.4 m.

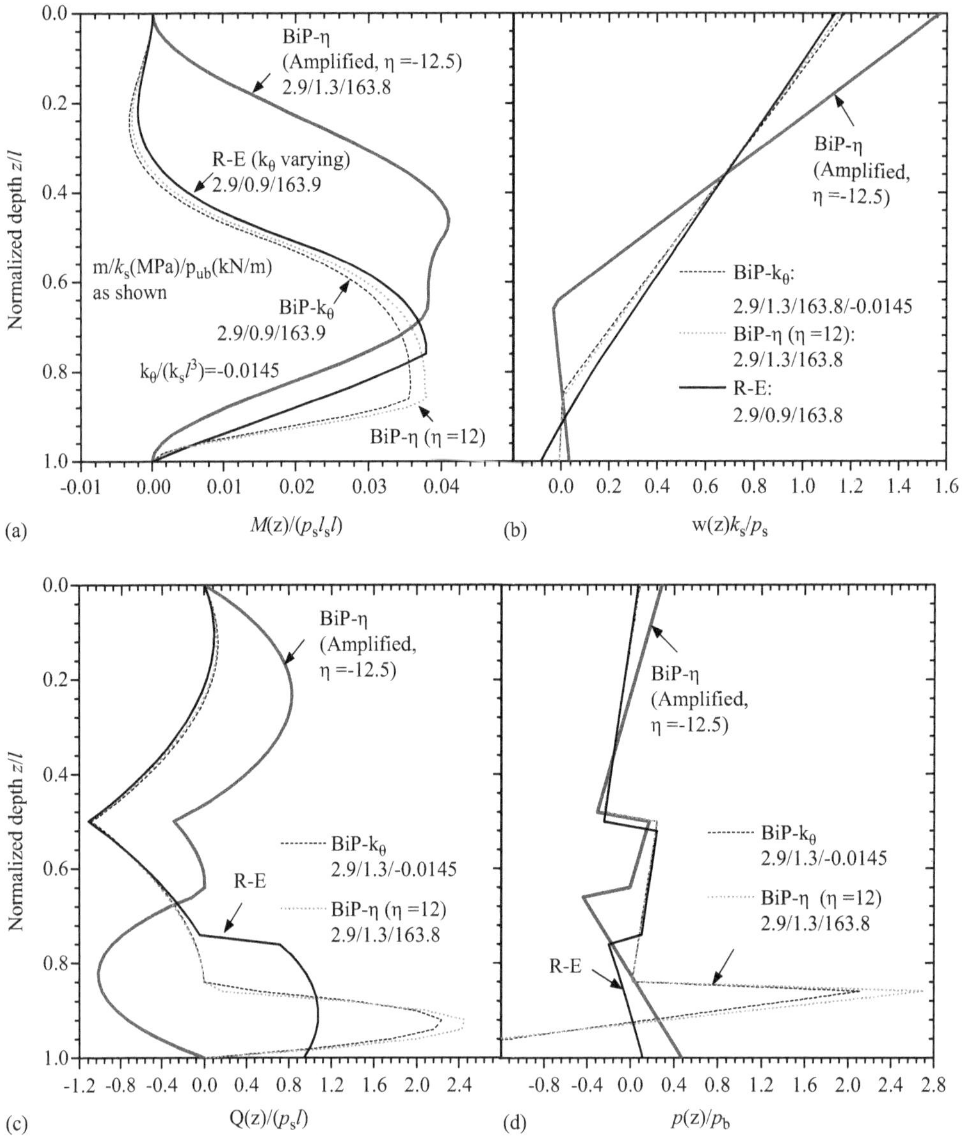

Figure 8.3 Profiles from three models: (a) bending moment; (b) displacement; (c) shear force; (d) on-pile force per unit length. [Adapted from Guo, W. D., *Int J Geomechanics*, **21** (1), 2021.]

With sufficient restraint around the pile-base by the 6 m-thick sand-layer, the T7 pile exhibits a rebound of 25% in deflection w_g and moment M_m from the peak (at l_{exc} = 1.2 m) to the final state. With a reduced restraint of 2.5 m-thick sand (41% of T7), Test 6 displays a rise-and-drop cycle of the M_m and 'brittle' failure. The M_m peaked at 238 kN·m (at l_{exc} =1.4 m), then reduced to 185 kN·m (at the final l_{exc} of 1.8 m) before ultimately degrading to 80 kN·m in about 10 months. The rise-and-drop cycle of M_m was first observed on piles subjected to lateral spreading (Dobry et al. 2003; Guo 2015b), which resembles the stress variation measured on wall embedded in sand (Milligan and Bransby 1976).

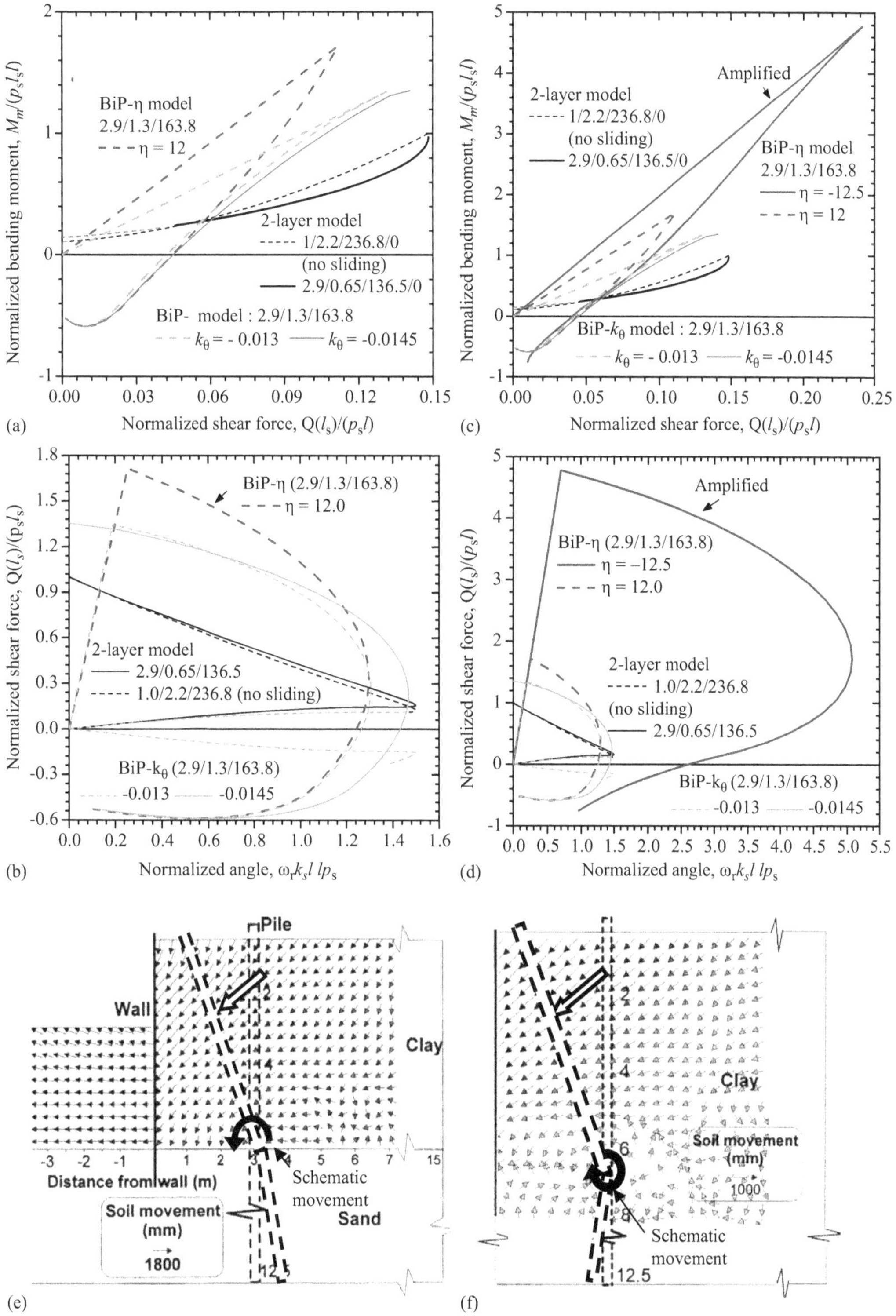

Figure 8.4 Normalized bending moment, shear force and rotation angle: (a and c) bending moment-shear force curves; (b and d) shear force-rotation curves ($\eta = -12.5$ for inward rotation); and sliding-rotational soil movement around; (e) Test 7; (f) Test 6 (Leung et al. 2006). [Adapted from Guo, W. D., *Int J Geomechanics*, **21**(1), 2021.]

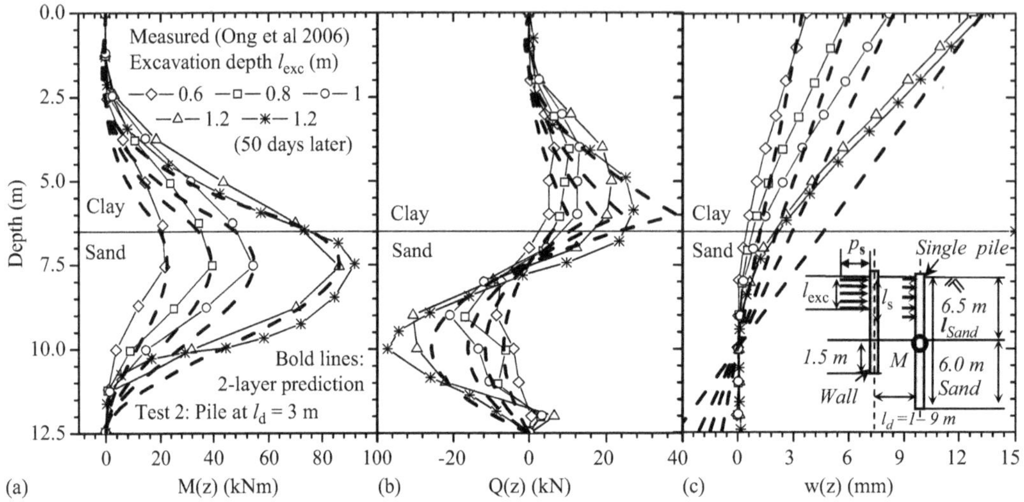

Figure 8.5 Predicted (2-layer model) versus measured (Ong et al. 2006) profiles of piles at excavation depths: (a) bending moment; (b) shear force; and (c) pile displacement. [Adapted from Guo, W. D., *Int J Geomechanics*, **21**(1), 2021.]

Table 8.5 Final response at end of tests (Guo 2021a)

Tests	w_s *(mm)*[c]	ω_s *(o)*[c]	w_g *(mm)*	ω_r *(o)*[d]	M_m *(kNm)*	$k_{e\theta}$[e] *(MNm/radians)*
T2 (WC1)[a]	100	0.818	26.0	0.184	225.0	70.1
T2 (WC2)[a]	30	0.307	12.7	0.097	85.7	50.6
T5[b]	83	0.76	24.5	0.193	163.0	47.08
T6[b]	210	2.0	100.5	0.544	219.2	23.08
T7[b]	1050	11.86	72.2	0.47	252.9	30.82

Notes:
[a] Piles in sand.
[b] Piles in clay-sand layer.
[c] Soil movement reduced linearly (at an angle ω_s) with depth from a ground level movement w_s.
[d] Rotation angle of upper rigid portion.
[e] $k_{e\theta} = M_m/\omega_r$ at the depth of M_m.

Example 8.1 Parameters for equivalent rigid piles

The 2-layer model (Guo 2015b) is adopted first to model the pile (l = 12.5 m, d = 0.63 m, s_u = 10 kPa, and H = 0). The input parameters were gained using the measured profiles with p_s (= $\alpha p_{ub} l_{exc}/l$), and $p_{ub} = 3.1 s_u d l_m$ (T2) or $4 s_u d l_m$) (T6 and T7), for which $p_{ub}/(\sigma_v d K_p)$ = 0.83 (T2) and 1.1 (T5, 6 and 7 with ϕ = 23°). At an invariable loading l_s (= l_m) of 6–6.25 m and with α ($\approx\beta_s/1.08$), the $M(z)$ and $w(z)$ profiles are matched at each known displacement w_g (see Table 8.6) as exemplified in Figures 8.5a through c for T2 pile (at l_d = 3 m), in Figures 8.7a and b for T6 pile, and in Figures 8.7c and d for T7 pile, respectively. The match resulted in a modulus $k_s^{profile}$ of $225s_u$ (T2, stable wall), $130s_u$ (T7, collapsed wall), and $(55–80)s_u$ (T6, for collapsed wall in thick, soft clay layer), respectively.

Salient features include a loading depth l_s ($\approx\beta_s l_{exc}$) of 5.6–6.25 m on the piles exerted by soil movement during excavation. The shift from stable to collapsed walls (movement modes) requires (i) α (see Figure 1.4b, Chapter 1) to increase from 1.2–2.3 to 2.5; (ii) the

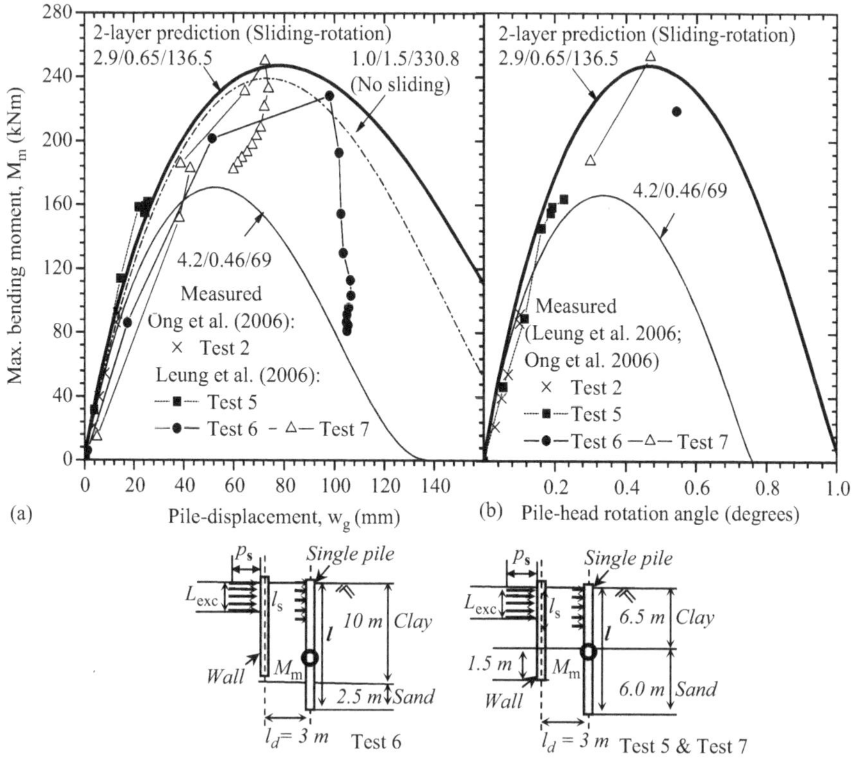

Figure 8.6 Predicted versus measured (Leung et al. 2006; Ong et al. 2006) response of piles (in clay-sand layers): (a) w_g–M_m; (b) rotation angle-M_m (Tests 2 and 5, stable wall; Tests 6 and 7, collapsed wall). [Adapted from Guo, W. D., *Int J Geomechanics*, **21** (1), 2021.]

modulus ratio m to increase from 2.7–2.9 to 4.0, and (iii) the frictional angle of sliding ϕ_i' to rise from $0.61\phi'$ (= 14°, T2) to $0.65\phi'$ (= 15°, T7) and to $0.83\phi'$ (= 19.5°, T6 in 10 m-thick clay) as deduced from the m variations, respectively. A high α value involves a fast approach to maximum response for a given excavation depth, while a high m value offers larger modulus (resistance) of underlying layer than the sliding layer and prompts rotation apart from sliding.

To model the overall response of flexible piles (E_p = 2.845 × 10^7 kPa) using the 2-layer (rigid pile) model (Guo 2021a), the p_{ub} and m values are reduced by up-to 20% and k_s is reduced to $[2 - (l/l_e)]k_s^{profile}$ ($k_s^{profile}$ written simply as k_s unless necessary). The length l_e (= l_{e1} + $0.5l_{exc}$, see Table 8.7) is the sum of the elastic portion l_{e1} (= 1.2 times the effective length of lateral piles) and the plastic portion of $0.5l_{exc}$ (at a distance l_d) (Guo 2012, 2013c). For sand with $E_s = 6z$ MPa (z in m), and for clay with $130s_u$ ($\approx k_s^{profile}$), the average E_s over an effective depth l_e of 8.35 m amounts to $8.4k_s^{profile}$ {i.e., 10.882 MPa = [44.55 × (8.35 – 6.5) + 1.3 × 6.5]/8.35 MPa}, or a shear modulus G_s of $3k_s^{profile}$ (for all tests). At k_s of $130s_u$, for instance, the l_{e1} is estimated as 7.33 m (= 1.2 × 6.11 m). The k_s is reduced subsequently to 0.46 MPa [= (2–12.5/10.7) × 0.55, see Figure 8.7b, l_e = 10.7 m and $k_s^{profile}$ = 0.55 MPa] (T6); and 0.739 MPa (l_e = 8.73 m, $k_s^{profile}$ = 1.3 MPa) (T7). The p_{ub} of all piles was deduced as 122.9–163.8 kN/m (see Table 8.6). Finally, a standard set of parameters m/k_s(MPa)/p_{ub}(kN/m)/k_θ is obtained as 2.9/0.65/136.5/0, while a residual set of 4.2/0.46/69/0 is also formed for collapsed piles. The 2-layer model with in particular k_θ = 0 offers sufficiently accurate predictions of the measured moment M_m and displacement w_g of all piles in the clay-sand

Table 8.6 Parameters for response profiles of piles behind excavation wall (2-layer model) (Guo 2021a)

Tests	l_{ex}[a]	α[a]	m/p_{ub} (kN/m)[a,b]	k_s (MPa)/d (cm)[a]	ϕ_i[a]	*Pred.*[c]/*Mea.* M_m (kNm)	*Pred. /mea.* w_g (mm)
		Piles embedded in clay-sand layer (Clay 6.5 m /sand 6 m) (Ong et al. 2006)					
T2@3m	0.6	1.2	2.7/122.9	2.25/63	14	22.5/21.6	3.5/3.7
l_s = 6 m	0.8	1.6	2.7/122.9	2.25/63		39.1/39.5	6.0/5.9
	1.0	1.8	2.7/122.9	2.25/63		54.7/54.3	8.4/8.1
	1.2	2.3	2.7/122.9	2.25/63		86.3/86.4	13/12.7
		Piles behind collapsed wall (Clay 6.5 m /sand 6.0 m) (Leung et al. 2006)					
T7@3m	0.6	2.5	2.9/163.8	1.3/63	15	60.2~	17/—
l_s = 6.25 m	1.2	3.9	2.9/163.8	1.3/63		185.3/184.6	52.2/38–44
	1.8	2.5	2.9/163.8	1.3/63		180.7/185 (?)	51/43(?)
	2.3	2.5	2.9/163.8	1.3/63		251/251.6	71/72
T5@3m	0.6	2.5	2.9/163.8	2.1/63		60.1/48	10/7.
	1.4	2.5	2.9/163.8	2.1/63		140.3/158–163	24/24
		Piles behind collapsed wall (Clay 10.0 m /sand 2.5 m) (Leung et al. 2006)					
T6@3m	0.6	2.89	4.0/163.8	0.55/63	19.5	96.3/99	43/20(?)
	1.2	2.89	4.0/163.8	0.55/63		192.6/—	87/90(?)
	1.4	2.89	4.0/163.8	0.55/63		224.8/220.8	101/101

Notes:
[a] The values of l_{exc}, l_s, l (=12.5m), m, p_{ub}, k_s, d (= 0.63m), α and ϕ_i are all input values.
[b] $p_s = \alpha_s p_{ub} l_{exc}/l$.
[c] Taking k_s = 18 MPa, the predicted w_g reduces to 0.86, 2.5 and 4.5 (mm) for l_{exc} of 1, 2 and 3 (m), respectively. $p_{ub}/(\sigma_v{}' dk_p)$ = 0.83 (Test 2) and 1.1 (Tests 5, 6 and 7 with ϕ = 23°).

Table 8.7 Modulus and effective pile lengths (Clay-sand Layer) (Guo 2021a)

Tests	$k_s^{profile}$ (MPa)	l_{el} (m)[a]	$0.5l_{exc}$ (m)	l_e (m)[b]	*Mea* (m)	k_s (MPa)[c]
T2	2.25	6.40	0.6	7.00	7.3–8.0	0.482
T5	2.1	6.51	0.9	7.41	7.5–8.0	0.657
T7	1.3	7.33	1.4	8.73	7.7–9.0	0.739
T6	0.55–0.65	9.1–8.7	0.9	9.6–10	10.7	0.38–0.49

Notes:
[a] $l_{el} = 1.26d(E_p/G_s)^{0.25}$ (Guo 2012) with $G_s \approx 3k_s^{profile}$.
[b] $l_e = l_{el} + 0.5l_{exc}$, length of rigid portion.
[c] $k_s = [2 - (l/l_e)]\, k_s^{profile}$.

layer behind the walls (see Figure 8.6). In summary, the model can roughly accounts for the flexible piles. The use of standard and residual sets of parameters allows for predicting the pile behaviour behind stable and collapsed walls.

Considering realistic k_θ ($\neq 0$), the predictions are made by using the BiP models and the unmodified parameters (for rigid piles) tabulated in Tables 8.5 and 8.6. Namely, the BiP-k_θ model for T6 and T7 piles (l = 12.5 m, d = 0.63 m, with s_u = 10 kPa and H = 0), adopts 2.9/1.5/163.8/-0.013 and 2.9/1.3/163.8/-0.0145, with corresponding $\overline{k}_T$ = –0.20 (T6) and $\overline{k}_T$ = –0.145 (T7). The BiP-η model employs 2.9/1.5/163.8 (T6) and 2.9/1.3/163.8 (T7), with η = 12 or –12.5, and a corresponding $\overline{k}_T$ = –0.17 (either T6 or T7). Given η = 12, for instance, the corresponding $-k_\theta$ value increases from 0.0083 to 0.015.

Example 8.2 Modelling T7 pile with elastic rebound

The BiP-k_θ model is adopted to simulate the behaviour of T7 pile. At l_s = 6.25 m (with an equivalent η = 10.95, Figure 8.7g), for example, the depth z_m was gained as 10.64 m using Equation (8.5). This offers M_L = 3,844 kNm, P_h = 511.88 kN, w_g = 0.0735 m, ω_r = -6.853×10^{-3} (radians), w_L = 0.0072 m, and ω_{rL} = -0.626×10^{-3} (radians) using expressions in Table 8.3. The predicted profiles of moment $M(z)$ and displacement $w(z)$ are plotted in Figures 8.7c and d, respectively. They are normalized and presented in Figure 8.3. The direction of rotation of the two segments is identical (at η = 10.95), which pushes the moment M_m downwards to occur in the middle of the hard layer, the maximum shear force and the maximum on-pile *FPUL* downwards as well (see Figure 8.3), compared to the 2-layer model (no hinge, see Figure 8.5). This downward trend alters the stability of the pile system.

Given $\bar{k}_\theta$ $(=-0.0145)$ and $\bar{k}_T$ $(=-0.145)$, the 'joint' depth z_m for a series of 'sliding' depths l_s is estimated (which corresponds to an equivalent η of 7.3 to 11). This offers the maximum bending moment M_m, the displacement w_g, and the rotation angle ω_r, which are plotted in Figure 8.8 along with the measured (triangular symbols) response. During unloading from point (72.24 mm, 244.4 kNm), the ω_r (and hence the rotational displacement $\omega_r l$) and M_m are obtained as with the loading process. This results in rebound (elastic)

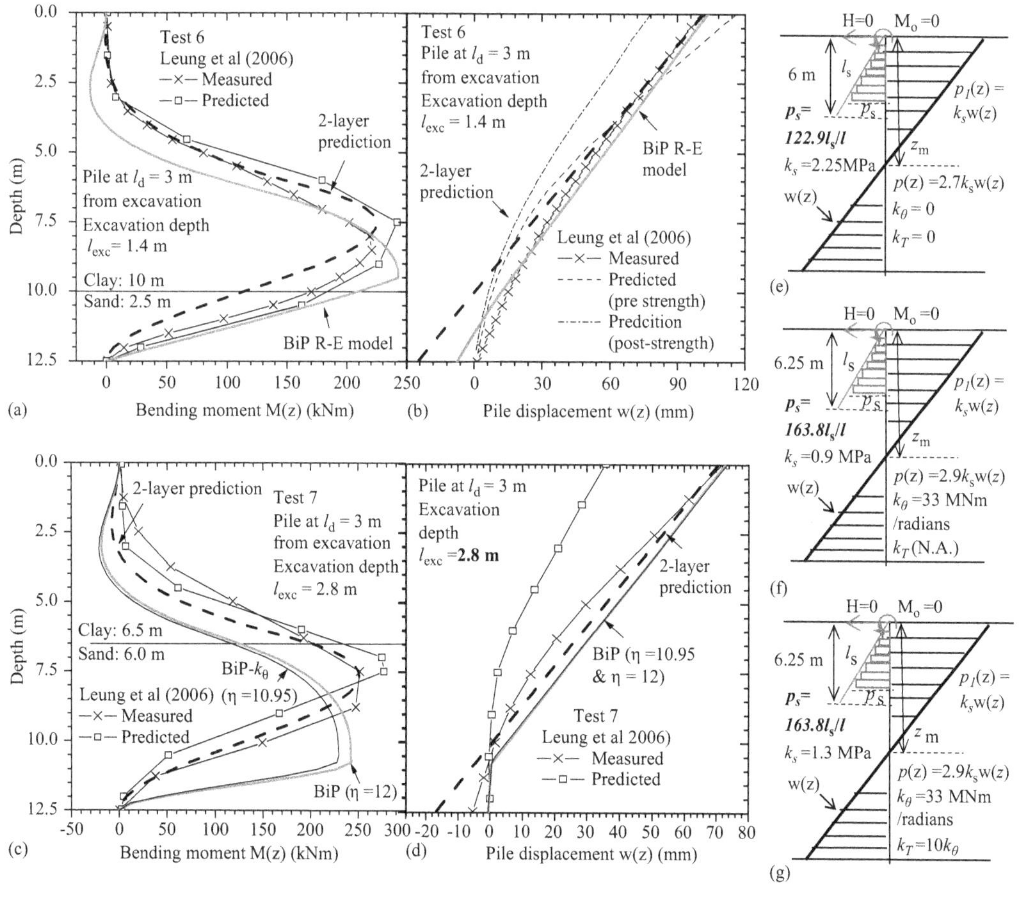

Figure 8.7 Predicted versus measured (Leung et al. 2006) profiles of bending moment and displacement (collapsed wall): (a–b) Test 6; (c–d) Test 7; Input parameters for (e) 2-layer model (T6); (f) R–E model (T6); (g) k_θ-model (T7). [Adapted from Guo, W. D., *Int J Geomechanics*, **21**(1), 2021.]

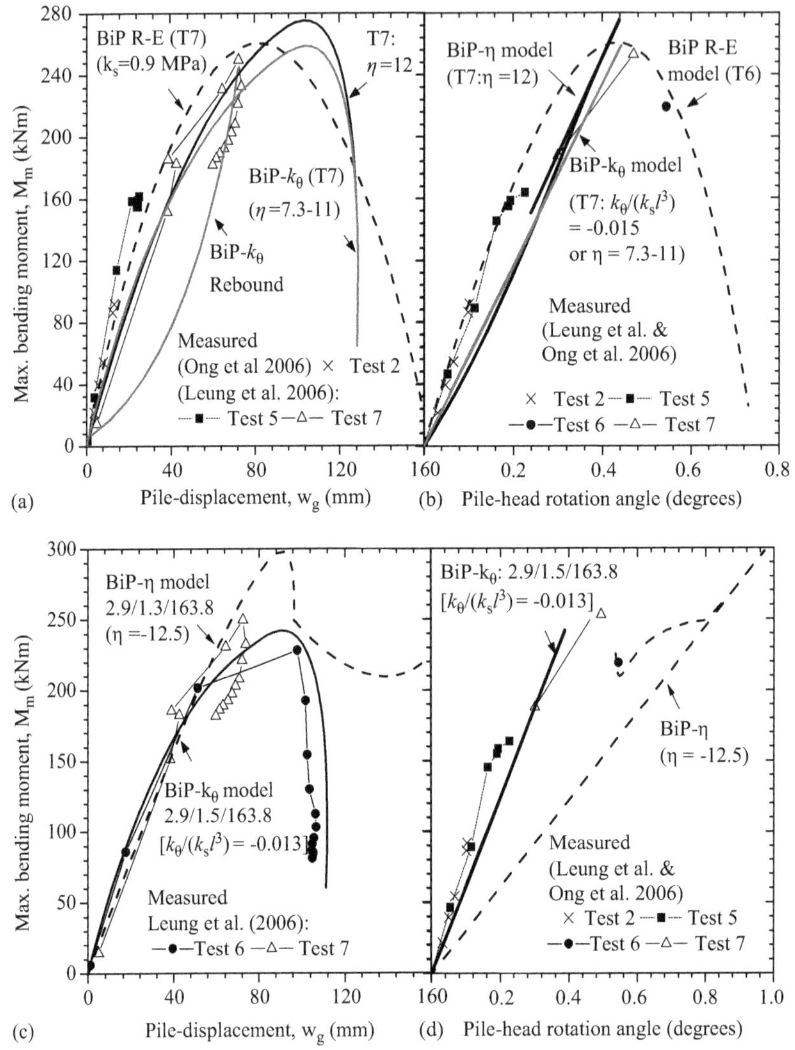

Figure 8.8 Predictions of BiP-k_θ, BiP-η and R-E models versus measured response of piles (Leung et al. 2006; Ong et al. 2006): (a and c) w_g–M_m; (b and d) rotation angles-M_m. [Adapted from Guo, W. D., *Int J Geomechanics*, **21**(1), 2021.]

displacement of $72.24 + \omega_r l$ (mm) and bending moment of 244.4-M_m (kNm). The predicted rebound-curve agrees well with the measured data (see Figure 8.8a).

Next, the BiP-η model was adopted to gain the pile response. Taking $\eta = 12$, and $\bar{k}_T = -0.17$, for instance, at $l_s = 0.5l$, it follows (i) $z_m = 10.72$ m using Equation (8.6); (ii) $\bar{k}_T = -0.014$ and $\bar{k}_\theta / \bar{k}_\theta^* \left(\approx \bar{k}_T / \bar{k}_T^* = -0.27 \right)$ using Equations (8.8) and (8.9); and (iii) $w_g = 0.073$ m, $w_L = 6.539 \times 10^{-3}$ m, $\omega_r = -6.758 \times 10^{-3}$ (radians), $\omega_{rL} = -0.563 \times 10^{-3}$ (radians), $M_L = 3,888$ kNm, and $P_h = 511.9$ kN using the expressions in Table 8.4. The predicted profiles of $M(z)$ and $w(z)$ are plotted in Figures 8.7c and d, and are normalized and presented in Figure 8.3. As excavation depth increases, the l_s increases from $0.02l$ to $0.62l$, which led to increase of $-\bar{k}_\theta$ from 0.0098 to 0.015, and increase of joint depth z_m from 4.53 m to the pile tip (12.5 m). The z_m is limited to l (of 12.5 m) and the $\bar{k}_\theta$ stays at -0.015 after $l_s \geq 0.62l$. The calculated moment M_m and rotation angle ω_r peak at a sliding depth l_s of $0.62l$ and decrease thereafter. The predicted M_m–w_g and M_m–ω_r curves align well with the measured data, as is seen in Figures 8.8a and b, which validate the BiP models.

Example 8.3 Modelling T6 pile exhibiting brittle failure

The response of T6 pile was simulated using both BiP models as with T7 pile. At a series of sliding depths ($l_s = 0.02l$–$0.62l$), the BiP-k_θ model has an equivalent η of 12.43–16.3(0.62l), whereas the BiP-η model has an equivalent $\bar{k}_\theta$ of –(0.0083–0.015) ($l_s = 0.02l$–$0.5l$). The predicted M_m–w_g and M_m–ω_r curves from both models largely agree with the respective measured data (see Figure 8.8). Specially, the BiP-k_θ model captures the 'brittle' drop in M_m at a near-constant displacement as measured, while the BiP-η mode suggests a constant residual rotation that matches the single measured point. The two mechanisms may occur but at different stages. The BiP-k_θ model predictions of T6 piles (at $l_s = 0.5l$) with $\bar{k}_\theta = -0.013$ offer a slightly larger maximum angle but the overall response was similar to that of the T7 piles (with $\bar{k}_\theta = -0.0145$ as shown in Figure 8.4.

Example 8.4 Response Amplification at typical $\eta < 0$

Soil movement may enforce an inward rotation (towards each other) of the upper and lower segments ($\eta < 0$), as well as lift the depth of M_m into a weaker layer (see Figure 8.3a, $\eta = -12.5$, BiP-η model, Guo 2021a), and resulting in amplified peak M_m and rotation angle (see Figure 8.8). This amplification is quantified herein using the singularity stiffness. Using the same values of m, k_s, and p_{ub}, as well as $\bar{k}_T = -0.17$, the T6 piles, may have a rotation ratio η of –12. At $l_s = 0.5l$, this results in $\bar{k}_\theta = 0.011$, $\bar{k}_\theta^* = 0.01$ and $\bar{k}_T^* = -0.159$, and leading to $\bar{k}_\theta \left(= 1.083\bar{k}_\theta^*\right)$ and $\bar{k}_T \left(= 1.068\bar{k}_T^*\right)$, which amplify the moment M_m to 955.8 kNm, and displacement w_h [$=w(z_m)$] to 0.121 m at the depth z_m. This M_m value is 321% (= 955.8/297.8) of the M_m (see Figure 8.8c) obtained using $\bar{k}_\theta \left(= 0.338\bar{k}_\theta^*\right)\left(\bar{k}_\theta^* = 0.021\right)$ and $\bar{k}_T = 0.361\bar{k}_T^* \left(\bar{k}_T^* = -0.296\right)$, which are measured as 228.43 kNm.

This calculation endorses the result of 'significant changes in computed wall deflections' from 'small variations in the undrained shear strength' (Whittle and Davis 2006). The amplified M_m is pulled up into the upper 'Lower Marine Layers', which unsurprisingly precipitated the Nicoll Highway collapse.

Example 8.5 R-E model and k_θ value

The R-E model (as listed in Table 4.6, Chapter 4) is adopted to simulate the behaviour of T6 piles using m (= 2.9) and p_{ub} (= 163.8 kN/m) (see Figure 8.8a). A 'reduced' modulus k_s of 0.9 MPa is adopted to gain a compatible stiffness $\bar{k}_\theta$ (via $\bar{k}_\theta = -1/I'_c$) of -0.013. The elastic parameters corresponds to the modulus of 2.9k_s (of the lower portion) are k = 2,453 kPa, N_p = 6,395 kN, α_c = 0.245, and β_c = 0.213 (see Tables 4.6 and 4.7, Chapter 4). The predicted $M(z)$ and $w(z)$ profiles agree well with the measured data (see Figures 8.7a and b), respectively. The nonlinear M_m–w_g and M_m–ω_r curves are largely well predicted (see Figures 8.8a and b), including the rotation angle at failure. The model, however, failed to capture the 'brittle' drop in M_m at a constant w_g. The model is capable of capturing non-zero shear force along pile base (see Figure 8.3c).

8.6 REMARKS

The $p_{ub}/(\sigma_v' K_p d)$ of 0.83–1.1 suggests that the limiting strength (of p_{ub}/d) varies within ±20% of the passive pressure. In modelling pile response, the on-pile strength p_b/d (= $\alpha p_{ub}/d$), being α times the limit strength, reflects the impact of residual strength S_r (≈0.5p_{ub}/d) or the ratio S_r/σ_v' (Ishihara 1993; Imamura et al. 2004; Armstrong et al. 2014). For stable and collapsed walls in clay-sand layers, the ratio $k_s/k_s^{profile}$ was found to be 0.3 and 1.1, respectively. It seems legitimate that the stress factor α (see Figure 1.4b) exceeds those deduced for 'transitional' piles in sliding slope (Guo 2015b) and translating-rotating piles under embankment loading (Guo 2016), as the soil movement mode alter with the offset distance of the piles behind the wall, leading to different values of α.

Guo (2021a) demonstrated that (i) the response profiles are largely well predicted for $\alpha \geq 1.0$. (ii) The p_{ub} cannot reach 330.8 kN/m, which invalidates the no-sliding (m = 1) set of parameters 1/1.5/330.8, despite its good match to the measured data (see Figure 8.6). (iii) The sliding-rotation mode, using the standard set, on the other hand, is confirmed in T6 and T7 piles. The mechanism for T5 piles (stable wall) is indeterminate, which warrants further investigation.

8.6.1 Sliding (shear band), sliding-rotation and pile-soil stability

The use of the angle of dilatancy ϕ_d' (= 5°) for piles in the sand layer stipulates that shear bands (slides) progressively evolve due to the 'softening' of the frictional soil during excavation (Lesniewska and Mroz 2000). Guo (2021a) demonstrates that

- All models exhibit a similar increase in shear force with lateral displacement (< 15 mm) of a stable wall.
- The shear-band model (sliding mechanism) is confined to stable walls and does not provide a force limit.
- The 2-layer model predicts a sustained shear force over a larger displacement for a high sliding friction (at m = 2.9) than no-sliding case (m = 1). The shear force sharply increases (after sliding depth > $0.5l$) prior to collapse.
- Based on the measured final displacement or rotation angle (see Table 8.5), T5 pile is deemed stable for reaching ~60% the maximum shear force, whereas T6 and T7 piles are relatively unstable at the zone of indefinite displacement and rotation (sliding–rotation mechanism) or beyond failure (no-sliding mechanism). The simple force and M_m locus may be used to identify stability of the pile-soil system.

Example 8.6 Anchored sheet piles

Anchored piles are typically adopted to retain excavation, which may rotate inward as bi-segments. Bourne-Webb et al. (2011) provide deflection, and bending moment of sheet piles (l = 0.26 m) with built-in hinges and pre-set yielding moments, as gained from centrifuge tests and 2-dimensional numerical simulations. Sheet piles (SPWFGs 13, 14, 15, and 16) were embedded in sand with a linearly increasing strength with depth (Gibson profile), and tested individually. The piles (embedded walls) were pulled by a single row of anchors at a depth a (m) with a measured tensile force H (kN/m), and H (a) = −1.89 (0.04), −1.85(0.04), −1.13(0.015), and −1.07(0.01) for SPWEG 13, 14, 15 and 16, respectively. The measured and numerically predicted profiles of displacement and bending moment for the tests are plotted in Figures 8.9 and 8.10.

The BiP-η model was adopted to predict the response profiles. The Gibson profile was simply modelled as a 2-layer soil with moduli k_s and $8k_s$ (m = 8), respectively. The k_s ($\approx E_s$) was obtained as 20.29 MPa (SPWEGs 13,14, 15), and p_{ub} = 10 kN/m. The $\eta\left(\overline{k}_T\right)$ values are determined as −1.125 (−0.027), −1.203(−0.023), −0.5(−0.0105), and −0.759(−0.02) for SPWEG 13, 14, 15 and 16, respectively. The predicted profiles of bending moment $M(z)$, and deflection $w(z)$ (at l_s = 0.16–0.17 m) are illustrated in Figures 8.9 and 8.10, for SPWGs 13 and 14, and for SPWGs 15 and 16, respectively. The degree of amplification (for η < 0) is assessed. The stiffness ratio $\overline{k}_\theta / \overline{k}_\theta^* \left(\overline{k}_\theta^*\right)$ of SPWEGs 13, 14, 15 and 16 tests was 0.198 (0.014), 0.186(0.0163), 0.213(0.0137) and 0.245(0.0136), respectively; with a corresponding $\overline{k}_T / \overline{k}_T^* \left(\overline{k}_T^*\right)$ of 0.529 (−0.051), 0.462(−0.054), 0.393(−0.027), and 0.525(−0.038). As previously discussed, these ratios incur amplification, which should have pulled anchors outward by a similar degree. The calculated $w(z)$ profiles were thus shifted outward by 1.0 mm (SP 13),

1.3 mm (SP 14), 1.3 mm (SP15), and 2.5 mm (SP16), respectively, resulting in a 'good' match with the measured $w(z)$ profiles.

The predicted depth of maximum bending moment (at zero shear force) was largely higher than the measured data owing to the built-in 'hinge' position (at shear force ≠ 0) and the stipulated 2-layer soil profile (to replace the Gibson profile). Despite this, the accuracy of the predictions, particularly with respect to amplification is still encouraging.

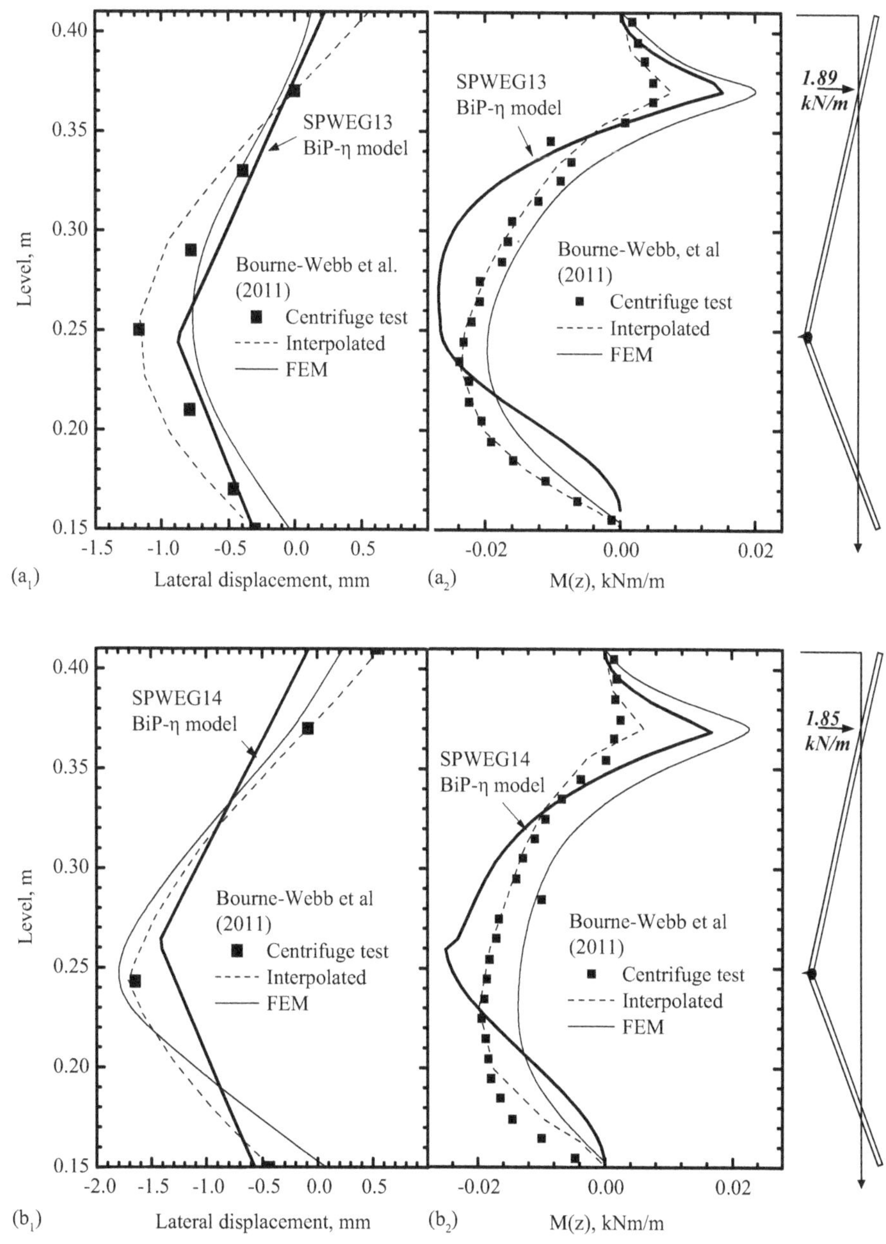

Figure 8.9 Comparison among the current BiP-η model, FEM and centrifuge tests (a_i) SPWEG13; and (b_i) SPWEG 14 (i = 1, displacement, i = 2 bending moment). [Adapted from Guo, W. D., *Int J Geomechanics*, **21**(1), 2021.]

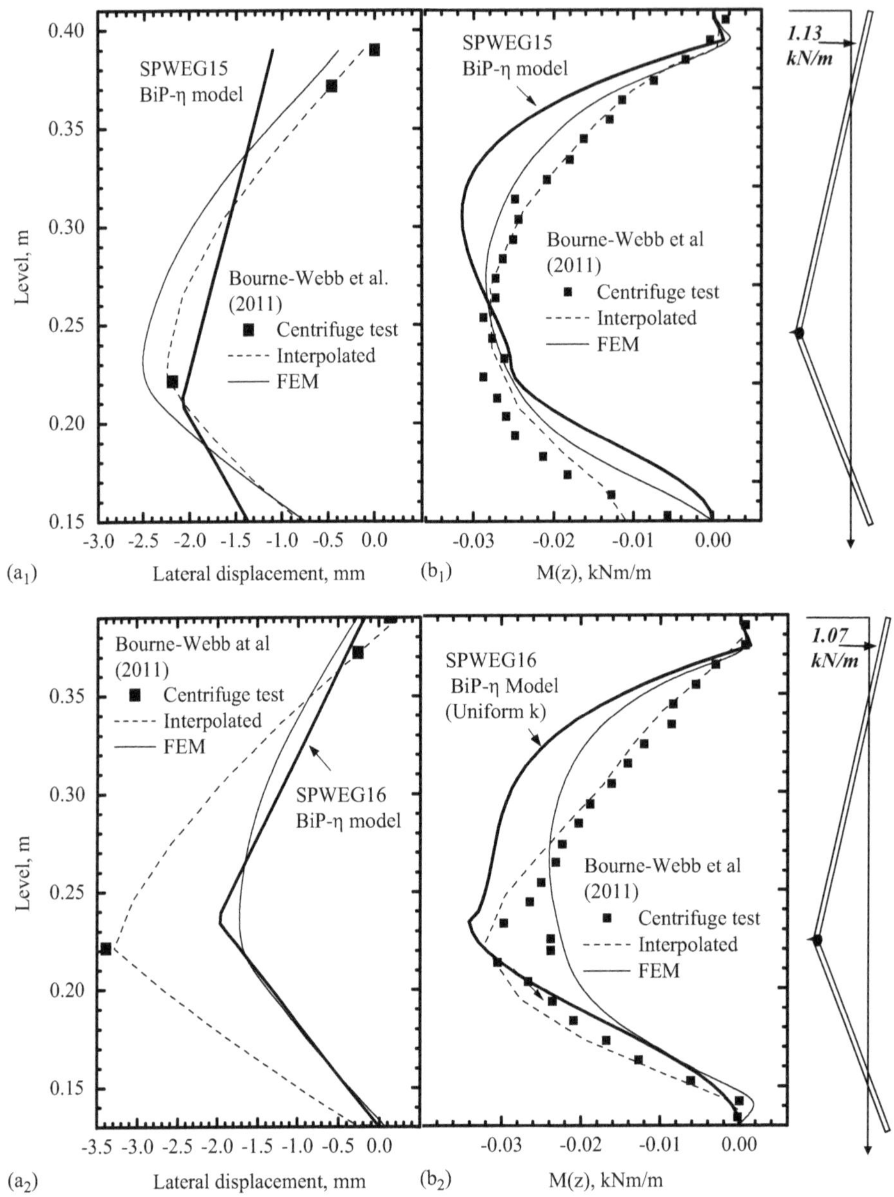

Figure 8.10 Comparison among the current BiP-η model, FEM and centrifuge tests: (a_i) SPWEG15; and (b_i) SPWEG 16 (i = 1, displacement, i = 2 bending moment). [Adapted from Guo, W. D., *Int J Geomechanics*, **21**(1), 2021.]

8.7 CONCLUSIONS

The BiP-k_θ and BiP-η models and solutions were developed (Guo 2021a) to capture the response of piles subjected to a mixed translation-rotational soil movement. A flexible pile is simplified as two-portions joined together by a rotational spring at the depth of maximum bending moment (z_m). The BiP-k_θ model assumes a constant rotational stiffness of the upper (rigid) portion about the depth z_m, which is capped by elastic restraint from the lower portion

(R-E model). On the other hand, the BiP-η model stipulates a constant rotation ratio between the two portions. The stiffness (k_θ) and rotation ratio (η) are rigorously interchangeable. The models reveal and quantify the response amplification for inward-rotating bi-portions. Simulations using the models are provided to capture the response of piles in centrifuge tests subjected to excavation loading. Discussion is made concerning the impact of five input parameters (i.e., limiting force per unit length, modulus of subgrade reaction, modulus ratio of stable over sliding layers, and rotational stiffness of each portion) on the predictions.

As piles in retaining excavation normally rotate inward as bi-segments, it is crucial to assess response amplification in deflection, rotation, bending moment and shear force, as well as shift of the depth of maximum bending moment to prevent unexpected failure. The BiP models and solutions arc useful to design of passive piles.

Chapter 9

Lateral piles under static through cyclic loading

9.1 INTRODUCTION

Piles are frequently subjected to lateral static through cyclic loading generated by waves, currents and winds etc. The pile behaviour under cyclic loading, in particular, has been extensively examined to date (Poulos 1982; Little and Briaud 1988; Long and Vanneste 1994; Lin and Liao 1999). Mono-piles are commonly used to support offshore wind turbines, and accurate estimation of accumulated pile rotation is essential for ensuring their safety performance (LeBlanc et al. 2010a, 2010b; Achmus et al. 2009; Choo et al. 2014).

Guo (2006, 2008) has developed models to capture nonlinear response of laterally loaded piles, in light of limiting force profiles (LFP) and the depth of mobilization (slip or gapping) along the pile. The impact of cyclic loading and number of cycles on the 'gapping' depth needs to be considered (Guo and Zhu 2005). This may be modelled using the BiP-k_θ model that simulates impact of increasing sliding pressure (e.g., excavation loading, Guo 2021a). Empirical expressions (e.g., LeBlanc et al. 2010a) are widely cited to estimate residual response, but the impact of eccentricity, levels, and number of cyclic loading on pile behaviour remains ambiguous.

This chapter presents a residual model, developed from load transfer model/solution (Guo 2008), to estimate residual displacement and rotation of piles over number of cyclic loading. The BiP-k_θ model (Guo 2021a) is recast into a lateral force form to capture pile response in stable through sliding soil. These models are adopted to investigate the nonlinear (including residual) response of piles embedded in sand and silt, and to deduce design parameters.

9.2 THEORETICAL MODELS

9.2.1 Guo08 model for laterally loaded pile

Table 9.1 presents the elastic-plastic solutions for a laterally loaded rigid pile (see Guo 2008). The model and solutions (hereafter referred to as Guo08; see Figure 9.1) are underpinned by the following stipulations.

- An ideal elastic-plastic p versus u relationship is adopted, with $p = kdu$ in elastic zone ($u \le u^*$, see Figure 9.1b), where p = force per unit length (*FPUL*), u = local pile displacement, d = pile outside diameter, k = subgrade modulus, which is related to the average shear modulus G_s over the pile embedded length l. The k (= $k_0 z^n$ [k_0, FL^{-n-3}]) may be referred to as constant k ($k = k_0$, $n = 0$, and in MN/m^3) and Gibson k ($k = k_0 z$, $n = 1$, in MN/m^4).
- Once the local displacement u exceeds a threshold u^* (i.e., $u \ge u^*$), the *FPUL* p reaches the limiting p_u and the pile-soil relative slip is initiated. The p_u largely increases linearly with depth z (n = 1, see the dashed line in Figure 9.1c) with $p_u = A_r zd$ (plastic zone)

 DOI: 10.1201/9781003315230-9

and $p_{ub} = A_r ld$, where $A_r z$ = net limiting on-pile pressure with $A_r = N_g \gamma_s' K_p^2$, and γ'_s = effective unit weight of the soil, i.e., bulk unit weight above water table and buoyant unit weight below, K_p[= $\tan^2(45° + \phi'/2)$] = coefficient of passive earth pressure, ϕ' = effective frictional angle of the soil, and N_g = non-dimensional parameter that can be back-estimated from measured pile response (Guo 2006, 2008).

- Tip yield occurs once the *FPUL* at pile tip attains the limiting p_{ub} (Figure 9.1c). Prior to and upon the *tip-yield* state, the *on-pile FPUL profile* follows the positive limiting LFP down to a slip depth of z_o, below which, it is governed by elastic interaction. Further increase in load beyond the tip yield state, the limiting force is fully mobilized from the tip as well to a depth of z_1. This actuates a new portion of the *on-pile FPUL profile* that follows the negative LFP over the depth l to z_1 (see Figure 9.1c). As the load continues to increase, the depths of z_o and z_1 approach each other and eventually tend to merge with the depth of rotation z_r. Achieving a state of $z_o = z_r = z_1$ is practically unattainable and is normally referred to as fully plastic (ultimate) state.

Table 9.1 Solutions for pre-tip and tip yield state (Guo 2008, 2012)

$u = \omega z + u_g$ and $z_r/l = -u_g/\omega l$

$p = kdu$, $p_u = A_r dz$, kd is the modulus of subgrade reaction, k is written as $k_0 z^n$.

Gibson k (n = 1)	Constant k (n = 0)
$\frac{H}{A_r dl^2} = \frac{1}{6}\frac{1+2\bar{z}_o+3\bar{z}_o^2}{(2+\bar{z}_o)(2\bar{e}+\bar{z}_o)+3}$	$\frac{H}{A_r dl^2} = \frac{\bar{z}_o}{2(2+3\bar{e}+\bar{z}_o)}$
$\frac{u_g k_o}{A_r} = \frac{3+2(2+\bar{z}_o^3)\bar{e}+\bar{z}_o^4}{[(2+\bar{z}_o)(2\bar{e}+\bar{z}_o)+3](1-\bar{z}_o)^2}$	$\frac{u_g k}{A_r l} = \frac{(2+3\bar{e})\bar{z}_o}{(2+3\bar{e}+\bar{z}_o)(1-\bar{z}_o)^2}$
$\omega\frac{k_o l}{A_r} = \frac{-2(2+3\bar{e})}{[(2+\bar{z}_o)(2\bar{e}+\bar{z}_o)+3](1-\bar{z}_o)^2}$	$\omega\frac{k}{A_r} = \bar{z}_o\frac{\bar{z}_o^2+3(\bar{z}_o-2)\bar{e}-3}{[2+3\bar{e}+\bar{z}_o](1-\bar{z}_o)^2}$
$\bar{z}_m = \sqrt{2H/(A_r dl^2)}$, $M_m = (2z_m/3 + e)H(z_m \le z_o)$	
$(\bar{z}_o^y)^3+(2\bar{e}+1)(\bar{z}_o^y)^2+(2\bar{e}+1)\bar{z}_o^y-(\bar{e}+1)=0$ (Solving numerically)	$\bar{z}_o^y = -(1.5\bar{e}+0.5)+0.5\sqrt{5+12\bar{e}+9\bar{e}^2}$
$A_\omega = 1-\frac{(3+4\bar{e})(1+2\bar{z}_o+3\bar{z}_o^2)}{3+2(2+\bar{z}_o^3)\bar{e}+\bar{z}_o^4}(1-\bar{z}_o)^2$	$A_u = 1-(1-\bar{z}_o)^2$
$A_\omega = 1+(1+2\bar{z}_o+3\bar{z}_o^2)(1-\bar{z}_o)^2$	$A_\omega = 1+\frac{3(1+2\bar{e})(1-\bar{z}_o)^2}{\bar{z}_o^2+3(\bar{z}_o-2)\bar{e}-3}$
$\zeta = \frac{1+2\bar{z}_o+3\bar{z}_o^2}{1+2\bar{z}_o^y+3(\bar{z}_o^y)^2}\frac{(2+\bar{z}_o^y)[(2\bar{e}+\bar{z}_o^y)+3]}{(2+\bar{z}_o)[(2\bar{e}+\bar{z}_o)+3]}$	$\zeta = \frac{H}{H_m} = \frac{\bar{z}_o(2+3\bar{e}+\bar{z}_o^y)}{\bar{z}_o^y(2+3\bar{e}+\bar{z}_o)}$

Note: $H, u, u_g, \omega, z, z_o, z_r$, e and l are defined in Figure 9.1. z_m is the depth of maximum bending moment M_m, z_o^y is the slip depth z_o at tip yield state. $\bar{z}_o = z_o/l$, $\bar{z}_m = z_m/l$, $\bar{e} = e/l$, $\bar{z}_o^y = z_o^y/l$.

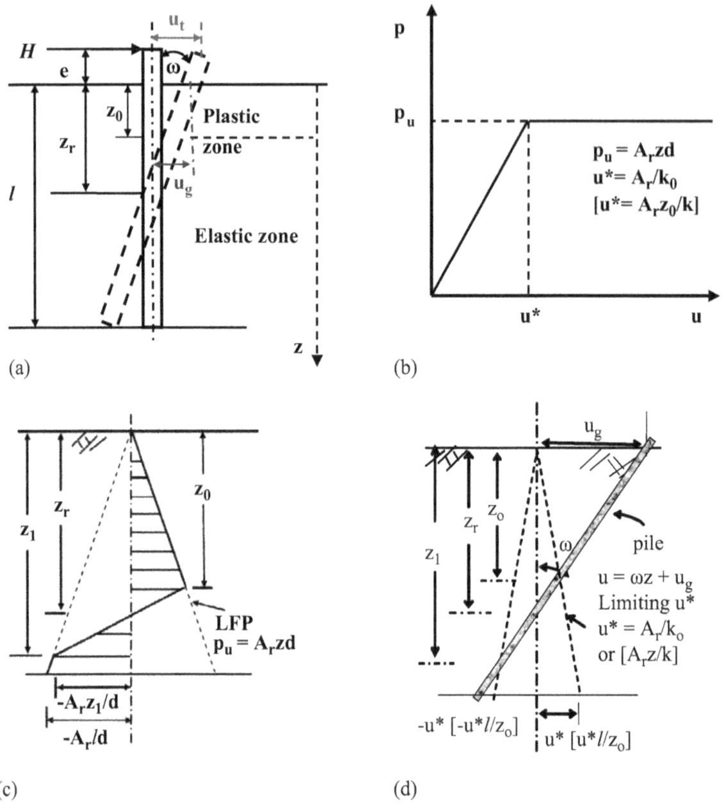

Figure 9.1 Schematic analysis for a rigid pile (a) pile-soil system; (b) load transfer model; (c) p_u (LFP) profiles; (d) pile displacement (after Guo 2008).

e = loading eccentricity above ground line;
H = lateral load; u_0 = pile displacement at ground line;
ω = angle of rotation (in radian); z = depth from ground line;
l = embedded length; z_0 = depth of slip; z_r = depth of rotation point;
p = soil resistance per unit length; p_u= ultimate soil resistance per unit length;
A_r = gradient of limiting force profile; d = outer diameter of the pile;
u = pile displacement; u^* = local threshold u above which pile soil relative slip is initiated;
k, k_0 = modulus of subgrade reaction, $k = k_0 z^m$, m = 0, and 1 for constant and Gibson k.

It is important to note that solutions underpinned by a single stipulated *on-pile FPUL profile* do not guarantee compatibility with altering *FPUL* profiles recorded at different loading levels for a single pile (see Figure 9.1c and d).

The measured response of 51 laterally loaded rigid piles in sand was matched using the Guo08 model (Qin and Guo 2014). The study delineates the impact of lateral load eccentricity on the normalized load-displacement and/or moment-rotation curves; as well as the capacity at tip-yield state and yield at rotation point. The Guo08 model and solution, based on a constant k and a linear increasing *LFP*, generally shows better agreement with the measured nonlinear response than the Gibson k. The following features are observed for the modulus G_s (for subgrade reaction k, or k_0) and gradient of limiting force profile factor N_g for A_r:

- kd/G_s = 3.27–6.91 (an average of 5.0) and G_s/G_{max} = (3–20)% for 11–16 full-scale field tests and 2 centrifuge tests; kd/G_s = 2.37–5.12 (an average of 3.7) and G_s/G_{max} = (0.8–2.6)% for 20–23 1g model tests; G_{max} = shear modulus at small strain, and N_g = $(0.4–1.8)(d/d_{ref})^{-0.25}$, where d_{ref} = 1.0 m.

9.2.2 Residual model for response under one-way cyclic loading

To estimate the residual displacement and rotation of piles, a new model is developed herein based on the lateral pile model (Guo 2008) and assuming a Gibson p_u and constant k. In light of the expressions of normalized force H and displacement u_g in Table 9.1, the total pile-displacement u_{gN} (at ground level, GL, Figure 9.2) at the Nth cycle is deduced as

$$\frac{u_{gN}kdl}{H} = \frac{2(2+3\bar{e})}{(1-\bar{z}_o)^2} \tag{9.1}$$

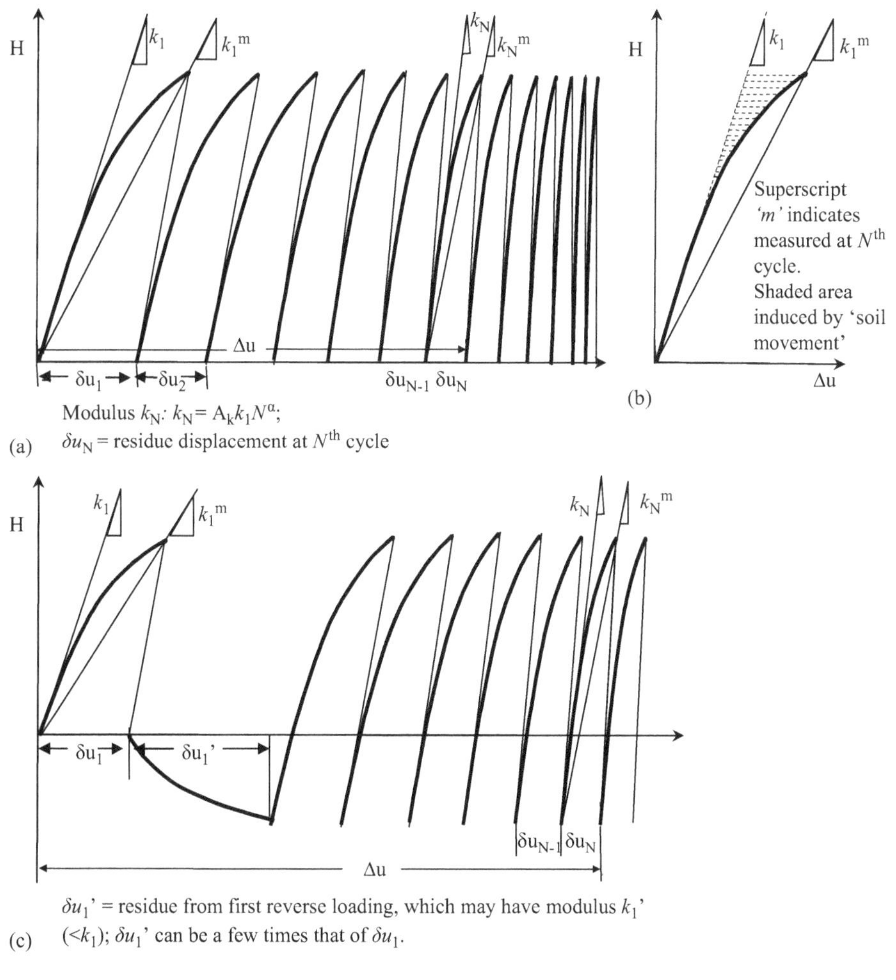

Figure 9.2 Schematic analysis for a rigid pile under cyclic loading: (a) One-way; (b) measure k_N; (c) residual δu_1^-.

where $\bar{z}_o = z_o/l$, normalized depth of slip; $\bar{e} = e/l$, normalized loading eccentricity e by l. Setting $\bar{z}_m = 0$ in Equation (9.1), the elastic displacement u_{gN}^e at the Nth cycle is deduced as $2(2+3\bar{e})H/(kdl)$. The modulus k_N at the Nth cycle is stipulated to be proportional to the power of N, with $k_1/k_N = A_k N^{-\alpha R}$, governed by a modulus factor A_k and a cyclic degradation factor α_R. Under one-way cyclic loading of $H \leftrightarrow H_{min}$ (= $\zeta_c H$), the ratio of residual displacement δu_N (= $u_{gN} - u_{gN}^e$) for the Nth cycle over the displacement under static loading u_g is determined as

$$\frac{\delta u_N}{u_g} = A_k A_u (1-\zeta_c) N^{-\alpha_R} \text{ and } \sum_{m=1}^{N} \frac{\delta u_m}{u_g} = A_k A_u (1-\zeta_c) \sum_{m=1}^{N} m^{-\alpha_R} \tag{9.2}$$

where $A_u = 1-(1-\bar{z}_o)^2$, average stress factor for displacement. The accumulated residual displacement after N cycles increases to a total of $\Delta u \left(= \sum_{m=1}^{N} \delta u_m \right)$, assuming a constant stress level or $\bar{z}_o$.

Likewise, the total pile-rotation ω_N at the Nth cycle is deduced as

$$\frac{\omega_N k d l^2}{H} = \frac{2\left[\bar{z}_o^2 + 3(\bar{z}_o - 2)\bar{e} - 3\right]}{(1-\bar{z}_o)^2} \tag{9.3}$$

Setting $\bar{z}_o = 0$ in Equation (9.3), the elastic rotation ω_N^e is obtained as $2(2+3\bar{e})H/(kdl^2)$. Under one-way cyclic loading of $H \leftrightarrow H_{min}$ (= $\zeta_c H$), the ratio of residual rotation $\delta\omega_N$ (= $\omega_N - \omega_N^e$) for the Nth cycle over the rotation under static loading ω is determined as

$$\frac{\delta \omega_N}{\omega} = A_k A_\omega (1-\zeta_c) N^{-\alpha_R} \text{ and } \sum_{m=1}^{N} \frac{\delta \omega_m}{\omega} = A_k A_\omega (1-\zeta_c) \sum_{m=1}^{N} m^{-\alpha_R} \tag{9.4}$$

where

$$A_\omega = 1 + \frac{3(1+2\bar{e})(1-\bar{z}_o)^2}{\bar{z}_o^2 + 3(\bar{z}_o - 2)\bar{e} - 3} \tag{9.5}$$

and A_ω = average stress factor for rotation. The total residual angle of rotation after N cycles is amount to $\Delta\omega \left(= \sum_{m=1}^{N} \delta\omega_m \right)$. The maximum lateral load is defined as the load H_m at tip-yield state, with a normalized slip depth of $\bar{z}_o^y$ given by

$$\bar{z}_o^y = -(1.5\bar{e} + 0.5) + 0.5\sqrt{5 + 12\bar{e} + 9\bar{e}^2} \tag{9.6}$$

This allows the level of cyclic load ζ (= H/H_m) to be defined by a normalized slip depth $\bar{z}_o$.

$$\zeta = \frac{H}{H_m} = \frac{\bar{z}_o(2 + 3\bar{e} + \bar{z}_o^y)}{\bar{z}_o^y(2 + 3\bar{e} + \bar{z}_o)} \tag{9.7}$$

Equation (9.7) demonstrates the variation of load level ζ with normalized eccentricity $(\bar{e})$ and normalized slip depth $(\bar{z}_o)$. It should be noted that existing test results for a given ζ_b may correspond to different $\bar{z}_o$ states at dissimilar $\bar{e}$ values.

Equations (9.1)–(9.7) are referred to as residual model, which is underpinned by the parameters α_R and A_k, as well as the existing parameters A_r and k (for estimating u_g, and ω). The model is characterized by the following

- The total residual displacement Δu and angle of rotation $\Delta\omega$ become *Riemann* zeta function of the cyclic number N as N moves towards infinity. The condition of 'α_R > 1.0' involves convergent residual displacement/rotation (or stabilized state), otherwise '$\alpha_R \le 1.0$', divergent displacement/rotation (instability) with cycles.
- The model is underpinned by $p = k_N du$, with a subgrade modulus k_N at Nth cycle (see Figure 9.1b). The impact of the zone highlighted in Figure 9.2b, between the k_1 (1st cycle) line and the bold curve, is neglected. The measured modulus k_N^{m} (= $\alpha_s k_N$) is only a fraction of k_N, with modulus factor α_s at a high level of soil strain (e.g., α_s = 1% for lateral spreading sand, Guo 2015b). Measured ratio $k_N^{\text{m}}/k_1^{\text{m}}$ and α_s offer a constant modulus factor A_k for a given type of test.

$$A_k = \alpha_s N^{\alpha_R} k_1^{\text{m}} / k_N^{\text{m}} \tag{9.8}$$

- The ratios A_u, A_ω and ζ_b all depend on the normalized slip depth $\bar{z}_o$ and loading eccentricity $\alpha_e \bar{e}$ (with the factor α_e).

9.2.3 BiP-k_θ models for cyclic loading with soil movement

Cyclic loading on lateral piles does propel upper soil movement, which is characterized by a subgrade modulus of k_s and underlying stable layer with a modulus of mk_s. The movement may bend pile body into two rigid segments connected by a rotational spring at depth z_m (Guo 2021a, see Figure 8.1, Chapter 8). The upper and lower segments of the pile rotate at angles ω and ω_{rL}, respectively, while the spring has a stiffness k_θ and a maximum bending moment M_m. The pile is also subjected to rotation constraint stiffness k_G at GL and k_T at pile-tip level (see Figure 8.1, Chapter 8). This pile-soil interaction mechanism is well captured by the BiP-k_θ model (Guo 2021a) as incited by excavation loading; or by the BiP-η model if a constant ratio η (= ω/ω_{rL}) of upper segment (ω) over the lower one (ω_{rL}) is stipulated.

Under constant cyclic loading, the depth z_m of maximum bending moment z_m is largely constant. The BiP-k_θ and BiP-η models offer an explicit expression of the lateral load H

$$H = \frac{0.5\bar{l}_m^2 - \bar{l}_m C_\theta}{C_\theta - \bar{a}} p_b l_m \tag{9.9}$$

$$C_\theta = \frac{\frac{12}{m}\left(6\bar{k}_\theta + \bar{z}_m^3\right)\bar{k}_T + \frac{6}{m}\left[m\left(1-\bar{z}_m\right)^3 + \left(3-\bar{z}_m\right)\bar{z}_m^2\right]\bar{k}_\theta + \bar{z}_m^3\left(1-\bar{z}_m\right)^3}{3\bar{z}_m\left[\frac{12}{m}\left(\bar{k}_T\bar{z}_m + \bar{k}_\theta\right) - \bar{z}_m\left(\bar{z}_m - 1\right)^3\right]} \tag{9.10a}$$

$$C_\theta = (-1)\frac{(m+3-2\eta)\bar{z}_m^3 - 3(m+1)\bar{z}_m^2 + 3m\bar{z}_m - m - 12\bar{k}_T}{6\left[(\eta-1)\bar{z}_m^2 + \bar{z}_m\right]} \quad (9.10b)$$

where $\bar{k}_\theta = k_\theta / (k_s l^3)$, $\bar{k}_T = k_T/(k_s l^3)$, $\bar{a}(= a/l)$, $\eta \neq 0.5(m + 3)$, $\bar{z}_m = z_m / l$, $p_b = \alpha p_{ub}$, a (= $-e$) depth (or eccentricity) of lateral load H, $m = K_{p1}/K_{a2}$ (subscripts '1' and '2' for upper layer and lower layer), and α = soil movement factor (Guo 2015b). $k_\theta = k_G + k_T$, total rotational stiffness along the pile. The k_θ has a maximum value of $k_{\theta e}$ ('e' for elastic state) given by

$$-k_{\theta e} == 2E_p I_p \lambda_2 \frac{\sinh^2\left[\lambda_2(l - z_{me})\right] + \sin^2\left[\lambda_2(l - z_{me})\right]}{\sinh\left[2\lambda_2(l - z_{me})\right] + \sin\left[2\lambda_2(l - z_{me})\right]} \quad (9.11)$$

where $\lambda_2 = [k_{s2}/(4E_pI_p)]^{0.25}$, $k_{s2} = mk_s$, and E_pI_p = flexural stiffness of pile. The model is underpinned by four soil-related parameters p_{ub} (via A_r, α), m, k_s and k_θ (with $k_T \approx 0$). Given a specific sliding depth ratio l_m/l, the on-pile *FPUL* p_s (= $\alpha p_{ub} l_m/l$) and the load H are obtained. Expressions in Table 9.2 (Guo 2021a) are subsequently adopted to gain the normalized pile-head displacement $\bar{w}_g (= w_g k_s / p_s)$, rotation angle $\bar{\omega}(= \omega k_s l / p_s)$, and $\bar{\omega}_{rL} (= \omega_{rL} k_s l / p_s)$; and the distribution of $M(z)$ and $w(z)$. Given a series of l_m/l ratios, nonlinear response of the pile is obtained (Guo 2016, 2021a). Studies against measured pile response show $k_\theta \approx k_{\theta e}$ gained using Equation (9.11), and that k_s (BiP model for passive loading, at a higher strain level) = (0.25–0.3)k_s (k_s for laterally loaded piles). As elaborated previously (Guo 2021a), a k_θ value incurs singularity at a negative value (back-rotation) of k_θ^*.

The non-dimensional response of passive piles has previously been discussed at length in light of the BiP-k_θ model using an exact depth z_m (Guo 2021a). Figures 9.3 and 9.4 present the general trajectory of the $\bar{w}_g$, $\bar{\omega}$, and $M_m/(p_s l_m l)$ response for a typical case. The figures also include the response calculated using a constant depth z_m. Based on either z_m, the solutions display similar correlations between normalized displacement and sliding depth, as well as between normalized displacement and rotation. In contrast, the constant z_m solutions do not have upper limits of displacement and rotation, and offer increasing maximum bending moment M_m with normalized pile formation w_g. Impact of these disparities is limited as shown in late examples.

Table 9.2 BiP-k_θ model for piles under passive loading

Moment about z_m: $M_L + k_\theta(\omega_{rL} - \omega_r)$	$\bar{M}_L = M_L / (p_s l^2)$ $= \bar{H}(\bar{z}_m - \bar{a}) + 0.5\bar{l}_s(2\bar{z}_m - \bar{l}_m)$	$\bar{H} = H/(p_s l)$ $\bar{P}_h = \bar{H} + \bar{l}_m$
Normalized displacement w_g and rotation ω_r at GL	$\bar{w}_g = 2\dfrac{\left[\bar{z}_m^3 + 6\bar{k}_\theta\right]\bar{P}_h - 3\bar{z}_m^2\bar{M}_L - 3\bar{z}_m^2\bar{k}_\theta\bar{\omega}_{rL}}{-\left(\bar{z}_m^3 - 12\bar{k}_\theta\right)\bar{z}_m}$	$\bar{\omega} = \dfrac{6\bar{B}_w\left(\bar{z}_m\bar{P}_h - 2\bar{M}_L\right)}{\bar{B}_w\left(\bar{z}_m^3 - 12\bar{k}_\theta\right) + 144\bar{k}_\theta^2}$
	$\bar{B}_w = m(1-\bar{z}_m)^3 + 12\left(\bar{k}_T + \bar{k}_\theta\right)$, $\bar{k}_T = k_T/(k_s l^3)$, $\bar{\omega}_{rL} = 12\bar{k}_\theta\bar{\omega}/\bar{B}_w$	

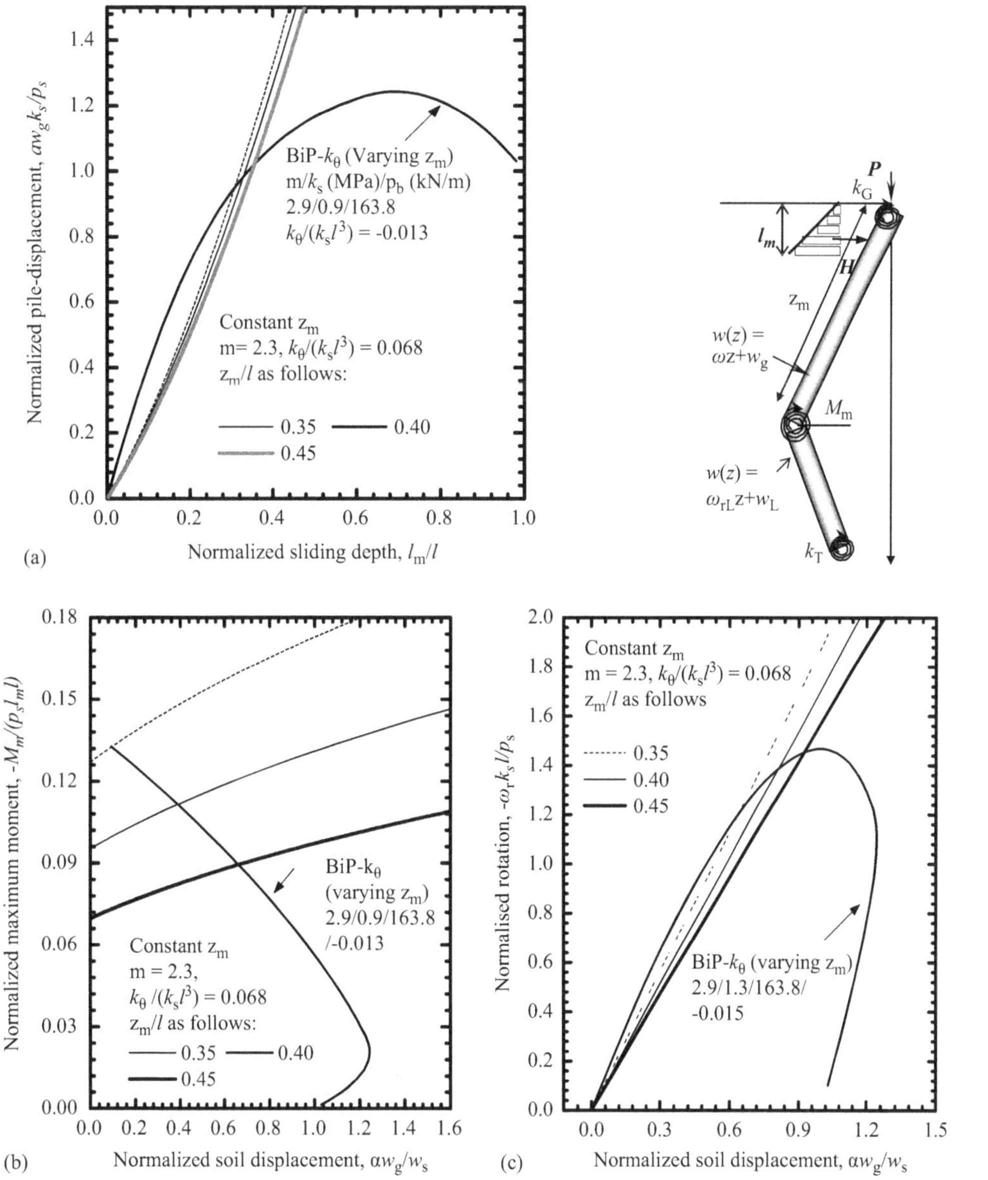

Figure 9.3 Normalized response of a rigid pile under two-way cyclic loading.

9.3 IMPACT OF CYCLIC DEGRADATION-SOIL MOVEMENT

Figure 9.5 shows 1g model tests on single piles in dry Queensland (Q) sand (Qin and Guo 2016a). The aluminium pipe pile has a length of 1.2 m, an outer diameter of 32 mm and a wall thickness of 1.5 mm. The Queensland quartz sand has a uniformity coefficient C_u of 2.92 and coefficient of curvature of C_c of 1.15. It has a relative density (D_r) of 89%, a unit weight γ_s of 16.27 kN/m^3, and an (residual) angle of internal friction ϕ of 38°. Each single pile was

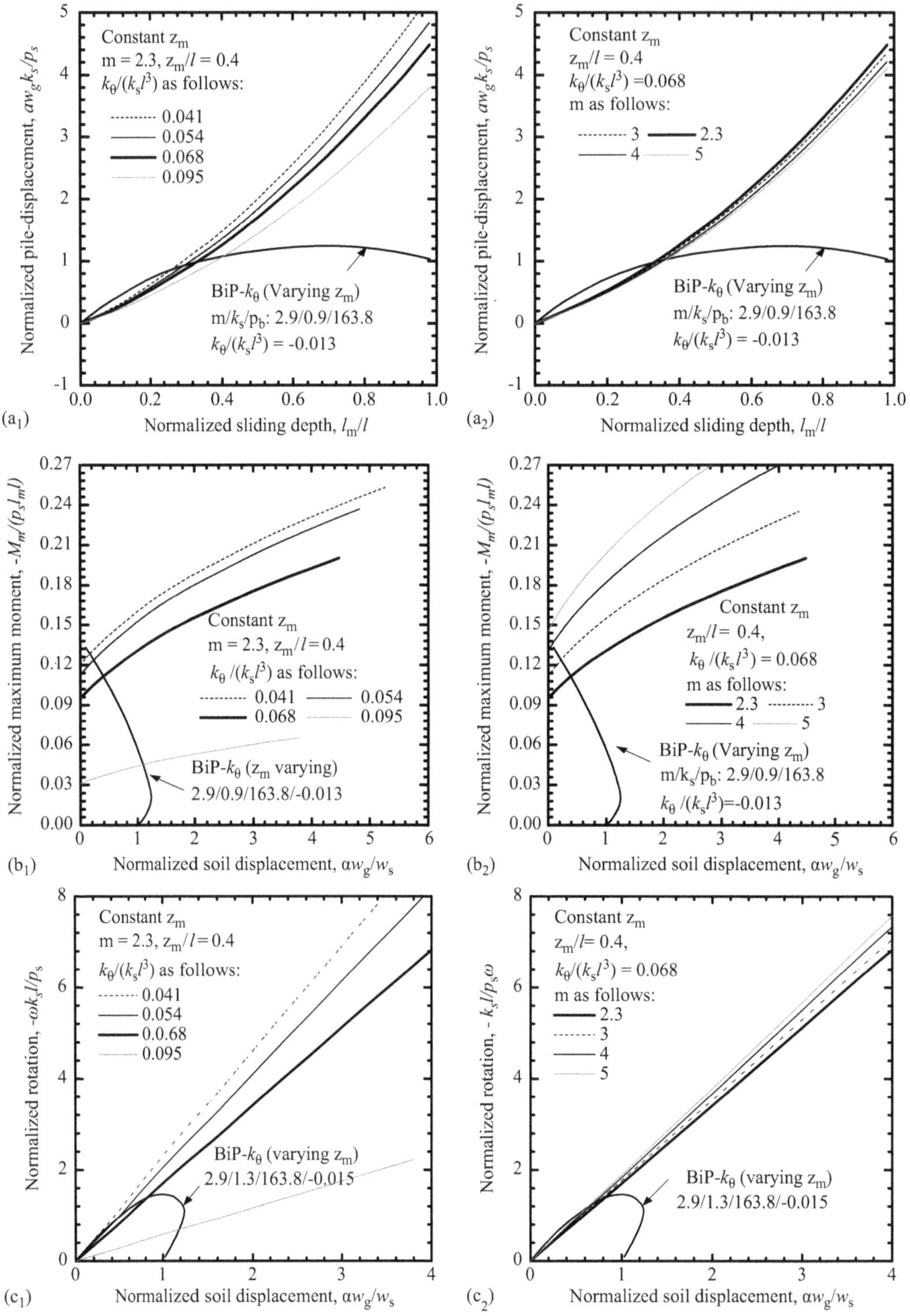

Figure 9.4 Normalized response of a rigid pile: (a_1–c_1) varying k_θ; (a_2–c_2) varying *n*.

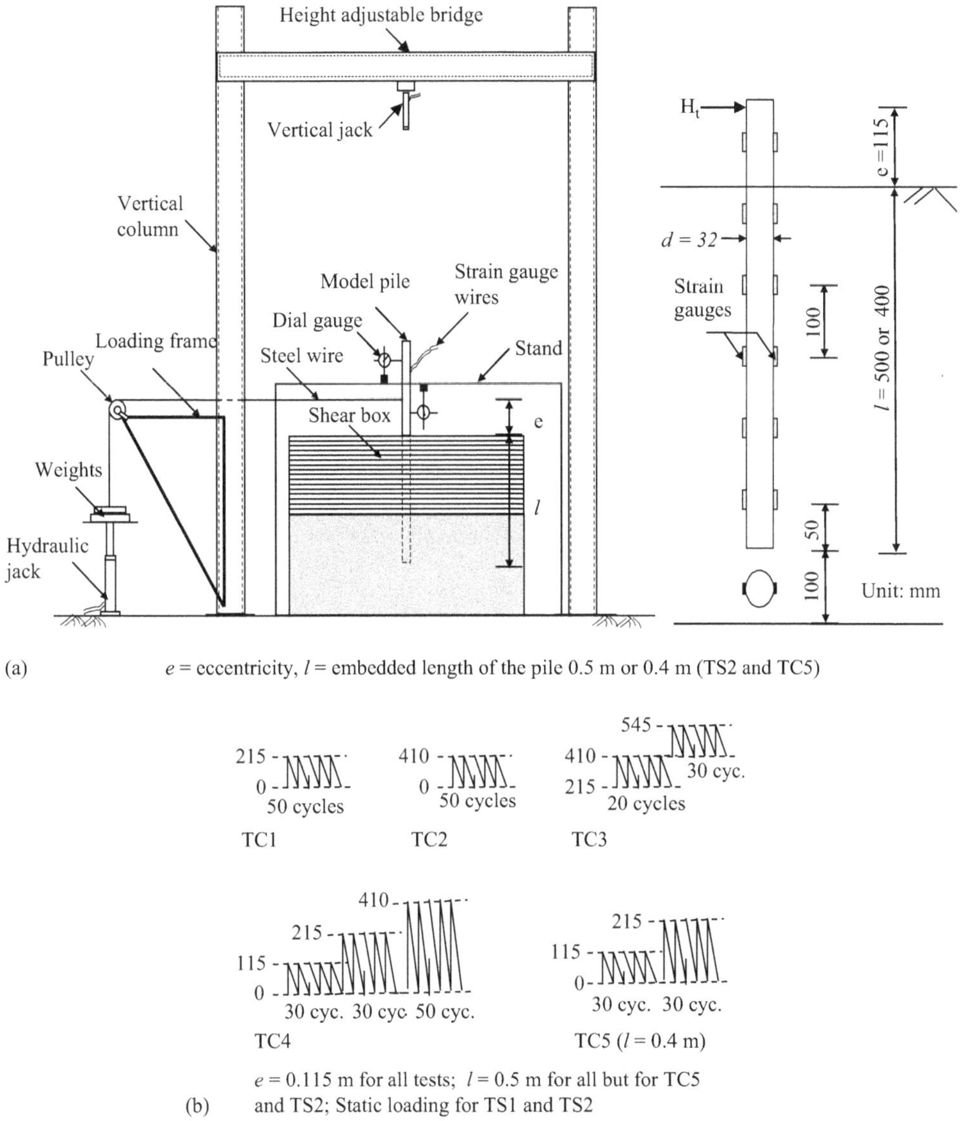

Figure 9.5 (a) Schematic experimental setting up and an instrumented pile; (b) loading patterns on five model test TC1-5.

installed to a depth l of 0.5 m (or to a depth of 0.4 m for TS2 and TC5 piles), and subjected to static loading (TS1 and TS2 piles) or cyclic loading (TC1, 2, 3, 4 and 5 piles) at a loading eccentricity e of 0.115 m above GL. The typical pile response is reviewed next to validate the new mechanism and models.

9.3.1 Static loading

The TS1 pile was gradually loaded to failure under static loading. Figure 9.6 provides the measured response, including (a) displacement u_g at GL, (b) rotational angle of pile ω; (c) maximum bending moment M_m with lateral load H; and distributions of (d) bending moment $M(z)$, and (e, f) pile displacement $w(z)$ with depth z. The load-displacement curves measured

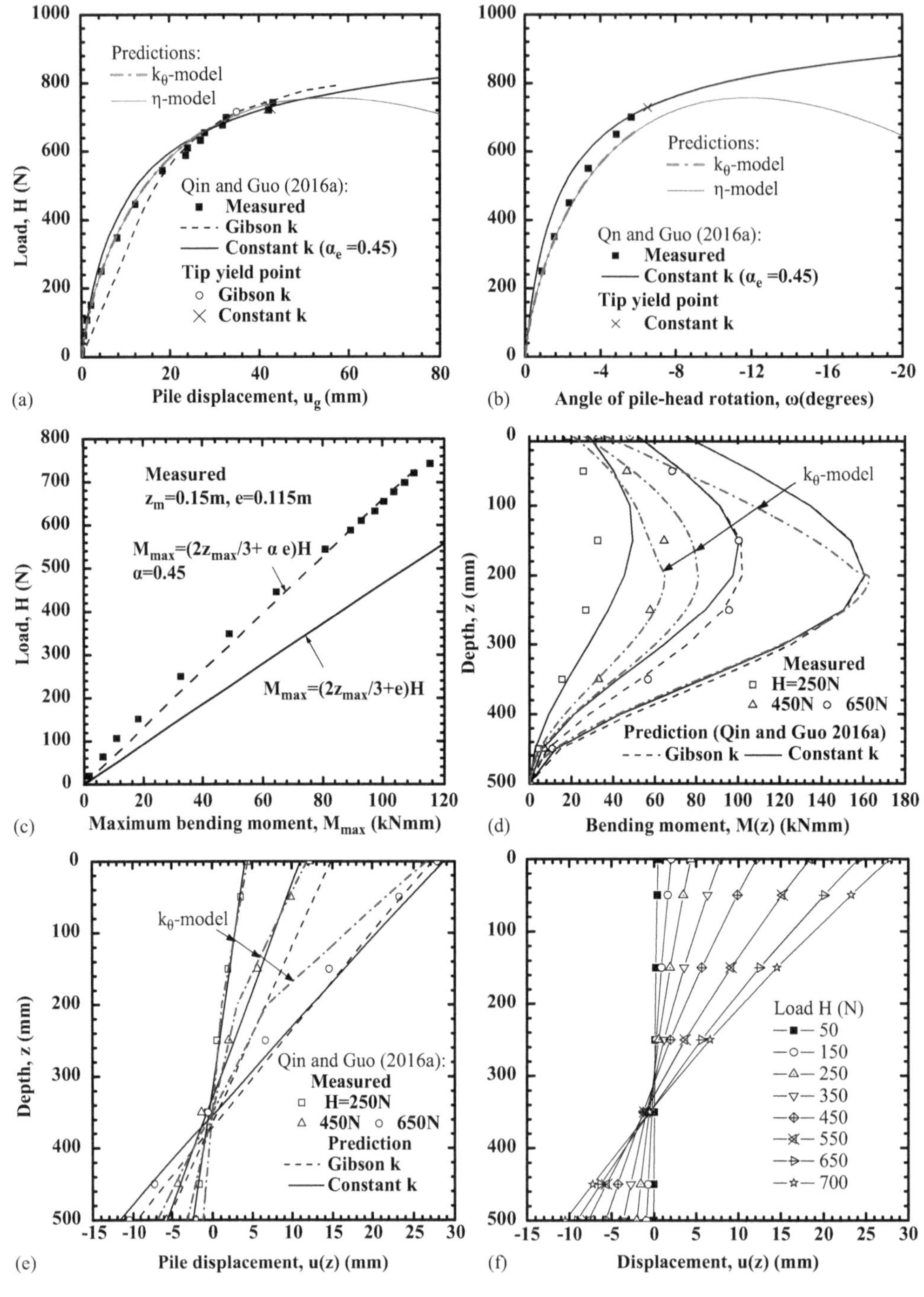

Figure 9.6 Predicted and measured response of test TS1 (a) H–w_g; (b) H–ω; (c) H–M_m; (d) $M(z)$; (e) $w(z)$; and (f) $u(z)$.

do not exhibit a distinguishable 'ultimate' load H_u, even at displacements u_g of $1.25d$. The piles primarily rotate rigidly about a depth of $0.85l$ with $z_m \approx 0.38l$. These features reinforce the use of nonlinear Guo08 model and k_θ-model (with a constant z_m) to simulate piles under static loading and in 'moving' soil.

9.3.2 One-way cyclic loading

Cyclic loading on piles can display diverse patterns, as induced by current and/or wind (McManus and Kulhawy 1994; Long and Vanneste 1994; Al-Douri and Poulos 1995). Imposing one-way cyclic loading (without reversal of load direction), TC1–TC5 piles were tested under uniform and non-uniform (storm) H. The cyclic load H_{min} (in N, see Figure 9.5) was raised to maximum H_m and then dropped to H_{min} ($H_{min} \leftrightarrow H_m$). The test results for TC1–TC5 piles are presented in Figures 9.6–9.9, respectively. For instance, the measured response (see Figure 9.8) includes H–u_g, H–ω, $M(z)$ and $w(z)$, as well as the M_m–H relationships plotted together with TS1, TC1 (load of 0 ↔ 215N). As shown in Figure 9.5, TC2 pile was subjected to 50 cycles of load H of 0 ↔ 410N, and subsequently loaded monotonically to failure. The TC3 pile was subjected to 20 cycles of load 215N ↔ 410N, and 30 cycles of 410N ↔ 545N. TC4 and TC5 piles were subjected to 30 cycles of 0 ↔ 115N, then 30 cycles of 0 ↔ 215 N. The TC4 was further subjected to 50 cycles of 0 ↔ 410 N, which was not applied to TC5 pile for a short embedment l of 0.4 m. For each test, readings of strain gauges and dial gauges were recorded for the first five cycles and every five cycles afterwards. The tests exhibit the following characteristics (Qin and Guo 2016a).

(1) The accumulated response was dominated by a low number of cycles at each load level. For instance, the cyclic load of 215 N on TC1 resulted in a slightly increased bending moment along the pile, with sand flowing around and separating from the pile. After 50 cycles, a densified (core) zone of 24 mm deep over a distance of up to 2.5d behind the pile was observed, beyond which the sand spreads laterally.
(2) Increasing the (uniform) cyclic load (= $\zeta_b H_m$) by 50% from 215 N (TC1) to 410 N (TC2) caused increase of the displacement u_g by three times and ultimate capacity (against static loading test TS1) by only 10% after 50 cycles. As expected, the displacement u_g of TC4 pile was limited at a relatively low load level ζ_b (of 0.243), but increases continuously with the loading cycles at a larger ζ_b (of 0.46). The impact of loading levels on piles, including stability, demands incorporating effect of slip depth and eccentricity.

The same mechanisms has been observed in the tests on piles in silt (Chen et al. 2015), thus further validating the residual model for storm loading. In particular, the pile behaviour is consistent with that observed at a relatively low ζ_b of cyclic loading through to incremental collapse (Poulos 1982; Swane 1983; Levy et al. 2009).

9.3.3 Depth z_m and Loading eccentricity modification α_e

It is worth emphasizing that the response of laterally loaded piles, regardless of preceding cyclic-loading, is well predicted using Guo's (2008) solution for 'static' loading (Qin and Guo 2016a). However, cyclic loading incurs permanent (residual) soil movement around each pile. The associated impact can be captured using the k_θ-model solutions (Guo 2021a) based on a constant value of z_m. The normalized $\bar{z}_m (= z_m / l)$ at a specific level of cyclic load (via $\bar{z}_o$) is given by (Guo 2008)

$$\bar{z}_m = \frac{1 + (\bar{z}_o + 3\bar{e})\bar{z}_o}{3(1 + 2\bar{e}) - (\bar{z}_o + 2\bar{e})\bar{z}_o} \quad \text{and / or} \quad \bar{z}_m = \sqrt{2\bar{z}_r^2 - 1} \tag{9.12}$$

The measured depth of pile rotation z_r allows an upper bound for the depth z_m at ultimate, impossible state to be estimated. Given $\bar{z}_r = 0.75 - 0.85$ (Zhu et al. 2015), the $\bar{z}_m$ is deduced

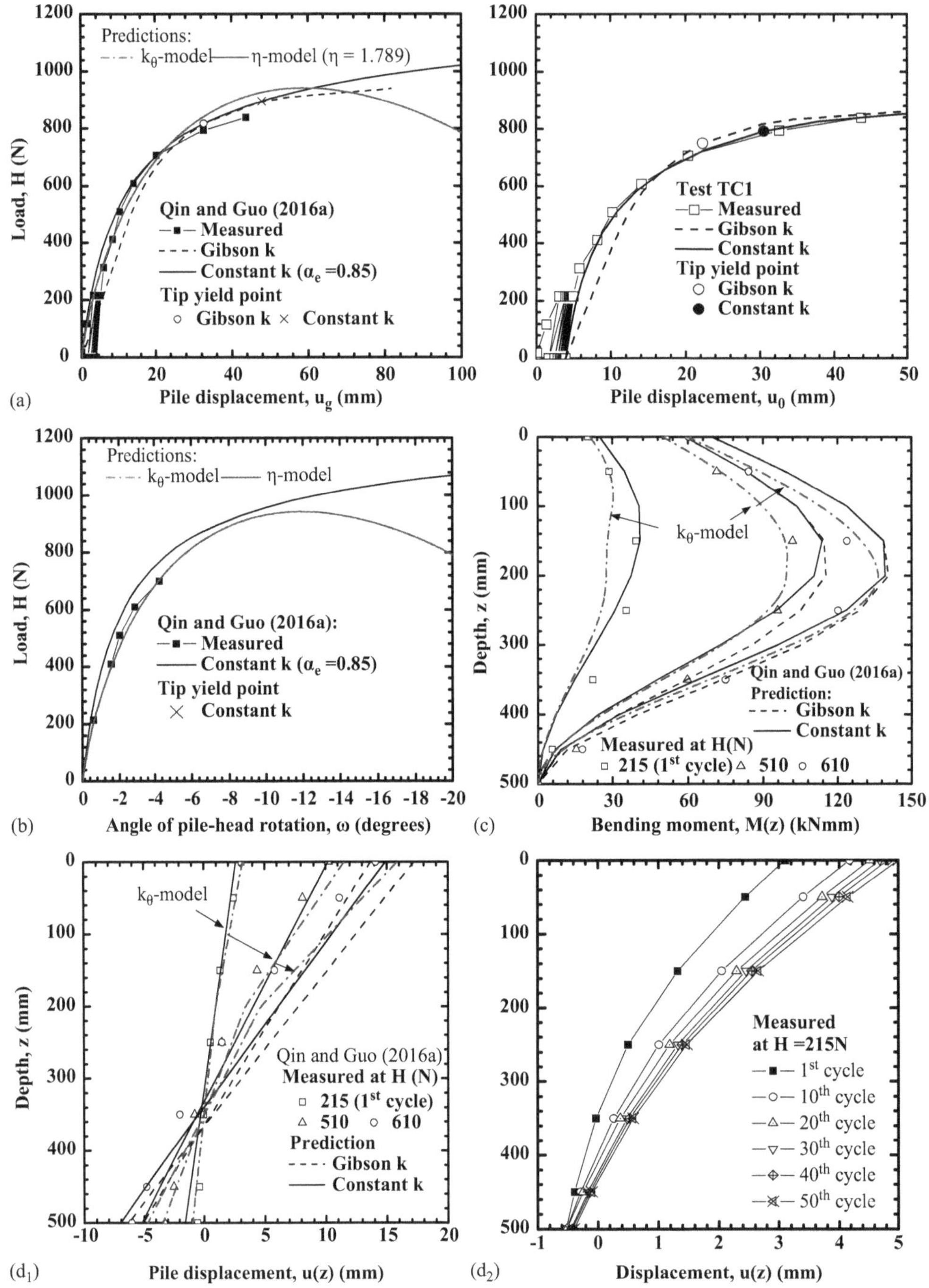

Figure 9.7 Predicted and measured response of test TC1 (a) H–w_g; (b) H–ω; (c) $M(z)$; and (d_1, d_2) $u(z)$.

as 0.322–0.479 with an average z_m/l of 0.4. Cyclic loading increases the maximum bending moment, M_m by ~20% (TC3), owing to the rising bending moment M_0 at GL, which is given by invariable depth z_m and cyclic H (Broms 1964, Guo 2008) (see Figure 9.8c) using

$$M_m = \frac{2}{3} H z_m + \alpha_e M_o \tag{9.13}$$

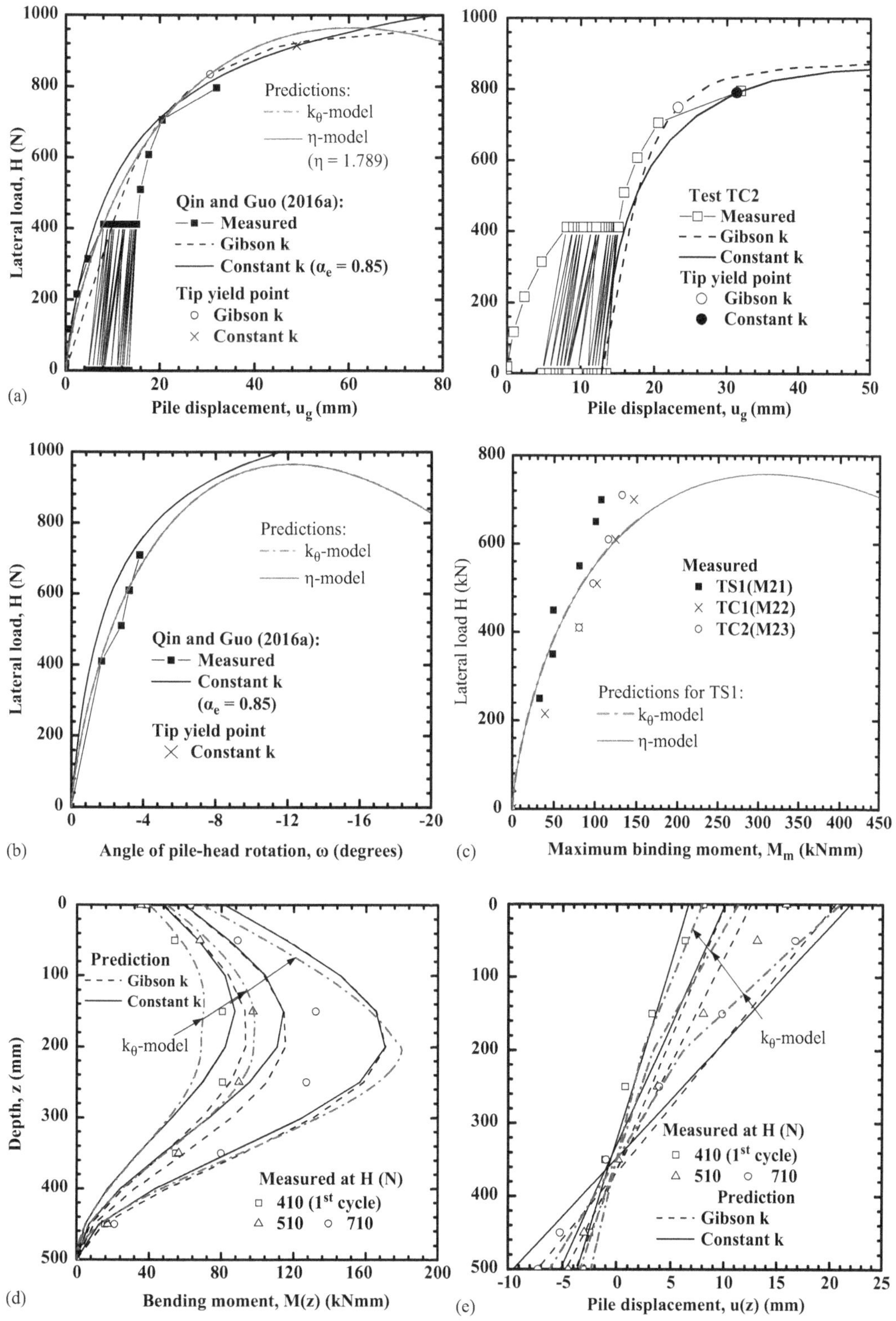

Figure 9.8 Predicted and measured response of test TC2 (a) H–w_g; (b) H–ω; (c) H–M_m; (d) $M(z)$ and; (e) $u(z)$.

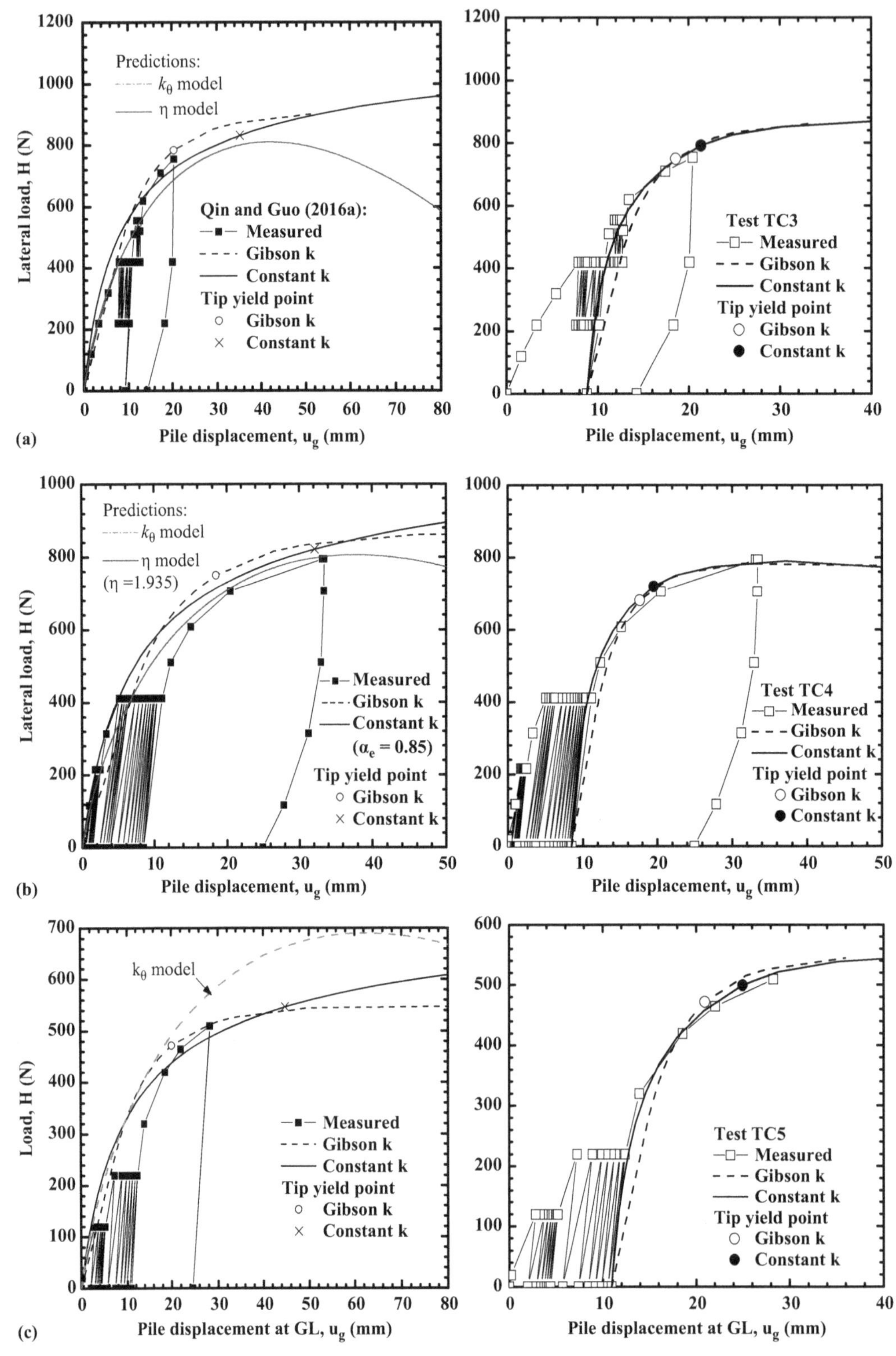

Figure 9.9 Predicted and measured H–u_g response of test (a) TC3; (b) TC4; (c) TC5.

where α_e = 0.45–1.0 with an average of 0.85. Incorporating the new factor α_e offers a better agreement between the predicted $\alpha_e M_o$ and the measured bending moment M_o at GL.

Example 9.1 Impact of eccentricity and soil movement

To use the residual model, it is desirable to obtain a reliable $\bar{e}$ value. Given a reduced eccentricity (α_e shown in figures) of laterally loaded pile, the solutions of Guo (2008) were matched visually with measured response, including load-displacement at GL, distributions of bending moment, and displacement. The match is presented in Figures 9.6 through to 9.9 for Test TS1 and TC1–5, and yielded the values of input parameters A_r (or N_g) and k (or k_0) (see Table 9.3). For instance, Test TC2 (pre-cyclic loading, or N = 1) involved A_r = 1,225 kN/m² (or N_g = 4.26) and k = 36.5 MM/m³ and k_0 = 121 MM/m⁴. The predicted load-displacement curves using a constant and a linear increasing modulus (Gibson) of subgrade reaction well bracket the measured data, but the bending moment is generally overestimated using a unit of α_e, as seen in Test TS1.

The cyclic loading leads to 3.1 times enlargement of the residual displacement and a 2.86 times enlargement of residual rotation. The impinging 'movement' is evaluated by using the modified k_θ-model and η models (with z_m = 0.4l), and the predictions are compared to the measured data of TS1, TC1 and TC2 in Figures 9.6–9.8, respectively. Herein, the prediction is exemplified for TS1 test. The pile had l = 0.5 m, d = 0.032 m, e = 0.15 m, and a flexible rigidity E_pI_p of 1.28 kNm². The soil is characterized as k_s = 288 kPa (= kd, sliding layer), m = 4.1, and k = 1,067 kPa (see Table 9.4). The coupled elastic parameters are obtained as N_p = 9.037 kN, α_N = 4.025 /m, β_N = 3.56/m, (Guo 2012), and λ_2 = 3.80/m, and k_θ = 4.285 kNm/rad (with z_{me} = 0.2 m). Given e/l = 0.104 (= 0.115α_e/0.5, with a low α_e of 0.45), it follows $\bar{z}_o^y$ = 0.593, and z_m/l = 0.29 using Equations (9.6) and (9.12). However, the upper bond z_m/l of 0.4 is taken as per Equation (9.13). For each l_m/l ratio, the load H was estimated using Equation (9.9) and w_g using Table 9.2 expression. For instance, at l_m/l = 0.16, the BiP-k_θ model offered –H = 123 N, and w_g = 1.57 mm, etc. Note the load is negative, as it is effectuated by an opposite direction of soil movement. The calculation is repeated for several l_m/l ratios (see Table 9.5), which is plotted in Figure 9.6 together with the measured response of H–u_g, H–ω, $M(z)$ and $w(z)$. The BiP-k_θ model offers a reduced loading capacity at large displacement/rotation, which is critical for designing piles under cyclic loading. Setting η = 1.62, and z_m/l = 0.4, the results from the BiP-η model were nearly 'identical' to those gained for the BiP-k_θ model (not repeated herein).

Equation (9.11) offers a stiffness k_θ of 4.859 kNm/radian, which exceeds the actual stiffness by 13.4%. The disparity, nevertheless, alters the pile response only by ±4% (with +3.8 for H, –1.8% for w_g, and –3.4% for ω_r). Equation (9.11) is therefore sufficiently accurate for practical design.

Table 9.3 Parameters for Guo08 model (Qin and Guo 2016a)

Tests	*Cycles*	*A_r (kN/m³) [N_g]*	*k (MN/m³)*	*k_0 (MN/m⁴)*
TS1	Static at H = 450 N	1050[3.65]	30	91
TC1	1 – 50	1200[4.17] – 1100	36.5 – 59.5	111 – 181
TC2	1 – 50	1225[4.26] – 1100	36.5 – 85.5	121 – 321
TC3	1 – 30(1st)[a], 20(2nd)	1150[4.00] – 1100[3.83]	48 – 126	170 – 340
TC4	1 – 30, 30, 50	1100[3.83] – 1000[3.48]	50 – 130	180 – 330
TC5	1 – 30 (1st), 30(2nd)	1150[4.00] – 1150[4.00]	32 – 92	170 – 340
D_r 4%	Static	166.0[0.754]	0.656	LeBlanc et al (2010a)
D_r 38%	Static	332.0[1.155]	1.547	

[a] parcel

Table 9.4 Parameters for BiP-$k_\theta(\eta)$ models

Tests	A_r *(kN/m³) [N_g]*	k_s *[kd]*[b] *(kPa)*	k_θ *(kNm/rad)*	*n/η*	*Reference*
TS1[a]	891 [3.10]	288.0	4.285	4.1/1.62	Qin and Guo (2016a)
TC1[a]	1077 [3.74]	389.3	5.549	5.0/1.68	
TC2[a]	1103 [3.83]	389.3	5.549	5.0/1.79	
TC3[a]	1035 [3.60]	512.0	6.303	5.0/1.91	
TC4[a]	990 [3.44]	533.3	6.418	5.0/1.94	
TC5[a]	1032 [3.59]	320	4.098	5.0/1.45	
2D[a]	141.1 [0.31]	1989	17,600	4.6/2.54	Hutchinson et al. (2005)
6D[a]	159.1 [0.35]	1059	10,230	2.3/1.706	
Test 1[a]	114.87 [0.5]	1320	594.3	5.38/1.16	Zhu et al. (2015)
	220.9 [0.96][b]	[2466][b]			

Note:
[a] $k_T = 0, k_s = (1/2–1/3)kd$.
[b] Guo08 model.

Table 9.5 BiP-k_θ (at $\bar{z}_m = 0.4$) and constant k models of TS1 pile

l_m/l[c]	*−H (N)*	w_g *(mm)*	M_m *(Nm)*	*ω (deg)*
0.16[a]/0.08[b]	123[a]/119[b]	1.57[a]/0.99[b]	8.8[a]/20.7[b]	−0.306[a]/−0.149[b]
0.22[a]/0.15[b]	215[a]/217	3.44/2.11	19.3/47.0	−0.674/−0.32
0.33[a]/0.30[b]	407/410	9.74/5.85	54.6/113.8	−1.907/−0.891
0.35[a]/0.32[b]	443/434	11.4/6.56	63.7/123.3	−2.226/−0.998

[a] BiP-k_θ model.
[b] Constant k model.
[c] $l_m < 0.59l$ ($= z_o^{tip\text{-}yield}$), stable cyclic loading.

Example 9.2 Multi-levels of cyclic load on piles in Queensland Sand

The closed-form solutions (Guo08 model) can be used to match the reloading-displacement curve at a specific cyclic loading for each test, as shown in Figures 9.7-right through to 9.9-right. This match is conducted for the post cyclic-loading state. The back-estimation results demonstrate that (1) the modulus k and the parameter k_0 increase with unloading-reloading cycles. For instance, TC2 test involves $k_N^m/k_1^m = 1.63–2.88$ ($k = k_1^m$) and $k_{0N}^m/k_{01}^m = 1.60–2.0$ ($k_0 = k_{01}^m$), e.g., $k_{50}^m = 85.5$ MN/m³ and $k_1^m = 36.5$ MN/m³. (2) The N_g, on the other hand, reduces to 90% of static loading, which is consistent with the practice in constructing $p–y$ curves for piles in sand by Murchison and O'Neill (1984) and API (1993). The N_g is 0.56–0.64 times that for static loading of six laterally loaded free-head flexible piles in calcareous sand (Guo and Zhu 2005). For instance, the A_r (TC2) decreases from 1,225 ($N = 1$) to 1,110 ($N = 50$) MN/m³.

The residual response accumulated over cyclic loading is determined for TC2 pile. With $A_r = 1{,}225$ kN/m³, $d = 0.032$ m, and $l = 0.5$ m, it follows $\alpha_e e/l = 0.196$, and $\bar{z}_o = 0.21696$ at $H = 0.41$ kN. Equations (9.2) and (9.3) offer $\delta u_N/u_g = 0.387A_k N^{-\alpha_R}$, and $\delta\omega_N/\omega = 0.413A_k N^{-\alpha_R}$. With $k_{50}^m/k_1^m = 2.34$ (= 85.5/36.5 at $N = 50$), and taking $\alpha_s = 1.1\%$, Equation (9.8) offers A_k of 0.5 ($= 0.011 \times 50^{1.2}/2.34$). Given $A_k = 0.5$ and $\alpha_R = 1.2$, the accumulated $\delta u_N/u_g$ and $\delta\omega_N/\omega$ ratios over 50 cycles amount to 2.56 and 2.73 respectively

$$0.387A_k\sum_{m=1}^{50}(1/m)^{\alpha_R} = 2.56, 0.413A_\omega\sum_{m=1}^{50}(1/m)^{\alpha_R} = 2.73 \tag{9.14}$$

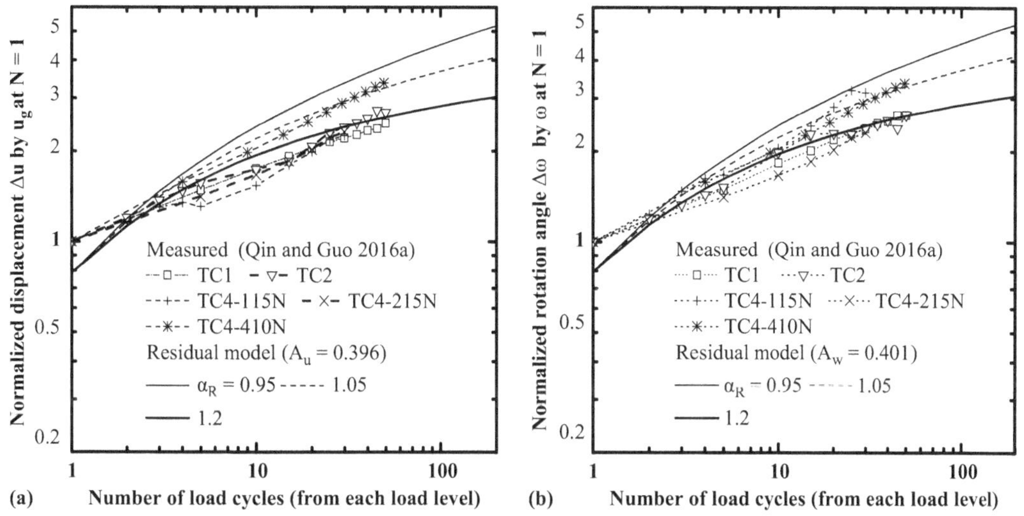

Figure 9.10 Predicted versus measured (Qin and Guo 2016a): (a) residual deformation; (b) residual angle of rotation.

As shown in Figure 9.10, both ratios agree with the measured data. Importantly, the degradation factor α_R is almost constant regardless of cyclic loading level, but it does vary with the type of test (one-way or two-way with unequal cyclic magnitude) and soil property, as evidenced subsequently.

The model and parameters (α_R and A_k via α_s) provide a good prediction for pile TC1 and TC4 concerning the evolution of normalized (residual) deformation $\Delta u/u_g$ and angle of rotation $\Delta\omega/\omega$ with the number of loading cycles. Importantly, u_g or ω is taken as the first cycle of each load level. For instance, the u_g or ω for TC4 under 215 N is taken as that at the 31st cycle; and the number of cycle is plotted as 1st–30th cycles (rather than 31st–60th cycle). The u_g or ω under 410 N is taken as that at the 61st cycle, and the 61st–110th cycle is plotted as 1st–50th cycles. This treatment is intended to minimize the impact of steep raise in loading level on the normalized behaviour. Given non-zero minimum load H, the model can be used for TC3 as well.

Example 9.3 Uniform cyclic load on piles in Leighton Buzzard Sand

Leblanc et al. (2010a) conducted 21 model tests on copper pipe pile (l = 0.36 m, d = 0.08 m, and 0.002 m in wall thickness) in Leighton Buzzard (LB) Sand. The sand has E_s (Young's modulus) of 4.64 MPa, ϕ (angle of internal friction) of 35° (D_r = 4%) and 43° (38%), and γ_s (density) of 14.7–17.6 kN/m^3. A lateral (static) load was exerted at a height of 1.19l (= 0.428 m) above GL. The load–displacement curves recorded are shown in Figure 9.11a. The k_s is obtained as 656 kPa (D_r = 4%) and 1547 kPa (D_r = 38%), respectively for e/l = 1.19. The *LFP* profile is described by A_r = 166.0 kN/m (N_g = 0.754) and A_r = 332.0 kN/m (N_g = 1.155). The parameters are tabulated in Table 9.3. The Guo08 model well predicts the measured response (Figure 9.11a), but for the hardening moment M_o (at GL) at large displacement. The k_0-model (not shown herein) shows a similar discrepancy from measured response to that presented in Figure 9.7 for TC1.

Leblanc et al. (2010a) established an empirical expression for estimating the accumulated angle $\Delta\omega/\omega$ over a number of loading cycles, as depicted in Figure 9.11b for ζ_b of 0.27, 0.4 and 0.52. The correlation for dense sand is well fitted using Equation (9.4), α_R = 0.75, and

$A_k = 5$ (at $D_r = 38\%$ and any ζ_b). A close fit is also observed for loose sand ($D_r = 4\%$, Figure 9.11c) albeit with A_k of 8 ($\zeta_b = 0.53$), 14 (0.27–0.4) and 35 (0.2). In particular, the current 'residual' model solution offers a close match to most test data at low number of cycles.

The tests have a $k_N{}^m/k_1{}^m$ ratio of 1.176 ($N = 100$ cycles) and 1.264 ($N = 10^3$ cycles). At $N = 10^3$ cycles, the tests in dense sand correspond to α_s of 3.55% (= $1.264 \times 5/1000^{0.75}$, $A_k = 5$). The α_s for loose sand is largely around 10.0% (= $1.264 \times 14/1000^{0.75}$, $A_k = 14$), and reaches 25% at a low load level ζ_b of 0.2. Overall, the α_s values are legitimate between static ($\alpha_s = 1.0$) and lateral spreading ($\alpha_s = 1\%$) cases.

Imposing a two-way cyclic loading, the normalized rotation angle $\Delta\omega/\omega$ (as measured) gradually reduces towards increasing number of cycles, especially in tests with near equal magnitudes of two-way cyclic loading (with ζ_c of −0.81 and −0.98). This behaviour is well captured using the BiP k_θ-model as shown in late example.

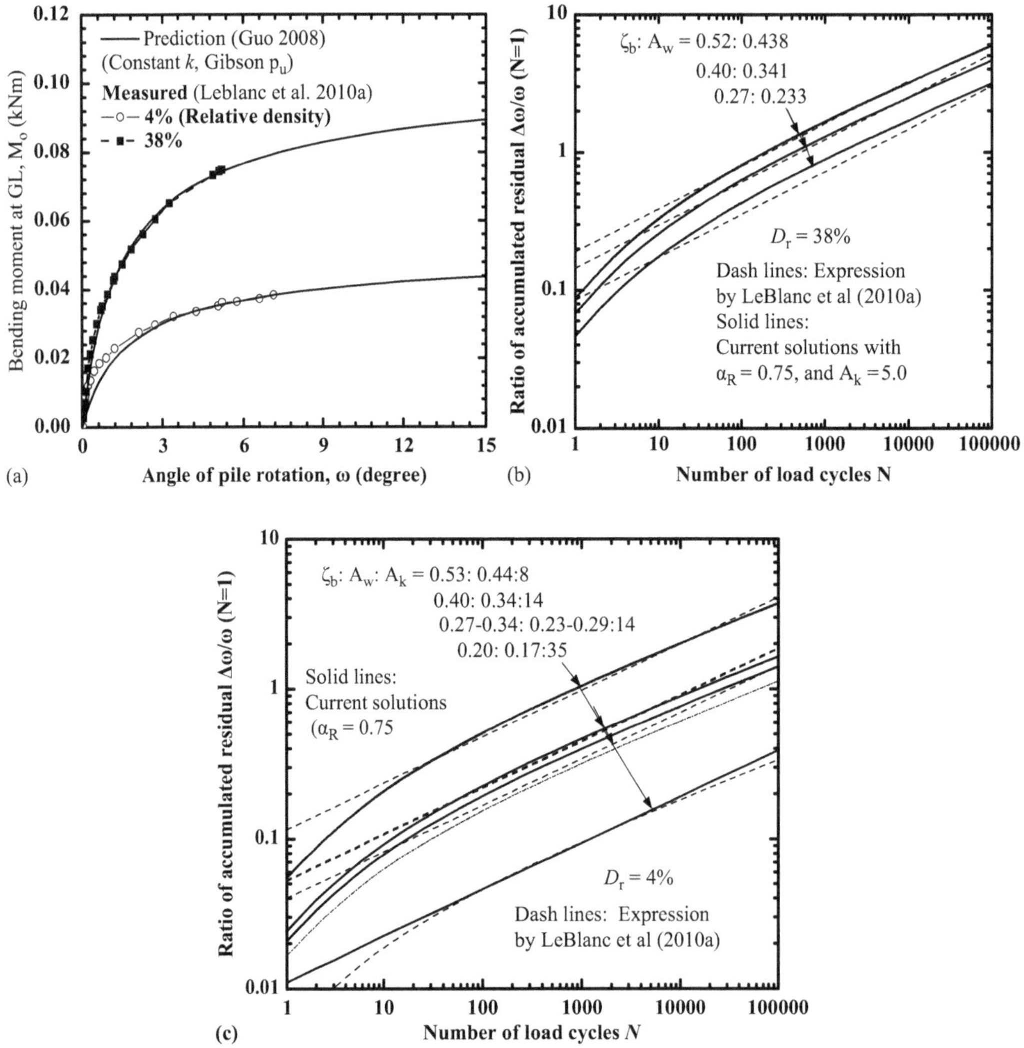

Figure 9.11 Predicted versus measured (Leblanc et al. 2010a) response of lateral piles.

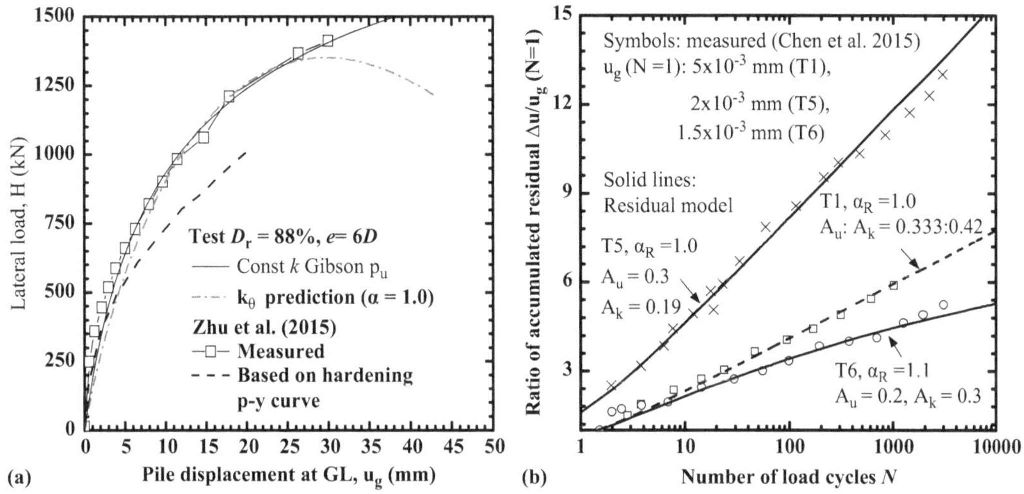

Figure 9.12 Predicted response of lateral piles versus (a) measured u_g–H (Zhu et al. 2015); (b) measured $\Delta u/u_g$ – number of cycles (Chen et al. 2015).

Example 9.4 Uniform cyclic loading on piles in Qiantang River silt

Zhu et al. (2015) reported six individual model tests on a single steel pipe pile (l = 0.915 m, d = 0.165 m, and 0.003 m in wall thickness) in Qiantang River (QR) silt. The pile has a flexible rigidity (E_pI_p) of 1.05 MN/m^2. The silt has D_r of 88%, E_s (Young's modulus) of 4.64 MPa, ϕ (angle of internal friction) of 35.5°, and γ_s (density) of 16.5 kN/m^3. In Test 1, lateral load was exerted at a height of 0.99 m (= 6d) above GL. The measured load–displacement curve is presented in Figure 9.12a, along with the predicted curve using the p–y curve of API (1993). The parameters k_s/G_s = 4.354 for e/l = 0.52 (Guo 2012) and $G_s = E_s/[2(1 + 0.3)]$ offer k = 7.77 MPa, and k_s = 1,282 kPa (= kd). The *FPUL* profile (i.e., *LFP*) is described by A_r = 220.9 kN/m (N_g = 0.962). The Guo08 model prediction, using constant k and Gibson p_u, agrees with the measured data. The BiP k_θ-model, with associated parameters (see Table 9.4), is also consistent with the measured response (Figure 9.12a), and predicts a force H softening at u_g > 26 mm.

Chen et al. (2015) conducted cyclic loading tests under similar conditions. The measured residual pile displacement Δu (accumulated) of tests T1, T5 and T6 is normalized by the displacement u_g (at N = 1) of 0.005 mm, 0.002 mm and 0.0015 mm. This provides the evolution of normalized residual displacement $\Delta u/u_g$ with loading cycles N, as shown in Figure 9.12b. The evolution route is well predicted using Equation (9.4), with α_R (= 1.0–1.1) and A_k (= 0.19–0.42). The A_u values are calculated for ζ_b = 0.37 (T1), 0.34 (T5), and 0.25 (T6). The tests in QR silt induce an average k_N^m/k_1^m ratio of 1.335 (at $N \geq$ 100 cycles). The modulus 'reduction' factor α_s is deduced as 0.25% (= 1.335 × 0.19/100, A_k = 0.19) and 0.56% (= 1.335 × 0.42/100, A_k = 0.42), which are up to 50% lower than those adopted for lateral spreading cases.

Example 9.5 Flexible piles subjected two-way cyclic loading

Chai and Hutchinson (2002) conducted four full-scale tests on single, reinforced concrete piles (d = 406 mm) under combined axial compression and reversed cyclic quasi-static lateral loading. The piles were embedded to a depth (l) of 13.5d in river sand, with friction angles of 42–44° for dense sand (of upper 3 m) and 37–38° for loose sand. Lateral (cyclic) load were applied at the top of the pile 2d and 6d (or a normalized $\bar{e}$ of 0.148 and 0.44) above GL in both loose and dense dry sand conditions. The piles tested in dense sand were previously simulated using the finite element method, adopting the API model and backbone p–y curve. The simulations were performed under displacement control with single cycles of reversed lateral loading applied to the same target displacements used during pile testing. The predictions are plotted in Figures 9.13a and b together with the envelope curves of the measured force and pile-head displacement relationships generated by a number of loading cycles.

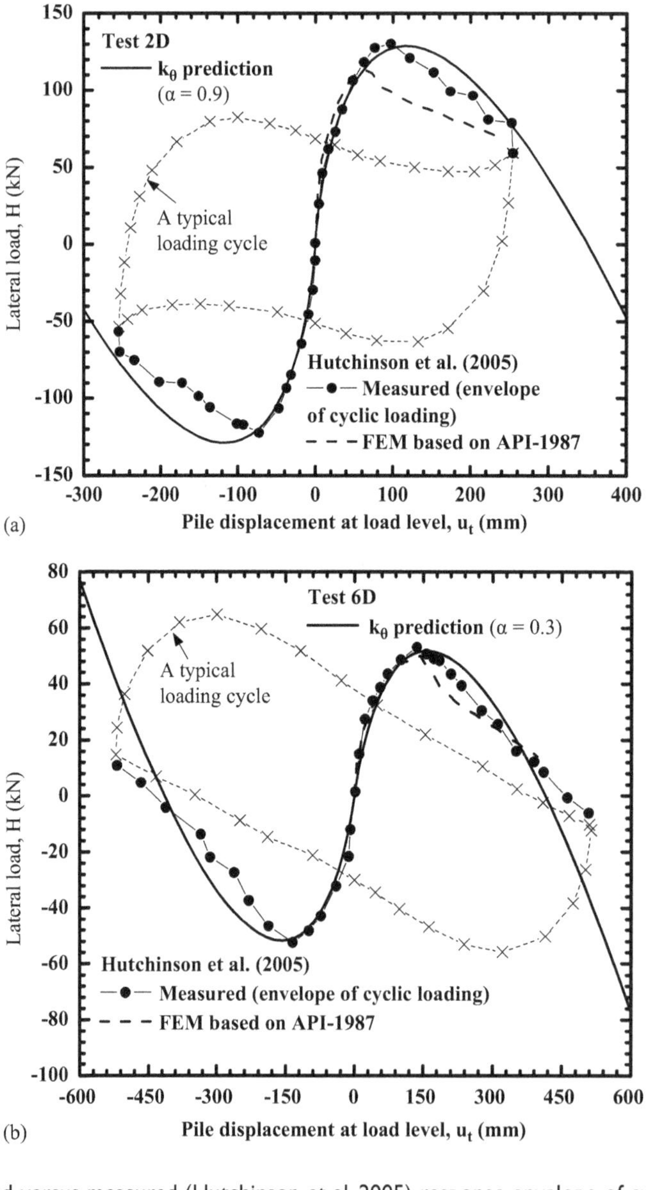

Figure 9.13 Predicted versus measured (Hutchinson et al. 2005) response envelope of cyclic degradation: (a) Test 2D; (b) Test 6D.

The BiP-k_θ model of the '2d' pile adopts $m = 4.6$, $k_s = 1.989$ MPa, $p_{ub} = 314.0$ kN/m, and $\bar{k}_\theta = 0.054$ and $\bar{k}_T = 0$ (written together as 4.6/1.99/314./0.054). The model corresponds to $\eta = 2.541$ using the BiP-η model. The sliding frictional angle was taken as 21.5° (= 0.5ϕ) and $m = K_{p1}/K_{a1}$ (subscript '1' for the entire soil, $K_{p1} = 2.157$, and $K_{a1} = 0.464$). The average CPT tip resistance q_c over the pile embedment was 14.6 MPa. The k was taken as $3q_c$ (MPa) for static loading (Guo 2012), and $kd = 17.78$ MPa (= $3 \times 14.6 \times 0.406$).

The sand exhibited similar properties to LB Sand with $\alpha_s = (3.6–25)\%$. Taking $\alpha_s = 11\%$ for cyclic loading, it follows $k_s (= \alpha_s kd) = 1.99$ MPa, $A_r = 141.1$ kN/m (with $N_g = 0.31$, $K_p = 5.289$, and $\gamma_s' = 16.27$ kN/m³), and $p_{ub}(= A_r dl) = 314.0$ kN/m. Given k_s and n, the subgrade modulus k_s (lower layer) was estimated (Guo 2012) as 9.839 MPa. Assuming a flexural stiffness $E_p I_p$ of 26.67 MPa, it follows $\lambda_2 = 0.551$, and $k_\theta = 17.6$ MNm/rad. The k_θ value is slightly higher than 15.81 MNm/radian estimated using Equation (9.11).

The prediction (see Figure 9.13a) compares well with the measured envelope curve using a soil movement profile factor α of 0.9. The cyclic load precipitated plastic hinge(s). To gain the associated bending moment and pile deformation, a reduced flexural stiffness (due to 'cracking') may be adopted. This calculation is similar to the example presented previously (Guo 2012), and is not pursued herein.

Likewise, the input parameters for the '6d' pile are determined and tabulated in Table 9.4. The BiP-k_θ model adopts $m = 2.3$, $k_s = 1.059$ MPa, $p_{ub} = 354.5$ kN/m, and $\bar{k}_\theta = 0.059$ and $\bar{k}_T = 0$. It corresponds to $\eta = 1.706$. Assuming a sliding frictional angle of 11.9° (= 0.276ϕ), it follows $m = 2.3$ (= 1.518/0.3659). The average q_c over l was 10.8 MPa, resulting in kd of 13.15 MPa (= $3 \times 10.8 \times 0.406$) for static loading, and 1.059 MPa for cyclic loading (taking $\alpha_s = 8\%$). The p_{ub} is estimated using $A_r = 159.3$ kN/m with $N_g = 0.35$, $K_p = 5.289$, and $\gamma_s' = 16.27$ kN/m³. The modulus k_2 (lower layer) is 2.357 MPa (with $m = 2.3$), $\lambda_2 = 0.386$ (with $E_p I_p = 26.67$ MPa), and $k_\theta = 10.23$ MNm/radian. In contrast, Equation (9.11) predicts a 5.4% higher k_θ value of 10.78 MNm/radian. The prediction using $\alpha = 0.3$ compares well with the measured data as well (see Figure 9.13b).

9.4 COMMENTS ON MODELLING

Under cyclic loading, subgrade modulus $k_N{}^m$ is only fraction (α_s times) of that evolved under static loading ($k_1{}^m$ of piles) with $\alpha_s = 1.1\%$ in dry Q sand, 3.6–25% ($A_k = 5–35$) in LB Sand, 0.25–0.56% ($A_k = 0.19–0.42$) in QR Silt, and 8–11% in river sand. The modulus reduction factor α_s is accord with strain levels right up to lateral spreading cases ($\alpha_s = 1\%$). The tests involve diverse normalized eccentricities, including $\bar{e}$ of 0.106–0.196 (dry sand), 1.19 (LB sand), 0.52 (QR silt) and 0.148–0.44 (river sand). It is important to note that a given load level (ζ_b) can mobilizes different slip depths ($\bar{z}_0$) [see Equation (9.7)]. Before using the results to other pile-soil systems, it is important to quantify the associated impact. For instance, in the dry Q sand and in the QR silt, the pile rotation approaches stable (with $\alpha_R > 1.0$) after the cycle loading. In LB sand, the rotation shifts from unstable state of Δu_g or $\Delta \omega$ (with $\alpha_R = 0.75$) to stable state ($\alpha_R > 1.0$) even after 10^4 cycles, for high $\bar{e}$ ratio of the tests.

It is worth noting that the assumption of $k_N = N^{\alpha R} k_1 / A_k$ is only valid within the depression zone around the pile, where the displacement and rotation are accumulated. Outside the zone, the modulus (as measured) adjusts itself to yield a balance force to support the evolution of the depression zone. After number of cycles (e.g., $N > 100$), the modulus stays as a constant of $\alpha_s N^{\alpha R} k_1$, as is evident from a constant force (pressure) along the pile embedment (Chen et al. 2015). A similar variation of the modulus k_N is evident in the tests in sand (Leblanc et al. 2010a).

Equation (9.4) offers an equivalent dimensional factor T_c (Belanc et al. 2010a).

$$T_c = 1 - \zeta_c \tag{9.15}$$

Under a reverse (two-way cyclic) loading, the rotation $\delta\omega$ for the first cycle (see Figure 9.2c) encompasses $\delta\omega_1$ (rebound) and $\delta\omega'_1$ (reverse loading) with $\delta\omega = \delta\omega_1 + \delta\omega'_1$. Ignoring the reverse loading component $\delta\omega_1$, the total residual rotation over N cycles is given by

$$\sum_{m=1}^{N} \frac{\delta\omega_m}{\omega} = A_k A_\omega \sum_{m=1}^{N} m^{-\alpha R} + A_{kc} A_{\omega c} \sum_{m=1}^{N} m^{-\alpha R} \tag{9.16}$$

where A_{kc} and $A_{\omega c}$ denote A_k and A_ω of reverse loading $\zeta_c\zeta_b$ (= H_{min}/H_m). The parameters depend on, for instance, a new normalized slip depth $\bar{z}_{oc}$

$$A_\omega = 1 + \frac{3(1+2\bar{e})(1-\bar{z}_{ob})^2}{(\bar{z}_{ob})^2 + 3(\bar{z}_{ob}-2)\bar{e} - 3} \text{ and } A_{\omega c} = 1 + \frac{3(1+2\bar{e})(1-\bar{z}_{oc})^2}{(\bar{z}_{oc})^2 + 3(\bar{z}_{oc}-2)\bar{e} - 3} \tag{9.17}$$

The equivalent dimensional factor T_c is now written as

$$T_c^u(\bar{z}_{oc}) = 1 + \frac{A_{kc}}{A_k}\frac{A_{\omega c}}{A_\omega} \tag{9.18}$$

The A_{kc}/A_k ratio is largely around unity. This yields the upper T_c (i.e., T_c^u)–ζ_c curve up to point A at $(\zeta_A, T_c^u(\bar{z}_{oc}))$ (see Figure 9.14). From point A, the T_c drops linearly with ζ_c (<0, lower line) until $T_c^l = 0$ at $\zeta_c = -1$, which is described by

$$T_c^l = T_c^u(\bar{z}_{oc})\frac{1+\zeta_c}{1+\zeta_A} \tag{9.19}$$

Tests conducted by Frick and Achmus (2020) involve $\bar{e}$ of 0.6, and $\bar{z}_o^y$ of 0.565. The loading level ζ_b of 0.35 is associated with $\bar{z}_{ob}$ of 0.18 using Equation (9.7). A reverse loading ζ_c of −0.25 extends T_c^u–ζ_c curve to point A (−0.25, 1.257) (see Figures 9.14a and b). Equations (9.18) and (9.19) are drawn as the solid lines through point A. Likewise, given $\bar{z}_o^y$ of 0.593, ζ_b of 0.2 ($\bar{z}_{ob}$ of 0.1055), and blue A (−0.2, 1.202) at $\zeta_c = -0.2$ (see Figure 9.14c), Equations (9.18) and (9.19) offer the dash lines. Given $\zeta_b = 0.5$ (i.e. $\bar{z}_{ob} = 0.275$), and black A of (−0.75, 1.759) at $\zeta_c = -0.75$, the solid lines are determined in Figure 9.14c from the equations. The calculated lines are in close agreement with the majority of measured data (e.g., Richards et al. 2020). It is worth noting that the model stipulates A_{kc}/A_k = 1.0 and $\delta\omega'_1 = 0$, while high measured T_c values may be predicted by using $A_{kc}/A_k > 1.0$ and/or $\delta\omega'_1 \neq 0$.

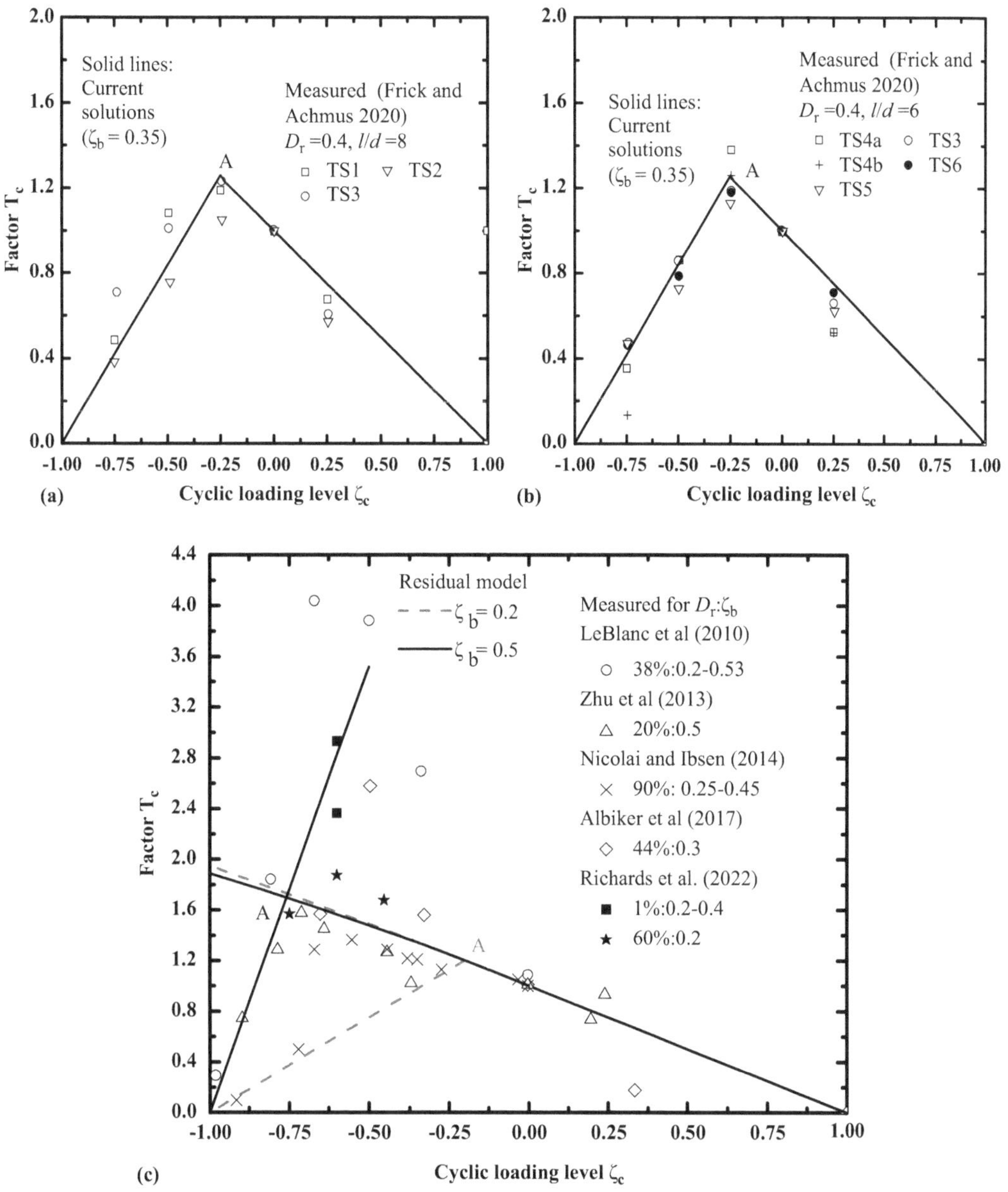

Figure 9.14 Predicted load factors T_c versus ζ_c against measured (a) l/d = 8; (b) l/d = 6 (Frick and Achmus 2020); (c) typical tests (Richards et al. 2020).

9.5 CONCLUSIONS

This chapter presents a model for estimating the residual response of piles under cyclic loading. The model builds upon the existing model for static loading and introduces new parameters to account for cyclic loading. Additionally, the BiP-k_θ model is reformulated to enable prediction of pile response under static through cyclic lateral loading in stable and sliding soil. The models are compared with four types of model tests in sand and silt, with identified salient features. The study provides impact of eccentricity, cyclic loading level and number, and stability of soil (via (α_R) and shear strain level (via A_r), and offering a range of input parameters. The model is illustrated via examples to support the use of model for practical design. Overall, the study justifies the need of the proposed expressions to incorporate the impact of loading eccentricity and shear strain levels on residual pile response.

Chapter 10

Torsional piles

Input parameters

10.1 INTRODUCTION

Laterally loaded piles are likely to undergo torsion owing to load eccentricity, as is evident in offshore mono-piles (Georgiadis 1987). Design of screw piles (Van Impe 1994) requires estimation of the torque to drive an auger into ground. Torque tests on SPT test rod (called SPT-T) may be conducted to gain soil properties (Decourt and Filho 1994; Decourt 1998; Lutenegger and Kelly 1998). Torsional tests (including on *SPT* rods) offer soil modulus and limiting strength to model vertically loaded piles (Poulos 1975; Randolph 1981, 1983), including set-up (capacity rise) behaviour (subsequent to installation) in conjunction of in situ loading tests (Axelsson 2000). In all cases, the torsional response is sensitive to the depth variation of stiffness (modulus) encapsulated as (Guo 2000a):

$$G = A_{gj}\left(\alpha_{gj} + z\right)^{n_j} \tag{10.1}$$

where G = shear modulus at depth z; α_{gj} = equivalent depth for the ground level modulus; A_{gj} = modulus constant; n_j = non-homogeneity factor; and 'j = 1' and 'j = 2' denote the upper-layer ($0 \le z \le l_s$), and the lower-layer ($z > l_s$), respectively.

This chapter provides closed-form solutions for elastic response of a single pile embedded in two-layered non-homogenous soil deposits, with each layer described by Equation (10.1) and $\alpha_{gj} \ge 0$. The solutions are presented in non-dimensional form, along with continuum-based solutions (Chow (1985) and the program GASPILE (Guo 1997; Guo and Randolph 1996, 1997a, b). Simplified expressions are also provided to define 'rigid' and 'flexible' piles, respectively. Two cases are investigated to gain insight into the differences between vertical and torsional loading piles, particularly regarding the influence of soil profile (e.g., GL stiffness and two-layers) and critical pile lengths.

10.2 CLOSED-FORM SOLUTIONS

Figures 10.1a and b depict a torsional pile embedded in a profile of single- and two-layered soil) for a constant average soil shear modulus, G_{avej} of each layer. The upper-layer has a thickness l_s and the pile has a total length l. Described by Equation (10.1), the profile at $n_j = 0$ is referred to as two-layered homogenous soil. The profile reduces to a single (lower) layer at $l_s = 0$ (see Figure 10.1a).

Numerical solutions are published for piles in single-layered soil ($n_1 = 0$, or 1, $\alpha_{g1} = 0$, and $z >$ embedment l) (Poulos 1975) and in two-layered elastic medium (described by $n_j = 0$, and

DOI: 10.1201/9781003315230-10

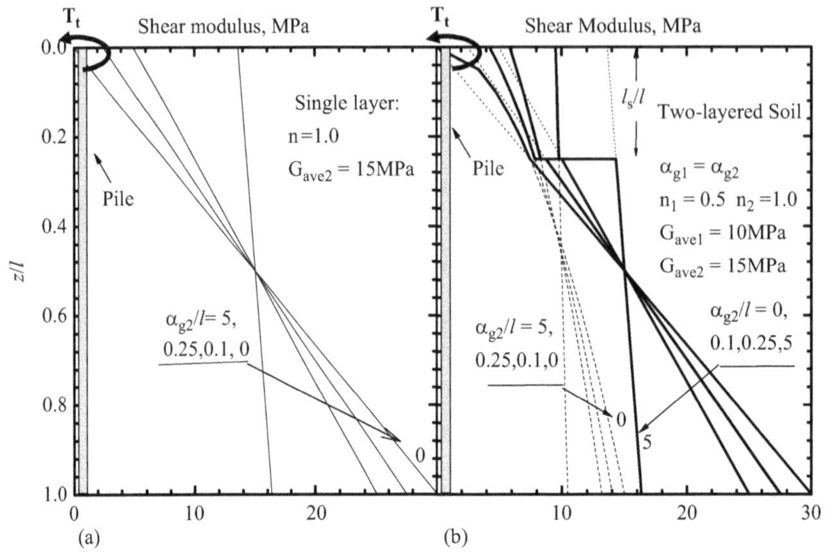

Figure 10.1 Typical soil profiles (solid line) addressed (a) single layer; (b) two-layer. [Adapted from Guo, W. D., Chow, Y. K. and Randolph, M. F., *Int J Geomechanics*, **7**(6), 2007.]

$\alpha_{gj} = 0$, Chow 1985; Hache and Valsangkar 1988). In the framework of load transfer model (with a series of independent springs distributed along the pile shaft), Table 10.1 provides the governing equation for the torsional pile in the single layer (Figure 10.1a), as well as the relationship between the angle of twist at any depth and the local shear stress. The associated closed-form solutions are available for elastic response characterized by $n_j = 0$, or 1 and $\alpha_{gj} = 0$ (Randolph 1981), and by any n_j, and $\alpha_{gj} = 0$ for elastic-plastic response (Guo and Randolph 1996). Given the general nature, the closed-form solutions for 2-layer soil by Guo et al. (2007) will be presented herein.

Table 10.1 Simplified solutions

$\dfrac{d^2\phi(z)}{dz^2} = \dfrac{\pi d^2 \tau_o}{2(GJ)_p}$	Equation (T10.1a): $\phi(z)$ = angle of twist at depth, z; τ_o = local shear stress on pile shaft; d = pile diameter; and $(GJ)_p$ = torsional rigidity of the pile.
$\phi = \tau_o/(2G)$	Equation (T10.1b): characterized by load transfer model (Guo and Randolph 1996)
$C_{\tau j}(z) = \dfrac{C_{1j}(z) + C_{2j}(z)\chi_j}{C_{3j}(z) + C_{4j}(z)\chi_j}\left(\dfrac{z+\alpha_{gj}}{l+\alpha_{gj}}\right)^{n_j/2}$	Equation (10.6): $C_{ij}(z)$ ($i = 1, 4$; and $j = 1, 2$) is expressed as modified Bessel functions, I and K, of fractional order, m_j and $m_j - 1$.
$C_{1j}(z) = -K_{m_j-1}I_{m_j-1}(y_j) + K_{m_j-1}(y_j)I_{m_j-1}$ $C_{3j}(z) = K_{m_j-1}I_{m_j}(y_j) + K_{m_j}(y_j)I_{m_j-1}$ $C_{2j}(z) = K_{m_j}I_{m_j-1}(y_j) + K_{m_j-1}(y_j)I_{m_j}$ $C_{4j}(z) = -K_{m_j}I_{m_j}(y_j) + K_{m_j}(y_j)I_{m_j}$ Equation (T10.1c)	$y_j = 2m_j\left(\pi_{\tau j}\dfrac{z+\alpha_{gj}}{l+\alpha_{gj}}\right)^{l/(2m_j)}$ (T10.1d) The base factor, χ_2 ($j = 2$) is given by $\chi_2 = \dfrac{4r_o}{3\pi}\dfrac{\pi_{\tau 2}^{l/(2m_2)}}{l+\alpha_{g2}}$ (T10.1e)

10.2.1 Solutions for piles in 2-layered soil (Guo et al. 2007)

10.2.1.1 Governing equations

Embedded in 2-layer soil [Equation (10.1)], governing equation for the torsional piles may be deduced from Equation (T10.1a) (see Table 10.1) and rewritten respectively, as

$$\frac{d^2\phi(z)}{dz^2} = \eta_g \left(\frac{\pi_{t2}}{l+\alpha_{g2}}\right)^{1/m_2} (z+\alpha_{g1})^{n_1}\phi(z)\ (0 \le z \le l_s) \tag{10.2}$$

$$\frac{d^2\phi(z)}{dz^2} = \left(\frac{\pi_{t2}}{l+\alpha_{g2}}\right)^{1/m_2} (z+\alpha_{g2})^{n_2}\phi(z)\ (l_s \le z \le l) \tag{10.3}$$

where $\phi(z)$ = angle of twist at depth, z; $m_j = 1/(2 + n_j)$ (j = 1, and 2 unless otherwise specified); $\eta_g = A_{g1}/A_{g2}$; and the pile-soil relative stiffness ratio, π_{tj} is defined by

$$\pi_{tj} = \left[\frac{4\pi r_o^2 A_{gj}}{(GJ)_p}\right]^{m_j} (l+\alpha_{gj}) \tag{10.4}$$

In Equation (10.2), the stiffness ratio for the upper-layer, π_{t1} is expressed through the lower-layer, π_{t2} and the ratio η_g.

10.2.1.2 Equivalent base factor at the depth of l_s

The influence of the lower layer is incorporated into the properties of the upper layer by an 'equivalent base factor'. The ratio of torque $T(z)$ over the angle of twist $\phi(z)$ at depth l_s is given by Guo and Randolph (1996) as

$$\frac{T(z)}{\phi(z)} = \frac{(GJ)_p}{l+\alpha_{gj}} \pi_{tj}^{1/(2m_j)} C_{tj}(z) \tag{10.5}$$

where $l_s \le z \le l$ (j = 2 concerning the lower-layer); $0 \le z \le l_s$ (j = 1 in handling the upper layer). The coefficient, $C_{tj}(z)$ is given by

$$C_{tj}(z) = \frac{C_{1j}(z) + C_{2j}(z)\chi_j}{C_{3j}(z) + C_{4j}(z)\chi_j} \left(\frac{z+\alpha_{gj}}{l+\alpha_{gj}}\right)^{n_j/2} \tag{10.6}$$

where $C_{ij}(z)$ (i = 1, 4; and j = 1, 2) is expressed as modified Bessel functions, I and K, of fractional order, m_j and $m_j - 1$, underpinned by argument, y_j and non-dimensional base factor, χ_2 (j = 2) (see Table 10.1).

Three points need to be emphasized: (i) At the pile-base ($z = l$), $T(l)/\phi(l) = 16r_o^3G_L/3$; (ii) χ_2 of Equation (T10.1e) synthesizes the relative effect of the base stiffness over the shaft stiffness; and (iii) The continuity of $T(l_s)/\phi(l_s)$ at depth l_s is maintained, which, based on Equation (10.5) for upper- or lower- layer, offers an equivalent 'base' factor, χ_1

$$\chi_1 = \frac{\dfrac{C_{t2}(l_s)}{\sqrt{\eta_g}} \dfrac{(l+\alpha_{g2})^{n_2/2}}{(l_s+\alpha_{g1})^{n_1/2}} C_{31}(l_s) - C_{11}(l_s)}{-\dfrac{C_{t2}(l_s)}{\sqrt{\eta_g}} \dfrac{(l+\alpha_{g2})^{n_2/2}}{(l_s+\alpha_{g1})^{n_1/2}} C_{41}(l_s) + C_{21}(l_s)} \tag{10.7}$$

where $C_{i1}(l_s)$ is $C_{ij}(z)$ ($i = 1...4$) at depth l_s for the upper layer ($j = 1$); $C_{t2}(l_s)$ stands for $C_{t2}(z)$ at depth l_s for the lower-layer. For a multi-layered soil, the equivalent factor for each layer can be assessed from the pile base until the χ_1 is found (see Guo et al. 2007).

10.2.2 Torsional influence factor

The torque (T_t) and the angle of twist at the pile head (ϕ_t) are correlated with a pile-head torsional influence factor, I_ϕ by

$$\phi_t = I_\varphi \frac{l+\alpha_{g1}}{(GJ)_p} T_t \tag{10.8}$$

With Equation (10.5), the influence factor I_ϕ is recast into

$$I_\phi = \left(\pi_{t1}^{1/(2m_1)} C_{t1o}\right)^{-1} \tag{10.9}$$

where C_{t1o} = limiting value of the $C_{t1}(z)$, as z approaches zero. The factor I_ϕ is computed using the overlying soil properties that incorporate non-dimensional parameters π_{tj}, α_{gj}/l, and n_j of each layer, and the ratio l_s/l. The underlying layer is embodied via the exclusive factor χ_1. Particularly, Equation (10.4) implies that the π_{tj} may also be represented through the A_{gj}, or via the ratios of π_{t2} and η_g.

10.2.3 Validation and salient features

Equation (10.9), based on the equivalent base factor χ_1, is sufficiently accurate for estimating I_ϕ compared to continuum-based numerical approaches (Guo and Randolph 1996). The I_ϕ reduces to $l_s/(l + \alpha_{g1})$ for very low modulus of upper layer (see Figure 10.2a). It agrees with the finite element analysis (FEA) and the discrete element analysis (DEA) (Chow 1985), as well as the load-transfer (GASPILE) analysis (Guo and Randolph 1996, 1997a,b) for piles in a single layer of low- or upper-layer properties (at $l_s/l = 0$ or 1.0) and a two-layered soil ($0 < l_s/l < 1.0$) with $\alpha_{gj} = 0$, $n_j = 0$ and $\pi_{t2} = 4.472$. The I_ϕ never exceed the 'upper limit'.

10.3 SIMPLE SOLUTIONS

Simplified expressions have been developed for modelling 'rigid and 'flexible' piles, respectively (Guo et al. 2007).

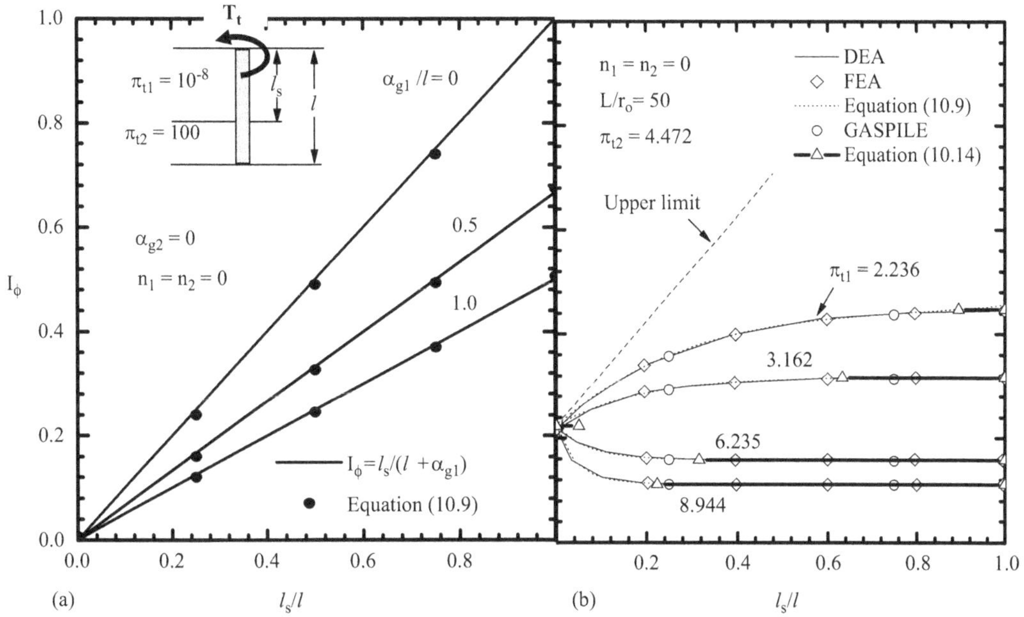

Figure 10.2 Verification of current solutions by (a) simplified; (b) numerical approaches. [Adapted from Guo, W. D., Chow, Y. K. and Randolph, M. F., *Int J Geomechanics*, **7**(6), 2007.]

10.3.1 Rigid piles

For rigid piles, the angle of twist is identical along pile length (see Table 10.2). Neglecting pile-base resistance (thus $\chi_2 = 0$), Equations (10.2) and (10.3) are integrated to offer

$$\frac{1}{I_\varphi} = \frac{\pi_{t1}^{2+n_1}}{1+n_1}\left(\left(\frac{\bar{\alpha}_{g1}+\bar{l}_s}{\bar{\alpha}_{g1}+1}\right)^{1+n_1} - \left(\frac{\bar{\alpha}_{g1}}{\bar{\alpha}_{g1}+1}\right)^{1+n_1}\right) + \frac{\pi_{t2}^{2+n_2}}{1+n_2}\left(\frac{\bar{\alpha}_{g1}+1}{\bar{\alpha}_{g2}+1}\right)\left(1-\left(\frac{\bar{\alpha}_{g2}+\bar{l}_s}{\bar{\alpha}_{g2}+1}\right)^{1+n_2}\right) + \chi_2\frac{\bar{\alpha}_{g1}+1}{\bar{\alpha}_{g2}+1} \quad (10.10)$$

where $\bar{\alpha}_{g1} = \alpha_{g1}/l$, $\bar{\alpha}_{g2} = \alpha_{g2}/l$, and $\bar{l}_s = l_s/l$.

Example 10.1 Rigid piles in a single-layered soil

A rigid pile is embedded in a single (lower) layer ($l_s = 0$). Its torsional I_ϕ can be estimated using Equation (T10.2b) provided in Table 10.2. The equation also offers a critical stiffness ratio π_t^* of Equation (T10.2c) at $I_\phi^* = 4\{1 - [\alpha_g/(\alpha_g + l)]^{(1+n)}\}$, e.g., at $\alpha_g = 0$, $\pi_t^* = 0.5$ ($n = 0$), 0.676 (0.5) and 0.794(1.0) (see dot symbols in Figure 10.3). Equation (10.10) or (10.2b) is sufficiently accurate (see Figure 10.3) for $\alpha_g/l = 0$, 1, 3, and 5 within $\pi_t < \pi_t^*$ (rigid pile); but it underestimate the I_ϕ [as inferred from Figure 10.3(b) and 10.3(c)] for $\pi_t > \pi_t^*$. Equation (10.9) excludes the fraction of base resistance (on safe side), compared to Equation (10.10) underpinned by a real χ_2.

Table 10.2 Simplified solutions

$I_\phi = \dfrac{(GJ)_p \phi_t}{T_t(l+\alpha_{g1})} = \dfrac{l_s}{l+\alpha_{g1}}$	Equation (T10.2a): $k \approx 0$ (Upper-layer) [see Figure 10.2a]
$I_\phi = \dfrac{1+n}{\pi_t^{2+n}}\left(1-\left(\dfrac{\alpha_g}{\alpha_g+l}\right)^{1+n}\right)^{-1}$	Equation (T10.2b): rigid piles in low layer: With $l_s = 0$, $\alpha_{gj} = \alpha_g$, $n_j = n$, and $\chi_2 = 0$ in Equation (10.10)
$\pi_t^* = \left(\dfrac{1+n}{4}\right)^{1/(n+2)}\left(1-\left(\dfrac{\alpha_g}{\alpha_g+l}\right)^{1+n}\right)^{-2/(2+n)}$	Equation (T10.2c): Critical stiffness ratio deduced from Equation (T10.2b)
$\pi_{t1}^* = \dfrac{\left(\dfrac{1+n_1}{4}\right)^{m_1}\left(1-\left(\dfrac{\bar{\alpha}_{g1}}{\bar{\alpha}_{g1}+1}\right)^{1+n_1}\right)^{-m_1}}{\left(\left(\dfrac{\bar{\alpha}_{g1}+\bar{l}_s}{\bar{\alpha}_{g1}+1}\right)^{1+n_1}-\left(\dfrac{\bar{\alpha}_{g1}}{\bar{\alpha}_{g1}+1}\right)^{1+n_1}\right)^{m_1}}$ (T10.2d)	$\pi_{t2}^* = \left(\dfrac{\dfrac{1+n_2}{4}\left(\dfrac{\bar{\alpha}_{g2}+1}{\bar{\alpha}_{g1}+1}\right)}{1-\left(\dfrac{\bar{\alpha}_{g2}+\bar{l}_s}{\bar{\alpha}_{g2}+1}\right)^{1+n_2}}\right)^{m_2}\dfrac{1}{\left(1-\left(\dfrac{\bar{\alpha}_{g2}}{\bar{\alpha}_{g2}+1}\right)^{1+n_2}\right)^{m_2}}$ (T10.2e)
Critical stiffness ratio deduced from Equation (10.10) and taking $\chi_2 = 0$	
$\dfrac{T_t}{G_c r_0^3 \phi_t} = \dfrac{\sqrt{2}\pi}{1+n}\sqrt{G_p/G_c}$ & $l_c \approx 2l/\pi_t$	Flexible piles with $l > l_c$, Equations (T10.2f), (T10.2g)

Example 10.2 Rigid piles in two-layered soil

For rigid piles in two-layered soil with $\pi_{tj} < \pi_{tj}^*$, Equation (10.10) is recommended, and for $\pi_t < \pi_t^*$ (a single layer). The corresponding π_{tj}^* for the I_ϕ^* at j-th layer is presented in Table 10.2, except for the singularity (at $l_s \neq 0$) or single layer ($l_s \neq l$). For example, given $l_s/l = 0.25$, $\alpha_{gj} = 0$, and $n_1 = n_2$, Equations (T10.2d and e) offer $\pi_{t1}^* = 1.0$ ($n_1 = 0$) and 2.0 ($n_1 = 1$), and $\pi_{t2}^* = 0.58$ ($n_2 = 0$), and 0.81 ($n_2 = 1$). Using Equation (10.10) and assuming $\chi_1 = \chi_2 = 0$, the associated I_ϕ^* for $\pi_{t2} = 0.4$ is estimated as 3.077 ($n_1 = 0$) and 3.571 ($n_1 = 1$); whereas the I_ϕ^* for $\pi_{t1} = 0.4$ is estimated as 3.42 ($n_1 = 0$) and 3.982 ($n_1 = 1$) (see dot symbols in Figure 10.4).

For n (= n_j) = 0, 0.5, and 1, Equation (10.9) offers the π_{t1}–I_ϕ curves at π_{t2} = 0.4, 1.0, 5.0 and 50 (see Figure 10.4a); and the π_{t2}–I_ϕ curves at π_{t1} = 0.4, 1.0, 5.0 and 15 (see Figure 10.4b). Taking $\chi_2 = 0$, Equation (10.10) offers accurate I_ϕ values for $\pi_{t2} = 0.4$ ($< \pi_{t2}^*$) and $\pi_{t1} < \pi_{t1}^*$, as well as π_{t1} = 0.4, and 1.0 ($\leq \pi_{t1}^*$) and $\pi_{t2} \leq \pi_{t2}^*$, respectively; otherwise, it underestimates I_ϕ values (not shown) at $\pi_{t2} = 1.0$ ($> \pi_{t2}^* = 0.58$–0.81).

10.3.2 Flexible piles

For a torsional, flexible pile, Equations (T10.2f and g) provide pile-head stiffness and a maximum (critical) length l_c ($\approx 2l/\pi_t$, or $\pi_t = \pi_t' = 2$) (Randolph 1981; Guo and Randolph 1996). A '$\pi_t > \pi_t'$ (= 2)' implies '$l > l_c$'. An equivalent factor, n_e for $\alpha_g/l \neq 0$ was introduced to Equation (10.1) (Guo 2000a):

$$n_e = (1+n)\left(1+\bar{\alpha}_g - \frac{\bar{\alpha}_g^{\,1+n}}{(1+\bar{\alpha}_g)^n}\right)^{-1} - 1 \qquad (10.11)$$

where $\bar{\alpha}_g = \alpha_g / l$. The torsional factor, I_ϕ at $\pi_t > \pi_t'$ is subsequently deduced from the stiffness as

$$I_\phi = \frac{1+n_e}{\pi_t^{(2+n)/2}} \left(\frac{1+\bar{\alpha}_g}{2/\pi_t + \bar{\alpha}_g} \right)^{n/2} \tag{10.12}$$

Equation (10.12) offers a low bound (at a large $\bar{\alpha}_g$, Figure 10.3) of Equation (10.9) as

$$I_\phi = \left(\pi_t^{(2+n)/2} \right)^{-1} \tag{10.13}$$

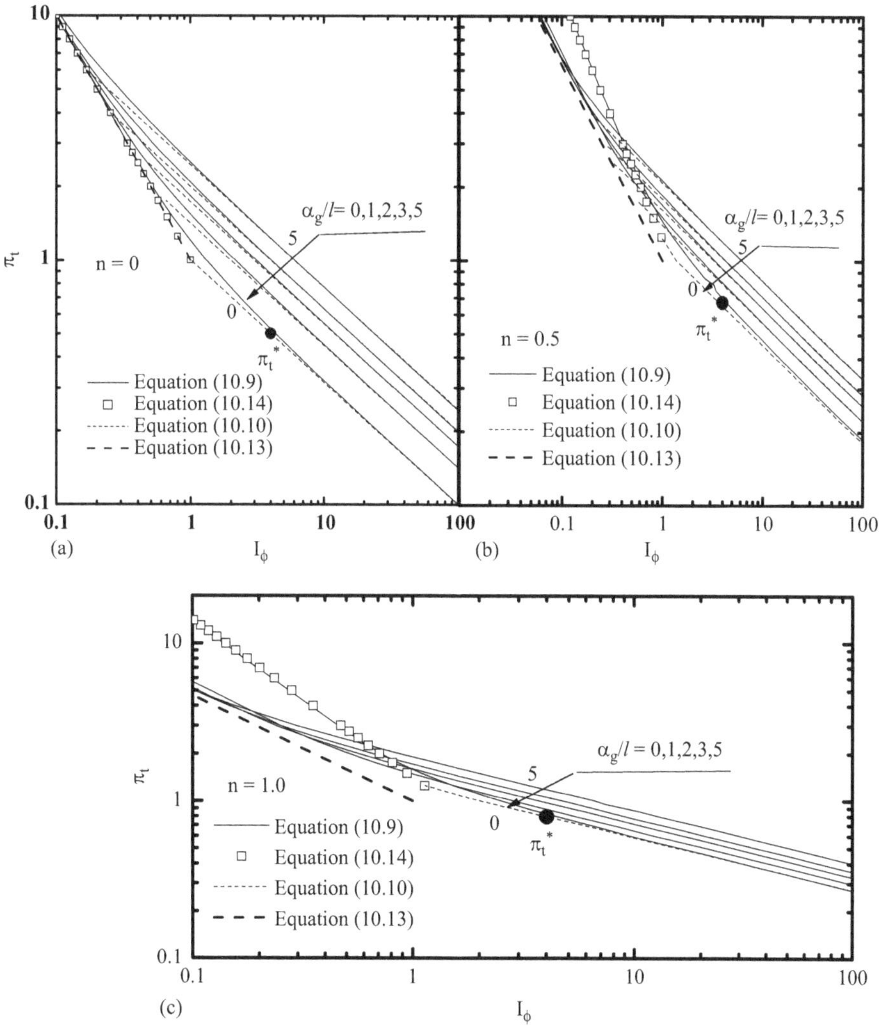

Figure 10.3 Elastic influence factor vs the relative stiffness relationship. [Adapted from Guo, W. D., Chow, Y. K. and Randolph, M. F., *Int J Geomechanics*, **7**(6), 2007.]

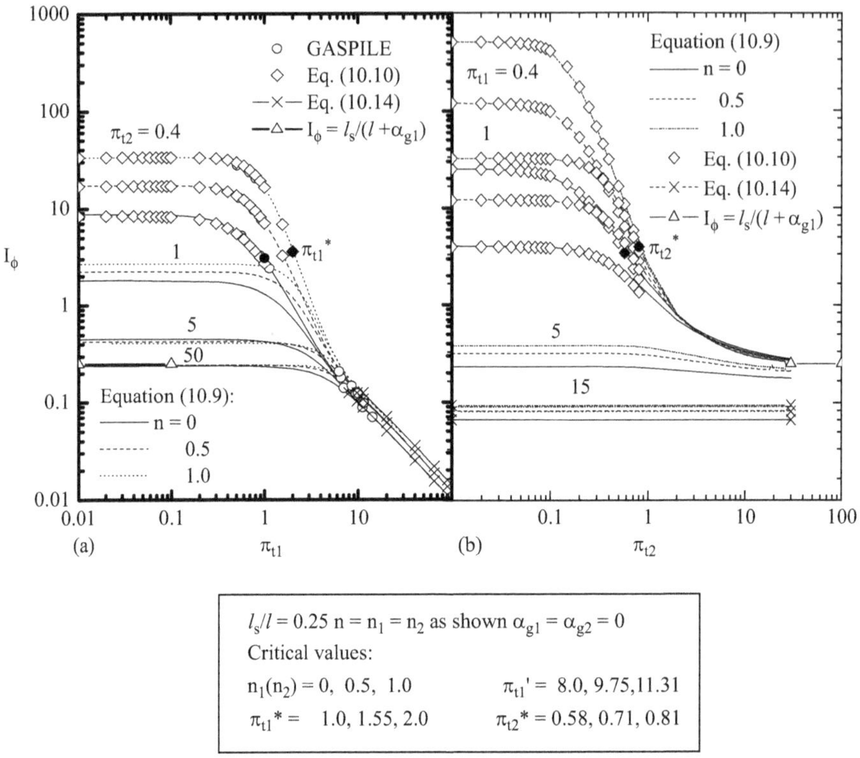

Figure 10.4 Torsional influence factor by different approaches ($l_s/l = 0.25$) [Adapted from Guo, W. D., Chow, Y. K. and Randolph, M. F., *Int J Geomechanics*, **7**(6), 2007.]

The I_ϕ–π_t curve of Equation (10.13) passes through point (1,1) at a gradient of $-(n + 2)/2$. At a low $\bar{\alpha}_g$ (Guo et al. 2017) with $n_e \approx n$, Equation (10.12) also offers an upper bound of

$$I_\phi = \frac{1+n}{2^{n/2}\pi_t} \tag{10.14}$$

Equations (10.14) and l_c (= $2l/\pi_t$) are consistent with the numerical results presented previously in Figure 10.2(b) and Table 10.3.

Table 10.3 Comparison of I_ϕ estimated using Equations (10.14) and (10.9)

Input parameters		*Equation (10.9)*	*Equation (10.14)*	*Notes*
$\pi_{t2} = 4.472$, $\alpha_{gi}/l = 0$, $n_j = 0$		0.2236	0.2236	I_ϕ at $l_s/l = 0$ in Figure 10.2b
$\pi_{t2} = 15.0$	$n_j = 0$	0.0667	0.0667	I_ϕ at $l_s/l = 0$ in Figure 10.4b
$\alpha_{gi}/l = 0$	$n_j = 0.5$	0.0827	0.0841	
	$n_j = 1.0$	0.0915	0.0943	
$\alpha_{gi}/l = 0$	$\pi_{t2} = 1.0$	1.313	1.313	Consistent with Figure 10.3a
$n_j = 0$	$\pi_{t2} = 3.0$	0.335	0.333	
	$\pi_{t2} = 6.0$	0.1667	0.1667	
	$\pi_{t2} = 10.0$	0.10	0.10	
	$\pi_{t2} = 15.0$	0.0667	0.0667	

Example 10.3 Flexible piles in two-layered soil and comments on l_c

The upper-layer thickness beyond the critical depth l_c (i.e., the portion of l_s-l_c) does not affect the head stiffness, as indicated by the symbol 'x' ($\pi_{t1} \geq 8.0$–11.3) in Figure 10.4a for $\pi_{t1} > 2l/l_s$. The impact of l_s/l on the torsional factor was examined extensively via Equation (10.9) (Guo et al. 2007).

For a single lower layer (at $l_s = 0$), Equation (10.14) is accurate (see Table 10.3). To avoid singularity, $l_s = 10^{-8}$ is adopted in using Equation (10.9). Given $\pi_t = 5$, and $n = n_j = 0.5$, Equations (10.13), and (10.14) offer lower and upper limits of I_ϕ of 0.1337 and 0.252, respectively. The two equations do bracket all the I_ϕ values at $l_s = 0$ and $\bar{\alpha}_{g1} = 0 - 1.0$ shown in Figure 10.5b.

10.3.3 Upper rigid, low flexible piles

A pile may behave extremely rigid in the upper-layer, but flexible in the lower-layer (i.e., an upper rigid, lower flexible pile). The associated I_ϕ involves a component of $\bar{l}_s/(1+\bar{\alpha}_{g1})$ and another one from Equation (10.12) (with parameters n_2, $\bar{\alpha}_{g2}$, and π_{t2}).

$$I_\phi = \frac{1}{1+\bar{\alpha}_{g1}}\left(\bar{l}_s + \frac{1+n_{e2}}{\pi_{t2}^{(2+n_2)/2}}\left(\frac{1+\bar{\alpha}_{g2}}{2/\pi_{t2}+\bar{\alpha}_{g2}}\right)^{n_2/2}\left(1+\bar{\alpha}_{g2}\right)\right) \tag{10.15}$$

where n_{e2} is given by setting $n = n_2$ and $\alpha_g = \alpha_{g2}$ in Equation (10.11). For a low π_{t2}, the pile involves an appreciable twist angle in the lower layer (see Figure 10.2a). Equation (10.15) agrees with Equation (10.9), provided $I_\phi < I_\phi'$ with I_ϕ' given by

$$I'_\varphi = \frac{0.5\left(\sqrt{1+n_{e2}}+\bar{l}_s\right)}{1+\bar{\alpha}_{g1}} \tag{10.16}$$

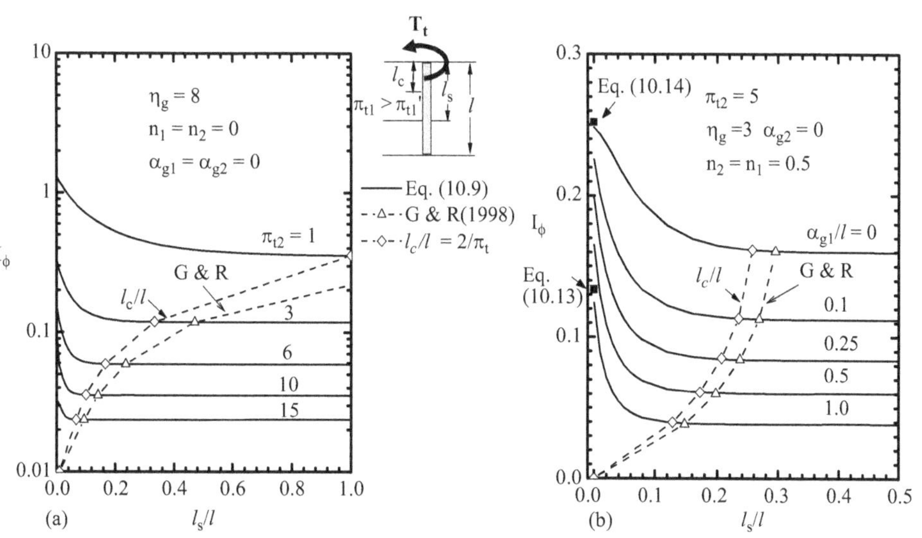

Figure 10.5 Effect of π_{t2} and α_{g1}/l on the critical depth l_s/l. [Adapted from Guo, W. D., Chow, Y. K. and Randolph, M. F., *Int J Geomechanics*, **7**(6), 2007.]

Equation (10.15) offers an *upper limit* of $I_\phi = 1/\pi_{t2} + \bar{l}_s$ for Figure 10.2b given $I_\phi \le 0.2236 + \bar{l}_s$; and an $I_\phi = \bar{l}_s/(1+\bar{\alpha}_{g1})$ at a high $\pi_{t2} = 50$ (see Figures 10.4a and b).

Rigid or flexible piles are classified using critical stiffness ratios, as it is not straightforward to estimate the critical lengths iteratively. Guo et al. (2007) provides example calculations of the critical length, l_1^* (rigid in upper-layer), l_2^* (rigid in lower-layer), and l_c (flexible). For instance, l_c (lower-layer) is obtained by using an extremely low π_{t1} for the upper-layer.

Example 10.4 Piles of intermediate stiffness ($\alpha_{gj} = 0$)

Equation (10.9) or GASPILE is adopted to estimate the torsional factor for a stiffness π_{t1} of $\pi_{t1}^* < \pi_{t1} < \pi_{t1}'$, e.g., $\pi_{t1} = 1.55–9.75$ at $n = (n_j =)\ 0.5$ (see Figure 10.4a). Equation (10.9) offers the I_ϕ for $\alpha_{gj} = 0$, three $\bar{l}_s$ ratios and five n_j values. The I_ϕ is plotted against A_{g1}/A_{g2} concerning $n_1 = n_2$ in Figure 10.6a; and various n_1 at $n_2 = 0$ in Figure 10.6b. The figures demonstrate: (i) Each logarithmic plot of I_ϕ versus A_{g1}/A_{g2} has a gradient of $-1/(2 + n)$ for flexible piles; (ii) Equation (10.14) may be used to estimate I_ϕ at $\pi_{t1}\left[=(A_{g1}/A_{g2})^{1/(2+n)}\pi_{t2}\right] > 2/\bar{l}_s$. e.g., when $A_{g1}/A_{g2} > 2.56$ ($n_j = 0$) or > 4.1 ($n_j = 1$) at $L/L_s = 0.25$ and $\pi_{t2} = 5.0$; and (iii) Equation (10.9) generally agrees well with the program GASPILE, except for $\pi_{t2} > 5$, $A_{g1}/A_{g2} > 1$, $\bar{l}_s \ge 0.5$, and $n_1 > 0$ ($n_2 = 0$). The exception leads to '$l_c < l_s$' or no effect from the low layer (Guo et al. 2007), for which Equation (10.9) is preferred to GASPILE.

Example 10.5 Effect of α_{gj}

Equation (10.9) is useful for estimating the factor I_ϕ, as shown in Figures 10.3a, b and c for a single-layered soil with $n = 0–1.0$ at $\bar{\alpha}_g = 0-5$; and in Figure 10.5b for two-layered soil ($n_j = 0.5, \bar{\alpha}_{g2} = 0$) with $\bar{\alpha}_{g1} = 0-1.0$. It also provides I_ϕ values for a pile ($l = 12.5$ m, and $r_o = 0.25$ m, and $\chi_2 = 0$) embedded in a soil profile ($\bar{l}_s = 0.25$, $n_1 = 0$, $n_2 = 1.0$, and $\bar{\alpha}_{g1} = 1.6$, see Figure 10.7) for varying lower layer $\bar{\alpha}_{g2}\ (=0-1.6)$ at π_{t2} of 0.4–15. The critical stiffness ratios are estimated as $\pi_{t1}^* = 2.6$, $\pi_{t1}' = 20.8$, and $\pi_{t2}^* = 0.59\ (\bar{\alpha}_{g2} = 0)$, 0.87 (0.8), and 0.95 (1.6), respectively.

- I_ϕ increases with an increase in $\bar{\alpha}_{g2}$ (see Figure 10.7) and/or n_1 ($= n_2$) (see Figure 10.6a). Otherwise, at a high $\pi_{t2} \ge 15$ and $\pi_{t1} < \pi_{t1}^*$ (a soft upper-layer), I_ϕ reduces with an increase in $\bar{\alpha}_{g2}$, as is the I_ϕ gained using Equation (10.15) and shown in Table 10.4. For instance, the I_ϕ at $\pi_{t2} = 15$ and $\bar{\alpha}_g\ (=\bar{\alpha}_{g2}) = 0$ is equal to 0.1324 [< I_ϕ' of 0.368 obtained using Equation (10.16)] with n_{e2} (= n_e) = 1.0 estimated using Equation (10.11).
- Equation (10.14) sufficiently accurate for $\pi_{t1} > \pi_{t1}'$ (i.e., $l_c < l_s$), and for $\alpha_{gj}/l = 0$ in Figure 10.4.

Equation (10.9) predicts increased I_ϕ with an increase in $\bar{\alpha}_{g2}\ (=0-1.6)$ (see Figure 10.8) for sets of $\bar{l}_s\ (=0.25-0.75)$, and A_{g1}/A_{g2} (= 0.1–10) at $\pi_{t2} = 0.4–15$ ($n_1 = 0, n_2 = 1, \bar{\alpha}_{g1} = 1.6$), but a reduced I_ϕ at $\pi_{t2} = 15$ ($>\pi_{t2}^*$) (see Figure 10.7) and a different range of π_{t1} and A_{g1}/A_{g2}. The prediction using Equations (10.10) and (T10.2b) ignored the pile-base stiffness (by taking $\chi_2 = 0$) (Guo et al. 2007)

10.4 ANALYSIS OF TORSIONAL TESTS

Stoll (1972) reported torsional load tests on pile A-3 and V-4 (out diameter 0.273 m and wall thickness 6.3 mm). These pipe piles, back-filled with concrete, had a torsional rigidity $(GJ)_p$

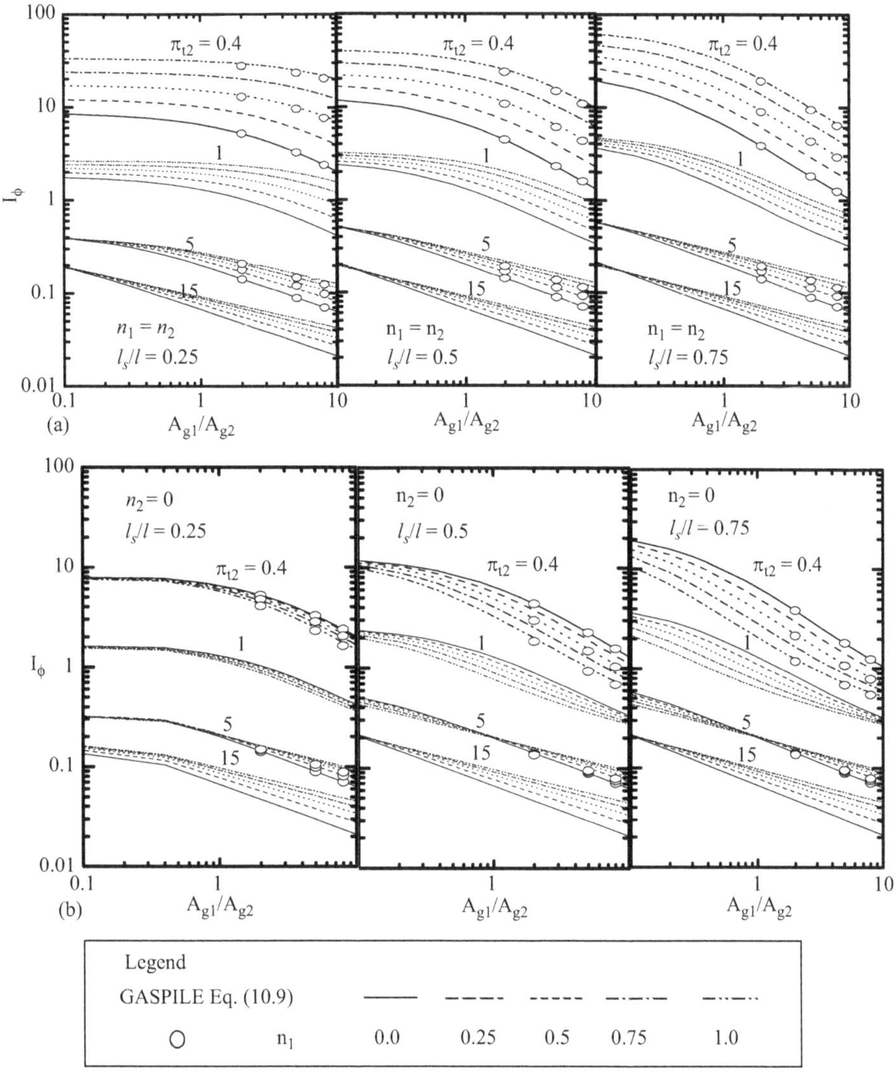

Figure 10.6 Effect of upper layer soil profile on torsional influence (a) n_1 (= n_2) different; (b) n_1 different (n_2 = 0). [Adapted from Guo, W. D., Chow, Y. K. and Randolph, M. F., *Int J Geomechanics*, **7**(6), 2007.]

Table 10.4 I_ϕ determined from Equation (10.15)

Input parameters		n_{e2}	I_ϕ	*Notes*
π_{t2} = 4.472, α_g/l = 0, n_j = 0		0	0.2236 + l_s/l	*Upper limit* of I_ϕ in Figure 10.2b
π_{t2} = 15.0, α_{g1}/l = 1.6, n_2 = 1.0, l_s/l = .25	α_{g2}/l = 0	1.0	0.1324	Figure 10.7 ($I\phi < I\phi'$)
	α_{g2}/l = 1.6	0.2381	0.1222	
π_{t2} = ∞, n_2 = 1.0, l_s/l = .25	α_g/l = 1.6	0.2381	0.0961	Equation (10.15) = Equation (T10.2a) in Figure 10.7

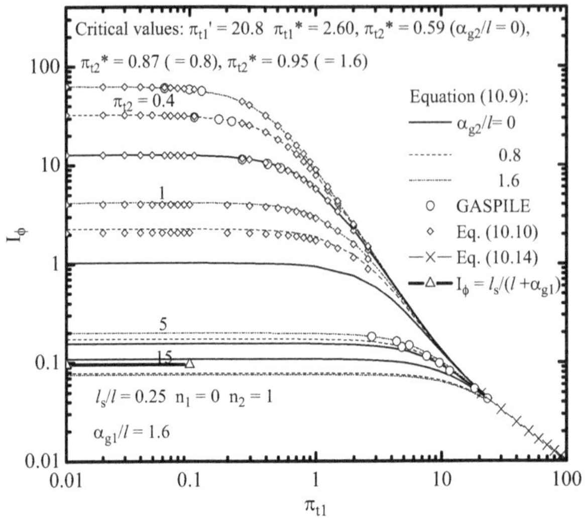

Figure 10.7 Influence of α_{g2}/l on $I\phi$ (l_s/l = 0.25, L/r_o = 50, n_1 = 0, n_2 = 1, α_{g1}/l = 1.6). [Adapted from Guo, W. D., Chow, Y. K. and Randolph, M. F., *Int J Geomechanics*, **7**(6), 2007.]

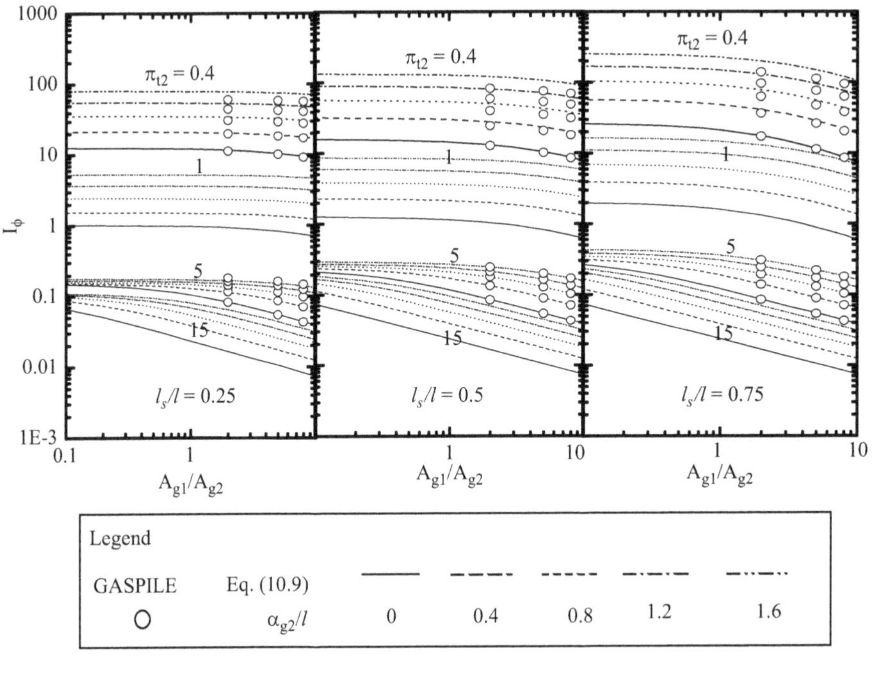

Figure 10.8 Effect of α_{g2} on relationship torsional influence factor (n_1 = 0, n_2 = 1, α_{g1}/l = 1.6, A_{g1}/A_{g2} different). [Adapted from Guo, W. D., Chow, Y. K. and Randolph, M. F., *Int J Geomechanics*, **7**(6), 2007.]

of 12.8 MNm^2, and G_p of 2.35 × 10^4 MPa. The tests were simulated using the solutions (Guo and Randolph 1996, Guo et al. 2007).

Pile A-3 was driven to a depth l of 17.4 m, over which $N \approx 1.38z$ (N = blow count of SPT, z = depth in m), and shear modulus increases linearly with depth (n = 1). On the other hand,

Pile V-4 was driven to l of 20.7 m, with $N \approx 0$ (z = 0–2.4), and $N \approx 2.62(z - 2.4)$. The pile involves $\pi_{t1} = 0$ for the upper layer (2.4 m), and a lower-layer stiffness being proportional to $(z - 2.4)$.

For pile A-3, the measured $T_t/(r_0\phi_t)$ was 20 MN/rad, which corresponds to a torsional influence factor I_ϕ of 0.26946 using Equation (10.8) and l = 17.4 m. Given $n = 1$ and $\alpha_g = 0$, Equation (10.14) offers $\pi_t = 5.248$, while Equation (10.4) gives A_g = 1.5 MPa/m. The l_c/r_0 of 48.6 [= 2.0 × 17.4/(0.273/2 × 5.2483)] is lower than 127.5 (of l/r_0). The pile was indeed flexible, which justify the use of Equation (10.14).

For Pile V-4, the measured $T_t/(r_0\phi_t)$ was 15 MN/rad, which gives I_ϕ (at pile-head level) = 0.302. Equation (10.15) offers an I_ϕ value (at l_s = 2.4 m) of $0.1860\left(= 0.302 - l_s, l = 20.7\,\text{m}\right)$. The π_t was estimated as 7.601 given $I_\phi = \sqrt{2} / \pi_t$ [with $\alpha_{gj} = 0$, see Equation (10.12)]. Subsequently, it follows A_g = 2.70 MPa/m, and l_c/r_0 = 40. Again, with $l_c/r_0 < l/r_0$ of 151.65 (= 18.3/0.273 × 2), the Pile V-4 is deemed as flexible. The results can also be deduced by using Equation (10.14) and the responses at the depth of 2.4 m (i.e., l = 18.3 m, I_ϕ = 0.2109, and π_t = 6.706, Guo and Randolph 1996).

The A_g value back-figured offers Young's modulus, E_s of ~$3N$ MPa (ν_s = 0.4, pile A-3), and $2.88N$ MPa (pile V-4), respectively. These values are comparable to $3N$ MPa for shallow foundations (Parry 1978), but are lower than $4N$ MPa reported for vertically loaded piles (Poulos 1975) and $7N$ MPa (Shibata et al. 1989). The length l^* for a rigid pile and effective pile l_c were 2.63 m and 6.63 m (pile A-3), and 2.17 m and 5.46 m (pile V–4). Both lengths l_c and l^* resemble influence depth of a shallow foundation.

10.5 PARAMETERS BETWEEN TORSIONAL AND VERTICAL LOADING

10.5.1 Set-up behaviour between torque rod and vertically loaded pile

Axelsson (2000) conducted a series of tests at a site located 20 km southwest of Stockholm to investigate the use of torque tests for evaluating the setup behaviour of vertically loaded piles following installation. The tests include CPT tests, static and dynamic vertical loading tests on rods and piles and a series of torque tests on driven (SPT) rods.

The test site consists of clay and sand layers. A clay layer with an average cone resistance q_c of 1 MPa extends to a depth z of 2.5 m below GL, and beneath it lies a silty sand layer that is sandwiched by a sand layer at z = 4.5–8 m (see Figure 10.9). The q_c gradually increases from 0.2 MPa at z = 2.5 m to 6 MPa at z = 6 m; and remains at approximately 2.5 MPa over z = 6–12 m.

A solid steel *SPT* rod (32 mm in diameter) was driven to a depth of 11.7 m. The rod has $(GJ)_p$ of 21.618 kNm^2 with G_p of 210 GPa, and $J_p = 1.02944 \times 10^{-7}$ m^4 ($= \pi d^4/32$). The rod was tested 1 hour, 1 day, 38 days, and 231 days, respectively subsequent to installation. The circumferential displacement of the rod (at GL) for applied torque shear were recorded, which are plotted in Figures 10.10a and b as dotted solid lines. The elastic torsional influence factor was evaluated using Equation (10.14), and is tabulated in Table 10.5. Adjacent to the torque rod, a square concrete pile D (0.235 m in width) was driven to a depth of 12.8 m and vertically loaded to failure 1, 8, 120, and 667 day(s) after installation, respectively. The measured load-displacement curves are plotted in Figure 10.10c.

The measured curves were matched using Equation (10.9) and/or program GASPILE (e.g., Guo 1997) by adjusting the time-dependent α_c and α_f for shear modulus G ($= \alpha_c q_c$, q_c in MPa), as well as limiting shaft friction τ_f ($= \alpha_f q_c$). The results are presented as G/G_{initial} and $\tau_f/\tau_{f\,\text{initial}}$ ratios ('initial' being taken as 1 hour and 1 day for the rod, and the pile D, respectively). The

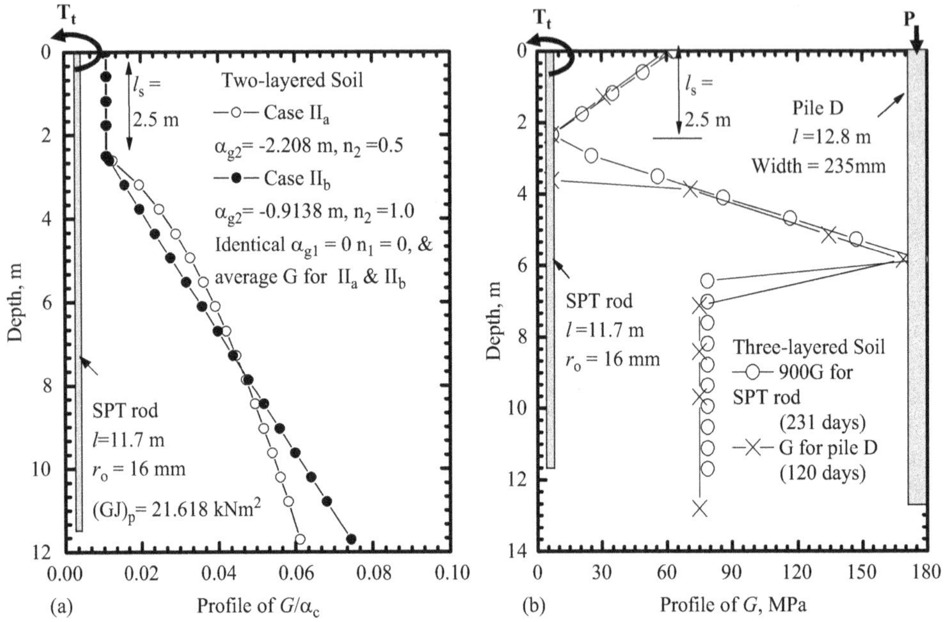

Figure 10.9 (a) Two-layered shear modulus profiles for SPT rod; (b) Profile of G for pile D and 900G for SPT rod.

back-estimation was conducted by stipulating the soil as a single-layered, two-layered and three-layered medium, respectively, in the form of five profiles of shear modulus G and τ_f profile.

10.5.2 Set-up effect on shear modulus from torsional rod

10.5.2.1 Single-layered soil – impact of l^* and l_c

The soil was first crudely treated as a homogeneous ($\alpha_g = 0$, $n = 0$) single-layered medium (Axelsson and Westin 2000). The match between Equation (10.9) and the measured I_ϕ of the rod ($r_o = 0.016$ m, $l = 11.7$ m) [see Figure 10.10a] offered a modulus G of 31.5–50.5 kPa (Table 10.5, with $\alpha_c = G/1{,}000$ MPa). For example, the one-hour test I_ϕ of 1.878 involves $\pi_t = 0.801$, in which $\chi_1 = 1.259 \times 10^{-4}$, $C_{t1o} = 0.665$, and $G = 31.5$ kPa. The base stiffness ratio T_b/ϕ_b (of 0.0028 kN/rad) is only 0.07% the head stiffness ratio T_t/ϕ_t (of 4.07 kN/rad). Similar calculation was undertaken for other tests and is provided in Table 10.5 as well.

It is crucial to justify methods of back-estimation to avoid misleading results. The rod tested in soil ($A_g = 31.5$ kPa) involves $\pi_t = 0.8011$, $l^* = 7.3$ m (using Equation T10.2c, Table 10.2), and $l_c = 29.2$ m ($= 2L/\pi_t$), respectively. With $l^* < l$ (= 11.7 m) $< l_c$ (see Table 10.5), the rod is neither rigid nor flexible.

Rigid pile/rod observes non-linear correlation between torque shear and displacement (Guo and Randolph 1996) of

$$\frac{T_t}{A_g l^n r_o^3 \varphi_t} = \frac{16}{3} + \frac{4\pi}{1+n}\left(\frac{l}{r_o}\right)\frac{-R_f \tau_o / \tau_f}{\mathrm{Ln}\left(1 - R_f \tau_o / \tau_f\right)} \tag{10.17}$$

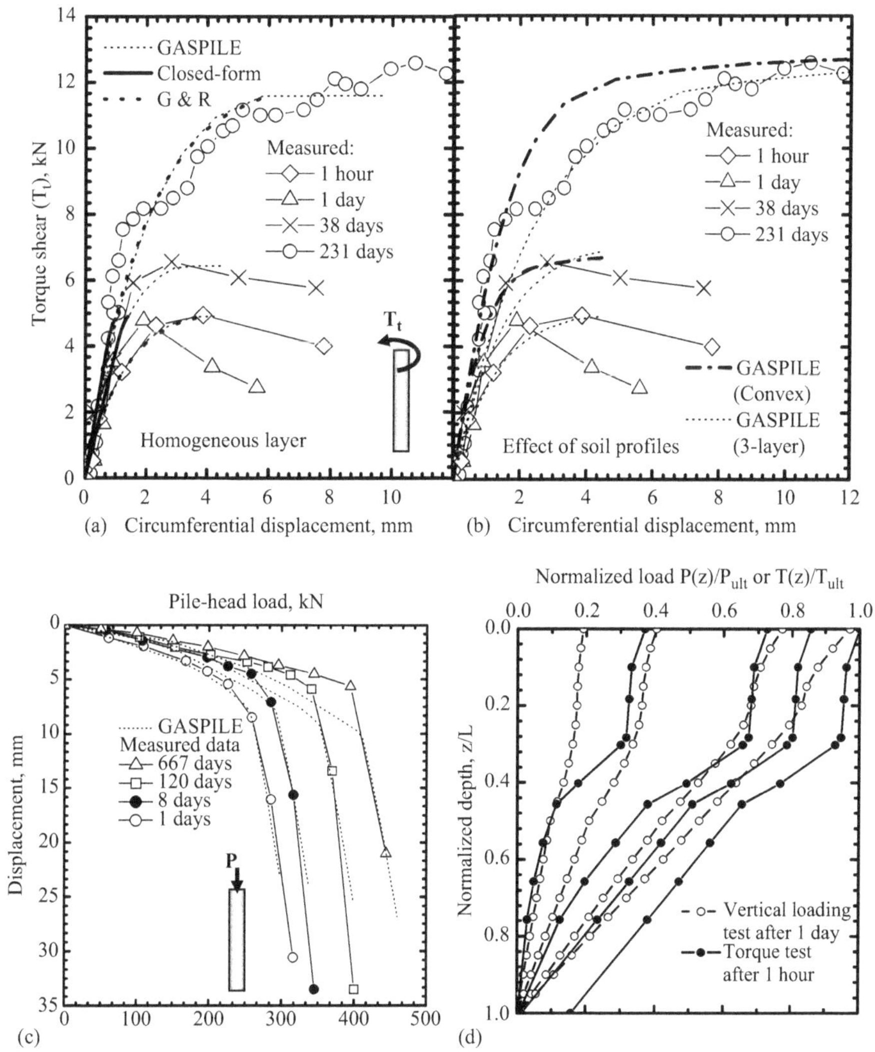

Figure 10.10 Comparison between the measured (Axelsson 2000) and the predicted (a)(b) load and circumferential displacement; (c)(d) load and displacement of Pile D and load distribution between the torque test and the vertical loading test (GASPILE).

Table 10.5 Parameters deduced from torque tests (homogeneous profiles)

Time (days)	*1/24 (= 1 hr)*	*1*	*38*	*231*
I_ϕ by Equation (10.8)	1.878	1.543	1.294	1.284
G^* (kPa) by Equations (10.9) and (10.14)	31.5(13.9)[a]	40(20.7)	50(29.3)	50.5(31.2)
τ_f (kPa) by GASPILE	4.16	4.10	5.46	9.83
$G/G_{\text{at 1 hr}}$	1	1.27	1.59	1.60
$\tau_f/\tau_{f \text{ at 1 hr}}$	1	1	1.31	2.36
l^*/l_c (m)	7.30/29.21	6.48/25.92	5.80/23.19	5.77/23.07

Notes:
[a] Values in parentheses are obtained using Equation (10.14). Using q_c = 1 MPa, τ_c = 1,000α_f kPa, $^*G$ = 1,000α_c kPa

Table 10.6 Predicted results of torque tests using G & R solutions

	Test at 1 hour (T_{ult} = 4.95 kN, G = 1.732MPa, R_f = 0.95)				Test on 231 days (T_{ult} = 11.57 kN[a], G = 2.64MPa, R_f = 0.95)			
T_t (kN)	0.204	2.656	3.882	4.904	5.301	8.193	10.121	11.567
$R_f\tau_o/\tau_f$	0.040	0.515	0.752	0.95	0.4354	0.673	0.831	0.95
$T_t/(\phi_t r_o)$	3997	2904	2200	1295	4734	3743	2905	1973
$\phi_t r_o$(mm)	0.05	0.92	1.76	3.79	1.12	2.19	3.48	5.86

Notes:

[a] This value is lower than the measured one of 13.06kN, in order to be consistent with GASPILE analysis. However, use of the latter would not lead to much different G.

and $\tau_o/\tau_f = T_t/T_{ult}$ (T_{ult} = ultimate T_t) at $\alpha_g = 0$, and $n = 0$. Given measured T_{ult} = 4.95 kN (1 hour test), and 13.06 kN (231-days), $R_f\tau_o/\tau_f = 0.04$ at T_t of 0.204 kN (of 1 hour test), Equation (10.17) offers $T_t/(\phi_t r_o) = 3997$ ($A_g = G = 1.732$ MPa), and $\phi_t r_o$ (circumferential displacement) = 0.05 mm. The calculation was repeated for a series of T_t (torque shear) to gain the displacement ($\phi_t r_o$, see Table 10.6), and the plot in Figure 10.10a as G & R. Likewise, the 'G & R' prediction for the test on 231 days is made using $G = 2.64$ MPa (i.e. available G/q_c ratio given by Lunne and Christoffersen 1985). These moduli and the 'G & R' prediction would be acceptable, but the 'rod' was not deemed as rigid with $l^* = 0.789$ m at $G = 2.64$ MPa. The deduced modulus of 1.73–2.64 MPa is, surprisingly, 52–55 times the 'true' values (see Table 10.5) deduced using Equation (10.9).

Taking the rod flexible, G is deduced as 13.9–31.2 kPa for 1/24–231 days using Equation (10.14). For instance, the one-hour test involves $I_\phi = 1.878$, $n = 0$, and $\pi_t = 0.5325$. Equation (10.14) offers A_g (= G) = 13.92 kPa; and the π_t gives $l_c = 43.9$ m. During set-up, the length l_c reduces from 43.9 m (1 hour) to 29.3 m (231 days), but due to the fact of '$l < l_c$' Equation (10.14) cannot be used. As a result, the values of 13.9–31.2 kPa are 40–60% of the 'real values' (Table 10.5) derived using Equation (10.9). This study warns against use of simplified expressions to deduce modulus.

10.5.2.2 Two-layered soil

Secondly, the soil was assumed as a two-layered medium with an interface located at $l_s = 2.5$ m. Given identical average q_c over the lower-layer, two shear modulus profiles, Case II$_a$ and II$_b$, are adopted (see Figure 10.9)

$$G = \alpha_c \ (0 \le z \le l_s), G = 2.935\alpha_c (z - 2.208)^{0.5} \quad (l_s \le z \le l)(\text{Case II}_a) \tag{10.18a}$$

$$G = \alpha_c \quad (0 \le z \le l_s), G = 0.63\alpha_c (z - 0.9138) \quad (l_s \le z \le l)(\text{Case II}_b) \tag{10.18b}$$

where G is in MPa, z is in m and α_c is dimensionless. Using Equation (10.9), the prediction is matched with the measured I_ϕ using either profile (as exhibited in Figure 10.10a) for each test to deduce the α_c (see Table 10.7). Compared to a single layer (Table 10.5), Table 10.7 shows that (i) α_c reduces by 65% [= (31.5–10.93)/31.5, Case II$_b$], which is misleadingly close to the invalid value deduced using Equation (10.14). (ii) G increases by 10–20%, e.g., 13.7% [= (35.8–31.5)/31.5] at one-hour test; (iii) l_c increases by 60–70%; and (iv) ratio α_c (= G/q_c) of 0.01093–0.01901 (Case II$_b$) is less than 1% what was suggested by Lunne and Christoffersen (1985).

Table 10.7 Back-estimated from torque tests for Cases II_a and II_b profiles

Time (days)	*1/24 (1 hr)*	*1*	*38*	*231*
$\alpha_c \times 10^3$	10.73˙/10.93@	14.07/14.40	18.26/18.78	18.46/19.01
$\alpha_f \times 10^3$	548.˙/548.@	550./550.	430./388.	235./225.
Average G (kPa)	35.2˙/35.84@	46.2/47.24	59.9/61.6	60.6/62.36
Average τ_f (kPa)	4.48˙/4.48@	4.47/4.47	5.71/6.33	10.46/10.92
π_{t1} by Equation (10.4)	0.468˙/0.472@	0.535/0.542	0.610/0.619	0.613/0.622
π_{t2} by Equation (10.4)	0.924˙/1.088@	1.03/1.192	1.143/1.303	1.148/1.308
χ_1 by Equation (10.7)	1.396˙/1.368@	1.77/1.72	2.33 /2.235	2.361/2.266
l_1^* by Equation (T10.2d)	62.63˙/61.51@	47.76/46.67	36.8/35.78	36.4/35.35
l_2^* by Equation (T10.2e)	2.21˙/8.35@	2.21/7.68	2.21/7.09	2.21/7.07
l_c ($\approx 2l/\pi_t$)#	50.05˙/49.60@	43.71/43.20	3837/37.83	38.16/37.60
$G/G_{at\ 1hr}$	1.0˙/1.0@	1.29/1.32	1.70/1.72	1.72/1.77
$\tau_f/\tau_{f\ at\ 1hr}$	1.0˙/1.0@	1.0/1.0	1.27/1.41	2.33/2.44

Notes:
1. ˙Case II_a and @ Case II_b using Equations (10.18a) and (10.18b).
2. $\pi_{t1}^*, \pi_{t1}' = 1.082, 9.36$ for either n_2, but $\eta_g = 0.54$˙/1.586@
3. $\pi_{t2}^* =$ NA˙/0.78@
4. χ_2 by Equation (T10.1e) = (6.48–8.5) × 10^{-4}•/(7.14–9.42) × 10^{-4}@
5. # using upper-layer properties.

It is important to note that the rod with $\pi_{t1} < \pi_{t1}^* = 1.082$ (i.e., $l < l_1^*$) is rigid in the overlying clay layer (<2.5 m), but neither rigid nor flexible (i.e., $l_2^* < l < l_c$) in the underlying low silty sand and sand layers. During the set-up period, the lower layer has a slightly reduced impact on the rod response, as factor χ_1 increase by 66% [e.g., from 1.368 to 2.266 Case II_b], and G and τ_f increase by 74 and 134%, while base contribution (χ_2 of an order of 10^{-3}) is negligible. To derive soil parameters, Equation (10.9) may be employed, with due consideration of layered soil properties.

10.5.2.3 Three-layered soil

Finally, sensitivity of the back-estimation to the profiles was examined using GASPILE. Taking the 3-layered CPT (q_c) profile (see Figure 10.9), measured T_t versus displacement curves (Figure 10.10b) were matched for both the convex portion (i.e., Case III_a) and overall trend (Case III_b) using GASPILE (Convex) and GASPILE (3-layer), respectively. The α_c was also deduced using the closed-form solutions for a multi-layered soil to gain the influence of the second and bottom layer via equivalent factors χ_2 and χ_3.

- For Case III_a the deduced α_c values were 0.033 (38 days) and 0.035 (231 days), and an average G (over the rod penetration) of 78.6 and 83.3 kPa, respectively.
- For Case III_b, the α_c values deduced were 0.014, 0.0145, 0.018 and 0.018, and average G of 33.5, 34.7, 43.1 and 43.1 kPa, for the four tests performed following the installation.

Compared to the results presented in Tables 10.7 and 10.5, the average G values for Case III_a are 30–40 and 65% higher; while those for Case III_b are 15–55 % lower. These results indicate the deduced results using the two-layered profiles (see Table 10.7) are typical.

Table 10.8 I_ϕ values from GASPILE and Equation (10.9) (two-layered profiles)

Time (days)	*1/24 (1 hr)*	*1*	*38*	*231*
a): $I\phi$ by Equation (10.9)	1.878[a]/1.878[b]	1.543/1.543	1.294/1.294	1.284/1.284
b): $I\phi$ by GASPILE	1.826[a]/1.802[b]	1.497/1.483	1.258/1.255	1.247/1.240
c): $(a-b)/a \times 100(\%)$	2.76[a]/4.0[b]	2.98/3.89	2.78/3.0	2.88/3.43

Notes:
[a] Case II_a.
[b] Case II_b.

10.5.2.4 Closed-form solutions and GASPILE

The agreement between Equation (10.9) and the non-linear GASPILE prediction is up to 50% ultimate torque, with only 3–4% difference in the predicted I_ϕ values for Cases II_a and II_b and use identical α_c (see Table 10.8). While the equation is efficient in assessing the impact of a lower layer via the equivalent factor χ_1, GASPILE provides additional information of limiting shaft stress τ_f and response of the rod right up to failure.

10.5.3 Set-up response in modulus from vertically loaded pile

The modulus values for vertically loaded pile D were deduced using GASPILE by assuming (1) a uniform set-up rate along depth (Axelsson 2000), and (2) proportional values of shear modulus and limiting shaft to the three-layered CPT profile (e.g., Figure 10.9). Given E_p = 20 GPa, the measured load-displacement curves were matched using GASPILE (Figure 10.10c), but for overestimation of pile-head displacement at high load-levels (or from 8 days onwards). The overestimation was ignored concerning stress, as it is attributed to stress hardening in lateral earth pressure, and a relatively higher set-up rate of stiffness and shaft friction of silty sand layer (z = 8–11 m) (Axelsson 2000). A typical profile of the modulus for 120 days has been plotted in Figure 10.9b as 'G for pile D'.

There is 900 times discrepancy between the deduced modulus for pile D and torsional SPT rod (see Table 10.7). The inconsistency is due to two facts: (i) The load was transferred to a distance of $3d$ from a rod/pile axis and within a depth of $0.3l$ under torsional loading, but to around $75d$ from the axis and along entire length under vertical loading (Guo and Randolph 1996) (see Figure 10.10d, note that the load $P(z)$ or torque $T(z)$ at depth z normalized by ultimate load P_{ult} or torque T_{ult} was determined from GASPILE.); and (ii) The elastic displacement limit of 1.0–1.5 mm (e.g., on 38-day test) for torsional test (see Figure 10.10a) was much lower than 5 mm observed for the vertical loading case (see Figure 10.10c), as is evident in other tests.

10.5.4 Limiting shaft stress τ_f from torque rod and pile

The GASPILE program was used to deduce the limiting stress τ_f (also α_f) for a single-layered medium (see Table 10.5) and a two-layered medium [Table 10.7, with $\alpha_c = \alpha_f$ and G as τ_f in Equations (10.18a) and (10.18b)]. In the three-layered Case III_b the τ_f (over the rod embedment) was deduced as 4.48, 4.48, 6.33 and 10.92 kPa, for the four tests with α_f of 0.548, 0.548, 0.388, and 0.225, respectively.

The deduced τ_f values fluctuate by 16% for each test with τ_f = 4.16–4.48 kPa (1 hour), 4.1–4.47 kPa (1 day), 5.46–6.33 kPa (38-days), and 9.83–10.92 kPa (231-days) subsequent

Table 10.9 Parameters deduced from vertically loaded pile D (3-layered profiles)

Time (days)	*1*	*8*	*120*	*667*
G (MPa)/α_c	27.63 /12.0	50.66/22.0	69.1/30.0	92.0/40.0
τ_f (kPa)/α_f	48.0/1.737	54.73/1.08	65.27/0.945	76.53/0.832
$G/G_{(at\ 1day)}$	1	1.833	2.501	3.330
$\tau_f/\tau_{f\ (at\ 1day)}$	1	1.152	1.392	1.632

to the installation. These values are 10–20% higher (for the one-hour test), or 100–200% lower than the respective τ_f values (i.e., 3.7, 15.0, 19.0 and 22.0 kPa, Axelsson 2000). The deduced ratio $\tau_{f\ pile}/\tau_{f\ rod}$ reaches 8–14, with τ_f ($\tau_{f\ pile}$) obtained from pile D (see Table 10.9) and $\tau_{f\ rod}$ from the torsional rod (Table 10.5). The $\tau_{fpile}/\tau_{f\ rod}$ ratio is far higher than unity (Lutenegger and Kelly 1998) and 2–4 (Axelsson 2000), highlighting the importance of using a justified back-estimation method.

Further comparison among Tables 10.5, 10.7, and 10.9 indicates that over the set-up period, the regain rate around vertically loaded pile was similar to that of the rod for the $\tau_f/\tau_{f\ initial}$ ratio, and 1.5–2 times faster for $G/G_{initial}$. The regain rate is provided Figure 10.11, in which the normalized increases in G, τ_f, and capacity are given by $(G - G_{initial})/(G_\infty - G_{initial})$, $(\tau_f - \tau_{f\ initial})/(\tau_{f\ max} - \tau_{f\ initial})$, and $(Q_s - Q_{s\ initial})/(Q_{s\ max} - Q_{s\ initial})$, respectively. Q_s is shaft capacity gained from the loading testing results shown in Figure 10.10c. The terms 'initial' and 'max' refer to those observed at 1 hour, and on 231 days for a torsional rod; or those on 1 day and 667 days for a vertically loaded pile, respectively. The time 't' refers to the time of tests; and t_{90} is simply taken as 231 days and 667 days for the torsional tests and vertical loading tests, respectively.

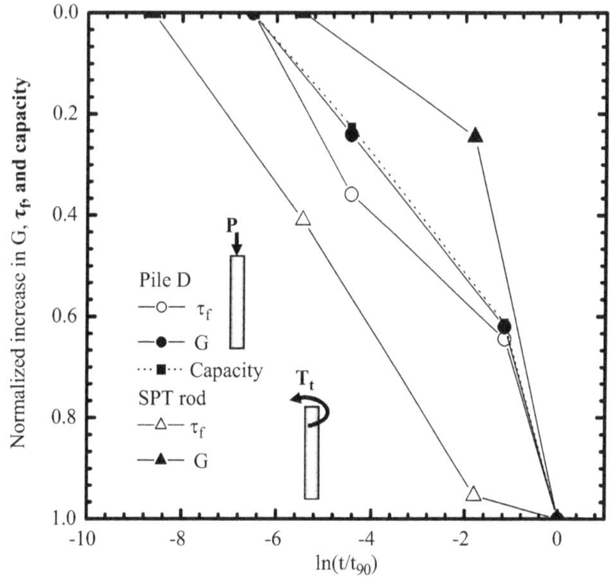

Figure 10.11 Set-up response for pile D and SPT rod.

The gradient of each load-displacement curve (of pile D) ahead of failure is quite similar for all tests. This implies a constant base resistance over the reconsolidation. The increase in total capacities was considered as the difference between Q_s and $Q_{s\ initial}$.

Overall, alternative back-estimation methods could end up with three orders and one order of magnitude difference in modulus and limiting stress. Equation (10.9) is sufficiently accurate for back-estimating the modulus and predicting elastic torsional response; while the GASPILE is flexible for estimating the limiting stress, incorporating nonlinear response.

10.6 CONCLUSIONS

The determination of input parameters such as limiting strength and subgrade modulus is critical to design of piles, and test data can be used to back-estimate them. However, the accuracy of the back-estimation under torsional loading must be justified by critical lengths. This chapter provides analytical solutions for a torsional pile embedded in a two-layered soil profile, which are expressed in terms of Bessel functions of non-integer order, and simplified expressions for rigid and flexible piles. In addition, a spreadsheet program GASPILE is also available for gaining numerical solutions. The solutions are presented in non-dimensional solutions, and can be used to deduce the modulus and strength (friction) between torsional and vertically loaded pile. Examples are provided for the back-estimating of the input parameters.

References

Abdoun, T., Dobry, R., O'Rourke, T. D., and Goh, S. H. 2003. Pile response to lateral spreads: Centrifuge modeling. *J Geotech Geoenviron Eng, ASCE* **129**(10): 869–878.

Abdoun, T., and Wang, Y. 2002. Modelling of pile foundations retrofitting strategies against seismically induced lateral spreading. *Int J Phys Model Geotech* **3**: 17–28.

Achmus, M., Kuo, Y. S., and Abdel-Rahman, K. 2009. Behavior of monopile foundations under cyclic lateral load. *Comput Geotech* **36**: 725–735.

ACI. 1993. Bridges, substructures, sanitary, and other special structures-structural properties. In *ACI Manual of Concrete Practice, Part 4*. Detroit.

Alba, P. D., and Ballestero, T. P. 2006. Residual strength after liquefaction: A rheological approach. *Soil Dyn Earthq Eng* **26**: 143–151.

Al-Douri, R. H., and Poulos, H. G. 1995. Predicted and observed cyclic performance of piles in calcareous sand. *J Geotech Eng, ASCE* **121**(1): 1–16.

Anderson, K. H., Murff, J. D., Randolph, M. F., Clukey, E. C., Erbrich, C. T., Jostad, H. P., Hansen, B., Aubeny, C., Sharma, P., and Supachawarote, C. 2005. Suction anchors for deepwater applications. In *Proc 1st Int Symp on Frontiers in Offshore Geotechnics*. Perth, Australia, Tayler and Francis.

API. 1993. *RP-2A: Recommended Practice for Planning, Designing and Constructing Fixed Offshore Platforms*, Washington, D. C, American Petroleum Institute.

Armstrong, R. J., Boulanger, R. W., and Beaty, M. H. 2014. Equivalent static analysis of piled bridge abutments affected by earthquake-induced liquefaction. *J Geotech Geoenviron Eng, ASCE* **140**(8): 04014046.

Aubeny, C. P., Han, S. W., and Murff, J. D. 2003. Inclined load capacity of suction caissons. *Int J Numer Anal Methods Geomech* **27**(14): 1235–1254.

Axelsson, G. 2000. *Long-term set-up of driven piles in sand.* Doctoral thesis, Royal Institute of Technology, Stockholm, Sweden.

Axelsson, G., and Westin, A. 2000. Torque tests on driven rods for prediction of pile set-up. In New technological and design development in deep foundations, *Proc GeoDenver 2000*, Denver, Colorado.

Bang, S., and Cho, Y. 2001. Ultimate horizontal capacity of sunction piles. *Proc 11th Int Offshore and Polar Engrg Conf.* Stavanger, Norway.

Bang, S., Jones, K., Kim, Y. S., and Cho, Y. 2011. Horizontal capacity of embedded suction anchors in clay. *J Offshore Mech Arct Eng, ASME* **133**(2): 0111041.

Berrill, J. B., Christensen, S. A., Keenan, R. P., Okada, W., and Pettinga, J. R. 2001. Case study of lateral spreading forces on a piled foundation. *Geotechnique* **51**(6): 501–517.

Bhattacharya, S., Bolton, M. D., and Madabhushi, S. P. G. 2005. A reconsideration of the safety of piled foundations in liquefied soils. *Soils Found* **45**(4): 13–24.

Boulanger, R. W., Kutter, B. L., Brandenberg, S. J., Singh, P., and Chang, D. 2003. Pile foundations in liquefied and laterally spreading ground during earthquakes: Centrifuge experiments and analyses. Rep. No. UCD/CGM-03/01, Center for Geotech. Modeling, Univ. of California, Davis. California, USA.

Bourne-Webb, P. J., Potts, D. M., Konig, D., and Rowbottom, D. 2011. Analysis of model sheet pile walls with plastic hinges. *Geotechnique* **61**(6): 487–499.

Brandenberg, S. J., Boulanger, R. W., and Kutter, B. L. 2005a. Discussion of "Single piles in lateral spreads: Field bending moment evaluation" by R. Dobry, T. Abdoun, T. D. O'Rourke, and S. H. Goh. *J Geotech Geoenviron Eng, ASCE* **131**(4): 529–534.

Brandenberg, S. J., Boulanger, R. W., Kutter, B. L., and Chang, D. 2005b. Static pushover analyses of pile groups in liquefied and laterally spreading ground in centrifuge tests. *J Geotech Geoenviron Eng Div, ASCE* **133**(9): 1055–1066.
Bransby, M. F., and Springman, S. M. 1996. 3-D finite element modelling of pile groups adjacent to surcharge loads. *Comput Geotech* **19**(4): 301–324.
Bransby, M. F., and Springman, S. M. 1997. Centrifuge modelling of pile groups adjacent to surcharge loads. *Soils Found* **37**(2): 39–49.
Brinch Hansen, J. 1961. *The ultimate resistance of rigid piles against transversal forces*. The Danish Geotech. Institute Bulletin No. 12: Copenhagen, Denmark.
Broms, B. B. 1964. Lateral resistance of piles in cohesive soils. *J Soil Mech Found Eng, ASCE* **90**(2): 27–63.
Brown, D. A., Morrison, C., and Reese, L. C. 1998, Lateral load behaviour of pile group in sand. *J Geotech Geoenviron Eng, ASCE* **114**(11): 1261–1276.
BSI. 1985. Structural use of concrete. BS 8110, London, Parts 1–3.
Cai, F., and Ugai, K. 2003. Response of flexible piles under laterally linear movement of the sliding layer in landslides. *Can Geotech J* **40**(1): 46–53.
Carrubba, P., Maugeri, M., and Motta, E. 1989. Esperienze in vera grandezza sul comportamento di pali per la stabilizzaaione di un pendio. *Proc 17th Convegno Nazionale di Geotechica, Assn. Geotec,* Italiana.
Chai, J. C., Shen, S. L., Ding, W. Q., Zhu, H. H., and Carter, J. P. 2014. Numerical investigation of the failure of a building in Shanghai, China. *Comput Geotech* **55**: 482–493.
Chai, Y. H., and Hutchinson, T. C. 2002. Flexural strength and ductility of extended pile-shafts. II: Experimental study. *J Struct Eng, ASCE* **128**(5): 595–602.
Chen, L. T., and Poulos, H. G. 1997. Piles subjected to lateral soil movements. *J Geotech Geoenviron Eng, ASCE* **123**(9): 802–811.
Chen, L. T., Poulos, H. G., and Hull, T. S. 1997. Model tests on pile groups subjected to lateral soil movement. *Soils Found* 7(1): 1–12.
Chen, L. T., Poulos, H. G., Leung, C. F., Chow, Y. K., and Shen, R. F. 2002. Discussion of "Behavior of pile subject to excavation-induced soil movement". *J Geotech Geoenviron Eng, ASCE* **128**(3): 279–281.
Chen, R. P., Sun, Y. X., Zhu, B., and Guo, W. D. 2015. Lateral cyclic deformation and pile-soil interaction studies on rigid model monopiles. *Geotech Eng (ICE)* **168**(2): 120–130.
Chmoulian, A. 2004. Briefing: Analysis of piled stabilization of landslides. *Proc Institution of Civil Engineers, Geotech Engrg* **157**(2): 55–56.
Choo, Y. W., Kim, D., Park, J. H., Kwak, K., Kim, J. H., and Kim, D. S. 2014. Lateral response of large-diameter monopiles for offshore wind turbines from centrifuge model tests. *Geotech Test J, ASTM* **37**(1): 1–14.
Chow, Y. K. 1985. Torsional response of piles in non-homogeneous soil. *J Geotech Eng Div, ASCE* **111**: 942–947.
Chow, Y. K. 1996. Analysis of piles used for slope stabilization. *Int J Numer Anal Methods Geomech* **20**(9): 635–646.
Coyle, H. M., and Reese, L. C. 1966. Load transfer for axially loaded piles in clay. *J Soil Mech Found Eng Div, ASCE* **92**(2): 1–26.
Cubrinovski, M., Haskell, J., Winkley, A., Robinson, K., and Wotherspoon, L. 2014. Performance of bridges in liquefied deposits during the 2010–2011 Christchurch, New Zealand, Earthquakes. *J Perform Constr Facil, ASCE* **28**(1): 24–39.
Cubrinovski, M., Kokushob, T., and Ishihara, K. 2006. Interpretation from large-scale shake table tests on piles undergoing lateral spreading in liquefied soils. *Soil Dyn Earthq Eng* **26**: 275–286.
De Beer, E., and Carpentier, R. 1977. Discussion on 'methods to estimate lateral force acting on stabilising piles' By Ito, T., and Matsui, T. (1975). *Soils Found* **17**(1): 68–82.
Decourt, L. 1998. A more rational utilization of some old in situ-tests. *Proc 1st Int Conf on Site Characterization*, Atlanta, Georgia, **2**: 913–918.
Decourt, L., and Filho, A. R. Q. 1994. Practical applications of the standard penetration test complemented by torque measurements, SPT-T; present stage and further trends. *Proc 13th Int Conf on Soil Mech and Found Engrg*, New Dehli, India, **1**: 81–112.

Dobry, R., Abdoun, T., O'Rourke, T. D., and Goh, S. H. 2003. Single piles in lateral spreads: Field bending moment evaluation. *J Geotech Geoenviron Eng, ASCE* **129**(10): 879–889.
Doyle, E. H., Dean, E. T. R., Sharma, J. S., Bolton, M. D., Valsangkar, A. J., and Newlin, J. A. 2004. Centrifuge model tests on anchor piles for tension leg platforms. *Proc Offshore Technology Conf, Paper No. OTC. 16845*. Houston, Texas, USA.
Duncan, J. M., Evans, L. T., and Ooi, P. S. K. 1994. Lateral load analysis of single piles and drilled shafts. *J Geotech Eng, ASCE* **120**(5): 1018–1033.
Duncan, J. M., Robinette, M. D., and Mokwa, R. L. 2005. Analysis of laterally loaded pile groups with partial pile head fixity. *Advances in Deep Foundations*, GAS 132, ASCE.
Ellis, E. A., and Springman, S. M. 2001. Full-height piled bridge abutments constructed on soft clay. *Geotechnique* **51**(1): 3–14.
Erbrich, C. T. 2004. A new method for the design of laterally loaded anchor piles in soft rock. *Proc Offshore Technology Conf, Paper No. OTC. 16441*. Houston, Texas, USA.
Esu, F., and D'Elia, B. 1974. Interazione terreno-struttura in un palo sollecitato dauna frana tipo colata. *Rivista Italiana di Geotechica* **111**(?): 27–38.
Frank, R., and Pouget, P. 2008. Experimental pile subjected to long duration thrusts owing to a moving slope. *Geotechnique* **58**(8): 645–658.
Franke, K. W., and Rollins, K. M. 2013. Simplified hybrid p-y spring model for liquefied soils. *J Geotech Geoenviron Eng, ASCE* **139**(4): 564–576.
Frick, D., and Achmus, M. 2020. An experimental study on the parameters affecting the cyclic lateral response of monopiles for offshore wind turbines in sand. *Soils Found* **60**: 1570–1587.
Fukumoto, Y. 1972. Study on the behaviour of stabilization piles for land-slides. *Soils Found* **12**(2): 61–73 (in Japanese).
Fukumoto, Y. 1976. The behaviour of piles for preventing landslide. *Soils Found* **16**(2): 91–103 (in Japanese).
Fukuoka, M. 1966. Damage to civil engineering structures. *Soils Found* **6**(2): 45–52.
Fukuoka, M. 1977. The effect of horizontal loads on piles due to landslides. *Proc 9th Int Conf on Soil Mechanics and Found Engrg*, Speciality session 10, Tokyo, 27–42.
Georgiadis, M. 1987. Interaction between torsional and axial pile response. *Int J Numer Anal Methods Geomech* **11**: 645–650.
Ghee, E. H. 2009. *The Behaviour of Axially Loaded Piles Subjected to Lateral Soil Movements*. School of Engineering. Gold Coast, Griffith University. Ph. D.
GNS Science. n.d. https://www.gns.cri.nz/Home/Our-Science/Natural-Hazards-and-Risks/Recent-Events/Canterbury-quake/Hidden-fault
Goh, A. T. C., Teh, C. I., and Wong, K. S. 1997. Analysis of piles subjected to embankment induced lateral soil movements. *J Geot Geoenviron Eng, ASCE* **123**(9): 792–801.
González, L., Abdoun, T., and Dobry, R. 2009. Effect of soil permeability on centrifuge modeling of pile response to lateral spreading. *J Geotech Geoenviron Eng, ASCE* **135**(1): 62–73.
Guo, W. D. 1997. *Analytical and numerical solutions for pile foundations*. PhD thesis, The University of Western Australia.
Guo, W. D. 2000a. Vertically loaded single piles in Gibson soil. *J Geotech Geoenviron Eng Div, ASCE* **126**(2): 189–193.
Guo, W. D. 2000b. Visco-elastic consolidation subsequent to pile installation. *Comput Geotech* **26**(2): 113–144.
Guo, W. D. 2003. A simplified approach for piles due to soil movement. *Proc 12th Panamerican Conf on Soil Mechanics and Geotech Engrg*, Cambridge, Massachusettes, USA, Verlag Gluckauf GMBH. Essen (Germany), **2**: 2215–2220.
Guo, W. D. 2005. Limiting force profile and laterally loaded pile groups. *Proc 6th Int Conf on Tall Buildings*, Hong Kong, China: World Scientific.
Guo, W. D. 2006. On limiting force profile, slip depth and lateral pile response. *Comput Geotech* **33**(1): 47–67.
Guo, W. D. 2008. Laterally loaded rigid piles in cohesionless soil. *Can Geotech J* **45**(5): 676–697.
Guo, W. D. 2009. Non-linear response of laterally loaded piles and pile groups. *Int J Numer Anal Methods Geomech* **33**(7): 879–914.

Guo, W. D. 2012. *Theory and Practice of Pile Foundations*. Boca Raton, London, New York, CRC Press.

Guo, W. D. 2013a. Simple model for nonlinear response of 52 laterally loaded piles. *J Geotech Geoenviron Eng Div, ASCE* **139**(2): 234–252.

Guo, W. D. 2013b. Laterally loaded rigid piles with rotational constraints. *Comput Geotech* **54**(2): 72–83.

Guo, W. D. 2013c. p_u based solutions for slope stabilising piles. *Int J Geomech* **13**(3): 292–310.

Guo, W. D. 2014. Elastic models for nonlinear response of rigid passive piles. *Int J Numer Anal Methods Geomech* **38**(18): 1969–1989.

Guo, W. D. 2015a. Nonlinear response of lateral piles with compatible cap stiffness and p-multiplier. *J Eng Mech* **141**(9): 06015002.

Guo, W. D. 2015b. Nonlinear response of laterally loaded rigid piles in sliding soil. *Can Geotech J* **52**(7): 903–925.

Guo, W. D. 2016. Response of rigid piles during passive dragging. *Int J Numer Anal Methods Geomech* **40**(14): 1936–1967.

Guo, W. D. 2020a. Instability of back-rotated piles with near singularity stiffness, *J Eng Mech* **146**(9): 06020005.

Guo, W. D. 2020b. Response amplification of back-rotated piles, *Can Geotech J* **57**(11): 1780–1795.

Guo, W. D. 2021a. Response and stability of piles subjected to excavation loading, *Int J Geomech* **21**(1): 04020238.

Guo, W. D. 2021b. Simulation of piles subjected to excavation or embankment loading. *Proc NZGS Symposium, Good Grounds for the Future*, 15–17 October, Dunedin, New Zealand.

Guo, W. D. 2022a. Ductile response of piles to stabilize landslide. *Int J Numer Anal Methods Geomech* **46**(7): 1292–1305.

Guo, W. D. 2022b. Sway, sliding and back-rotation of piles subjected to pulse-like soil movement, *Int J Phys Model Geotech* **22**(4): 169–191.

Guo, W. D., Chow, Y. K., and Randolph, M. F. 2007. Torsional piles in two layered non-homogenous soil. *Int J Geomech* **7**(6): 410–422.

Guo, W. D., and Ghee, E. H. 2004. Model tests on single piles in sand subjected to lateral soil movement. *Proc 18th Australasian Conf on Mechanics of Structures and Materials*, Perth, Australia, Taylor and Francis Group, **2**: 997–1003.

Guo, W. D., and Ghee, E. H. 2005a. A preliminary investigation into the effect of axial load on piles subjected to lateral soil movement. *Proc 1st Int Symp on Frontiers in Offshore Geotechnics*, Perth, Australia, Taylor and Francis.

Guo, W. D., and Ghee, E. H. 2005b. Response of axially loaded pile groups subjected to lateral soil movement – An experimental investigation. *Proc 6th Int Conf on Tall Buildings (ICTB-VI), Hong Kong*, 333–338.

Guo, W. D., and Ghee, E. H. 2006. Behaviour of axially loaded pile groups subjected to lateral soil movement. *Proc Geoshanghai, Found Analysis and Design*, Shanghai, China. GSP **153**: 147–181.

Guo, W. D., and Ghee, E. H. 2010. Model tests on free-standing passive pile groups in sand. *Proc 7th Int Conf Physical Modelling in Geotechnics (ICPMG 2010)*, **2**: 873–878.

Guo, W. D., and Lee, F. H. 2001. Load transfer approach for laterally loaded piles. *Int J Numer Anal Methods Geomech* **25**(1): 1101–1129.

Guo, W. D., and Qin, H. Y. 2006. Vertically loaded piles in sand subjected to triangular profiles of soil movements. *Proc 10th Int Conf on Piling and Found*, Amsterdam, Netherland, **1**: paper no. 1371.

Guo, W. D., and Qin, H. Y. 2010. Thrust and bending moment of rigid piles subjected to moving soil. *Can Geotech J* **47**(2): 180–196.

Guo, W. D., Qin, H. Y., and Ghee, E. H. 2006. Effect of soil movement profiles on vertically loaded single piles. *Proc Int Conf on Physical Modelling in Geotechnics*, Hong Kong, China, Taylor and Francis Group, London, UK. **2**: 841–846.

Guo, W. D., Qin, H. Y., and Ghee, E. H. 2017. Modelling single piles subjected to evolving soil movement. *Int J Geomech* **17**(4): 04016111.

Guo, W. D., and Randolph, M. F. 1996. Torsional piles in non-homogeneous media. *Comput Geotech* **19**(4): 265–287.

Guo, W. D., and Randolph, M. F. 1997a. Vertically loaded piles in non-homogeneous media. *Int J Numer Anal Methods Geomech* **21**(8): 507–532.

Guo, W. D., and Randolph, M. F. 1997b. Non-linear visco-elastic analysis of piles through spreadsheet program. *Proc 9th Int Conf of Int Association for Computer Methods and Advances in Geomechanics – IACMAG 97*, Wuhan, China, **3**: 2105–2110.

Guo, W. D., and Zhu, B. T. 2005. Static and cyclic behavior of laterally loaded piles in calcareous sand. *Frontiers in Offshore Geotechnics*: ISFOG 2005, Gourvence and Cassidy (Eds), London, Taylor and Francis Group, 373–379.

Guo, W. D., and Zhu, B. T. 2011. Structure nonlinearity and response of laterally loaded piles. *Aust Geomech* **45**(3): 41–52.

Hache, R. A. G., and Valsangkar, A. J. 1988. Torsional resistance of single pile in layered soil. *J Geotech Eng Div, ASCE* **114**(2): 216–220.

Hamada, M, and O'Rourke, T. D. 1992. Case studies of liquefaction and lifeline performance during past, *Earthquakes* **1**. Japanese case studies, Technical report NCEER-92–0001.

Hansen, B. J. 1961. The ultimate resistance of rigid piles against transversal forces. Copenhagen, Denmark, *The Danish Geotech. Institute Bulletin* No. 12.

Haskell, J. J. M., Madabhushi, S. P. G., Cubrinovski, M., and Winkley, A. 2013. Lateral spreading-induced abutment rotation in the 2011 Christchurch earthquake: Observation and analysis. *Geotechnique* **63**(15): 1310–1327.

Hayashi, K., Fujita, K., Tsuji, M., and Takewaki, I. 2018. A simple response evaluation method for base-Isolation building-connection hybrid structural system under long-period and long-duration ground motion. In Takewaki, I., ed., *Performance of Innovative Controlled Buildings Under Resonant and Critical Earthquake Ground Motions*. Lausanne, Frontiers Media.

He, L., Elgamal, A., Abdoun, T., Abe, A., Dobry, R., Hamada, M., Menses, J., Sato, M., Shantz, T., and Tokimatsu, K. 2009. Liquefaction-induced lateral load on pile in a medium Dr sand layer. *J Earthq Eng* **13**: 916–938.

Hetenyi, M. 1946. *Beams on Elastic Foundations*. Ann Arbor, University of Michigan Press.

Houlsby, G. T., and Martin, C. M. 2003. Undrained bearing capacity factors for conical footings in clay. *Geotechnique* **53**(5): 513–520.

Huang, F.-Y., He, L.-F., Shan, Y.-L., Hu, C.-X., and Zhou, Z.-M. 2021. Experiment on interaction of soil-abutment-RC pile in integral abutment jointless bridges (IAJBs). *Rock Soil Mech* **42**(7): 1803–1814.

Idriss, I. M., and Boulanger, R. W. 2007. SPT- and CPT-based relationships for the residual shear strength of liquefied soils. *Proc 4th Int Conf on Earthquake Geotech Engrg*, K. D. Pitilakis, ed., Dordrecht, Netherlands, **1**: 1–22.

Imamura, S., Hagiwara, T., Tsukamoto, Y., and Ishihara, K. 2004. Response of pile groups against seismically induced lateral flow in centrifuge model tests. *Soils Found* **44**(3): 39–55.

Ishihara, K. 1993. Liquefaction and flow failure. *Geotechnique* **43**(3): 351–415.

Ishihara, K., and Cubrinovski, M. 1998. Soil-pile interaction in liquefied deposits undergoing lateral spreading. *Proc 11th Danube-European Conf*, May, Croatia.

Ishihara, K., and Cubrinovski, M. 2004. Case studies of pile foundations undergoing lateral spreading in liquefied deposits. *Proc 5th Int Conf on Case Histories in Geotech. Engrg*, New York, NY.

Ito, T., and Matsui, T. 1975. Methods to estimate lateral force acting on stabilizing piles. *Soils Found* **15**(4): 43–59.

Jakrapiyanun, W. 2002. *Physical modeling of dynamic soil-foundation-structure-interaction using a laminar container*. *Civil Engrg* San Diego, La Jolla, University of California. Ph.D.

Jeong, S., Seo, J. L. D., and Park, J. 1995. Time-dependent behavior of pile groups by staged construction of an adjacent embankment on soft clay. *Can Geotech J* **41**: 644–656.

JGJ. 1994. (Chinese) Technical code for building pile foundations. JGJ 94-94, Beijing, 252.

JRA 2002 Specification for highway bridges, in Japan Road Association, *Prepared by Public Works Research Institute (PWRI) and Civil Engrg Research Laboratory*: Japan.

Juirnarongrit, T., and Ashford, S. A. 2006. Soil–pile response to blast-induced lateral spreading. II: Analysis and assessment of the p-y method. *J Geotech Geoenviron Eng, ASCE* **132**(2): 163–172.

Kagawa, T., Sato, M., Minowa, C., Abe, A., and Tazoh, T. 2004. Centrifuge simulations of largescale shaking table tests: Case studies. *J Geotech Geoenviron Eng, ASCE* **130**(7): 663–672.

Karim, M. R. 2013. Behaviour of piles subjected to passive subsoil movement due to embankment construction – A simplified 3D analysis. *Comput Geotech* **53**: 1–8.

Kim, H., Kim, Y. S., Kim, K. O., and Cho, Y. 2006. Centrifuge model tests on embedded sunction anchor pullout. *Proc 6th Int Conf on Physical Modelling in Geotechnics*, Hong Kong.

Knappett, J. A., and Madabhushi, S. P. G. 2009. Influence of axial load on lateral pile response in liquefiable soils. Part I: Physical Modelling. *Geotechnique* **59**(7): 571–581.

Kourkoulis, R., and Gelagoti, F. I. 2011. Anastasopoulos, and G. Gazetas, Slope stabilizing piles and pile-groups: Parametric study and design insights. *J Geotech Geoenviron Eng, ASCE* **137**(7): 663–677.

LeBlanc, C., Byrne, B. W., and Houlby, G. T. 2010b. Response of stiff piles to random two-way lateral loading. *Geotechnique* **60**(9): 715–721.

LeBlanc, C., Houlby, G. T., and Byrne, B. W. 2010a. Response of stiff piles to long-term cyclic lateral loading. *Geotechnique* **60**(2): 79–90.

Lesniewska, D., and Mroz, Z. 2000. Limit equilibrium approach to study the evolution of shear band systems in soils. *Geotechnique* **50**(3): 521–536.

Leung, C. F., Chow, Y. K., and Shen, R. F. 2000. Behaviour of pile subject to excavation-induced soil movement. *J Geotech Geoenviron Eng, ASCE* **126**(11): 947–954.

Leung, C. F., Ong, D. E. L., and Chow, Y. K. 2006. Pile behavior due to excavation-induced soil movement in clay II: Collapsed wall. *J Geot Geoenviron Eng, ASCE* **132**(1): 45–53.

Levy, N. H., Einav, I., and Hull, T. 2009. Cyclic shakedown of piles subjected to two-dimensional lateral loading. *Int J Numer Anal Methods Geomech* **33**: 1339–1361.

Lin, S. S., and Liao, J. C. 1999. Permanent strains of piles in sand due to cyclic lateral loads. *J Geotech Geoenviron Eng, ASCE* **125**(9): 798–802.

Little, R. L., and Briaud, J. L. 1988. *Cyclic horizontal load tests on six piles in sands at Houston Ship Channel.* Research Report 5640 to USAE Waterways Experiment Station. Civil Engrg, Texas A&M University.

Long, J. H., and Vanneste, G. 1994. Effects of cyclic lateral loads on piles in sand. *J Geotech Eng, ASCE* **120**(1): 225–244.

Lunne, T., and Christoffersen, H. P. 1985. Cone penetrometer interpretation for offshore sands. In *Proc Offshore Technol Conf*, Houston, U.S.A., OTC4464, 181–192.

Lutenegger, A. J., and Kelly, S. P. 1998. Standard penetration tests with torque measurement. *Proc 1st Int Conf on Site Characterization*, Atlanta, Geogiria, **2**: 939–945.

Matlock, H., Wayne, B., Kelly, A. E., and Board, D. 1980. Field tests of the lateral load behaviour of pile groups in soft clay. Paper No. OTC. 3871, *Proc 12th Ann Offshore Technol Conf*, Houston, Texas, 671–686.

McGann, C. R., Arduino, P., and Mackenzie-Helnwein, P. 2011. Applicability of conventional p-y relations to the analysis of piles in laterally spreading soil. *J Geotech Geoenviron Eng, ASCE* **137**(6): 557–567.

McManus, K. J., and Kulhawy, F. H. 1994. Cyclic axial loading drilled shafts in cohesive soil. *J Geotech Eng, ASCE* **120**(9): 1481–1497.

Merifield, R. S. 2011. Ultimate uplift capacity of multiplate helical type anchors in clay. *J Geotech Geoenviron Eng, ASCE* **137**(7): 704–716.

Mesri, G. 2007. Yield strength and critical strength of liquefiable sands in sloping ground. *Geotechnique* **57**(3): 309–311.

Meyerhof, G. G., Yalcin, A. S., and Mathur, S. K. 1983. Ultimate pile capacity for eccentric inclined load. *J Geotech Geoenviron Eng Div, ASCE* **109**(3): 408–423.

Militano, G., and Rajapakse, R. K. N. D. 1999. Dynamic response of a pile in a multi-layered soil to transient torsional and axial loading. *Geotechnique* **49**(1): 91–109.

Milligan, G. W. E., and Bransby, P. L. 1976. Combined active and passive rotational failure of a retaining wall in sand. *Geotechnique* **26**(3): 473–494.

Miwaa, S., Ikedaa, T., and Satob, T. 2006. Damage process of pile foundation in liquefied ground during strong ground motion. *Soil Dyn Earthq Eng*, **26**, 325–336.

Mokwa, R. L., and Duncan, J. M. 2003. Rotational restraint of pile caps during lateral loading. *J Geotech Geoenviron Eng, ASCE* **129**(9): 829–837.

Motamed, R., and Towhata, I. 2010. Shaking table model tests on pile groups behind quay walls subjected to lateral spreading. *J Geotech Geoenviron Eng, ASCE* **136**(3): 477–489.

Muraro, S., Madaschi, A., and Gajo, A. 2014. On the reliability of 3D numerical analyses on passive piles used for slope stabilisation in frictional soils. *Geotechnique* **64**(6): 486–492.

Murchison, J. M., and O'Neill, M. W. 1984. Evaluation of p-y relationships in cohesionless soils. In J. R. Meyer, ed., *Analysis and Design of Pile Found*, ASCE. San Francisco, National Convention, 174–191.

Murff, J. D., and Hamilton, J. M. 1993. P-Ultimate for undrained analysis of laterally loaded piles. *J Geotech Geoenviron Eng Div, ASCE* **119**(1): 91–107.

Oliveira, J. R. M. S., Almeida, M. S. S., Almeida, M. C. F., and Borges, R. G. 2010. Physical modelling of lateral clay-pipe interaction. *J Geotech Geoenviron Eng, ASCE* **136**(7): 950–956.

Olson, S. M., and Stark, T. D. 2002. Liquefied strength ratio from liquefaction flow failure case histories. *Can Geotech J* **39**(2): 629–647.

Ong, D. E. L., Leung, C. F., and Chow, Y. K. 2003. Piles subject to excavation-induced soil movement in clay. *Proc 13th European Conf on Soil Mechanics and Geotech Engrg*, Prague, Czech Republic, **2**: 777–782.

Ong, D. E. L., Leung, C. F., and Chow, Y. K. 2006. Pile behavior due to excavation-induced soil movement in clay. I: Stable wall. *J Geotech Geoenviron Eng, ASCE* **132**(1): 36–44.

Ooi, P. S. K., Chang, B. K. F., and Wang, S. 2004. Simplified lateral load analyses of fixed-head piles and pile groups. *J Geotech Geoenviron Eng, ASCE* **130**(1): 1140–1151.

Park, R., and Paulay, T. 1975. *Reinforced Concrete Structures*. John Wiley and Sons, Inc.

Parry, R. H. G. 1978. Estimating foundation settlements in sand from plate bearing tests. *Geotechnique* **28**(1): 107–118.

Poulos, H. G. 1975. Torsional response of piles. *J Geotech Eng Div, ASCE* **101**(10): 107–118.

Poulos, H. G. 1982. Single pile response to cyclic lateral load. *J Geotech Eng, ASCE* **108**(3): 355–375.

Poulos, H. G. 1995. Design of reinforcing piles to increase slope stability. *Can Geotech J* **32**(5): 808–818.

Poulos, H. G., Chen, L. T., and Hull, T. S. 1995. Model tests on single piles subjected to lateral soil movement. *Soils Found* **35**(4): 85–92.

Poulos, H. G., and Davis, E. H. 1980. *Pile Foundation Analysis and Design*. New York, John Wiley and Sons.

Qin, H. Y. 2010. *Response of pile foundations due to lateral force and soil movements*. School of Engineering. Gold Coast, Griffith University. Ph.D.

Qin, H. Y., and Guo, W. D. 2014. Nonlinear response of laterally loaded rigid piles in sand. *Geomech Eng Anal Int J* **7**(6): 679–703.

Qin, H. Y., and Guo, W. D. 2016a. Response of static and cyclic laterally loaded rigid piles in sand. *Mar Georesour Geotechnol* **34**(2): 138–153.

Qin, H. Y., and Guo, W. D. 2016b. Response of piles subjected to progressive soil movement. *Geotech Test J, ASTM* **39**(1): 1–20.

Rajashree, S. S., and Sitharam, T. G. 2001. Nonlinear finite-element modelling of batter piles under lateral load. *J Geotech Geoenviron Eng Div, ASCE* **127**(7): 604–612.

Randolph, M. F. 1981. Piles subjected to torsion, *J Geotech Eng Div, ASCE* **107**(8): 1095–1111.

Randolph, M. F. 1983. Design consideration for offshore piles. *Proc Conf on Geot Practice in Offshore Engrg Div*, ASCE, Austin, Texas, 422–439.

Randolph, M. F., and Houlsby, G. T. 1984. The limiting pressure on a circular pile loaded laterally in cohesive soil. *Geotechnique* **34**(4): 613–623.

Rao, S. N., and Veeresh, C. 1994. Influence of pile inclination on the lateral capacity of batter piles in clays. *Proc 4th Int Offshore and Polar Engrg Conf*, Osaka, Japan.

Reese, L. C. 1973. A design method for an anchor pile in a mooring system. *Proc Offshore Technol Conf, Paper No. OTC 1745*.

Reese, L. C. 1997. Analysis of laterally loaded shafts in weak rock. *J Geotech Geoenviron Eng, ASCE* **123**(11): 1010–1017.

Reese, L. C., Cox, W. R., and Koop, F. D. 1974. Analysis of lateral loaded piles in sand. *Proc 6th Ann Offshore Technol Conf OTC 2080*, Dallas Texas.

Richards, I., Byrne, B. W., and Houlby, G. T. 2020. Monopile rotation under complex cyclic lateral loading in sand. *Geotechnique* **70**(10): 916–930.

Rollins, K. M., Gerber, T. M., Lane, J. D., and Ashford, S. 2005. Lateral resistance of a full-scale pile group in liquefied sand. *J Geotech Geoenviron Eng, ASCE* **131**(1): 115–125.

Scott, R. F. 1981. *Foundation Analysis*. Englewood Cliffs, NJ, Prentice Hall.

Seed, R. B., and Harder, L. F. 1990. SPT-based analysis of cyclic pore pressure generation and undrained residual strength. *Proc H. Bolton Seed Memorial Symp*, BiTech, Richmond, BC, Canada, **2**: 351–376.

Senpere, D., and Auvergne, G. A. 1982. Suction anchor piles—A proven alternative to driving or drilling. *Proc Offshore Technol Conf, Paper No. OTC 4206*, Houston, USA.
Sharp, M. K., and Dobry, R. 2002. Sliding block analysis of lateral spreading based on centrifuge results. *Int J Phys Model Geotech* **2**: 13–32.
Shibata, T., Yashima, A., and Kimura, M. 1989. Model tests and analyses of laterally loaded pile groups. *Soils Found* **29**(1): 31–44.
Smethurst, J. A., and Powrie, W. 2007. Monitoring and analysis of the bending behaviour of discrete piles used to stabilise a railway embankment. *Geotechnique* **57**(8): 663–677.
Springman, S. M. 1989. *Lateral loading on piles due to simulated embankment construction*. Ph. D Thesis, University of Cambridge.
Stewart, D. P., Jewell, R. J., and Randolph, M. F. 1993. Numerical modelling of piled bridge abutments on soft ground. *Comput Geotech* **15**(1): 21–46.
Stewart, D. P., Jewell, R. J., and Randolph, M. F. 1994. Design of piled bridge abutment on soft clay for loading from lateral soil movements. *Geotechnique* **44**(2): 277–296.
Stoll, U. W. 1972. Torque shear test of cylindrical friction piles. *Civ Eng, ASCE* **42**: 63–64.
Swane, I. C. 1983. *The cyclic behavior of laterally loaded piles*. PhD Thesis, The University of Sydney.
Tasiopoulou, P., Gerolymos, N., Tazoh, T., and Gazetas, G. 2013. Pile-group response to large soil displacements and liquefaction: Centrifuge experiments versus a physically simplified analysis. *J Geotech Geoenviron Eng Div, ASCE* **139**(2), 223–233.
Van Impe, W. F. 1994. Developments in pile design. *Proc 4th Int DFI Conf*, Stresa, Italy, A. A. Balkema.
Viggiani, C. 1981. Ultimate lateral load on piles used to stabilise landslide. *Proc 10th Int Conf Soil Mech and Found Engrg*, Stockholm, Sweden, **3**: 555–560.
Vivatrat, V., Valent, P. J., and Ponterio, A. A. 1982. The influence of chain friction on anchor pile design. *Proc Offshore Technol Conf, Paper No. OTC 4178*. Houston, USA.
Wang, S.-T., and Reese, L. C. 1998. Design of pile foundations in liquefied soils. In P. Dakoulas, M. Yegian, and R. Holtz, eds., *Proceedings of Geotechnical Earthquake Engineering and Soil Dynamics III*, Reston, VA. **2**: 1331–1343.
Wang, Y. F. 2019. *Dynamic modeling of pulse-like earthquakes and ground motions*. A dissertation submitted in partial satisfaction of the requirements for the degree, Doctor of Philosophy in Geophysics, University of California San Diego.
White, D. J., Thompson, M. J., Suleiman, M. T., and Schdafer, V. R. 2008. Behaviour of slender piles subjected to free-field lateral soil movement. *J Geotech Geoenviron Eng Div, ASCE* **134**(4), 428–436.
Whitney, C. S. 1937. Design of reinforced concrete members under flexure or combined flexure and direct compression. *J ACI* **33**: 483–498.
Whittle, A. J., and Davis, R. V. 2006. Nicoll Highway collapse: Evaluation of geotechnical factors affecting design of excavation support system. *Proc Int Conf on Deep Excavations*, Singapore.
Xiao, Y., Wu, H., Yaprak, T. T., Martin, G. R., and Mander, J. B. 2006. Experimental studies on seismic behavior of steel pile-to-pile-cap connections. *J Bridg Eng, ASCE* **11**(2): 151–159.
Yang, Z., and Jeremic, B. 2002. Numerical analysis of pile behaviour under lateral load in layered elastic-plastic soils. *Int J Numer Anal Methods Geomech* **26**(14): 1385–1406.
Young, A. G., Phu, D. R., Spikula, D. R., Rivette, J. A., Lanier, D. L., and Murff, J. D. 2009. An approach for using integrated geoscience data to avoid deepwater anchoring problems. *Proc Offshore Techn Conf, Paper No. OTC 20073*.
Yu, H. S., Herrmann, L. R., and Boulanger, R. W. 2000. Analysis of steady cone penetration in clay. *J Geotech Geoenviron Eng Div, ASCE* **126**(7): 594–605.
Yu, L., Liu, J., Kong, X.-J., and Hu, Y. 2011. Numerical study on plate anchor stability in clay. *Geotechnique* **61**(3): 235–246.
Yun, G., and Bransby, M. F. 2007. The horizontal-moment capacity of embedded foundations in undrained soil. *Can Geotech J* **44**(4): 409–424.
Zhang, L. M., McVay, M. C., and Lai, P. W. 1999. Centrifuge modelling of laterally loaded single battered piles in sands. *Can Geotech J* **36**: 1074–1084.
Zhu, B., Sun, Y. X., Chen, R. P., Guo, W. D., and Yang, Y. Y. 2015. Experimental and analytical models on laterally loaded rigid mono-piles with hardening p-y curves. *J Waterw Port Coast Ocean Eng* **141**(6): 04015006.

Index

Pages in *italics* refer to figures and pages in **bold** refer to tables.

For Product Safety Concerns and Information please contact our EU representative GPSR@taylorandfrancis.com
Taylor & Francis Verlag GmbH, Kaufingerstraße 24, 80331 München, Germany

www.ingramcontent.com/pod-product-compliance
Lightning Source LLC
Chambersburg PA
CBHW061352210326
41598CB00035B/5956
* 9 7 8 1 0 3 2 3 2 4 7 7 7 *